D1562990

MATHEMATICAL METHODS OF OPERATIONS RESEARCH

BY THOMAS L. SAATY

University Professor
University of Pittsburgh

Dover Publications, Inc.
NEW YORK

To Bernadine

Copyright © 1959, 1988 by Thomas L. Saaty.
All rights reserved under Pan American and International Copyright Conventions.

Published in Canada by General Publishing Company, Ltd., 30 Lesmill Road, Don Mills, Toronto, Ontario.
Published in the United Kingdom by Constable and Company, Ltd., 10 Orange Street, London WC2H 7EG.

This Dover edition, first published in 1988, is an unabridged, enlarged republication of the work first published by the McGraw-Hill Book Company, Inc., New York, 1959. Chapter 13 and a new Preface have been specially written by the author for this Dover edition.

Manufactured in the United States of America
Dover Publications, Inc., 31 East 2nd Street, Mineola, N.Y. 11501

Library of Congress Cataloging-in-Publication Data

Saaty, Thomas L.
    Mathematical methods of operations research / by Thomas L. Saaty.
        p.   cm.
    Reprint. Originally published: New York : McGraw-Hill, 1959. With a new preface and a new chapter.
    Bibliography: p.
    Includes index.
    ISBN 0-486-65703-5 (pbk.)
    1. Operations research. 2. Mathematical analysis. I. Title.
T57.6.S15 1988
658.4′03′4—dc19   88-10265
                                                                         CIP

PREFACE TO THE DOVER EDITION
A Generation Later

Although operations research has legitimized the use of quantitative common sense in Everyman's everyday operations, some people have written that operations research is dead. It may be early to make this assertion but it could become true if operations research (O.R.) remains in the same state it is in today. Some O.R. people give the impression of a coterie dedicated to applying a very limited set of tools to a very big world of problems.

Since I wrote this book in the middle of the century, several things have happened. There has been a burgeoning of methods and their advocates. But O.R. contributions have declined in creativity. Most of the effort has gone into extending and refining old methods, such as those presented in this book. Successful O.R. people tend to become managers, but do not add much to the understanding of the field. There have been some excellent contributions to queueing operations, inventory, scheduling, and other mechanical (as distinct from social) problems of society. Applications of optimization have been extended to include optimization of thousands of variables all treated in the same way. Disappointingly, people have fallen for this method of quick and automatic treatment of complexity in a homogeneous way as if it were a potion that will cure every disease. Some people have acted as if they were challenging infinity this way, but infinity remains very remote. Some problems have been "solved"—if one can call these dispositions "solutions." Others have been aggravated by the mindless or mechanical use of modeling or by trying to solve related problems without looking at the entire system. Systems thinking has been spreading in the operations-research and management-science professions, but most technicians find it easier to deal with concrete and simple models than with philosophical abstractions and interpretations characteristic of the systems approach. Still, philosophy is needed to give quantification greater depth and meaning. But a good O.R. philosophy has not emerged.

By studying the mathematics literature one can conclude that the core of mathematics has not been perturbed in the least by operations-research methods. Most mathematicians do not know about O.R. Unlike the mathematics introduced under the influence of physics, the O.R. mathematics developed so far consists of stopgap techniques with little structural consequence for mathematics as a whole. More important, although there has been a shifting of attention in society away from the study of objects and machines in themselves to people and their behavior in industries and organizations, O.R. has not been able to rise to the occasion with new and effective models. This is symptomatic of too close an attachment to techniques at a high cost to creative thinking. How people work together and what (if any) law drives their psyches have become major concerns in our competitive society. One thing we are finding is that people can be brainwashed to think of as their own things and ideas that have actually been unconsciously borrowed. But if we accept the goal that people themselves must participate in deciding what they really want, then no abstract model alone can decide for people on the basis of pure principles. We need expert support systems to facilitate the process of interaction and learning in individuals and in groups working together.

Three problems face our world today, the survival-and-conflict problem, the population-and-food problem, and the lifetime-learning-and-profitable-occupation problem, all involved with our attempt to pass our lives more pleasantly on this earth. The world is threatened with annihilation and most people are hungry and despondent. How can our approach to problem solving help? To this Dover edition we have added a new chapter (Chapter 13) to offer the reader a modern approach to multicriteria decision making. This is a relatively young area and the work presented here is primarily that of the author. It amounts to assisting people to structure the unstructured problems they experience, to provide paired comparison judgments, and to create relations within this structure. These judgments are then transformed into ratio scales and synthesized over the entire structure to obtain a relative scale of priority of the elements represented in the structure, be they goals, criteria, subcriteria, or alternatives.

The theory has been validated in a number of applications. It is useful when there are intangible criteria and a variety of judgments and observations are to be made. It operates on the principle that there are no permanently best decisions, only good decisions in light of what we know. If we know more, we are likely to change our minds. It incorporates also a gradual realization that modeling is a metaphor rather than a direct path to the truth.

It is of interest to people deeply involved in mathematical modeling to know that the quantitative approach to reality is one to which many of us are introduced at an early age, and we tend to take it for granted as an innate way of thinking rather than as a powerful tool that we learn to use through painstaking effort over a long period of time. The social scientist does not have the illusion that the complex problems he encounters can necessarily be subjected to this method of analysis. He regards other problem-solving approaches as more powerful and more natural because his problems are less rigidly structured, more amorphous than those to which we are accustomed. Of course there are many different metaphors of reality, and the versatile problem solver tries to learn to choose them appropriately. Artificial-intelligence people tend to take this very broad view of problem definition and solution.

We need to remember that when we are faced with an unstructured problem it is we who *create* the model in the form of a quantitative metaphor; there is no correct model waiting in the wings for us to call onstage. Two quantitative minds may differ substantially as to which of two dramatically different models serves best to represent a given problem. Anthropologists, psychologists, musicians, and poets do not subscribe to our way of constructing metaphors; they have their own ways, and they humor us, as we do them, when it is a question of who is in closer contact with the "real" reality.

In the book "Einstein's Space and Van Gogh's Sky: Physical Reality and Beyond" by the psychologist Lawrence LeShan and the physicist Henry Margenau (Macmillan, New York, 1982), the see-touch-and-measure sensory way of the scientist is contrasted with three other recognized ways of looking at the world. Together, the four ways are as follows: (1) sensory; (2) clairvoyant; (3) transpsychic; and (4) mythic. In the first (the scientific way), individuals construct reality as if they were detached observers of a larger whole; in the second, as if they were extensions of the whole without separation, as in dancing, music, and meditation; in the third, as if they were reciprocals of the whole, with their wishes and desires able to urge on the forces of nature, as people do in prayer; and in the fourth, as if they were identical with the whole, as happens in dreaming and play, where curiosity and creativity become means of apprehending reality.

A basic limiting principle of the sensory mode is the assumption that all phenomena are sequentially continuous in space and time, related by linear cause and effect. Despite disproof of this principle in quantum mechanics, "it has not been clearly understood that its abandonment means the complete collapse of the system of one rationality ruling the entire universe. A completely consistent cosmos cannot be inconsistent in one area. One exception collapses it all." Causation, a basic tenet of science, works only for isolated systems such as we are able to bring about or selectively observe in the physical world. But causal interpretation does not work for all aspects of consciousness. Consciousness goes beyond the immediately sensory; it constantly reaches into the past and future, to distant places and to possibilities of places. Only a noncausal perspective of numerous alternative realities allows us to understand this domain adequately. The challenge is to develop methods of studying nonsensory modes of reality, methods that are as effective as those used in the sensory mode and that are as scientific as possible in the noncausal realm. The authors explore several possibilities in art, ethics, parapsychology, and the study of human consciousness.

The modern scholar and philosopher Earl R. McCormac, in his book "A Cognitive Theory of Metaphor" (MIT Press, Cambridge, Mass., 1985), has given particular attention to the role of metaphor (a noncausal mode) in all human understanding. He refers to the writing of the great economist John Maynard Keynes, who described Newton's attempts to reconcile his mathematical physics with his intuitions about the world as follows:

"Newton was not the first of the age of reason. He was the last of the magicians, the last of the Babylonians and Sumerians, the last great mind which looked out on the visible intellectual world with the same eyes as those who began to build our intellectual inheritance rather less than 10,000 years ago.

"Why do I call him a magician? Because he looked on the whole universe and all that is in it as a riddle, as a secret which could be read by applying pure thought to a certain evidence, certain mystic clues which God has laid about the world to allow a sort of philosopher's treasure here to the esoteric brotherhood. He believed that these clues were to be found partly in the evidence of the heavens and in the constitution of the elements (and that is what gives the false suggestion of his being an experimental philosopher), but also partly in certain papers and traditions handed down by the brethren in an unbroken chain back to the original cryptic revolution in Babylonia. He regarded the universe as a cryptogram set by the Almighty...."

A recent book, "The Mathematical Experience" by Philip J. Davis and Reuben Hersh (Birkhauser, Cambridge, Mass., 1981), a best-seller among mathematicians, tackles this entire theme from the standpoint of the mathematical tradition, opening further avenues of exploration.

In closing, operations research is not dead. It does, however, need a broader philosophy and new methods of dealing with problems in more versatile ways. To remain alive, operations research must open its eyes to fresh possibilities; it must confront its tasks with better spirit, with creativity, and with more thorough preparation than it does at present.

T. L. S.

University of Pittsburgh

PREFACE TO THE FIRST EDITION

Operations research—the scientific approach to operational problems for the greater fulfillment of objectives—is a parallel development to an increased realization of the usefulness of staff planning and analysis functions. The implementation of these functions, under the name of operations research or otherwise, has led to rapid development of new methodology and application of old methodology to new problems. The books so far available in the field have concentrated mainly on case histories. The fact that operations research utilizes a great variety of mathematical methods has been an incentive for writing this book.

The practice of operations research involves more than mathematical techniques. However, problems usually require analytical treatment, which then enables one to form conclusions with greater confidence.

The material in this book was gradually developed, over a period of several years, as a set of notes for a course in operations research adapted to the widely different backgrounds and specialities of the students taking the course. Compromise in the presentation had to be made on many occasions. Some elements of this compromise remain in the book because it was felt that there would be a similar variety of background among the readers.

For those readers who are interested in operations research but lack adequate mathematical training, many problems are posed in elementary fashion in order to give, if possible, better insight and to stimulate suggestions for probable solutions.

The mathematical methods of operations research encompass most of applied mathematics and some pure mathematics, e.g., basic set theory, lattice theory, etc. To include all these methods would be outside the scope of the book.

The book requires for background a course in calculus, with some elements of advanced calculus and rudimentary knowledge of matrix theory. There are a few places in which a degree of analytical experience would be helpful.

Chapters 1 to 4 provide some background material and also serve to complete the "perspective." A chapter on some mathematical models is included to illustrate, by brief discussions and examples, how the need for such tools arises and some types of theoretical questions associated with them which also require answers. This chapter will serve as a starting point for expanding these methods wherever desirable.

Because of the variety of methods, a greater degree of completeness has been attempted in those subjects which it was felt needed such unity as far as the literature is concerned. The Part on optimization is an illustration of this.

Although the material in the two chapters on probability and statistics is well documented in the existing literature, it seemed best to give a unified treatment of the analytical methods most commonly used in operations research and to provide some of the needed background for the chapters on applications. The brief presentations on probability and statistics are meant for the beginning student, and only as guides—not as last words. The interested reader has many references to use in expanding his knowledge of the subject. Examples which were available and illustrative have been used from the examined literature.

Although it is easier to present systematically only some of the analytical methods available, an attempt was made to include material on creative imagination. In the final analysis, the individual himself, his talents and methods of approaching problems, form the fundamental qualities required in solving problems. Those analytical methods studied give a reliable and powerful approach to problems.

However, many of man's problems have been successfully solved by a combination of intuitive talent and a desire to accept the challenge presented by a new problem. There is a place for both the analytical and the intuitive methods of attack. Neither can succeed completely without the other. Both require creative ability. It is difficult to write about general creativity, but since the approach to it used with the class was successful, it is included here in Chapter 12 as an essay.

We wish to warn the reader at the outset that even though we may attempt partly to inform and partly to stimulate him, however short we may fall of this goal, we have little intention of settling here questions which should not and cannot be settled in the few pages of one book. These remarks apply in the same measure to the mathematical methods given to be used primarily in increasing understanding of problems and secondarily in giving solutions to these problems.

Many friends have contributed to the preparation of this book by suggestions, criticism, and proofreading. My indebtedness to various authors, journals, and books is clear in the references. Here, my thanks to Dr. George Shortley, editor of *Operations Research,* for useful suggestions and permission to use my paper from the journal as Chapter 11. My gratitude for encouragement and guidance in the general structure of the book goes to my Operations Evaluation Group colleague Dr. Robert

S. Titchen and to my friend Dr. George G. O'Brien. In occasional discussions, many useful ideas and examples and criticisms came from my colleagues Dr. Richard H. Brown, Dr. John Danskin, and Robert P. Smith.

In particular, I wish to thank Dr. Allan J. Goldman, of the National Bureau of Standards, for his careful reading of and valuable suggestions on the optimization section, and Dr. Joan Rosenblatt for illuminating discussions on probability and statistics. To my tireless friend Theodore Wehe go my thanks for proofreading and frequent help in assembling and writing the first and second chapters, and to Monte Bourjaily, for encouragement and editorial help on the essay on creativity. To my Yale professor Carl Hempel (now at Princeton) go my thanks for lending me ideas and material from a lecture on symbolic logic.

I am indebted to Professor Sir Ronald A. Fisher, Cambridge, to Dr. Frank Yates, Rothamsted, and to Oliver & Boyd, Ltd., Edinburgh and London, for permission to reprint Tables 10-5 to 10-10 from their book "Statistical Tables for Biological, Agricultural, and Medical Research."

Other valuable technical discussions have been held with my friends Saul I. Gass, Professor Glen D. Camp, Dr. Philip Wolfe, Professor Philip M. Morse, Dr. Herbert P. Galliher, Dr. John Y. Barry, Dr. Isidor Heller, Dr. George Suzuki, Otto Rauschwalbe, David Bourland, Donald Y. Barrer, Dr. Edwin O. Elliott, and Dr. Herbert Glazer.

T. L. S.

London

CONTENTS

Preface to the Dover Edition   *iii*
Preface to the First Edition   *vii*

1. Operations Research—History and Concepts   1

Part 1. Scientific Method—Truth; Mathematics and Logic—Validity; Outlines of Some Useful Mathematical Models

2. Some Remarks on Scientific Method in Operations Research   21
3. Mathematical Existence and Logical Proof   41
4. Elementary Classical Methods Used in Model Formulation   65

Part 2. Optimization, Programming, and Game Theory

5. Optimization   99
6. Linear and Quadratic Programming   165
7. The Theory of Games   209
Some Optimization Problems   228

Part 3. Probability, Applications, Statistics, and Queueing Theory

8. Probability   235
9. Some Applications   275
10. Fundamental Statistics   302
11. Résumé of Queueing Theory   331
Some Probability and Statistics Problems   375

Part 4. An Essay

12. Some Thoughts on Creativity   381
Elementary Analytical Problems and Projects   400

Part 5. Decision Making

13. Multicriteria Decision Making: The Analytic Hierarchy Process   415

Index   451

Index to Chapter 13   459

CHAPTER 1

OPERATIONS RESEARCH—HISTORY AND CONCEPTS

The main purpose of this book is to present some mathematical methods essential to operations research. This chapter on the historical background of operations research sets forth general ideas on the subject, to provide perspective. The remaining eleven chapters pursue the task of examining specific ideas and illustrating specific methods. The mathematics of this book is divided essentially into two parts—one on methods of optimization and another on the theory of probability and on statistical methods. The last chapter, which is an essay, challenges the reader to approach problems creatively and includes a collection of elementary problems as illustrations. The ideas presented in this book will serve as useful auxiliary guides to the individual interested in operations research. There has been an attempt to present a variety of material within a workable structure, and it is hoped that the reader will pursue the subject at greater length through the indicated references.

1-1. Historical Remarks

The scientific method has been man's outstanding asset in his technological development. Its success has been manifest not only in his greater understanding of the natural world and his ability to predict its behavior, but also in material advances which have lengthened his life, increased his leisure time, and enabled him to pursue a far greater variety of activities.

In the field of applications man has pragmatically accepted these advances as sufficient evidence of the usefulness of the scientific method. Yet, perhaps for historical reasons, there remained a stubborn resistance to admitting the scientific method into the more personal and less materialistic affairs of man himself. As the structure of an industrial society becomes progressively more complex and far-flung, the acts of the statesman produce deeper and more decisive influences on the course of human events [20]. These influences are felt by the members of society in a more direct, personally important way than in the days of older and looser social organization. Hence, the consequences of the acts of the statesman should be calculated with greater care. We must try to make available to him the best tools to predict the results of any course of action he may propose and to analyze dispassionately his aims and objectives. Accurate prediction, of course, is one of the primary concerns of science. Thus there would appear to be obligations both on the part of the statesman to make greater use of scientific knowledge and on the part of the scientist to convince the statesman of the effectiveness of his predictive techniques in human affairs as well as in technology. Despite the possibility of moralization, essentially a scientist remains a scientist, regardless of the domain to which his efforts are applied, since his tool is the scientific method.

It has often been observed that war accelerates and intensifies the development of new devices and methods to an extent difficult to imagine under the conditions of peace. In times of grave national crises, all available talents and techniques are needed to outwit and conquer the enemy. Thus, early in World War II, the barriers were lifted, and the scientific method, after centuries of confinement to technical problems, was at last called forth to struggle with the operations involved in human enterprises. At that time, more than ever before, teams of scientists were organized in an intensified effort to use their scientific discipline in warfare. This type of activity has become known as *operations research*. Since this initial breakthrough, the recognition of operations research has spread beyond the confines of national emergency. The success of the scientific method has been adequately demonstrated in both military and industrial operations almost as fully as it has been in the domain of the natural sciences.

However, it could not be expected that science, having left the comfortable control of the laboratory and the cool indifference of the cosmos, could enter into the fiercely competitive and somewhat chaotic world of human affairs without making certain sacrifices. Mathematical precision in prediction and description had to be replaced by a hard-headed acceptance of the complexity, transiency, and emotional conflict of the world of human beings. One immediate consequence of this change in outlook was a great increase in the use of nondeterministic techniques, such as probability and statistics, for the description of phenomena. Yet as one looks back, it is significant to note that this trend had already been anticipated in the natural sciences, the notion of probability being by this time already well entrenched in such fields as atomic physics and statistical mechanics. Perhaps this development—the increase in flexibility and nondeterministic quality of scientific techniques—was essential before science could be adapted to the needs of the manager and the statesman.

1-2 Definitions

As an organized field, operations research is in its early growth. The inability of those who conduct research in operations to state succinctly their function in the research world has had both good and bad effects. On the negative side, an unsought aura of mystery has grown up around the name, clouding the picture of what is really going on and preventing an informed evaluation. On the positive side, lack of confinement to a conventional domain has meant that no group has been kept out and, consequently, that contributions have come from a wide variety of sources [5].

Descriptive Definitions

In the absence of previous exploration of underlying ideas to provide motivation, simplified definitions are rarely of any real value. However, for whatever use they may be to the reader, the following definitions of "operations research" are offered:

Of the large number of definitions with which the author is familiar, he prefers, "Operations research is the art of giving bad answers to problems to which otherwise worse answers are given." This definition is both scientifically and philosophically appealing; it lacks the finality which makes definitions of difficult activities rigid and ultimately useless. Another favored definition is that which refers to operations research as "quantitative common sense," giving it the spirit of modern science. Still another, due to J. Steinhardt, calls it "research into operations." This is not tautological, as might at first appear, since both the concepts of research and of operations immediately remind the individual of familiar ideas, e.g., "Operations are (usually) orderly and purposeful activities of organizations . . . through the men and materiel which compose the personnel and plant . . . ." [18]. Although much has been learned about operations in the past few years, we do not as yet have a formal definition which characterizes the structure of operations in a few words.

We finally give the "usual" definition of operations research: an aid for the executive in making his decisions by providing him with the needed quantitative information based on the scientific method of analysis [10, 14]. What is meant by the "scientific method" is left to the imagination of the reader and to the discussion in Chapter 2.

A Working Definition, or Procedure

Keeping in mind as a guide the various foregoing descriptions of operations research, it is possible to outline a step-by-step account of the processes undertaken by the operations analyst, i.e., the individual doing operations research, in the solution of a problem. The individual first *examines the objectives* which are presented to him to be fulfilled. He then recognizes for himself, or is presented with, a problem. This may be a goal, for which he must find the means of attainment, or almost as frequently, but usually in industry rather than in the military, the analyst *may be given the means and be asked to discover a goal, or use, for those means.* In either case, the formulation of the problem is necessary, its relation to the goal or to the means should be considered, and the correct type of data gathered and applied in an analysis of the problem. *He then determines the type of data* necessary to the solution of the problem, their availability, and how readily they are obtainable. There may be more than one way to approach the problem, and the analyst's next step is to generate and *examine the various practical methods* to decide which may lead to the desired ends. Having *selected and applied those methods* that seem best to attain the objectives (provided that an answer or specific result is possible), the analyst finally communicates his findings or results, without bias, to the decision maker from whom the original objectives were derived. It may be noted that the process starts with the objectives and ends with the analysis of the operation, followed by recommendations as to improved operation in *the light of analysis.* The formulation of the objectives and *the action* on the recommendations are not properly included within the activity of the operations analyst, being literally the boundary conditions imposed on the freedom of his operation.

In the course of attaining a set of desired objectives, the evolution of an entirely new theory or technique concerning a solution may occur. An integral part of the creative analysis of a problem is the making of a hypothesis. Ideas for advancing hypotheses may possibly be derived by noting the contiguity, similarity, or contrast to other tested and successful hypotheses or methods. When such similarities, etc., have been noted and examined, the analyst should indulge in a period of reflection to assure himself that what he has noted actually exists. An eager analyst on the track of the elusive connection between his hypothesis and its bearing on or application to the problem may otherwise discover only that he has made erroneous judgments and untenable assumptions. When he has assured himself of the validity of his ideas, he then proceeds with a synthesis of his solution, using in his equations or techniques known data leading to known results in order to test his hypothesis, particularly as to its predictive content. When, after adjustment and alteration, the equations or techniques satisfactorily reproduce the known results, they are verified whenever possible by application to test situations, using new data. If the results obtained are as predicted, the analyst has successfully completed a solution of a problem. His final act is the communication of this verified theory or technique to those with the means to act upon or use his results. This process of lucid communication is one of the more difficult aspects facing the individual.

A BRIEF HISTORICAL ACCOUNT OF OPERATIONS RESEARCH

1-3. Pre-World War II

History is replete with examples of man's ingenuity in solving operational problems. Using many elements of the scientific method, often unconsciously, man has brought the forces of imagination and resourcefulness to bear on matters closely related to his existence. The entire course of progress might be regarded, in at least one sense, as a continuing solution of the largest of all operational problems—that of living. Man's ingenuity and creativeness have taken him into a multitude of fields and in many directions [3].

One of the fields in which man seems to have excelled is that of war. He has evinced a very decided talent for devising weapons and other means for his own destruction. Nearly all the weapons of today have had their crude counterparts in antiquity. Sharpened sticks, stone-tipped spears, daggers, and swords were the forerunners of modern bayonets and commando knives. "Greek fire" and boiling pitch evolved into flame throwers and napalm bombs. All modern projectiles, whether rifle bullets, rockets, or atomic shells, reflect man's desire to extend his destructive reach beyond the length of his arm or the range of his slingshot.

The Trojan horse was a tactical weapon psychologically used. The infiltration among civilian refugees by North Korean and Communist Chinese soldiers during the Korean War or the use of submarines to surprise and sink enemy shipping would indicate that surprise, stealth, and camouflage used as tactics have changed but little.

Another field in which man's ingenuity has played an important part is that of industry. At one time, man clothed himself in animal skins (some, of course, still do); carried his burdens on his own back; walked when he wished to go somewhere; and grew, made, or caught all the bare necessities of life. Creativity and invention, trial and error, methodical consideration and evaluation by multitudes of men have advanced the state of industry to the point today where there seems to be no limit to the number and type of goods and services available for the amelioration of raw, physical existence.

In the past, research and development often have been confined to narrow and separate paths. Only one or two factors at a time have been considered. But as mathematical techniques were developed and refined, more detailed aspects of larger numbers of factors could be handled on a broader scale. It might be said that operations research began to emerge from its shell when prediction techniques advanced beyond those involved with throws of a die. Such men as Taylor, Erlang, Lanchester, Edison, and Levinson all contributed to the early development of operations research, although not under that name [12, 1].

In the late 1800s, pioneers in management consulting and industrial engineering were proving the value of scientific techniques in the fields of production and planning—techniques which operations research was to refine and extend. Frederick W. Taylor campaigned for the application of scientific analysis to methods of production as early as 1885 when he published the results of his own experiments in this field. Perhaps the most familiar example of Taylor's work is that involving a simple shovel. Taylor's objective was to determine the weight load of a shovelful of material that would provide a maximum of material moved with a minimum of labor fatigue. After many experiments with varying weights, Taylor arrived at a weight load considerably lighter than that commonly used but which proved the most effective over the period of a day. Taylor had brought an open mind to bear on an age-old assumption—that the laborer who carried the most material on his shovel was the most efficient worker. His contribution was not the development of new facts but the reevaluation of traditional and common knowledge in the impersonal light of science and his subsequent compilation of his discoveries into the basic laws of a new science.

In 1917, A. K. Erlang, a Danish mathematician working with the telephone company, published his most important work, "Solution of Some Problems in the Theory of Probabilities of Significance in Automatic Telephone Exchanges," containing his formulas for loss and waiting time which he had developed on the basis of the principle of statistical equilibrium; these now well-known formulas are of fundamental importance to the theory of telephone traffic. A few years after its appearance his formula for the probability of loss was accepted by the British Post Office as the basis for calculations respecting circuit facilities.

Erlang grappled with various technical problems, the first of which was to measure stray currents in the manholes of the streets of Copenhagen. He aided the engineers of the company in solving problems of a physical and mathematical nature in a Socratic manner. Instead of giving a direct solution of the problem posed, he preferred to enter into a somewhat lengthy discussion elucidating the subject from every conceivable point of view, thus aiding the inquirer to solve the problem independently. Erlang's ideas and work in telephony anticipated by almost half a century the modern concepts of waiting-line theory [1].

In Great Britain, an aeronautical pioneer, Frederick W. Lanchester, experimented with the problem of translating complex military strategies into mathematical formulas. "Aircraft in Warfare," published in 1916, set forth Lanchester's views on the effectiveness of quantitative analysis as applied to military strategy. Perhaps Lanchester's most significant expression of this type of analysis was the so-called "N-square" law.

concerned with the relationship of victory, numerical superiority, and weapon superiority. The law applies in a situation in which all units of both forces engage simultaneously and states that (1) numerical superiority, under the assumed conditions, is relatively more important than accuracy (kill probability, etc.) and (2) it is of utmost importance to deploy available combatant units in a single force and, if possible, to split up the enemy force.

The reader will find in Ref. [15] Lanchester's chapter on mathematics in warfare which treats in some detail Nelson's strategy at the Battle of Trafalgar, which may be considered a classic example of the "N-square" law.

In the United States, Thomas A. Edison, as chairman of the Naval Consulting Board during World War I, used a "tactical game board" to plot and analyze the effectiveness of zig-zagging and other techniques by merchant ships in order to evade enemy submarines.

During the 1930s, Horace C. Levinson began to apply scientific analysis to the problems of merchandising. His work for L. Bamberger and Company involved, among other things, a study of customer buying habits, the response to advertising, and the relation of environment to the type of articles sold. In the 1920s, Levinson had completed a study for a mail-order house on the effect of speeded-up shipment on the customer acceptance or rejection of c.o.d. packages. His success in predicting general human reactions from the collection and analysis of great quantities of data led him later to initiate a study for Bamberger on the effectiveness, in terms of added sales, of keeping department stores open at night.

Levinson not only extended the sphere of operations research to include an examination of the effectiveness of advertising, placement of merchandise, and other techniques of merchandising, such as the innovation of night openings, which lend themselves to quantitative analysis; he also refined the use of the model, employing higher mathematics.

1-4. World War II

Much has already been said about operations research during World War II, and it is not the purpose here to add to the already abundant and published material on this subject. Examination of the references at the end of this chapter will supply many of the details to the interested reader. It has been felt sufficient to present here two examples of the application of operations research to the prosecution of the war in order to give a representative idea of the problems encountered.

Early in the war, when the Allies were struggling to gain command of the skies over Europe, the number of participating aircraft in bombing missions, as well as of missions themselves, was increasing. Improvement of bombing accuracy had always been recognized as being of primary concern, but few recommendations had been made that led to anything approaching the desired accuracy. Following the assignment of several operations analysts to the Eighth Bomber Command in England in 1942, accuracy studies were made that led to more precision. In comparison with the existing accuracy in 1942, when less than 15 per cent of bombs dropped landed within 1000 feet of their aiming point, the results of analysis increased accuracy to greater than 60 per cent in 1944 [2].

It was immediately obvious that many factors governed the accurate placement of bombs on the ground. These included such items as the size and type of aircraft formation, the number of independent aimings per formation, the selection of lead bombardiers, their training, tactics used on the bomb run, etc. As a measure of the effectiveness of bombing, it was decided to use the percentage of bombs falling within 1000 feet of the aiming point. To gain an insight into previous bombing results and as a check on the improvements in accuracy resulting from changes in bombing methods, cameras were placed on several planes in each formation to provide a record of the results achieved by that formation. In addition, other vital statistics—number of aircraft in the formation, dimensions of the formation, number and type of bombs carried, altitude and speed over target, weather conditions, enemy opposition, etc.—were gathered.

Analysis of the photographs seemed to justify the assumption that the bombs could be considered as being distributed evenly in rectangular patterns over the target area. From consideration of the variations of these rectangles with the use of various operational suggestions, further recommendations were derived. Recommendations such as the nearly simultaneous release of their bombs by all the bombardiers instead of the practice of each bombardier aiming and releasing his own bombs, the salvoing of bombs instead of presetting them to release in a string, the decrease in the number of aircraft per formation from 18 to 36 to the smaller amount of 12 to 14, and greater attention to precision flying all resulted in quadrupling the delivery effectiveness over a period of two years.

Offensive patrols by Allied aircraft against surface shipping and enemy submarines originally were carried out on a rather random basis. It was believed that a study of the factors relating to systematic patrol would lead to greater chances of sighting, and thus of sinking, enemy ships than had thus far been achieved. Through consideration of the contact ranges, flying speeds, distances and areas to be covered, etc., it was believed possible to devise flight paths across a body of water and to arrange schedules in such a sequence that there always would be a patrolling aircraft in the air prepared to intercept any shipping or submarines that might attempt to pass through the area under surveillance.

The contact range of a patrolling aircraft was determined by the effective "sighting" distance of its radar. Half of this distance marked at either side of the flight line delineated a "swept" path across the particular body of water. The flying speed of the aircraft determined the scheduling of succeeding departures, whereas the area to be patrolled decided the number of paths to be flown. Both of these last considerations, in turn, along with maintenance and overhaul time on the ground, determined the number of planes required for the operation.

One such devised flight schedule was applied to the patrol of the "waist" of the South Atlantic with the intent to intercept German shipping of war materiel [17]. The success of this method of patrol is illustrated by the fact that three German blockade runners were caught on their way to Germany as they returned from Japan with raw materials. A fourth escaped by virtue of an elaborate disguise, and a fifth, while penetrating this particular blockade, was trapped by a similar patrol of the Bay of Biscay.

These are but two examples of the effectiveness of operations research as applied to the military during the war.

1-5. Post-World War II

Since World War II, operations-research groups have been continued in all branches of the armed forces. These groups are now known as the Operations Evaluation Group (Navy), the Operations Analysis Group and the RAND Corporation (Air Force), and the Operations Research Office (Army). A group called the Weapons Systems Evaluation Group, reporting to the Joint Chiefs of Staff, continues to function under the Institute for Defense Analysis. In addition, there are several smaller groups of analysts working within the military-service framework. Similar groups also exist in the United Kingdom.

At the close of the war, the industrial climate on both sides of the Atlantic was ripe for the introduction of operations research into business planning. Operations-research work during the war years had been "top secret." As some indications of its value were released and, later, more detailed descriptions of actual studies accomplished, its possible contributions to the business world became more apparent. There was no denying the importance that the military continued to attach to operations research after the war ended. Warfare, per se, seemed to be undergoing a rigorous reappraisal at all three levels—technological, tactical, and strategic. Millions of dollars had been set aside for this type of research and analysis. Industry needed to revamp its production and organization to service peacetime needs quickly. In the United States, it was a question of competition. In Great Britain, a critical economic situation demanded drastic increase of efficiency in production and in the development of new markets.

Industrial operations research in Great Britain and in the United States developed along different lines. In Britain the nationalization of a few industries provided a fertile field for experimentation with operations-research techniques in industries as a whole. The pressures of the economic situation, unrelieved by the end of the war, resulted in the application of operations research to government, social, and economic planning. For example, a "Survey of Sickness," implemented during the war years under the Governmental Social Survey, provided a comprehensive picture of the general health of the population. Operations-research studies for the Ministry of Food surveyed food consumption and expenditure patterns to predict the effects of government food and price policies on nutrition and the family budget.

In the United States, the application of operations-research techniques to business operations was slower because many executives, already accustomed to management consulting and industrial engineering, believed that operations research was merely a new application of an old technique. Heightened competition also slowed down the development of the state of the art because companies had no desire to aid competitors by releasing the results of studies undertaken.

It is not unlikely that many of the objectives of operations research in industry had already been pursued by persons called management consultants, specialists in quality control, time-and-motion experts, marketing analysts, efficiency engineers, and design engineers. However, the intensity and purpose of the new approach have provided broader, more precise, and always quantitative studies of industrial problems, often closer to the objectives of the entire operation. The executive can, more than before, base his decisions on analyses having these objectives and values as their primary interest.

Many industrial concerns are now utilizing operations analysts in an effort to reduce costs, increase production, and to speed their products to the consumers. The aircraft industry was, perhaps, one of the first to feel the need for the advanced techniques and broad approach of operations research in order to keep step with Air Force developments. Convair and Lockheed are among those who have established their own internal units to conduct complementary research on facets of new developments, such as human reactions to new high-altitude and jet-propelled planes and the development and integration of new capabilities with new tactics. U.S. Rubber, Sun Oil, and Du Pont have either placed operations analysts directly on their own staffs or obtained their services through contracts with private industrial-operations consulting firms.

A large number of similar groups are presently employed by industry to do operations research on broad areas such as transportation, communications, agriculture, merchandising, and various aspects of manufacturing. In the study of production and inventory control problems, traffic and flow problems, sales tactics and optimum effort in sales competition, many operations-research techniques are directly applicable to industry. The study of optimum economic measures, optimum transportation costs, traffic congestion, securities transactions, the utilization of natural resources, shipping, etc., is directly applicable to business and society.

As a concrete example of operational techniques applied to problems of industry, consider the situation in which a large manufacturing concern wished to determine how many additional salesmen were needed to handle its expanding line of products and how the effort of the existing salesmen could be improved [21]. It was desired to examine three factors relating to the problem: the allocation of sales time by salesmen to their accounts (current and prospective), the optimum mixture and number of accounts assigned to each salesman, and, time permitting, the personal characteristics required of a salesman.

In the examination of the first factor, the problem was to relate the time spent on an account to the dollar volume of sales derived from the account and to classify the types of accounts handled, since some types required more time than others for fruition. However, after the collection and tabulation of data deemed pertinent to the problem, it was found that there appeared to be no clearly defined connection between the number of sales calls and the amount spent by the account. Then it was recognized that there was no difference in the average annual dollar volume between accounts that received an increased number of calls and those that received a decreased number over several years. It was suggested that some reduction in the number of calls per account could be made without causing sales to be reduced, and this was verified, on an average, by reference to the data. This reduction in calls on established customers allowed the salesman more time to make new contacts and thus to increase his total sales. It also gave a basis for determining the number of salesmen needed to reach certain sales goals. Further, by limiting the number of calls per new prospect, the actual increase in the number of new accounts could be expected to reach 11 per cent. In a study of how salesmen spent their time, the ratio of the time spent on paper work connected with their accounts to the time spent calling on customers was discovered to be about 1/1, a ratio considered to be overly high.

On the basis of these studies, a number of recommendations were made: (1) that plans for hiring additional salesmen be curtailed; (2) that time spent in the office be reduced; (3) that the number of calls to prospects be limited, and (4) that, after further study, the office paper work and sales effort in the field perhaps should be separated. As a result of action on these recommendations, the sales-volume objectives for the first year of a five-year plan were met. The annual saving in not acquiring additional salesmen was equal to about 25 times the cost of the study.

This is just one example, of many available, to illustrate the problems and procedures of operations research that may be encountered in industry. The techniques used in this solution were simple. A variety of other techniques have also been in use. Thus, for example, linear programming has found wide application in solving such diverse problems as determining the optimum utilization of machines, most efficient scheduling procedures, optimum salary-scale systems, minimum-cost transportation routing in the oil-refining industries, etc.

Queueing theory has had appreciable success in handling congestion problems in the transportation and industrial media. An outstanding application of this theory has been made by the Port of New York Authority in an attempt to speed up traffic operation at the New York tunnels and routes leading to them.

Attention to inventory-control problems by a large number of people has produced a variety of models.

The interest generated in operations research by industry, the military, and the government has influenced several colleges to include courses on its methodology in their curricula. Among the first schools to do so were the Massachusetts Institute of Technology, Columbia University, The Johns Hopkins University, Case Institute of Technology, and the University of Michigan.

Another indication of the interest in operations research has been the formation, in the early 1950s, of two considerably active societies, The Operations Research Society of America (with its journal *Operations Research*) and The Institute of Management Sciences (with a journal *Management Science*). In England, similar interest has been evinced by the formation of the Operational Research Society (with its journal, the *Operational Research Quarterly*). More recently, the Operational Research Society of France (Société Française de Recherche Opérationelle) was formed and has published a journal, *La Recherche Opérationelle*. In Italy, the Center of Operations Research (Centro per la Ricerca Operativa) publishes a bulletin of the same name, and in Austria the publication *Unternehmensforschung* recently made an appearance.

A rough survey of a large share of published case-history literature in operations research yields the following outline of some industrial applications. The main journals examined include those previously mentioned and, in addition, the *Naval Research Logistics Quarterly, Journal of the Society for Industrial and Applied Mathematics, Econometrica, Harvard*

Business Review, Journal of the American Statistical Association, Fortune Magazine, and the *Proceedings of the Institute of Radio Engineers.*

A. Concern with large problems of industry and their relation to the entire economy (top-level planning and control)
   1. Optimum utilization of capital
   2. Expansion and growth—geographic distribution of plants in multiple-plant operation
   3. Changing markets and effect of evolution of technology
B. Production
   1. Costs
   2. Scheduling (linear programming)
      a. For a single period
      b. For an arbitrary number of periods
   3. Machines—capital-equipment analysis
      a. Capacity—optimum loading
      b. Reliability and maintenance of equipment
      c. Defectives
      d. Preventive sampling
      e. Repair as a queueing problem
   4. Efficient utilization of personnel
   5. Design of products
   6. Congestion in service operations
   7. Capacity
   8. Quality control
C. Inventory control
   1. When the demands are known
   2. When the demands are unknown
D. Sales—marketing
   1. Promotional effort and effectiveness of advertising
   2. Price forecasting
   3. Forecasting of manufacturing needs
   4. Analysis of consumer buying habits
   5. Optimum geographic distribution of publicity
E. Purchasing
   1. Optimum size of order
   2. Selective purchasing
   3. Consumer buying habits
F. Transportation
G. Computers in industrial operations research

THE SCOPE OF OPERATIONS RESEARCH

In its recent years of organized development, operations research has entered successfully many different areas of research for the military, the government, and industry. At what point does operations research cease to be effective? What are its limitations? How broad is its scope?

It is possible, at least from one standpoint, to classify and outline four major areas of research which have proved amenable to the peculiar techniques of operations research [11]. (A somewhat different classification will be given in Chap. 2.) Very briefly, they cover broad problems of prediction for the future, questions of strategy in competition, problems that the military would classify as tactical, and equipment-systems problems. Studies in all these areas have proved equally valuable to both industry and the military. Governmental activities have been able to make some use of prediction for the future in developing economic and social policies.

1-6. Predictions

Broad problems of prediction for the future in industry generally concern either an industry's future demand for a product or the possibility of a technological advance creating a market for a new product. The fluctuation of demand with time is the crucial factor in both of these problems.

To illustrate an industry's future demands for a product, suppose that a chemical company is considering diversification. It might well conduct an operations-research study of several possible new products to determine the one best suited to the company's setup. This would involve both an analysis of the ease with which the company could add the production of any one of the products to its organization and a study of the future demand for such products in the industry as a whole. A prediction of the future demand for any one product can be made on the basis of its past use, of the increase or decrease of its usefulness in the past decade, and the protracted expansion of the entire industry. Estimated competition from other companies within the industry must also be taken into consideration before a recommendation for any one product can be made.

1-7. Competitive Strategy

Game-theory analysis of strategy can be applied to industry. A simple example would be the offer of a $2000 fur coat in August for $1400. If this is considered simply as an action by the merchandiser, the store would incur a $600 loss, since the customer attracted by the August sale might well buy the coat in October. However, if a competing store offered the same sale in August and drew the customer away, the first store would suffer the loss of any possible profit, since it had not offered the coat at a sale price. This is an example of the minmax theory; the original store did not try to get the maximum price for the coat but preferred to minimize the possible strategy of a competitor.

1-8. Tactical Optimization

Operations research has proved an invaluable aid in solving problems that the military would classify as tactical. The problem of the operations-research team is, basically, to observe a given operation in action and, once having understood its basic components, to construct a model simulating it. If the operation is thus fundamentally comprehended, it is possible to determine the outcome of various tactics or courses of action applied to it to attain a given objective—in industry, a greater return in dollar profits or the minimizing of a loss; in the military, victory or the effective countering of an enemy's strategy.

For example, the railroad industry must constantly reroute empty freight cars to take care of freight shipments at various points across the country. An operations-research study showed that this appreciable loss could be minimized by routing and scheduling empty cars in advance. It is often possible to estimate a day or so ahead of time the requirements of various shippers for cars as well as available empties and their location. If cars are distributed in terms of individual-day requirements, a car will be routed to the nearest requirement point. The next day it may have to travel double the distance, incurring considerable expense over the two-day period. If requirements for the next day are estimated in advance, it becomes possible to chart the first day's distribution in terms of the demands for the same cars at different locations on the second day. Thus, a car may have to travel 25 per cent farther on the first day but may reduce its traveling distance by 75 per cent on the second day. Linear programming, when applied to this problem, does not always reduce loss by 50 per cent, but the total cost of distribution for the two days will usually be considerably less than that incurred by the method of daily distribution.

1-9. Equipment and Systems Problems

Operations research can also be applied effectively to methods of inventory in the military, industrial, and government fields.

An example of governmental tactics in maintaining inventories was the enactment of the legislation for the stockpiling of strategic materials to ensure preparedness for possible national emergencies. This policy was undertaken after a thorough investigation of the state of manufacture of such materials, the possible immediate increase in their production and the long-range expansion of production, and the predicted need at a given time for strategic materials. The policy of stockpiling was implemented because the study showed that we would not have sufficient inventories of these materials on hand to meet the demand of a national emergency.

One of the many problems of inventory that confront the military is the maintenance of a sufficient supply of equipment to cope with an emergency, even though developments in the state of the art and production of new equipment are constantly rendering these large stocks obsolete. Taking into consideration the many other factors involved, such as the policies and military potential of our government and that of an opponent and predictions of the development of the state of the art, operations research can arrive at some point of inventory maintenance which will leave the country sufficiently prepared with a practicable outlay. This specific point is arrived at through rational analysis and prediction, not through whim or guesswork.

1-10. Future Applications

Some future applications of operations research may lie in the fields of education and sociology. For example, an operational approach to the organization of general school curricula better suited to the requirements of our present society might produce a well-integrated program, emphasizing creativity and motivation in the individual. In the field of medicine, much of the student's education in the early years involves memorizing. With the application of ideas of symmetry, etc., to the study of general anatomy and with the introduction of simplified nomenclature, study might be facilitated, with a reduction in learning time.

A scientific approach could also be applied in the social field. For example, the expansion of population has caused some to demand birth control by legislative fiat. An operational analysis of the threat of overpopulation might suggest other less culturally and religiously controversial solutions.

PERSPECTIVES

1-11. Qualities of the Operations Analyst

Since operations research often requires different, sometimes unorthodox, approaches to problems it may be desirable to discuss some of the helpful qualities of an operations analyst. The qualities briefly discussed below are common to any competent scientist. The distinguishing feature is that in operations research the individual usually is faced with an operation about which little, if any, theoretical knowledge is available; hence considerable originality and enterprise for solving specific problems are required.

Foremost in the array of desirable individual attributes is the *desire to investigate*, to solve the many problems that may be present and to anticipate many of these problems before they officially emerge for solution. Broad *imagination* which covers many fields of interest is another important quality which frequently plays a major part in the approach and solution of a problem and in seeing the full scope of a problem and the subsequent problems that may arise from the utilization of a particular solution.

Usually there are several major and minor parts to a problem. Another desirable quality is the *ability to distinguish the important considerations* from the usual maze of facts, figures, and opinions constituting the minor considerations. When the practitioner tests a model of his problem, the inclusion of many minor constraints often makes the model too cumbersome to work with, in that there may be more variables than he can handle conveniently. *Ingenuity and resourcefulness* are the main aspects of originality, desirable in several phases of operations-research work: in the formulation of the problem, in the acquisition of the necessary information for its solution, and in the meaningful application of these data to the analysis of the problem. The operations analyst is not told what data to seek in order to solve a problem, nor how to seek them; these decisions are his own, and often they require trips into the field as well as observation of many seemingly unrelated factors.

Once he has acquired the information that he believes to be necessary to the problem, the practitioner should have the *technical skill and knowledge* to apply and relate the information so that it may yield useful results. He should be able to make further measurements and experiments to verify the sufficiency of his data and to manipulate at will these data and the mathematical techniques in order to achieve working hypotheses and relative predictions. In all his operations, a *sense of duty and responsibility* is essential, one that will take him out of his comfortable office or laboratory into the field when necessary.

Frequently, more than one skill or specialty is required for the successful solution of a problem. It is relatively impossible for one man to have at his command detailed technical skills and knowledge in more than one or two fields. Thus, when the situation so demands, several scientific specialists may be called together to work as a team, utilizing their various skills toward the desired objective. A necessary attribute in each of these men is, therefore, the *ability to work as part of a team*, neither trying to dominate it nor allowing himself to be subordinated by it.

When the operations analyst, as an individual, or the team collectively, arrives at the solution of a given problem, he, or it, should have the courage of his convictions and the *integrity* to accept the results of his research and to defend these results even if they may be unpopular with those who posed the problem. Finally, before any solution of a problem can be accepted and implemented, it must be communicated and explained to those who have the power of implementation and action. The day is not yet here when ideas may be transmitted clearly and instantaneously from one person to another, and so *facility in verbal and written expression is highly desirable.*

REFERENCES

1. Brockmeyer, E. H., L. Halstrom, and A. Jensen: The Life and Works of A. K. Erlang, *Transactions of the Danish Academy of Technical Sciences*, no. 2, 1948.
2. Brothers, L. A.: Operations Analysis in the United States Air Force, *Journal of the Operations Research Society of America*, vol. 2, no. 1, pp. 1–16, February, 1954.
3. Camp, G. D.: Operations Research: The Science of Generalized Strategies and Tactics, *Textile Research Journal*, vol. 25, no. 7, pp. 629–634, July, 1955.
4. Churchman, C. W., R. L. Ackoff, and E. L. Arnoff: "Introduction to Operations Research," John Wiley & Sons, Inc., New York, 1957.
5. Fair, W. R.: Operations Research at Stanford Research Institute, Seminar lecture, Menlo Park, March 10, 1954.
6. Goodeve, C.: Operations Research as a Science, *Journal of the Operations Research Society of America*, vol. 1, no. 4, August, 1953.
7. Herrmann, C. C., and J. F. Magee: Operations Research for Management, *Harvard Business Review*, vol. 31, no. 4, pp. 100–112, July–August, 1953.
8. Horvath, W. J.: Operations Research—A Scientific Basis for Executive Decisions, *The American Statistician*, October, 1948.
9. Johnson, E. A.: The Scope of Operations Research, Johns Hopkins University lecture.
10. Kittel, C.: The Nature and Development of Operations Research, *Science*, vol. 105, no. 2719, pp. 1–4, February, 1947.
11. Lamar, E.: Lectures on Operations Research, Operations Evaluation Group, Massachusetts Institute of Technology, Washington, D.C.
12. McCloskey, J. F., and F. N. Trefethen (eds.): "Operations Research for Management," Johns Hopkins Press, Baltimore, 1954.
13. McCloskey, J. F., and J. M. Coppinger (eds.): "Operations Research for Management," vol. 2, Johns Hopkins Press, Baltimore, 1956.
14. Morse, P. M., and G. E. Kimball: "Methods of Operations Research," John Wiley & Sons, Inc., New York, 1951.
15. Lanchester, Frederick William: Mathematics in Warfare, in J. B. Newman, "The World of Mathematics," vol. 4, Simon and Schuster, Inc., New York, 1956.
16. Rapoport, A.: "Operational Philosophy," Harper & Brothers, New York, 1954.
17. Steinhardt, J.: The Role of Operations Research in the Navy, *U.S. Naval Institute Proceedings*, vol. 72, no. 519, May, 1946.
18. Steinhardt, J.: Speech, Operations Evaluation Group, Massachusetts Institute of Technology, Washington, D.C.
19. Steinhardt, J.: Terminal Ballistics, *Journal of the Operations Research Society of America*, vol. 3, no. 3, pp. 231–238, August, 1955.
20. Stone, M. H.: Science and Statecraft, *Science*, vol. 105, no. 2733, pp. 507–510, May 16, 1947.
21. Waid, C., D. F. Clark, and R. L. Ackoff: Allocation of Sales Effort in the Lamp Division of the General Electric Company, *Operations Research*, vol. 4, no. 6, pp. 629–647, December, 1956.

PART 1

**SCIENTIFIC METHOD—TRUTH
MATHEMATICS AND LOGIC—VALIDITY
OUTLINES OF SOME USEFUL MATHEMATICAL MODELS**

CHAPTER 2

SOME REMARKS ON SCIENTIFIC METHOD IN OPERATIONS RESEARCH

2-1. Introduction

In order to balance the material presented in this book, the well-explored subject of scientific method will be briefly examined to distinguish between mathematical validity and scientific truth. Another purpose is to give some perspective for the fundamental notions needed in operations research and already discussed in the first chapter. A structure is provided which can be expanded from material in the literature, if so desired. A thorough understanding of concepts of this sort is conducive to good habits of mind which are, consequently, an asset in attacking and solving problems. There are many questions to be asked before, during, and after the solving of a problem, if indeed, there should be need for a solution at all. The classification and relating of ideas into a well-integrated structure provide a powerful instrument which can be used to penetrate more deeply into those domains where the mind interacts with external factors. The success of these methods has encouraged man to provide solutions of problems, to prevent others from arising, and to predict future occurrences and methods of handling them according to systematic, often called "reliable" procedures.

As yet, no one has advanced a closed system of ideas which outline and delineate the scientific method. Such a task, by the very nature of its subject, is doomed to failure. As long as the human mind expands its capacity to obtain and handle purposively information regarding the world of experience, the scientific method will continue to increase in refinement. Its limits are pushed farther out to encompass as many aspects of human life as can be rationally included under the scientific domain.

2-2. Operationalism

From the definitions of operations research given in the previous chapter, it is not difficult to discern the fact that the operations analyst is basically a scientific eclectic. He chooses those methods which are suitable for his work. He does not hesitate to adopt any creative methods which enable him to arrive at reasonable answers, "bad answers to improve on existing worse answers." Because of urgency, he cannot always wait to develop the theoretical tools to solve a simple problem, particularly since a distinguishing characteristic of his work is the frequent emphasis on end results.

Often it is essential that the operations analyst be capable of completing the analysis of a problem, even if it requires thorough familiarization with subjects in which he has little interest. This is a crucial point, since many individuals find it difficult to reconcile themselves to such demands. Yet, because of the large number and variety of problems encountered, it is possible to make good use of individuals who are not inclined to look into problems outside their specialties.

The human mind, as Comte expressly asserts, demands unity of method and doctrine. The construction of a scientific framework requires a certain respect for the facts of experience, open-mindedness, an experimental trial-and-error attitude, and the capacity for working within an incomplete framework. Einstein, in attempting to differentiate among the various mental attitudes which may be found in a scientist, points out that he may appear as a realist, an idealist, a positivist, or a Platonist. As a realist, he attempts to describe objectively the facts of experience. As an idealist, he depends on his theories and ideas as means for constructing convenient concepts. As a positivist he does not speak of realities at all and regards nothing as ascertainable or meaningful beyond sense experience. Even his theories must stand the rigorous test of logically justifying these sense data. As a Platonist, he considers logical simplicity as an essential tool for his research.

In "The Logic of Modern Physics," published in 1927, P. W. Bridgman [3] advanced the idea of *operationalism*. In this important work in the philosophy of science, Bridgman developed a theory for the definition of scientific concepts. This theory has, in fact, broad applications to other areas of thinking [1]. Operationalism emphasizes the role of operations and the measurements associated with them in the definition of concepts. Since all the processes by which concepts are ascertained are relative, notions of absolutes, such as absolute time and space, have no place in scientific thinking. Operationalism suggests that concepts always should be defined in terms of operations. Thus, for example, the concept of time is meaningless apart from the concrete and specific operations which are employed in its measurement [22].

Of interest are the operational interpretations of definitions and of truth. Operationally, a definition is a process of making a word usable to the individual interested in the definition. Definitions are not always absolute. A child may be introduced to the concept of time by means of a watch, a calendar, day and night, etc. The fact that the theoretical physicist may define time as the independent variable in the differential equations of motion cannot serve any useful purpose for a beginner; hence, the notion of adaptable definitions. This notion does not conflict with the need to define a concept before proceeding into lengthy discussions about it. Moreover, it allows for tolerance and the admittance of a diversity of experience.

The ability to communicate information is strongly associated with the background experience of the communicator. Therefore, it would be difficult for two people who have not shared a common experience to communicate. Actually, even though experiences may be shared, language and vocabulary differ, and the meanings of words are not always the same to different people.

Generally, a definition either assigns names to objects and phenomena, enabling reference to them, or elicits an experience for which a word stands. Verbal definitions save the trouble of simulating the experience or the object involved in the definition, a task rather impossible without the symbolism of language. An operational definition describes operations—physical or mental—to be performed in order to observe the thing defined or its effect.

A distinction is made by operationalism between the concepts of validity and of truth. Through correct use of logic, one validly obtains new assertions from given ones. Validity requires inner consistency with other assertions. Truth requires verification by examining things other than the assertion. It relates language with experience. With mathematics, one associates the concept of validity; with empirical science, truth. This distinction is not new and dates back to Kant, who distinguished between analytic propositions of validity proper and synthetic propositions of truth proper [14]. Validity of propositions belongs to the subject matter of logic, where there are formal rules for resolving problems. This topic will be partly illustrated through the use of truth tables in the next chapter. As for truth proper, there are no fixed rules for determining truth. An assertion is operationally true if all three of the following conditions are satisfied:

1. The assertion implies predictions, explicit or implicit, to be tested by conceivable operations. (If no conceivable operations are discerned or suggested, the assertion is operationally meaningless.)

2. The operations have been carried out to test the predictions of the assertion. (Otherwise the assertion is indeterminate.)

Remark: An indeterminate assertion is illustrated by the case of individuals who are judged neither guilty nor innocent; e.g., "not proven" is permissible under Scottish though not under English law. Nearly a century ago, Madeline Smith was tried in a Scottish court as a suspected murderer of her lover, Emile L'Angelier. Strong circumstantial evidence was brought forth. It was thought that she had reason and opportunity but the case could not be proved, despite the fact that she was believed guilty. The verdict was "not proven" since action on belief alone could not be justified in a court of law. Thus the proposition that Madeline Smith had or had not murdered her lover was left indeterminate. This example taken from human affairs, where the life of an individual was involved, shows the near impossibility of obtaining a positive answer to some propositions.

3. Finally, the predictions have been verified by the operations. If the operations fail to verify the predictions, the assertion is said to be false.

This definition of truth does not solve all problems in one stroke. However, it is suggestive of a type of patience and tolerance useful in scientific exploration. Frequently, elaborate discussion and analysis are required to determine the operations to be carried out. Assertions should not be dismissed as meaningless when operations for testing them are not immediately perceived. Each situation must be handled according to its own merits and within the scientific framework.

An apparent dichotomy in the scientific method may be discerned when attempting to distinguish between mathematics, on the one hand, and the empirical sciences, which include the natural and behavioral sciences, on the other hand [1]. An interesting method of observing the distinction between the two may be found in the vocabularies used to describe these sciences. In mathematics, the vocabulary includes words such as axiom, definition, theorem, corollary, deduction, proof, consistency, etc. In the empirical sciences, the vocabulary includes the words data, hypothesis, theory, law, classification, measurement, experiment, and verification.

In mathematics, proof is based on the deductive method, whereas nothing in the nature of strict proof enters into the natural and behavioral sciences, whose method is inductive, attempting to generalize on observations. The concept of probability is used in the latter to indicate the degree of confidence established by the proof. Careful observation, in order to form conjectures regarding a physical phenomenon under study, is an integral part of the empirical sciences but is not needed in the rational sciences. Verification decreases doubt and increases the probability of the truth of hypotheses. Even though the latter may then be taken as accepted theories, they are always subject to disconfirming predictions. Pragmatically, then, a proposition is true if it solves the problem in an acceptable way. Acceptability depends on those who must be convinced by the individual advancing an idea.

The deductive method of logic does not require that postulates be true in the empirical sense or even that they correspond to anything real. The *validity* of properties deduced from axioms depends only upon the correct use of the rules of logic. However, the properties are *true* if the axioms and postulates are true and can thus be applied to a certain subject matter in which the truth of the axioms may be shown by correspondence between the facts and the axioms.

In general, a scientist must make use of both the deductive and the inductive method. Thus logical deduction, a tool of the mind, has enabled the formation of theories which, when subjected to experimental tests, were found to succeed in predicting the occurrence of phenomena. One is usually forced to observe a few events and formulate hypotheses based upon them. Then after the probable truth of these hypotheses is confirmed by means of data, deductive methods may be applied to arrive at some logical consequences of the hypotheses. The predictions of the latter are, in turn, checked for verification. Note that there is a delicate difference in the two methods. The rational scientist may adopt any assumptions to serve his purpose; the natural scientist must make careful observations to make assumptions, which are only tentative and are subject to change on his obtaining new evidence. From a creative standpoint, the method of making arbitrary—not necessarily unreasonable—assumptions is a useful method even for the empirical scientist, particularly when he is not pressed for time. For then very frequently he can apply his deductive powers to obtain useful predictions. The choice of assumption must, of course, be made on the basis of some experience.

It is clear that an operational approach to operations-research problems is a powerful tool which must be employed in the stages immediately following hypothesis making. As for the latter, it is preferable that the point of view adopted be flexible since what may appear operationally meaningless or indeterminate may, with additional facts and analyses, become quite reasonable. That is to say, the hypotheses formed with regard to a specific problem should be posed from a variety of viewpoints in order to precipitate an operationally acceptable one, if such is at all possible.

2-3. The Scientific Method in Operations Research

The scientific method in operations research may be subdivided into three phases: the judgment phase, the research phase, and the action phase. Of the three, the research phase is the largest and longest, but the other two are just as important since they provide, respectively, the basis for and implementation of the research phase [13].

The judgment phase includes (1) a determination of the operation; (2) the establishment of the objectives and values pertinent to the operation; (3) the determination of suitable measures of effectiveness, and (4) the formulation of the problem relative to the objectives.

The research phase utilizes (1) observation and data collection for a better understanding of what the problem is (What, for example, does traffic congestion mean and how may one formalize the problem to make it amenable to analysis?), (2) formulation of hypotheses and models; (3) observation and experimentation to test the hypotheses on the basis of additional data; (4) analysis of the available information and verification of the hypotheses, using the preestablished measures of effectiveness; and finally (5) prediction of various results from the hypotheses, generalization of the results, and consideration of alternative methods.

The action phase in operations research consists of making recommendations based on the results of the above procedure and the subsequent decision process by those who first posed the problem for consideration or by any one in a position to make a decision influencing the operation in which the problem occurred.

2-4. The Judgment Phase

The Operation

The question, to which a brief answer was attempted in the first chapter, may be asked: What is an operation, and how can the scientific method aid in the implementation of an operation? [26]

An operation may be considered as a collection of few or many separate actions or functions dealing with raw materials, e.g., men and machines, that, when united, form a coherent structure from which action with regard to broader objectives is attained. The act of assembling a watch is an operation consisting of many separate motions and steps contributing to the completed timepiece on the wrist. A factory may be considered an operation that is a collection of many smaller operations; similarly, a military establishment or maneuver may be so considered. These examples are broad and quite self-evident, but it often helps to consider the various types of operations, both large and small, that may occur and do occur continuously. It is also interesting to remember that any conceivable operation has had in the past, currently has, and will have in the future, problems connected with its successful completion. How well the operation is completed depends on the understanding of the operation itself, of its problems, and particularly of the methods used in its execution.

Objectives and Values

The values assigned to something vary with the outlook and framework of association of the individual (or organization). The individual, in turn, has acquired his outlook and association from his experience and environment. Thus it is often helpful to be familiar with the experience of one who poses a problem relating to an operation in which he has a vested interest as well as to evaluate carefully the problem itself in order to arrive at realistic values and objectives for the problem.

The problem, or even the operation itself, may be motivated by individually assigned values, such as a desire for power, recognition, or money. A person may be seeking personal satisfaction, security, or independence, or he may be acting from aesthetic, altruistic, or religious motives. An individual may desire to save time by gaining speed and efficiency in the operation. This desire, in turn, may be motivated by competition affecting the individual or by his own intellectual drive, curiosity, or energy. The objectives of the problem, therefore, will, to a large extent, reflect the type of values set upon it.

In the judgment phase, care must be exercised in defining the frame of reference of the operation in question. Basic determinations should be made as to the type of situation, i.e., manufacturing, engineering, tactical, etc.; the degree of boldness acceptable in the solution; the range of risks involved; the impact of a solution on other areas; and even the degree of management sophistication. Objectives and values, whether economic, ethical, aesthetic, etc., should be examined with sufficient care to facilitate a clear approach to the solution of the problem. It is wise to ascertain the characteristics of the situation under study to discover if it is isolated or recurrent, how much information may be immediately available concerning it, and what the time limits might be for reaching a solution. It is helpful to know whether reversibility is necessary in the resultant solution. Separate from, and in addition to, the need to know desired time limits is the question of what degree of accuracy might be required in the results. It is desirable to know whether the feedback of acquired information into the operation is necessary.

Methods of attack should be decided upon when the problem is formulated. These may be either speculative, in which the solution of the immediate problem is put aside in the search for alternative approaches and for fundamental causes, or they may be direct in that the immediate problem is solved without an attempt to eliminate its causes. This latter method suboptimizes (i.e., it is not concerned with the broader view) more than the former and is subject to greater pitfalls. However, it is often more expedient and useful than the former, particularly when time limits are involved.

With the recognition of the values and objectives pertinent to a problem, there remains the question of determining the proper terms in which the solutions of the problem may be interpreted, so that the objectives are correctly expressed. Such terms are called *measures of effectiveness*.

Measures of Effectiveness

A measure of effectiveness is a criterion which measures the extent of success of a solution, as related to the objectives, when applied to a problem arising in an operation. More generally, it may be taken as a measure of the success of a model in representing a problem and providing its solution. It is the connecting link between the objectives and the analysis designed to provide corrective action.

By analogy with the physical sciences, where a study of errors and confidence limits is an integral part of judging how well a theory describes the existing facts and predicts successfully new future outcomes, operations research uses measures of effectiveness to test the success of a solution. Because of the required urgency and the higher values placed on problems dealing with the lives of people and their affairs, measures of effectiveness are of extreme importance in enabling tests for improvements or in changing the method of attack and altering the solution. However, unlike the physical sciences which can, in time, generate new methods of solution for their problems, operations research frequently cannot wait to advance a new theory. The existing solution must supply test criteria. Any solution which improves the situation perceptibly, although imperfectly, is desirable. Frequently, because of lack of appropriate knowledge, excessive effort spent in producing an elaborate solution is not advisable. In parallel with the urgency in the approach to problems, there is a developing and indispensable interest in basic research in the field, where new methods are developed and new concepts formulated.

It was observed during the war that merchant ships were attacked and sunk by enemy aircraft. In the search for a remedy, these questions were posed:

Is it worthwhile to install AA guns on merchant ships?
Is there a possibility of injurious suboptimization?
Would the guns introduce new dangers?
If such guns were installed, what effects should be looked for?
Should the effect be measured by the number of enemy aircraft shot down or by the number of ships saved?

This speculative approach (study of the problem in its broadest generalities) enables an accounting for more outcomes than the alternatives listed above. It might also be possible to avoid the problem of gun installation by considering the fact that ships are used for transportation and finding other methods of transportation. If and when this speculative approach is found to be futile, then it is necessary to resort to the direct approach.

Where the existence of a choice of methods allows the selection of the appropriate measure of effectiveness, research into the operation will clarify the relations between the alternatives and the results and will help in selecting the appropriate measure which is sensitive to changes. Measures of effectiveness may be expressed as ratios, or rates. For example, such ratios might be as follows: in traffic study at a specific intersection, cars per hour, cars per accident, or delay per car; in bombing ships, the number of hits per bomb, or number of ships sunk per bomb or per load; in aircraft interception, the number of planes shot down per number of interceptions attempted.

Measures of effectiveness are also frequently expressed in probabilities per unit effort, since many operations have two possible outcomes, either success or failure, with their associated frequencies. Then, too, not all the factors may be controlled or even known, making probability methods more desirable.

Very often failure to choose the proper measures of effectiveness can lead to completely wrong conclusions about the problem. This frequently stems from the original omission to determine the current values and objectives for the problem. However, even with the establishment of proper values and objectives, the use of the first plausible ratio or probability that springs to mind causes the criterion of selecting the proper "payoff" or "objective functions" to be neglected. Unless one develops methods of evaluating criteria and choosing effective ones, his quantitative methods may prove to be of little value in their application.

The Problem in the Operation

Since there are always problems connected with operations, part of the judgment phase should be concerned with the type of problem under study, how it originated, and, if time permits, its causes. Problems may fall into several basic categories. For instance, an operational problem may be of a *remedial* nature and have its origins in either actual or threatened accidents (whether airplane crashes or job-performance hazards). It may be remedial concerning the failure of a control system or be relative, perhaps, to the improvement of the inefficient performance of an operation.

Accomplishing a task efficiently may also be considered as *optimization*, another type of operations problem. A job may be optimumly completed with a given amount of effort, or, if the task is given, ways usually may be found to finish it with less effort. With an abundance of effort available, ways also may be sought to expand tasks profitably to make optimum use of this effort.

With new advances, improvements, and inventions in one field, there arises the problem of the *transference* of these new techniques to other fields of endeavor. As an illustration, radioactive isotopes now find uses in the field of medicine and in manufacturing, where they may be used to determine rates of absorption of substances in certain parts of the human body, on the one hand, or to test for wearing qualities in automobile tires or in machinery, on the other.

An operational problem may be that of *prediction*, of forecasting the problems associated with future developments and inventions along a certain line. In addition, the results of the introduction of new developments may be ascertained. Another type of operational problem may be one dealing with purely *hypothetical* considerations in which a specific danger, use, by-product of an invention, or operational function has not yet arisen. Man's years of consideration of space travel amply illustrate this type of problem, and the old and useful motto "Be prepared" has much meaning here.

While separate types of operations problems may be stated in theory, it is difficult to separate them in practice. Remedial and optimization problems are closely related, as are predictive and hypothetical problems, and transference and predictive problems. It is rarely possible to work solely with one of these types without overlapping or being forced to consider another. However, it might be noticed that the types of operations problems are presented above probably in order of increasing difficulty of solution. A new member of an operations-research team might be given an indoctrination problem requiring a solution in terms of remedial action, whereas he would need greater experience in operations research before tackling predictive problems.

Prior to the selection of problems for investigation, in order to anticipate a possible request for a solution, careful consideration is required as to whether a problem truly exists. Hasty selection of problems often has led to a waste of time on the part of the analyst and to erroneous conclusions. An illustration of such oversight occurred when an examination of artillery gun crews was made, culminating with the recommendation that three men be used to man the guns rather than six. The gunners, upon hearing this recommendation, stated that they, in fact, never did use more than three men and had been struggling for 20 years to be allotted three additional men to handle cooking and miscellaneous chores.

2-5. The Research Phase

Exploratory Observation, Data, and Other Information for Better Understanding of the Problem

This early process has the purpose of understanding and defining the problem arising in the operation. It usually has the function of improving the scope of the model. Actual observations by trained observers at the scene of the operation may be difficult and often dangerous to obtain. Nevertheless, such has been done in the field during combat. Pertinent data may also be obtained from action reports, track charts, dispatches, battle narratives, and from interviews and briefings, as well as from first-hand observation. In situations where ample time is available, it is often possible to set up operational experiments simulating the actual problem, from which most of the necessary data may be obtained. In the case of the military and the development of new weapons or missiles, proving-ground experiments facilitate the determination of technical information and the quality of improvements.

Where it is not possible to obtain information concerning a problem, the analyst should try to infer from the incomplete data available what the missing parts of the problem may be. Since nearly every problem, in one or more aspects, lacks some vital information, such inferences are commonplace, but they must be made on the basis of scientific procedures. For example, during the war, the RAF Coastal Command wished to determine whether enemy submarines were using search receivers to pick up signals from airborne radar. Intelligence reports did not help, and the number of submarine contacts per flying hour was inadequate to answer the question. It was decided to study the record of submarine evasive actions on the basis of both visual and radar sightings by the RAF patrols. The ratio of the number of times in which a submarine took evasive action following visual sighting to the number of times it did not, corresponded to the ratio of action to nonaction following radar sighting. It was correctly concluded that the enemy was indeed able to detect the signals emitted by the airborne radar.

The Formulation of Relevant Hypotheses; Models [2, 15, 24]

On the basis of previous knowledge, the significant factors of a problem may be selected. In the search for order among the facts, a rational explanation of the cause of the problem is often necessary in order to pursue a solution of the problem. Sometimes this order may be intelligible only to the individual doing the work. The reasoning involved frequently takes the form of rationalization. *Rationalization* is reasoning employed to justify a group of existing impulses or desires. The common understanding of the word "rationalization" implies the improvising of arguments to justify a point of view without the proper weighing of the facts or the use of rigor. Rationalization is at its best when it uses intricate logical arguments and scientific techniques. In chaotic types of experience, it is often necessary to set things in a proper light through rationalization with the purpose of introducing the order so vital for coping with problems.

Tentative explanations, when formulated as propositions, are called *hypotheses*. It is very essential to state the hypothesis and its anticipated consequences before any attempt is made at verification. In developing a hypothesis, frequently some facts must be regarded as more significant than others. The formulation of a good hypothesis depends largely upon sound knowledge of the subject matter. Methods of hypothesis formulation are based on a search for a general rule from specific facts and provide "order" among the facts.

A hypothesis must be formulated in such a way as to make possible some deductions from it. Frequently, while an immediate verification of the hypothesis may not be possible, the verification of its consequences is possible; e.g., only the effect of gravity can be observed. Unless each of the constituent terms is experimentally verifiable, the entire hypothesis is not empirically verifiable.

A hypothesis must provide an answer to the problem in question. Thus it must be capable of verification; otherwise, it can be refuted on the basis of empirical evidence. It should reveal propositions whose truth was not initially known or suspected. A good hypothesis can predict what will happen and explain or infer what has happened.

Hypotheses may be found in a number of ways. They are suggested by the problem and are needed throughout the inquiry to supply connections for the facts. The simpler hypothesis is to be preferred, and experiment will narrow down the choice. While no hypothesis stating a general proposition can be demonstrated as absolutely true, it gives interpretation to that which may otherwise appear as random observations. With this interpretation, extraneous observations (ones that do not suit the purpose) may be pointed out. Note that facts may be regarded as hypotheses for which there is a substantial amount of evidence.

Facts are certain distinguished and discriminated aspects in sense perception. They may also be regarded as an interpretation of sense data. Invariant properties, observed and experienced, are taken as facts. Events existing in space and time with their relations are also regarded as facts. Facts which are mostly spontaneous in experience function as criteria against which scientific theories and all other kinds of conjectures are ultimately tested [6].

An essential objective of operations research is consideration of an operation as an entity rather than as a series of disconnected suboperations. The fundamental conceptual device which enables one to regard the operation as a whole is a *model*, which is essentially a hypothesis. A model is an objective representation of some aspects of a problem by means of a structure enabling theoretical subjective manipulations aimed at answering certain questions about the problem. This representation, which attempts to establish a correspondence between the problem and rational thought, may be realized by forming a physical model, such as a wind tunnel for testing aircraft, or a theoretical model, such as the equations immediately related to an operation in queueing theory. Methods for manipulating a model to obtain answers are contained in the theory that attempts the solution. In this sense, the concept of theory includes that of a model. However, a model may be present without one's knowing all its theoretical implications.

On the other hand, a disadvantage of a model frequently lies in the identification of the theory with the model. The behavior of hydrogen atoms, in certain respects, as if they were solar systems makes the latter a useful model but not an adequate one for studying the properties of hydrogen atoms. Thus there is always the danger of transferring the logical necessity of the entire model or parts of it to parts of the theory. Differently stated, this says that the assumption of the model may impose limitations on the theory which, in fact, should not be in the theory. A remedy here is to keep the assumptions always in mind while developing the theory. Thus when a model is used to advance a theory it will make a difference whether the premises are logically necessary or logically contingent. Hertz points out the need for distinguishing between what one thinks is necessary, what is actually necessary, and what one wishes to appear necessary [2].

In scientific thinking, alternative theories can frequently be developed to explain the same empirical generalizations. It is also interesting to note that theoretical concepts in one scientific deductive system, such as optics, may be included in another more comprehensive system, such as electromagnetic theory.

The formulation of models is motivated by a belief in some order existing in the mind and in the real world. The success and adoption of a model hinge on several factors. In the absence of any pattern, it is better to start with a simple model than to have none. There is no conflict between this concept and that of viewing an operation with an open mind.

In the chapter on methodology, some mathematical models will be provided and applications either mentioned or illustrated.

The incentive to form models in operations research may be the desire and ability to discover problems and provide urgent solutions or the need to define and describe the problem. Listing the relevant contingencies requires a model to integrate the facts and to find interrelations among the parameters, using both theoretical tools and operational data. This results in isolating invariant quantities to work with, such as the entropy (a basic physical concept) of a system. Finally, suitable predictions are made from the model which allow solution of the problem.

Among the advantages of considering general models in operations research, such as those provided by queueing theory, are the following:

A logical and systematic approach

The provision of grounds for imitation in specific problems

The indications of the scope and limitation of the activity

The incorporation of useful tools which are then made available in a systematic manner, and the prevention of duplication in searching for theoretical tools to solve specific problems

The enabling of the detection of grounds for research and improvement in model making

The indication of the type of measurable quantities in a problem

Models may be formed by purely theoretical considerations or by building around hypotheses derived from known facts and data. The recommended successive-approximation procedure may be briefly described as follows:

Data   theory   prediction   further data   adjustment of the theory   further prediction . . .

Here one might mention the fact that models are attempts at understanding an operation and should not be taken as absolutes in any sense. Pitfalls of this kind encountered in philosophical interpretation of mathematical models have led to dead ends. The development of non-Euclidean geometries taught the lesson that with logic alone one cannot establish conclusive facts about nature. Experiments and data are necessary to test and verify one's outlook on reality through a model. In addition, models should be "ultraflexible"—flexible in the class of available models to represent a problem. This may be regarded as external flexibility. Probabilistic models have led to greater success in handling situations with imperfect information than have deterministic models, such as differential equations. The former, by their nature, may be regarded as ultraflexible.

Among the desirable properties of models which facilitate manipulation is economy in the number of parameters. This is best accomplished by studying the primary factors first. The law of parsimony, that is, using as few assumptions as possible, is a useful rule. An example of this is the operational approximation of using the normal distribution, which has only two parameters, to describe a large number of situations. An excessive number of parameters, some of which are vaguely determined simply to produce a better fit, is a hindrance to the use of a model. The same is true when parameters which are impossible to test by means of operational data, such as studying several aspects of the stock market, are included.

A model should also be "intraflexible"—have flexibility within—in order to allow for new information. It is desirable to have a model apply to more than the particular problem for which it is formulated. The general applicability of linear programming and queueing theory is an example of this. The concept of intraflexibility may also be illustrated by passing a polynomial through a set of n points of data as compared with using a least-square fit. New data would most likely not be on the curve in the first case but would be allowed for in the second case, indicating greater flexibility. Finally, the faithfulness of a model is tested by its success in prediction. There is little value to models which are regularly in error. This difficulty is illustrated by the present state of the theory of elementary particles.

The introduction of fine structure into models set up to study large effects often produces results which are chaotic and false; e.g., it is known in the social sciences that the study of mob behavior is not applicable to individual behavior.

The presence of several disjointed parts with no apparent connecting links produces poor models. Logic and clear thinking are often lost in the technical details of such a model. The consequence is a lack of specific results and clear prediction objectives.

The time devoted to the development of a model is a serious factor to consider. When urgent answers are needed, there is little advantage in developing elaborate models. A compromise as to size and detail is necessary. Economy should be considered in applying models to a problem, whether in exact form or by approximation methods.

It often happens that a model is needed simply for rough estimates. The consideration of great detail is then wasteful. The law of diminishing returns operates even in this sophisticated domain.

Upper and lower bounds for parameter values derivable from experiment are useful and should accompany a model, if possible. This enables the making of crude, theoretical estimates.

A model may be examined to see if it provides a solution of the problem. The wrong facts may have been considered. It may also point out other problems or even that the wrong problem has been considered. The nature of the model often dictates the type of data required to test the model, as well as their amount and refinement. The outstanding criterion for judging a model is its use in forecasting a solution and in providing a pattern for decision making. A simple and coherent model is not only easy to manipulate but may be convincing to the executive because it is readily grasped.

Observation, Experimentation, Further Data for Analysis and Testing of Hypotheses [34]

Free observation may have the purpose of learning, whether to verify a hypothesis or not. For example, one may observe the activity of a beehive to obtain some notions as to how the bees conduct their work. If these notions suggest a hypothesis, the truth of which the investigator wishes to test, then often a controlled observation is required. In other words, it may be desired to verify a statement of the form "if A, then B." Thus, by making A occur, one looks for the occurrence of B. Either A may be observed to occur naturally, and then one looks for the occurrence of B, or by a controlled experiment A is made to occur and the occurrence of B is then noted. Hence, experiments are guided by hypotheses or even determined by them. Often prior observation is required to determine the type of controls which would be effective in an experiment. The latter supplies data and information to verify the hypothesis. The question always arises as to whether the information obtained from an experiment is correct. This usually suggests repetition of the experiment, to confirm or to deny the correctness of the information. When it is decided that the information is correct, it is then used to confirm or deny the hypothesis. Without such information the testing of hypotheses becomes a speculative matter and is frequently in danger of being at the mercy of the individual's whims and biases.

Data, their collection, and processing are an essential part of the analysis of operations. Sampling will be discussed in the chapter on statistics. The types of data required are determined from the model to be used and from the experiments designed to test the model. This may be done by isolating the main factors in the problem, examining the purposes of each factor and their interrelation, and by describing the ones that will depict various degrees of accomplishment. The units of measurement, also necessary for the determination of desirable types of data, will become evident from the nature and scope of the problem and, to a certain extent, from the type of model to be constructed.

Before a model can be of much use to its builder, it is necessary for him to have collected the pertinent information concerning the problem from outside sources and observation and from direct controlled experimentation. However, usually there is never a sufficient amount of data to make the model detailed enough to cover all contingencies.

Analysis of Available Information and Verification of Hypothesis

Most of the time that a scientist spends in training is devoted to learning how to analyze and interpret information. There are qualitative methods of plausible reasoning as well as quantitative, which are mostly statistical methods of analyzing the data. If, for example, it is desired to determine the period T of a specific simple pendulum, then a hypothesis is made. This is based upon knowledge of physics, giving $T = 2\pi \sqrt{l/g}$ for a small angle of oscillation, where l is the length of the pendulum and g is the acceleration due to gravity. Measurements of length at different temperatures should enable one to determine the type of compensation required to keep the period constant, particularly if the nature of the metal from which the pendulum is made is not known. Perhaps a chemical experiment is first required to determine the nature of the metal. The coefficient of expansion of that metal is then used.

The model may require that statistical distributions be determined, such as the input and service distributions of times between arrivals and services in a waiting-line congestion study. Then the data gathered are grouped according to standard methods and often approximated by a known distribution. A significance test is then performed to discover how well the hypothetical distribution fits the data (see Chap. 11 on queueing).

A hypothesis does not have to be proved for every possibility in order to be acceptable. Sampling methods enable tests of hypotheses on the basis of a few, well-selected cases.

In any case, the verification of a hypothesis must be associated with the excellence of rigor. Only in this manner can one avoid the semblance of "begging the issue." When logical rigor and evidence are lacking, any one of the following mistakes may be committed:

1. Proof of a fact which at some point assumes the truth of the fact.

2. Confusing an implication with its converse. Thus to show that A implies B, one shows, instead, that B implies A.

3. Because of lack of enough repetitions of an experiment, conclusions drawn without accounting for chance occurrences.

4. The interference of individual bias in the reasoning and in the conclusions, sometimes twisting the facts by clever rationalization to make the worse seem the better cause.

5. The asking of questions which predetermine the answer, such as why does a child look more like its father than its mother?

6. Adopting the simpler of two hypotheses only on the basis that simplicity is preferable. This can lead to using the wrong hypothesis [6].

Inquiries arising from problems must be well formulated with the objective of discovering facts.

Some types of invariants are (1) the common properties of different objects, e.g., mass, chemical composition, etc.; (2) the regularities in nature, such as the law of gravitation, Boyle's law, etc.; (3) the variations in regularities when the latter are subjected to external forces, as, for example, the change in physical appearance of some biological organisms due to age.

The familiar methods of agreement, difference, agreement and difference, concomitant variation, and residues are used as aids to discovery; however all are subject to error [34].

Analogy is useful in interpretation when the example used for analogy explains some past behavior of the actual system. It should also be possible to use it for forecasting future behavior. Thus, analogy can be a powerful theoretical tool for explanation, e.g., Bohr's use of the solar system as a model for the atom, with the electrons corresponding to the planets and the nucleus to the sun. Useful results were obtained on the basis of this simple analogy. However, it was initially realized that this model would not provide all the answers.

Prediction, Generalization, and Alternative Methods

When a model has been verified by means of deductive analysis of the data, a theory is then developed from the model with the purpose of obtaining a complete description of the circumstances of the problem by studying variations in the parameters of the model, subject to the original assumptions. The next step is to use the theory which is based upon present and past information to extrapolate into the future, not only describing varied possibilities contingent on the operation, but also anticipating occurrences and accounting for remedial action. In this fashion, a solution is provided for the specific problem under study, and often, because of the extra information going into the model, other problems can be anticipated and corrective action recommended. Inductive reasoning aimed at prediction and generalization plays a major role in this part of research. Ability to predict and to use inductive reasoning correctly is highly dependent upon experience. In this manner, confidence is increased, and the necessity for induction becomes as urgent with the individual as the need for deductive rigor. Little of one's school training accounts for this aspect of research. More is derived from initial practical training and indoctrination and later experience.

In the last stages of research the analyst usually sees other methods by means of which the problem could have been attacked and will then be able to recommend, or even pursue, new research based on revised hypotheses. Imaginative analyses of problems are rarely confined to a single, narrow approach, since the information gathered may be used to account for a variety of questions not required in the consideration of the specific problem. To make efficient use of the information, it is desirable to make initially a variety of hypotheses which in the final analysis lead to alternative solutions. The advantages of this procedure are obvious and require little justification if economic considerations do not pose a problem.

2-6. The Action Phase

Recommendation and the Decision Process

In the action phase the communication (see Chap. 12) of remedial action to the decision maker should include a recapitulation of the situation, a statement of the assumptions underlying the presentation of the situation, the scope and limitations of this presentation, a presentation of the alternative courses of action, and a description of the impact or consequence of each alternative, as well as the recommendation for specific action. With this complete statement of the situation and its recommended solution the decision maker is better able to arrive at and justify (if necessary) the required course of action.

Briefly, the decision maker's reliance upon scientific knowledge is closely related to the fact that a scientist is, by profession, trained to reject unsupported statements and has an instinctive desire to rest all decisions on some quantitative basis, even if the basis is only a rough estimate. This enables a scientist to detect the existence of problems and questions of which the regular staff may be unaware. In addition, a scientist, through his research experience, is trained to arrive at the fundamentals of a question, to seek out broad underlying principles through a mass of sometimes conflicting and irrelevant data. He may be assumed to know how to handle data and how to guard against fallacious interpretations of statistics.

One of the most useful features of operations research is that it may bring to light unrealistic management goals. It also helps to single out the critical issues which require executive appraisal and analysis, and it provides factual bases to support and guide executive judgment. Further, the outstanding function of the operations-research staff is to increase the understanding of the administrator as to the implications of the decisions which he may make. The scientific method supplements the decision maker's ideas and experiences and is no substitute for them. But in order to attain his goals more fully, the decision maker, as a rational individual, cannot afford to ignore the utility of the scientific method.

REFERENCES

1. Benjamin, A. C.: Philosophy of the Sciences, in V. Ferm (ed.), "A History of Philosophical Systems," Philosophical Library, Inc., New York, 1950.
2. Braithwaite, R. B.: "Scientific Explanation," Cambridge University Press, London, 1955.
3. Bridgman, P. W.: "The Logic of Modern Physics," The Macmillan Company, New York, 1927.
4. Carnap, R.: "Testability and Meaning," Graduate Philosophy Club, Yale University, New Haven, Conn., 1950.
5. Clifford, W. K.: in K. Pearson (ed.), "The Common Sense of the Exact Sciences," newly edited by J. R. Newman, Alfred A. Knopf, Inc., New York, 1946.
6. Cohen, M. R., and E. Nagel: "An Introduction to Logic and Scientific Method," Harcourt, Brace and Company, Inc., New York, 1934.
7. D'Abro, A.: "The Evolution of Scientific Thought," 2d ed., Dover Publications, New York, 1950.
8. Doyle, A. C.: "The Adventures and Memoirs of Sherlock Holmes," Modern Library, Inc., New York.
9. Schlipp, P. A.: "A. Einstein, Philosopher—Scientist," Library of Living Philosophers, Evanston, Ill., 1949.
10. Frank, P.: "Modern Science and Its Philosophy," Harvard University Press, Cambridge, Mass., 1950.
11. Hempel, C. G.: "Fundamentals of Concept Formation in Empirical Science," mimeographed lecture notes, Yale University, New Haven, Conn., 1951.
12. Hocking, W. E.: "Types of Philosophy," Charles Scribner's Sons, New York, 1929, 1939.
13. Hurni, M. L.: Observations on Operations Research, *Journal of the Operations Research Society of America*, vol. 2, no. 3, pp. 234–248, August, 1954.

14. Kant, I.: "Critique of Pure Reason," Willey Book Co., New York, 1943.
15. Margenau, H.: The Competence and Limitations of Scientific Method, *Journal of the Operations Research Society of America*, vol. 3, no. 2, pp. 135–146, May, 1955.
16. Morse, P. M.: Measures of Effectiveness, in "Notes from MIT Summer Course on Operations Research," Technology Press, Massachusetts Institute of Technology, Cambridge, Mass., 1953.
17. Nagel, E.: The Methods of Science, *Scientific Monthly*, vol. 70, pp. 19–23, 1950.
18. Max, Otto: "Science and the Moral Life," New American Library of World Literature, Inc., New York, 1949.
19. Poincaré, H.: "Science and Method," Dover Publications, New York, 1955.
20. Poincaré, H.: "The Foundations of Science," Science Press, New York and Garrison, N.Y., 1913.
21. Poincaré, H.: "Science and Hypothesis," Dover Publications, New York, 1955.
22. Rapoport, A.: "Operational Philosophy," Harper & Brothers, New York, 1954.
23. Reichenbach, H.: "The Rise of Scientific Philosophy," University of California Press, Berkeley, Calif., 1951.
24. Saaty, T. L.: Generalized Models in Operations Research, Paper presented before the Operations Research Society of America, May, 1956.
25. Schrödinger, E.: "Science and Humanism," Cambridge University Press, London, 1952.
26. Solandt, O.: Observation, Experiment, and Measurement in Operations Research, *Journal of the Operations Research Society of America*, vol. 3, no. 1, pp. 1–14, February, 1955.
27. Sullivan, J. W. N.: "The Limitations of Science," New American Library of World Literature, Inc., New York, 1933.
28. Von Mises, R.: "Positivism, a Study in Human Understanding," Harvard University Press, Cambridge, Mass., 1951.
29. Weaver, W., and others: Fundamental Questions in Science, *Scientific American*, vol. 189, no. 3, September, 1953.
30. Weyl, H.: "Philosophy of Mathematics and Natural Science," Princeton University Press, Princeton, N.J., 1949.
31. Whitehead, A. N.: "Science and the Modern World," New American Library of World Literature, Inc., New York, 1948, 1949.
32. Whitehead, A. N.: "Essays in Science and Philosophy," Philosophical Library, Inc., New York, 1948.
33. Wightman, W. P. D.: "The Growth of Scientific Ideas," Yale University Press, New Haven, Conn., 1951.
34. Wilson, E. B., Jr.: "An Introduction to Scientific Research," McGraw-Hill Book Company, Inc., New York, 1952.
35. Wittgenstein, L.: "Tractatus Logico-philosophicus," Harcourt, Brace and Company, Inc., New York, 1922.

CHAPTER 3

MATHEMATICAL EXISTENCE AND LOGICAL PROOF

3-1. Introduction

Poincaré distinguished between two types of mathematical minds. One sort are logicians and analysts, the other intuitionists and geometers. Logicians do not always proceed from the general to the particular, as the rules of formal logic would seem to require of them. In fact, they have extended the boundaries of science by generalization from particulars.

Generalizations in mathematics are often obtained by formally analyzing specific problems and then observing a property common to all of them. A hypothesis is then made, and a proof of the existence of a wider class with well-defined characteristics having this property is supplied. In this sense the methods of mathematical discovery have the earmarks of scientific research.

3-2. Remarks on Existence Proofs and Methods of Solution

The general study of problems in mathematics may be divided, roughly, into the quest for existence theorems and the construction of solutions. The importance of existence and uniqueness theorems is that they ensure, when applied to physical systems, for example, that the mathematical model chosen to represent a phenomenon actually has a solution which can be unambiguously applied. A simple example to illustrate this point is that of the existence of a circle passing through three given points in a plane, not all of which lie on a straight line. One cannot prove an analogous general theorem for the case of four arbitrary points. Proving the existence of a unique solution simplifies matters for application. Uniqueness proofs indicate the amount of information needed to produce the desired solution, such as is the case in boundary-value problems in differential equations. On the other hand, these proofs do not always provide information on how a solution may be obtained. It is sometimes true, as in differential equations, that an existence proof is carried out by actually constructing the solution. Existence proofs are often discovered through generalization from results of special cases, where solutions are known to exist. It may be important to know how many solutions of a problem exist and the degree of arbitrariness, as specified by the number of constants in an ordinary differential equation, for example. Information of this type enables the individual to proceed with confidence that he has not overlooked other feasible solutions of the problem. Frequently, prior to proving existence, one may be able to place "a priori bounds" on the set from which the solutions are to be obtained. This will be illustrated below.

Next in the process is finding conditions which solutions may satisfy (provided that they exist) although one may proceed directly to construction, having previously proved existence. It is often possible to obtain (i.e., construct) solutions, if they exist, through necessary conditions, e.g., the vanishing of a derivative for a maximum or a minimum. Then, in order to verify that the desired solution has been obtained, sufficiency conditions may be imposed, e.g., the second-derivative test for a maximum or a minimum. The latter conditions are sometimes very difficult to obtain, and they generally require deep and detailed analysis. Such is particularly the case in the calculus of variations.

The constrained-optimization problems studied later show that it is usually easier to prove that a solution exists than actually to prescribe a method for obtaining one. A fortunate characteristic of the simplest problems of this type (e.g., linear programming) is that systematic procedures for obtaining solutions can be given. In addition, some procedures, such as the simplex process, obtain the exact solution in a finite number of steps. In general, convergence theorems must accompany methods which require infinite numbers of steps for solution; when the number of steps is finite, a proof that the process terminates is required. When an electronic computer is used to solve such a problem, it is desirable to be sure that the process will terminate; furthermore, it is useful to know, within reasonable limits, how much time will be required to solve the problem.

Many problems which occur in operations research involve logarithmic and exponential functions which present calculation difficulty but, on replacement by a series expansion using the first few terms, become manageable. The adequacy of the answer must be checked by estimating remainder terms and upper bounds to the errors committed.

In order to find a solution of a problem, one must first determine carefully what kinds of information are needed. Answers to the following questions supply useful guides.

1. Is a number needed for the solution? If so, with what accuracy? Or, alternatively, is it satisfactory to give merely upper and lower bounds to the desired answer? Or, if predictions are to be made, with what degree of accuracy are they required?

2. Is the solution needed in order to describe the behavior of a physical system? Then, for example, a steady-state solution (independent of time) may suffice.

Such questions are related to the following domains of investigation:
A priori bounds
Existence and uniqueness theorems
Characterization theorems
Methods for obtaining solutions
Convergence theorems
Bounds and degree of approximation

Several of the above ideas will now be illustrated. The examples selected should be instructive and useful either in material or in method. It is not considered desirable in this chapter to give a complete mathematical background in order to justify the material presented. There are ample examples in the literature further to illustrate the ideas.

3-3. Illustrations

Because of the relationship between linear constraint sets of optimization problems and polyhedra, a topic which will be further explored in this book, the following elementary illustrations have been chosen. They will be used to demonstrate some of the mathematical questions discussed above.

A Priori Bounds

A regular polyhedron in three dimensions is bounded by congruent regular polygons in such a way that any side of each polygon lies on two and only two polygons, its solid angles are equal, and it has no holes [9].

The following theorem asserts that there can be, at most, five such polyhedra, if any. It does not prove their existence.

Theorem 3-1: There are, at most, five regular polyhedra in three dimensions.

Proof: Using a Euclidean proposition which asserts that "any solid angle is contained by plane angles less than four right angles," we point out that five identical equilateral (and hence regular) triangles do contain a solid angle. The sum of their plane angles at the vertex is, of course, $5 \times 60 = 300$ degrees. As will be seen later, each vertex of an icosahedron is the meeting point of five such triangles. Similarly, four identical regular triangles enclose a solid angle of an octahedron, and three enclose a solid angle of a tetrahedron. Thus, at most, three regular polyhedra are constructable from regular triangles. Again in accordance with Euclid's proposition, only one regular polyhedron, the cube, can be constructed from squares and one, the dodecahedron, from pentagons. None can be constructed from hexagons or higher-sided polygons, since at least three such are required to define a solid angle and the sum of the plane angles at that vertex is equal to or exceeds four right angles. Thus, there can be, at most, five regular polyhedra.

An alternative proof of this fact, which supplies some properties (characterization) which can be used in existence and construction proofs, will now be given. We shall use the following lemma due to Euler as a basis for proof. Note that every regular polyhedron is a simple polyhedron, the latter defined in an analogous fashion to the former without requiring regular polygons for faces or equality for solid angles.

Lemma: The number of vertices V, edges E, and faces F of a simple polyhedron are related linearly as follows:

$$V - E + F = 2 \qquad (3\text{-}1)$$

Briefly, the proof of the lemma is carried out by taking out one of the faces of an arbitrary simple polyhedron and laying the remaining faces flat without overlapping any faces. (The reader may find it helpful to visualize an elastic polyhedron to be stretched flat through the gap.) It is then reduced to a triangle by operations such as adding a vertex and joining it to two other vertices or joining two vertices by a segment which does not cross the rest of the figure. In either case, the Euler characteristic $V - E + F$ is not changed, although it has one less face which has been cut out to enable flattening. Just as one may add an edge between two vertices, thus introducing an additional face which cancels the added edge in $V - E + F$, one may remove an edge and thus remove a face with it, and the expression remains unchanged. If the reverse operation of omitting instead of adding the above-mentioned elements is performed, the Euler characteristic remains the same. Therefore, by addition and omission as described above, a triangle is finally obtained. The characteristic for the triangle $V = 3$, $E = 3$, and $F = 1$ is obviously equal to 1. For the polyhedron it is equal to 2 since a face has been cut out to flatten the figure.

Proof: Here is the proof. Note that for a regular polyhedron the following two relationships hold:

$$nV = 2E \qquad (3\text{-}2)$$
$$mF = 2E \qquad (3\text{-}3)$$

where n and m are integers whose values are to be determined. The first holds since there are n faces attached to each vertex and thus n edges meet at each vertex. However, each edge is used twice. The second holds since each face is bound by m edges and each edge bounds two faces.

Substituting into Eq. (3-1) yields

$$\frac{2E}{n} - E + \frac{2E}{m} = 2 \tag{3-4}$$

or

$$\frac{1}{n} + \frac{1}{m} = \frac{1}{2} + \frac{1}{E} \tag{3-5}$$

This is meaningful only if $m \geq 3$, $n \geq 3$. If both m and n are greater than 3, then

$$\frac{1}{E} = \frac{1}{n} + \frac{1}{m} - \frac{1}{2} \leq \frac{1}{4} + \frac{1}{4} - \frac{1}{2} = 0 \tag{3-6}$$

which is absurd.

Let $n = 3$; then m can take on only the values 3, 4, and 5. In that case, $E = 6, 12, 30$. Because of the symmetry of Eq. (3-5) in m and n, the same values are obtained for E if $m = 3$ and $n = 3, 4$, and 5.

Thus, there are six possibilities, of which one is repeated, namely, $n = 3$, $m = 3$. Hence, there are, at most, five regular polyhedra for which the number of faces, edges, and vertices may be easily calculated, using the above information.

As an exercise in upper bound it may be shown that a quadratic equation can have no more than two roots.

Existence

Existence proofs may be obtained by two methods, as follows:

1. Existence by construction. The reader may prove by construction that there exists a unique circle passing through three noncollinear points in a plane. Or again, by completing the square, one may prove by substitution that a quadratic equation has at least one root and, in fact, two.

2. Existence without construction. Simple examples of this type of existence proof are found in studying certain games where it can be shown that there exists a way in which the first player can win, yet it is not known how. Another example is that of showing that an infinite series has a limit to which it converges, without actually constructing this limit [12].

Remark: If one does not succeed in giving a counterexample to invalidate a conjecture, it does not mean that the conjecture is true. Goldbach's conjecture, which asserts that every even integer which is not a prime or the square of a prime (this rules out the integers 2 and 4) is the sum of two odd primes, has not been proven, nor are there counterexamples to disprove it. To illustrate this, $60 = 7 + 53$ is the sum of two odd primes, although one also has, for example, $60 = 5 + 55$, and $60 = 3 + 57$. The conjecture says nothing about the last two possibilities.

In science a conjectural general statement becomes more credible if it is verified in a new particular case. Note that credibility does not imply truth. However, it indicates an increase in the determinateness of an assumption or a conjecture. Disproving a conjecture by counterexample is a powerful tool for tempering wild speculations. Frequently it enables one gradually to refine the generality of the conjecture until it becomes a provable theorem. In any case, one counterexample should be adequate to deal with a situation. As to whether a conjecture can always be proved or disproved or whether it is decidable, Kurt Gödel has demonstrated, in works which brought him the Albert Einstein prize in 1951, that there are mathematical propositions which are undecidable within the framework of formal mathematics. The implications of these findings can best be grasped by reading Gödel's work and several papers which have appeared as expositions on the subject [3, 5, 15].

Characterization

"Necessary and sufficient" or "if and only if" are conditions which characterize certain situations. Some analysis of "necessary and sufficient" conditions as implications is given later. It has been previously mentioned that, for a certain property to hold, a condition may be necessary but not sufficient, and conversely. The vanishing of the first derivative is a necessary condition for a stationary point; however, it is not sufficient. A familiar sufficiency condition is provided by the second-derivative test. Frequently, it is not possible to find a condition which is at the same time necessary and sufficient. However, exploration is needed in order to determine whether this would or would not be possible. An illustration of a necessary and sufficient condition will now be given.

Theorem 3-2: A necessary and sufficient condition that $u = u(x,y)$, $v = v(x,y)$, which are functions of x and y, be connected by an identical relation $f(u,v) = 0$ is the vanishing of the Jacobian or functional determinant of u and v with respect to x and y, provided that f as a function of u and v has no stationary value in the domain considered.

Note that the Jacobian J, also written as $\partial(u,v)/\partial(x,y)$, is the determinant

$$\begin{vmatrix} u_x & u_y \\ v_x & v_y \end{vmatrix} = u_x v_y - v_x u_y \qquad (3\text{-}7)$$

The subscripts indicate the variable with respect to which the first derivative is taken.

Proof: (a) The condition is necessary. In order to show this, the equation $f(u,v) = 0$ is assumed to hold and is differentiated with respect to

x and y to obtain

$$\frac{\partial f}{\partial u}\frac{\partial u}{\partial x} + \frac{\partial f}{\partial v}\frac{\partial v}{\partial x} = 0 \quad \text{or} \quad f_u u_x + f_v v_x = 0 \qquad (3\text{-}8)$$

$$\frac{\partial f}{\partial u}\frac{\partial u}{\partial y} + \frac{\partial f}{\partial v}\frac{\partial v}{\partial y} = 0 \quad \text{or} \quad f_u u_y + f_v v_y = 0 \qquad (3\text{-}9)$$

It is not difficult to see (also discussed later) that, in order for these two equations to have solutions in f_u and f_v which are not zero, by the non-stationarity assumption the coefficient determinant, which is the Jacobian in this case, must be zero.

(b) The condition is sufficient. In order to show this, assume that

$$u_x v_y - v_x u_y = 0 \qquad (3\text{-}10)$$

Suppose that v does not depend on y, in which case $\partial v/\partial y = 0$. From the fact that the Jacobian is zero, it follows that either $\partial v/\partial x = 0$ or $\partial u/\partial y = 0$. In the former case, v is a constant, and the functional relation may be given by $v = a$, where a is a constant. In the latter case, u and v depend on x alone, and the desired relation is obtained by eliminating x between $v = v(x)$ and $u = u(x)$.

If v actually depends on y, then $\partial v/\partial y \neq 0$, and $v = v(x,y)$ gives y as a function of x and v, that is, $y = f(x,v)$, which may be substituted in $u(x,y)$ to yield $u = F(x,v)$.

Thus in the Jacobian, $\partial u/\partial x$ is replaced by $\partial F/\partial x$ and $\partial u/\partial y$ by either $\partial v/\partial y = 0$ or $\partial F/\partial x = 0$. The former case contradicts the hypothesis that $\partial v/\partial y \neq 0$, and the latter case states that F is a function of v only, from which the desired relation $u - F(v) = 0$ is obtained, and the proof is complete [16].

Remark 1: Among other uses of Jacobians is their occurrence in transformations from one coordinate system to another. For many practical problems it is easier to formulate a problem in rectangular coordinates. However, spherical coordinates, for example, also occur. Transformation from one coordinate system to another frequently simplifies the resulting expressions and consequently also their manipulations.

Remark 2: Note that the last part of the above proof involved a contradiction, a *reductio ad absurdum*. The following example illustrates the power of this type of proof.

Example 1: Prove that e (the base of the natural logarithm) is irrational; i.e., it cannot be expressed as the ratio of two relatively prime integers (without a factor in common).

Solution: Suppose that e is rational and can therefore be expressed as $e = a/b$, where a and b are relatively prime positive integers. Then, from the series expansion of e^{-x} at $x = 1$, one has

$$\frac{b}{a} = e^{-1} = \sum_{n=0}^{\infty} \frac{(-1)^n}{n!} = \sum_{n=0}^{a} \frac{(-1)^n}{n!} + \sum_{n=a+1}^{\infty} \frac{(-1)^n}{n!} \qquad (3\text{-}11)$$

Multiplying by $(-1)^{a+1}a!$ and transposing terms yield

$$(-1)^{a+1}\left[b(a-1)! - \sum_{n=0}^{a} (-1)^n a!/n!\right] = \frac{1}{a+1} - \frac{1}{(a+1)(a+2)}$$
$$+ \frac{1}{(a+1)(a+2)(a+3)} - \cdots \qquad (3\text{-}12)$$

The right side has a value between zero and unity, since it is an alternating series converging to a value between its first term and the sum of its first two terms, while the left side is an integer. This is a contradiction. Thus, e is irrational.

As an exercise, using this method, show that $\sqrt{2}$ is irrational. Also show that the number of primes is infinite.

Remark 3: Constructive proofs, in contrast to proofs by contradiction, usually require more imagination, and their pursuit suggests new ideas and permits discovery. The following example illustrates a constructive proof. It literally enlarges the framework of the stated problem and obtains the conclusion independently of the added agents which serve as "catalysts" enabling quick and convenient proofs.

Example 2: Prove that a plane section of a right-circular cylinder is an ellipse or a circle [9].

Proof: Drop into the cylinder two spheres equal in radius to the generating circle. They touch the cylinder in two circles. Let F_1 and F_2 be the two points of tangency of the spheres with the plane section of the cylinder. Let B be a point on the curve of intersection of the plane with the cylinder. Draw the straight-line generator of the cylinder which passes through B and suppose that it intersects the two circles of tangency of the spheres with the cylinder at P_1 and P_2, respectively. Note that $BF_1 = BP_1$ since they are both tangents to the sphere from the same point B. Similarly, $BF_2 = BP_2$.

Thus, $BF_1 + BF_2 = BP_1 + BP_2 = P_1P_2$ which is the same, regardless of where B lies on the intersection curve. The curve must, by definition, be an ellipse with foci at F_1 and F_2. When F_1 coincides with F_2, the section is a circle.

Remark 4: Proof using finite induction (to be distinguished from transfinite induction extended to cases involving infinite classes) is briefly discussed in order to make the presentation complete. The principle of finite induction is an axiom which is added to the axioms for the number system to provide an additional method of proof.

The principle of finite induction requires that, in a sequence of elements, if a property holds for the first element and if it is shown that every time it holds for the nth element it also holds for the $(n + 1)$st, then the property may be assumed to hold in general.

A simple example will suffice to illustrate the method. Prove that

$$1 + 2 + 3 + \cdots + n = \frac{n(n + 1)}{2} \qquad (3\text{-}13)$$

The sequence here is $1, 1 + 2, 1 + 2 + 3, \ldots$, and the property states that the value of each element is obtained by substituting the number of terms in it in the formula on the right-hand side. It is clear that the quantity on the right is satisfied for the first element. Assume now that it is true for the nth element. To show that it is true for the $(n + 1)$st, one must have

$$1 + 2 + \cdots + n + (n + 1) = \frac{(n + 1)(n + 2)}{2} \qquad (3\text{-}14)$$

It can be easily verified that the quantity on the right is also obtained by adding $n + 1$ to both sides of the equation in n terms. Hence, by the principle of finite induction, the quantity on the right may be used to obtain the value of any term in the sequence.

Construction of Solutions

Although the idea of construction is common to constructive proofs and construction of solutions, the two are distinct in that the former refers to a method of logical proof whereas the latter is concerned with providing a solution of a problem after having proved or assumed that a solution exists. A solution of a problem may, in fact, be obtained by contradiction or even by induction. Since a major part of this book is concerned with this type of problem, no illustrations are given.

Convergence Proofs

Often the process of constructing a solution requires the use of successive steps, each closer then the previous one without necessarily coinciding with the solution. Such an iterative process may require a finite or an infinite number of steps to obtain the solution; hence the notion of convergence. Does the procedure converge to a solution? For example, does Newton's method of root approximation converge and what are the conditions under which convergence is produced? Does the simplex process for solving a linear-programming problem converge? Does it converge to a solution in a finite or an infinite sequence of steps? It is frequently convenient to use series representations of a function in some range of values of the variables if convergence is assured in this range. These are a few of the important ideas in this field.

Approximations and Bounds for Errors

Obviously, even if the procedure converges, one cannot carry out an infinite number of iterations. There follows, therefore, the idea of approximation and the error incurred by it. The examples of determining approximations and bounds for errors which are presented here are by no means a complete representation of the many possibilities. However, they illustrate the point adequately.

The trapezoidal method of approximation to the value of an integral gives approximately

$$\int_a^b f(x)\, dx \simeq \frac{b-a}{n}\left[\frac{f(b)+f(a)}{2} + f(x_1) + f(x_2) + \cdots + f(x_{n-1})\right] \quad (3\text{-}15)$$

where $x_0 = a, x_1, x_2, \ldots, x_{n-1}, x_n = b$ are the abscissas of division points of the interval (a,b) into intervals of equal lengths each equal to $(b-a)/n$. An upper bound to the error committed if the absolute value of the second derivative satisfies the condition $|f''(x)| < M$ is given by

$$\frac{(b-a)^3 M}{12 n^2} \quad (3\text{-}16)$$

Simpson's rule for the approximate evaluation of an integral gives

$$\int_a^b f(x)\, dx \approx \frac{h}{3}(f_0 + 4f_1 + 2f_2 + 4f_3 + \cdots + 2f_{n-2} + 4f_{n-1} + f_n) \quad (3\text{-}17)$$

where $h = (b-a)/n$ and n is an even integer and where

$$f_i = f(a + ih) \qquad i = 0, 1, 2, \ldots, n$$

This approximation is exact if $f(x)$ is a polynomial of third degree or less. If the calculation is done with n subdivisions and another with $2n$ subdivisions the error in the latter may be estimated by dividing the difference of the two results by 15. This easily follows from the fact that the error in the approximation is proportional to $1/n^4$.

Of frequent use is the Euler-Maclaurin theorem given by [23]

$$\sum_{i=0}^{N} y(x+ih) = \frac{1}{h}\int_x^{x+Nh} y(x)\, dx + \frac{1}{2}[y(x+Nh) + y(x)]$$

$$+ \sum_{k=0}^{\infty} \frac{B_{2k+2}}{(2k+2)!} h^{2k+1}[y^{2k+1}(x+Nh) - y^{2k+1}(x)] \quad (3\text{-}18)$$

It relates the sum of values of a function at equally spaced values of its argument to a definite integral of the function, together with correction terms involving Bernoulli numbers, powers of the spacing between argument values, and odd derivatives of the function at the end points of the interval. For any "well-behaved function," it is very rapidly convergent since the Bernoulli coefficients $\dfrac{B_{2k+2}}{(2k+2)!}$ ($k = 0, 1, 2, \ldots$) decrease very rapidly, the first five being $1/12$, $-1/720$, $1/30{,}240$, $-1/1{,}209{,}600$, and $1/47{,}900{,}160$, respectively.

The error committed in the Euler-Maclaurin formula, truncated at the terms involving the third derivatives, is given by

$$\frac{h^7}{6!} \int_0^1 \varphi_6(t) y^{\text{vi}}(x + ht)\, dt \tag{3-19}$$

where $\varphi_6(t)$ is the sixth of the Bernoulli polynomials $\varphi_n(t)$ which are the coefficients of $s^n/n!$ in the expansion of $s[(e^{ts} - 1)/(e^s - 1)]$ in ascending powers of s. For example,

$$\varphi_1(t) = t \qquad \varphi_2(t) = t(t - 1) \qquad \varphi_3(t) = t^3 - \frac{3t^2}{2} + \frac{t}{2}$$

3-4. Symbolic Logic and Truth Tables

Logic may be used not only to obtain correct conclusions from empirically derived premises but also to point out the *possible* conclusions which may be inferred from them. Just as it was initially possible to abstract from nature truths which were mathematically expressible and which were later used to produce predictions, so has it been possible to use logic in order to abstract from the specific, concrete meaning of propositions those elements which are purely formal. Thus, its subject matter is independent of the particular instance to which it is applied. Any valid deduction from premises about symbols used in logical composition sentences would be equally valid if these premises referred to any other situation. Thus, its virtue lies not only in the advantage resulting from the objectivity inherent in a formal symbolic system but also in the analytical methods it presents for the testing of correct reasoning in specific instances. It has been possible to apply with great success symbolic logic to situations which do not readily yield to intuitive understanding [13, 14].

Because of its formalism, symbolic logic supplements and does not replace intuitive methods. It provides mechanical checks of critical points. In addition, it also provides assistance with symbolic operations in complex situations, which increases precision and enables generality. Only the calculus of propositions which offers methods of symbolizing connectives such as "and," "or," etc., is studied here. The study of statements involving a variable x and "quantifications" of such statements by phrases such as "there exists an x . . . ," etc., belongs to the domain of propositional functions and will not be discussed [7, 19].

The Concept of Statement and Statement Connectives [6]

A statement is a sentence which asserts something to be the case. It is thus called an assertive sentence. If what a statement asserts is the case, the statement is called true. Otherwise it is false. Conversely, every sentence which is either true or false is a statement. With this characterization of statements among all sentences, it can be seen that only statements can serve to express any kind of knowledge, scientific or otherwise. The subject content of all knowledge must be expressible in a sentence for which truth can be claimed.

As an illustration, consider the following sentences:
1. Copenhagen is the capital of Denmark.
2. There are no centaurs.
3. Any two bodies in the universe attract each other with a force directly proportional to the product of their masses and inversely proportional to the square of their distance.
4. Two plus five equals seven.
5. America for the Americans.
6. Three cheers for Jim.

The first four of these are statements since each of them asserts something to be the case and is either true or false, even though, as in the second case, it is not known which of the two alternatives holds. On the other hand, the last two sentences make no assertions; thus the terms "true" and "false" cannot be meaningfully applied to them. They are not statements.

By means of *statement connectives* such as "and," "or," "not," "if . . . , then," "if and only if," it is possible to combine given statements to construct other more complex ones. This construction process is referred to as *statement composition*. The sentence "if he is ill or tired he will not come" is a compound statement formed by means of the connectives "if . . . , then" and "or" from the component statements "he is ill," "he is tired," and "he will not come." It is possible by means of so-called *truth tables* to explain concisely the use of statement connectives.

The word "and" in its connective use is usually symbolized by a dot. The result of combining two component sentences by the word "and" or by a dot is called the *conjunction* of those component sentences. Suppose that it is desired to investigate the conditions under which a conjunction is true, considering the truth value of its components in the sentence "tomorrow will be warm · tomorrow will be sunny." To decide about its truth, one must wait for tomorrow's weather. However, the following can be decided today: Each of the two components will be true or false so that altogether there are four different possibilities which are represented in the first two columns of the following table. It is obvious that the conjunction of the two components will be true only when each of the components is true and in no other case. This is expressed in the third column of the table. Analogous considerations apply to any other conjunction of two sentences to express this generality. Thus, one obtains the following rule for conjunction: A conjunction is true if and only if both of its components are true.

Note that the notion of truth tables arises from the fact that a statement has two possible truth values, namely, truth T or falsity F.

A	B	$A \cdot B$
T	T	T
F	T	F
T	F	F
F	F	F

The connective "or" has at least two different meanings. In the sentence "Beethoven was born in 1770 or in 1771," it is meant to assert that exactly one of the two components is true to the exclusion of the other. When used in this sense, the word "or" is said to be the exclusive "or" and is symbolized by a double wedge $\veebar$. Consider now the sentence "John is young or he is inexperienced." Here the possibility that John might be both young and inexperienced is not precluded. Thus, one or the other or both of the components are true. This is an illustration of the nonexclusive "or" which is symbolized by a wedge $\vee$. The phrase "and/or" often used in legal language is an alternative for the nonexclusive "or."

Depending on whether the two components of a sentence are connected by a single or a double wedge, the compound statement is referred to as the *nonexclusive* or the *exclusive alternation* of the two components, respectively. In agreement with general usage, when the term "alternation" is used alone it will be understood in the nonexclusive sense. By analogy with the truth table for conjunction, one has for the two types of alternation truth table 3-1.

TABLE 3-1

A	B	$A \vee B$	$A \veebar B$
T	T	T	F
F	T	T	T
T	F	T	T
F	F	F	F

The rules in this case are that a nonexclusive alternation is true if at least one of its components is true, and an exclusive alternation is true if and only if exactly one of its two components is true.

The word "not" is a connective term by courtesy only. It generates a new statement not from two components but from only one. A reliable procedure of forming the denial of a sentence is to prefix it by the phrase "it is not the case that." This phrase is usually represented by a tilde $\sim$ which is a degenerate N. The truth table for the negation is given by

A	$\sim A$
T	F
F	T

with the rule that the negation of a statement is true if and only if the statement itself is false.

When two components are connected by means of the phrase "if . . . , then," the result is a conditional sentence, or, briefly, a *conditional*. The component following "if" is the *antecedent*, and the word following "then" is the *consequent*. In the conditional "if you eat tomatoes, then you will have hives," the antecedent is "you eat tomatoes," and the consequent is "you will have hives." Note that the same assertion is made by the statement "you won't eat tomatoes or you will have hives," with the nonexclusive alternation. Symbolically, this may be expressed as "$\sim$ you eat tomatoes $\vee$ you will have hives."

The second sentence may be used as a definition of the first. Thus, if the symbol $\supset$, called "horseshoe," is used for the phrase "if . . . , then" and if the antecedent and consequent of the conditional are denoted by A and B, respectively, then $A \supset B$ means the same as $\sim A \vee B$.

Frequently an arrow is used instead of the horseshoe, yielding "$A \to B$" instead of "$A \supset B$." The truth table for the conditional is

A	B	$A \supset B$
T	T	T
F	T	T
T	F	F
F	F	T

Note that a conditional is false if and only if its antecedent is true and its consequent is false.

The converse of the conditional "$A \supset B$" is "$B \supset A$." Its inverse is "$\sim A \supset \sim B$," and its contrapositive is "$\sim B \supset \sim A$." Do all four associated conditional sentences make different assertions, or are there any two among them which assert the same thing? As will be seen later, the question can be answered easily by means of truth tables.

The conjunction of a conditional and its converse is called a *biconditional* and is denoted by the symbol $\equiv$ or by $\leftrightarrow$. Thus, "$A \equiv B$" is short for "$(B \supset A) \cdot (A \supset B)$." The statement "if Jack's grade is above 59, then he passes the course," and conversely, "if he passes the course, then his grade is above 59," becomes, by means of connectives, "(Jack's grade is above 59 $\supset$ Jack passes the course) $\cdot$ (Jack passes the course $\supset$ Jack's grade is above 59)." The first of the two conjoined conditionals may obviously be read thus: "Jack passes the course *if* his grade is above 59." The second conditional may be rendered in different ways, one of which is the following: "Jack passes the course *only if* his mark is above 59" (indeed, the second conditional asserts that if Jack passes the course his grade must be better than 59, i.e., that he can pass only if that condition is met).

The truth table for the biconditional is

A	B	$A \equiv B$
T	T	T
F	T	F
T	F	F
F	F	T

A biconditional is true if and only if its two components have the same truth value. Note that the statement "$A \to B$," which in words says that "if A, then B" or "only if B, then A," is equivalent to saying "A is sufficient for B" or "B is necessary for A." The statement "$A \leftrightarrow B$" gives A as a necessary and sufficient condition for B, and conversely.

Computation of Truth Tables for Complex Expressions

The reasoning leading to the truth table of the biconditional may be illustrated in steps as shown in Table 3-2.

TABLE 3-2

A	B	$B \supset A$	$A \supset B$	$(B \supset A) \cdot (A \supset B)$
T	T	T	T	T
F	T	F	T	F
T	F	T	F	F
F	F	T	T	T

where the truth values of the third and fourth columns depend only on those of the first two columns and that of the fifth column is the conjunction of those in the third and fourth columns. In a similar way, more complicated examples may be studied. As an illustration, the definition of $\vee$ by "$(A \vee B) \equiv (A \cdot \sim B) \vee (\sim A \cdot B)$" may be verified by Table 3-3.

TABLE 3-3

A	B	$A \cdot \sim B$	$\sim A \cdot B$	$(A \cdot \sim B) \vee (\sim A \cdot B)$
T	T	F	F	F
F	T	F	T	T
T	F	T	F	T
F	F	F	F	F

The correctness of the above definition is checked by comparing the last column with the truth-table column of "$A \vee B$." The two must have the same truth values.

3-5. Applications of Truth Tables

Test for Validity of Inferences

In an inference one may distinguish between (1) a set of sentences from which something is inferred—these are called the *premises*—and (2) the sentence which is inferred, the so-called *conclusion*. Frequently, the premises are written, in some order, above a horizontal line, and the conclusion below it. The horizontal line thus serves to indicate the claim that the last sentence can be inferred from the preceding ones, and the logically crucial question is whether the alleged inference is valid or not. It is one of the foremost objectives of logical theory to make clear what is meant by saying that an inference is valid and to develop general criteria by means of which the validity of an inference may be checked.

Consider, for example, the following two inferences:

$$\frac{\text{The gas tank is empty} \supset \text{the motor stalls}}{\sim\text{The gas tank is empty}} \qquad (a)$$

$$\frac{\begin{array}{l}\text{The gas tank is empty} \supset \text{the motor stalls}\\ \sim\text{The gas tank is empty}\end{array}}{\sim\text{the motor stalls}} \qquad (b)$$

Reflection will show that (a) is a logically valid inference, while (b) is invalid, or fallacious. In (a) it is not sure whether both premises are actually true; but certainly if they are then it is impossible that the conclusion could be false. In (b), however, it is conceivable that the premises might both be true and yet the conclusion false; thus, for example, the motor might stall because the battery is exhausted, even though the gas tank is not empty.

Generally, an inference is logically valid if it is impossible that its premises should be true and yet its conclusion false; otherwise it is logically invalid. In the first case the conclusion is said to follow (logically) from, or is a (logical) consequence of, the premises, that the conclusion can be validly inferred from the premises, and that the premises logically imply the conclusion. Note that the distinction "valid" or "invalid" is applied exclusively to inferences, the distinction "true" or "false" exclusively to statements.

For inferences of the kind illustrated by (a) and (b) above, the truth tables offer a simple mechanical and decisive method of testing for validity. The method has made it possible to build a logical decision machine whose mechanism embodies the rules of the truth tables just as a multiplication machine embodies the general rules of multiplication. The achievement of such results would have been impossible without the use of a symbolic notation of the kind explained here.

What was said previously about the validity of the inferences (a) and (b) can be expressed more clearly by reference to their "forms." In a manner analogous to that of replacing symbols by numbers to evaluate an algebraic statement, one can represent inference (a) by means of the letters A and B, which may be replaced by any desired statement. This yields the scheme

$$\frac{A \supset B}{\sim B} \qquad (c)$$
$$\overline{\sim A}$$

By substituting the statements "the gas tank is empty" and "the motor stalls" for A and B, respectively, (a), which is said to constitute a particular substitution instance of the inference form (c), is obtained. However, (b) is an instance of the following compositional form:

$$\frac{A \supset B}{\sim A} \qquad (d)$$
$$\overline{\sim B}$$

Note that in any substitution instance of (c), i.e., in any specific inference of the form (c), in which the premises are true sentences, the conclusion is always a true sentence. To show this, construct the truth table of the premises and the conclusion of (c) (see Table 3-4).

TABLE 3-4

A	B	$A \supset B$	$\sim B$	$\sim A$
T	T	T	F	F
F	T	T	F	T
T	F	F	T	F
F	F	T	T	T

Then check to see whether in all the rows in which both premises are marked T the conclusion is likewise marked T. Note that, in this particular case, there happens to be only one possible case in which both premises are true together (and thus are both marked T); it is the case represented by the fourth row, and in that row the conclusion is also marked T. This shows that there cannot possibly be a substitution instance of (c) in which the premises are true but the conclusion is not. For this reason, schema (c) is said to represent a valid form of inference. The specific argument (a) is now said to be a valid inference simply because it is a substitution instance of a valid form of inference.

Consider now the form (d). Its truth table is given by Table 3-5.

TABLE 3-5

A	B	$A \supset B$	$\sim A$	$\sim B$
T	T	T	F	F
F	T	T	T	F
T	F	F	F	T
F	F	T	T	T

The premises are both marked T in the second and fourth rows, but in the former of these, the conclusion is marked F. Hence, it is possible (though not necessary, as shown by the fourth row) that in a specific instance of form (d) both premises may be true and yet the conclusion false. For this reason, (d) is called an invalid form of inference.

For the use of truth tables to test the validity of more complex examples consider the following argument:

> If it rains, I take my umbrella
> If I take my umbrella, I am likely to lose it
> ---
> If it rains, I am likely to lose my umbrella

MATHEMATICAL EXISTENCE AND LOGICAL PROOF

This is a substitution instance of the form

$$\frac{\begin{array}{c} A \supset B \\ B \supset C \end{array}}{A \supset C} \tag{e}$$

There are three components involved in this schema. This means that there are altogether eight different possibilities for the truth values of the components: Each of the four possibilities distinguished with regard to A and B in the previous tables gives rise to two new possibilities, since, in each, C may be T or F.

The truth table for the premises and conclusion of (e) is given in Table 3-6.

TABLE 3-6

A	B	C	$A \supset B$	$B \supset C$	$A \supset C$
T	T	T	T	T	T
F	T	T	T	T	T
T	F	T	F	T	T
F	F	T	T	T	T
T	T	F	T	F	F
F	T	F	T	F	T
T	F	F	F	T	F
F	F	F	T	T	T

Note that both premises are marked T in the first, second, fourth, and eighth rows and that in all these rows the conclusion is likewise marked T. Hence, (e) is a valid form of inference, and the above argument about the umbrella, being a substitution instance of it, is a valid inference.

Consider now the following argument:

If Dick has missed the train he would have phoned or wired
Dick has not phoned
Dick did not miss the train

It is of the form

$$\frac{\begin{array}{c} A \supset (B \vee C) \\ \sim B \end{array}}{\sim A} \tag{f}$$

The truth tables of premises and conclusion are combined in the schema shown in Table 3-7.

TABLE 3-7

A	B	C	$B \vee C$	$A \supset (B \vee C)$	$\sim B$	$\sim A$
T	T	T	T	T	F	F
F	T	T	T	T	F	T
T	F	T	T	T	T	F
F	F	T	T	T	T	T
T	T	F	T	T	F	F
F	T	F	T	T	F	T
T	F	F	F	F	T	F
F	F	F	F	T	T	T

The premises are both marked T in the third, fourth, and eighth rows. However, in the third row the conclusion is marked F. Hence, (f) is an invalid form of inference. And indeed, in the concrete example given above, it is quite possible that the premises should both be true and yet the conclusion be false: Dick might have missed the train and not phoned but wired. Note that exactly this possibility is indicated by the third row of the last schema.

Observe that an inference need not have exactly two premises. Thus, for example,

$$\frac{\sim\sim A}{A} \qquad \frac{A \equiv B}{A \supset B} \qquad \frac{\begin{array}{c} A \supset B \\ B \supset C \\ \sim C \end{array}}{\sim A} \qquad \frac{A}{A \vee B}$$

are all valid forms of inference. (The last one shows that from any one given sentence one may validly derive infinitely many different consequences!)

For reference, a list of some simple valid forms of inference and some formal fallacies is given here.

$\dfrac{\begin{array}{c} A \supset B \\ A \end{array}}{B}$ (modus ponens) $\qquad \dfrac{\begin{array}{c} A \supset B \\ B \end{array}}{A}$ (fallacy of affirming the consequent)

$\dfrac{\begin{array}{c} A \vee B \\ A \end{array}}{B}$ (alternative syllogism) $\qquad \dfrac{\begin{array}{c} A \supset B \\ \sim B \end{array}}{\sim A}$ (modus tollens)

$\dfrac{\begin{array}{c} A \supset B \\ \sim A \end{array}}{\sim B}$ (fallacy of denying the antecedent) $\qquad \dfrac{\begin{array}{c} A \supset B \\ B \supset C \end{array}}{A \supset C}$ (chain rule)

Test of Logical Equivalence, Tautologies, and Contradictions

Consider the sentence (1) "Dick is in Chicago ⊃ Dick is in Illinois" and its contrapositive (2) "∼Dick is in Illinois ⊃ ∼Dick is in Chicago." The latter can be validly inferred from the former. The inference is of the form

$$\frac{A \supset B}{\sim B \supset \sim A} \qquad (g)$$

and Table 3-8 shows that in all cases where the premise is marked T so is

TABLE 3-8

A	B	$\sim B$	$\sim A$	$A \supset B$	$\sim B \supset \sim A$
T	T	F	F	T	T
F	T	F	T	T	T
T	F	T	F	F	F
F	F	T	T	T	T

the conclusion. But the same table also shows that

$$\frac{\sim B \supset \sim A}{A \supset B} \qquad (h)$$

is a valid form of inference as well.

Consequently, sentences (1) and (2) can be inferred from each other. Such sentences are said to be logically equivalent, and the rule for their equivalence is that, if two statements are corresponding instances of compositional forms with identical truth tables, then the statements are logically equivalent. Logically equivalent statements convey the same information in (possibly) different words. The above analysis shows that any conditional is logically equivalent with its contrapositive. This rule is known as the *principle of contraposition*. The equivalence of (1) and (2) is just a special instance of it.

The term "equivalent," as explained and used thus far, is applicable only to statements: Two statements are said to be (logically) equivalent if they make exactly the same assertion (in other words, if they express the same proposition). Such equivalence requires that the two statements be true under exactly the same conditions, and this consideration provided the basis for the use of truth tables in establishing certain equivalences.

It is convenient, however, to extend the use of the term "equivalent" (and similarly of the terms "incompatible" and "contradictory") to compositional forms. Two compositional forms are equivalent if their truth tables are identical.

The truth table corresponding to the compositional form

$$A \supset (B \lor {\sim}B)$$

is given by Table 3-9.

TABLE 3-9

A	B	B	$B \lor {\sim}B$	$A \supset (B \lor {\sim}B)$
T	T	F	T	T
F	T	F	T	T
T	F	T	T	T
F	F	T	T	T

The table shows that the above compositional form cannot be false under any circumstances and is true by virtue of its logical form.

A statement which is a substitution instance of a compositional form whose truth table contains exclusively *T*'s is called a *truth-table tautology*. The truth-table tautologies belong to the wider class of logically true, or analytic, statements, i.e., of those statements whose truth can be established by logical analysis alone and hence without any need to collect empirical data as evidence.

On the other hand, the compositional form

$$A \cdot {\sim}A$$

has the truth table

A	${\sim}A$	$A \cdot {\sim}A$
T	F	F
F	T	F

As this table shows, any statement of the form "$A \cdot {\sim}A$"—no matter what the subject matter and the truth value of its components may be— must necessarily be false, by virtue of its form alone.

A statement for which it can be shown, by means of logical analysis, that it must be false no matter what the facts may be (i.e., under all possible circumstances) is said to be *self-contradictory*, or logically false. Logically false statements include those statements which are substitution instances of compositional forms which have exclusively *F*'s in their truth tables. Logically false statements of this particular kind are also called *truth-table contradictions*.

It is now easy to establish the following laws by reference to truth tables:

Any statement of the form "$A \supset A$" is logically true (principle of identity).
Any statement of the form "$A \lor \sim A$" is logically true (principle of excluded middle).
Any statement of the form "$A \cdot \sim A$" is logically false (principle of contradiction).
These three principles are often referred to as the *laws of thought*.

REFERENCES

1. Belgodere, P.: "Aide-mémoire de mathématiques générales," and "Aide-mémoire de calcul différentiel et intégral," Centre de Documentation Universitaire, Tournier et Constans, Paris.
2. Carroll, L.: "Symbolic Logic," London, 1897.
3. DeSua, F. C.: Metamathematics: A Non-technical Exposition, *American Scientist*, pp. 488–494, 1955.
4. Euclid: "The Thirteen Books of Euclid's Elements," vol. 3, Dover Publications, New York, 1956.
5. Gödel, K.: "On Undecidable Propositions of Formal Mathematical Systems," mimeographed, Princeton University, Princeton, N.J., 1934.
6. Hempel, C. G.: Notes for Philosophy 10, Yale University, New Haven, Conn., 1951.
7. Hilbert, D., and W. Ackerman: "Gründzuge der theoretischen Logik," 2d ed., Berlin, 1938.
8. Hilbert, D., and P. Bernays: "Grundlagen der Mathematik," vol. 1, 1934; vol. 2, 1939, Berlin. Second printing, J. W. Edwards, Publishers, Inc., Ann Arbor, Mich., 1944.
9. Hilbert, D., and S. Cohn-Vossen: "Anschauliche Geometrie," Dover Publications, New York, 1944.
10. Hille, E.: Recent Trends in Mathematics, *American Scientist*, vol. 41, p. 106, January, 1953.
11. Kershner, R. B., and L. R. Wilcox: "The Anatomy of Mathematics," The Ronald Press Company, New York, 1950.
12. Landau, E.: "Differential and Integral Calculus," Chelsea Publishing Company, New York, 1951.
13. Ledley, R. S.: Mathematical Foundations and Computational Methods for a Digital Logic Machine, *Journal of the Operations Research Society of America*, vol. 2, no. 3, August, 1954.
14. Ledley, R. S.: Digital Computational Methods in Symbolic Logic, with Examples in Biochemistry, *Proceedings of the National Academy of Sciences*, vol. 41, no. 7, pp. 498–511, July, 1955.
15. Nagel, E., and J. R. Newman: Gödel's Proof, *Scientific American*, vol. 194, no. 6, June, 1956.
16. Phillips, E. G.: "A Course of Analysis," Cambridge University Press, London, 1946.
17. Poincaré, H.: "Science and Hypothesis," Dover Publications, New York, 1955.
18. Poincaré, H.: "Science and Methods," Dover Publications, New York, 1955.
19. Quine, W. V. O.: "Methods of Logic," Henry Holt and Company, Inc., New York, 1950.

20. Rosser, J. B.: "Logic for Mathematicians," McGraw-Hill Book Company, Inc., New York, 1953.
21. Tarski, A.: "Introduction to Logic," 2d rev. ed., Oxford University Press, New York, 1946.
22. Whitehead, A. N., and B. Russell: "Principia Mathematica," 2d ed., vols. 1, 2, and 3, Cambridge University Press, London, 1925 1927.
23. Whittaker, E. T., and G. Robinson: "The Calculus of Observations," 3d ed., Blackie & Son, Ltd., Glasgow, 1940.

CHAPTER 4

ELEMENTARY CLASSICAL METHODS USED IN MODEL FORMULATION

4-1. Introduction

The purpose of this chapter is to discuss, by means of examples, some representative mathematical methods used in problem formulation. Often physical or geometrical analysis of a problem gives rise to simple proportionality relationships or points out the need for more elaborate techniques such as differential and difference equations, integral equations, and statistical methods. Dimensional analysis is another useful device which occurs in physical considerations of a problem and which can also be applied to establish relations among the variables of a physical problem.

Before proceeding to the formal methods presented in the next sections of this chapter, two elementary examples are given to illustrate geometric and physical arguments used in formalizing a problem. Many such illustrations are provided in the problems of the final chapter.

Example 1: If a rectangle is inscribed in an isosceles triangle of base b and altitude h, express the area A of the rectangle as a function of one of its sides y.

Solution: If y is the base of the rectangle and x is its altitude, then by similarity of triangles one has

$$\frac{x}{h} = \frac{(b-y)/2}{b/2}$$

from which one obtains

$$x = \frac{h(b-y)}{b}$$

and the area can be simply expressed as

$$A = \frac{hy(b-y)}{b}$$

Example 2: In the "ride and tie" plan (where the first man rides a horse for a stated time and then leaves it for a second man, who is walking, to ride for a like time while the first man walks on) find the rate of progress and for what fraction of the time the horse can rest.

Solution: Let r and w be the riding and walking rates, respectively. At the beginning and end of a "cycle" of time duration t, the two men are together and the horse has rested for a length of time T. The distance traversed by walking equals the distance traversed by riding, that is, $w(t + T) = r(t - T)$, whence the fraction of time the horse can rest is $T/t = (r - w)/(r + w)$. The rate of progress is thus given by

$$\frac{w(t+T)}{t} = w\left[1 + \frac{r-w}{r+w}\right] = \frac{2rw}{r+w}$$

For the case of m men nd one horse the rate of progress is

$$\frac{mrw}{(m-1)r + w}$$

and the fraction of time the horse can rest is given by

$$\frac{(m-1)(r-w)}{(m-1)r + w}$$

4-2. Differential Equations

Other problems often require an approach using differential equations. In this section a brief presentation is made of existence theorems and methods of solutions of differential equations of frequent occurrence. See also Chap. 11.

Differential Equations of First Order; Existence and Construction of Solutions [2, 19]

In the first-order differential equation $dy/dx = f(x,y)$ let $f(x,y)$ be analytic in the neighborhood of (x_0,y_0); that is, it can be expanded in series in the neighborhood of the point (x_0,y_0). Then an integral $y(x)$ of the differential equation exists; is analytic in the neighborhood of (x_0,y_0), yielding y_0 at $x = x_0$; and has the series expansion

$$y - y_0 = \left(\frac{dy}{dx}\right)_0 (x - x_0) + \left(\frac{d^2y}{dx^2}\right)_0 \frac{(x - x_0)^2}{2!} + \cdots$$
$$+ \left(\frac{d^ny}{dx^n}\right)_0 \frac{(x - x_0)^n}{n!} + \cdots \quad (4\text{-}1)$$

where $(d^iy/dx^i)_0$ is the ith-order derivative at (x_0,y_0).

Note that

$$\frac{d^2y}{dx^2} = \frac{\partial f}{\partial x} + \frac{\partial f}{\partial y}\frac{dy}{dx}$$
$$\frac{d^3y}{dx^3} = \frac{\partial^2 f}{\partial x^2} + 2\frac{\partial^2 f}{\partial x\, \partial y}\frac{dy}{dx} + \frac{\partial^2 f}{\partial y^2}\left(\frac{dy}{dx}\right)^2 + \frac{\partial f}{\partial y}\frac{d^2y}{dx^2}$$

Seven classical types of first-order differential equations will be briefly discussed here.

1. An equation with separable variables is of the form

$$f(x)\,dx + g(y)\,dy = 0 \tag{4-2}$$

Integration leads to

$$\int f(x)\,dx + \int g(y)\,dy = C \tag{4-3}$$

To this type belong equations which do not contain y, such as $f(x,y') = 0$. These are resolved by obtaining $y' \equiv dy/dx = h(x)$, which is an equation with separable variables having $y = \int h(x)\,dx + C$ as a solution. Equations which do not contain x, such as $f(y,y') = 0$ also belong to this type. On setting $dy/dx = h(y)$, the solution is given by

$$x = \int \frac{dy}{h(y)} + C \tag{4-4}$$

2. A homogeneous equation (one which is not changed when kx is substituted for x and ky for y) of the form $dy/dx = f(y/x)$ is integrated by means of the substitution $y = tx$. This gives

$$\frac{dy}{dx} = t + x\frac{dt}{dx} = f(t) \tag{4-5}$$

from which $dx/x = dt/[f(t) - t]$ is obtained. Integration and multiplication by x yield y as a function of x, alone. The substitution $x = r\cos\theta$, $y = r\sin\theta$, can also be used to obtain the solution.

3. The linear equation $dy/dx = f(x)y + g(x)$ is solved by replacing

$$y = h(x)e^{\int f(x)dx} \quad \text{with} \quad h'(x) = g(x)e^{-\int f(x)dx}$$

The solution is then given by

$$y = e^{\int f(x)dx}\int g(x)e^{-\int f(x)dx}\,dx \tag{4-6}$$

as can be verified by substitution.

4. Bernoulli's equation, $dy/dx = f(x)y + g(x)y^n$, is linear in y, y', and y^n, n designating an arbitrary constant. It is reduced to a linear equation by the change of variables

$$y^{-n+1} = z$$

5. Clairaut's equation is $y = xy' + f(y')$. The general integral of this equation is given by the family of straight lines, $y = ax + f(a)$.

6. Lagrange's equation, $y = xf(y') + g(y')$, is linear in x and y. To integrate this equation, let $y' = p$, from which it follows that $y'' = dp/dx$. By differentiating with respect to x, the equation becomes

$$p = f(p) + [xf'(p) + g'(p)]\frac{dp}{dx}$$

which may be written as

$$[p - f(p)]\frac{dx}{dp} = xf'(p) + g'(p) \tag{4-7}$$

This is a linear equation yielding p as a function of x. Then Lagrange's equation itself provides y as a function of x.

7. The last of these equations is Riccati's equation, $y' = ay^2 + by + c$, where a, b, and c are functions of x. The solution of this equation is not generally obtainable by quadratures, as in the other cases. If a particular integral y_1 is known, then substituting $z = y - y_1$ leads to Bernoulli's equation. If two particular integrals, y_1 and y_2, are known, the substitution $u = 1/z$, $z = y_2 - y_1$ reduces the equation to a linear differential equation.

The classical process of solving the linear differential equation of nth order,

$$\frac{d^n y}{dx^n} + a_1 \frac{d^{n-1} y}{dx^{n-1}} + \cdots + a_{n-1} \frac{dy}{dx} + a_n y = 0 \tag{4-8}$$

where $a_i (i = 1, \ldots, n)$ are constants, is carried out by assuming a solution of the form e^{rx}. This leads to the equation

$$r^n + a_1 r^{n-1} + \cdots + a_n = 0 \tag{4-9}$$

called the *characteristic equation*. To each simple root r_k of this equation corresponds a particular integral $e^{r_k x}$ and to each multiple root r_k of order k correspond k particular integrals $e^{r_k x}, xe^{r_k x}, \ldots, x^{k-1} e^{r_k x}$. The general solution is then given as the sum of the particular integrals each multiplied by an arbitrary constant.

The General Linear Differential Equation of Second Order with Constant Coefficients

Such an equation is given by

$$a\frac{d^2 y}{dx^2} + b\frac{dy}{dx} + cy = f(x) \tag{4-10}$$

The general integral y is the sum of the integral y_1 of the equation with $f(x) = 0$, called the *homogeneous part*, and a particular integral y_2 of the nonhomogeneous equation with

$$f(x) \not\equiv 0 \quad \text{that is,} \quad y = y_1 + y_2$$

Now y_1 is obtained as follows: By assuming a solution of the form e^{rx} one has the characteristic equation

$$ar^2 + br + c = 0$$

If this quadratic equation has two real roots the solution is given as

$$y_1 = c_1 e^{r_1 x} + c_2 e^{r_2 x} \qquad (4\text{-}11)$$

If it has a double root r_1, then

$$y_1 = e^{r_1 x}(c_1 + c_2 x) \qquad (4\text{-}12)$$

Finally, if it has two conjugate imaginary roots $\alpha + i\beta$ and $\alpha - i\beta$, then

$$y_1 = e^{\alpha x}(c_1 \cos \beta x + c_2 \sin \beta x) \qquad (4\text{-}13)$$

In each case c_1 and c_2 are arbitrary constants to be determined by the initial conditions.

The calculation of y_2 depends on the form of the function $f(x)$. If

$$f(x) = a_0 x^n + a_1 x^{n-1} + \cdots + a_n \qquad (4\text{-}14)$$

then y in the left side of the differential equation is replaced by a polynomial of degree n if $c \neq 0$ and by a polynomial of degree $n + 1$ if $c = 0$ and $b \neq 0$, and the two sides of the equation are identified in order to determine the coefficients. In this manner a polynomial $g(x)$ is obtained as a particular integral of the equation. If, on the other hand, $f(x) = Ae^{\alpha x}$, the integral y_2 depends on the characteristic equation of the homogeneous part. If the characteristic equation does not vanish at $r = \alpha$, then

$$y_2 = \frac{A e^{\alpha x}}{a\alpha^2 + b\alpha + c} \qquad (4\text{-}15)$$

If the characteristic equation vanishes at $r = \alpha$ but its derivative does not, then

$$y_2 = \frac{A x e^{\alpha x}}{2\alpha a + b} \qquad (4\text{-}16)$$

Finally, if both the characteristic equation and its derivative vanish at $r = \alpha$, then

$$y_2 = \frac{A x^2 e^{\alpha x}}{2a} \qquad (4\text{-}17)$$

If $f(x)$ involves trigonometric or hyperbolic functions they can be expressed in terms of exponentials and the above procedure is then applied.

Remark 1: Euler's equation $ax^2 y'' + bxy' + cy = 0$ may be reduced to a linear equation with constant coefficients in y and t by substituting $x = e^t$.

Remark 2: Of interest are two types of second-order differential equations for which it is known how to find solutions. There are $f(x, y', y'') = 0$, in which y does not appear and which becomes a first-order equation $f(x, p, p') = 0$ by substituting $y' = p$, and $f(y, y', y'') = 0$, in which x

does not appear and which is reduced to the previous type by substituting

$$y' = p \qquad y'' = \frac{dp}{dx} = \frac{dp}{dy}\frac{dy}{dx} = p\frac{dp}{dy} \qquad (4\text{-}18)$$

This gives
$$f\left(y,p,p\frac{dp}{dy}\right) = 0 \qquad (4\text{-}19)$$

Systems of n Differential Equations of First Order

An example of a system of n differential equations of the first order is as follows:

$$\begin{aligned}\frac{dy_1}{dx} &= f_1(x,y_1,y_2,\ldots,y_n) \\ \frac{dy_2}{dx} &= f_2(x,y_1,y_2,\ldots,y_n) \\ &\cdots\cdots\cdots\cdots\cdots \\ \frac{dy_n}{dx} &= f_n(x,y_1,y_2,\ldots,y_n)\end{aligned} \qquad (4\text{-}20)$$

If the functions f_i are analytic in the neighborhood of x_0, $(y_1)_0$, ..., $(y_n)_0$, the system has a set of analytic integrals in the domain of x_0 and yields $(y_1)_0, \ldots, (y_n)_0$ at $x = x_0$. Consequently, such a system depends on n arbitrary constants and may be replaced by an equivalent system of a single differential equation of order n, and by $n - 1$ relations among the variable x and the functions $(y_1, \ldots, y_n)$. Such a relation between the variables is called a *first integral* of the system. A set of k distinct first integrals can replace any k equations of the system and hence lower the order of the differential system. Thus the resolution of such a system requires the finding of first integrals.

An nth-order Differential Equation

The equation

$$\frac{d^n y}{dx^n} = f\left(x,y,\frac{dy}{dx},\ldots,\frac{d^{n-1}y}{dx^{n-1}}\right) \qquad (4\text{-}21)$$

may be replaced by the equivalent system of n equations of the first order:

$$\frac{dy}{dx} = y_1 \quad \frac{dy_1}{dx} = y_2 \quad \cdots \quad \frac{dy_{n-2}}{dx} = y_{n-1} \quad \frac{dy_{n-1}}{dx} = f(x,y,y_1,\ldots,y_{n-1}) \qquad (4\text{-}22)$$

It follows that the equation has an analytic solution in the domain of x_0 such that together with its $(n - 1)$st derivatives it takes the values $y_0, y_0', \ldots, y_0^{(n-1)}$ at $x = x_0$. Thus, f must be analytic in the neighborhood of this point.

An nth-order Linear Differential Equation

The equation

$$\frac{d^n y}{dx^n} + a_1 \frac{d^{n-1} y}{dx^{n-1}} + \cdots + a_{n-1} \frac{dy}{dx} + a_n y + a_{n+1} = 0 \quad (4\text{-}23)$$

with coefficients a_i as functions of x assumed analytic in a circle C_0 of radius R with center at x_0 has an analytic integral in the circle C_0 and assumes the value y_0 at $x = x_0$ while its $(n-1)$st derivatives take on arbitrary constant values y_0', y_0'', ..., $y_0^{(n-1)}$ for $x = x_0$.

The following examples are interesting although they are not a complete illustration of the type of equations mentioned above.

Example 1: Suppose that N_1 units of one force A, each of hitting power α, are engaged with N_2 units of an enemy B, each of hitting power β. Suppose further that the engagement is of such a kind that the fire power of force A is directed equally against all units of B, and vice versa. Characterize the strength of a force.

Solution: the rate of loss of the two forces is given by

$$\frac{dN_1}{dt} = -k\beta N_2 \quad \text{and} \quad \frac{dN_2}{dt} = -k\alpha N_1 \quad (4\text{-}24)$$

where k is a constant.

The strength of the two forces is defined as equal when their fractional losses are equal, i.e., when

$$\frac{1}{N_1}\frac{dN_1}{dt} = \frac{1}{N_2}\frac{dN_2}{dt} \quad (4\text{-}25)$$

on dividing the first equation in (4-24) by the second and integrating one has two hyperbolas defined by

$$\alpha N_1^2 - \beta N_2^2 = C \quad (4\text{-}26)$$

By taking $C = 0$ one has Lanchester's N^2 law [8, 21] which states that the strength of a force is proportional to the fire power of a unit multiplied by the square of the number of units.

Example 2: Consider the problem of conflict between a host population and its parasite opponent[1] [22]. Let N_1 be the number of the population and N_2 that of the parasites. Let b_1 and b_2 be the birth rates in the population and in the invading parasites, and let d_1 and d_2 be their respective death rates; d_1 does not include deaths by parasites. Let kN_1N_2 be the death rate in the population due to parasite invasion, where the coefficient k may be a function of N_1 and N_2 but here is assumed to be a constant. The birth of a parasite follows the laying of a single egg in a host and

---

[1] From A. Lotka, "Elements of Mathematical Biology," reprinted by permission of Dover Publications, New York.

ultimately killing him. Thus one also has $b_2 = k'kN_1N_2$, where k' indicates that only a fraction of the eggs hatch. Describe the variations in the two populations.

Solution: With the above assumptions one has the time rates of change of the two populations described by the following differential equations:

$$\frac{dN_1}{dt} = (b_1 - d_1)N_1 - kN_1N_2$$
$$\frac{dN_2}{dt} = kk'N_1N_2 - d_2N_2 \quad (4\text{-}27)$$

Dividing the first equation by the second, integrating, and substituting

$$N_1 = x + \frac{d_2}{kk'} \equiv x + p$$
$$N_2 = y + \frac{b_1 - d_1}{k} \equiv y + q \quad (4\text{-}28)$$

yield $\quad d_2 \log(x + p) + (b_1 - d_1)\log(y + q) - kk'x - ky = C \quad (4\text{-}29)$

where C is the integration constant. After expanding in series about the origin and neglecting terms of higher order, one has integral curves given by ellipses which describe the periodic variations in the host population and in the parasites. These curves are defined by

$$\frac{d_2x^2}{p^2} + \frac{(b_1 - d_1)y^2}{q^2} = D \quad (4\text{-}30)$$

where D is a constant. The period of oscillation near the origin is given by

$$\frac{2\pi}{\sqrt{d_2(b_1 - d_1)}}$$

If there is an additional host population of number N_3 with birth and death rates per head b_3 and d_3, respectively, and with the death rate due to parasites given by hN_2N_3 then the first equation remains the same, whereas the term $hh'N_2N_3$ is added to the right side of the second, where h' is the fraction of parasite eggs hatching in the second host population. A third equation

$$\frac{dN_3}{dt} = (b_3 - d_3)N_3 - hN_2N_3 \quad (4\text{-}31)$$

is adjoined to the first two in order to describe the rate of change of the new population. Similar interesting conclusions to the two population cases may be obtained by solving the system of equations near the origin. (Vito Volterra [32] studied similar systems and obtained well-known equations named after him.)

Example 3: Given a series of cups of equal capacity filled with water and arranged one below another. Pour into the first cup an equal quantity of wine at a constant rate and let the overflow in each cup go into the cup

METHODS USED IN MODEL FORMULATION 73

just below it. Assuming that complete mixture takes place instantaneously, find the amount of wine in each cup at any time t and at the end of the process, i.e., at time T [1].

Solution: Let q denote the capacity of each cup, so that the rate of flow is q/T.

Let x_k be the amount of wine in the kth cup and x_{k-1} that in the $(k-1)$st cup. Then the rate of flow of wine into the kth cup is x_{k-1}/T and out of it is x_k/T. This gives

$$\frac{dx_k}{dt} = \frac{x_{k-1}}{T} - \frac{x_k}{T} \tag{4-32}$$

Any equation in the sequence can be solved if the one before it has been solved. Thus,

$$\frac{dx_1}{dt} = \frac{q}{T} - \frac{x_1}{T}$$

which gives

$$x_1 = q\left(1 - \frac{1}{e^{t/T}}\right)$$

$$\frac{dx_2}{dt} = \frac{q(1 - 1/e^{t/T})}{T} - \frac{x_2}{T}$$

which gives

$$x_2 = q\left[1 - \frac{1 + (1/1!)(t/T)}{e^{t/T}}\right]$$

and so on.

For the final amounts in the successive cups one has

$$x_k = q\left(1 - \frac{e_k}{e}\right) \tag{4-33}$$

where e_k is the sum of the first k terms of

$$1 + \frac{1}{1!} + \frac{1}{2!} + \frac{1}{3!} + \cdots$$

as $k \to \infty$, $x_k \to 0$.

Note that the rate of flow is not essential. If x, the amount of wine poured into the first cup, is the independent variable, then

$$\frac{dx_1}{dx} + \frac{x_1}{q} = 1 \cdots \frac{dx_k}{dx} + \frac{x_k}{q} = \frac{x_{k-1}}{q} \tag{4-34}$$

and $x_k = q\left\{1 - \left[1 + \frac{x}{q} + \frac{1}{2!}\left(\frac{x}{q}\right)^2 + \cdots + \frac{1}{(k-1)!}\left(\frac{x}{q}\right)^{k-1}\right]e^{-x/q}\right\}$ (4-35)

The final result is obtained by setting $x = q$.

Example 4: The equation describing the trajectory of a projectile subject to frictional forces is obtained by solving the system of differential equations

$$\frac{d^2y}{dt^2} = -k\frac{dy}{dt} - g \quad \text{and} \quad \frac{d^2x}{dt^2} = -k\frac{dx}{dt} \tag{4-36}$$

where k is the coefficient of friction and g is the acceleration due to gravity.

Solution: If one writes

$$v_y = \frac{dy}{dt} \quad \text{and} \quad v_x = \frac{dx}{dt}$$

the above equations become

$$\frac{dv_y}{dt} = -kv_y - g \quad \text{and} \quad \frac{dv_x}{dt} = -kv_x \qquad (4\text{-}37)$$

The second of these two equations has the solution

$$v_x = (v_x)_0 e^{-kt} \qquad (4\text{-}38)$$

where $(v_x)_0$ is the initial velocity at time $t = 0$.

Also, by combining the two equations one has

$$\frac{dv_y}{dv_x} = \frac{v_y + g/k}{v_x} \qquad (4\text{-}39)$$

from which

$$v_y + \frac{g}{k} = Av_x = A(v_x)_0 e^{-kt} \qquad (4\text{-}40)$$

is obtained, where A is the integration constant. By putting $t = 0$ one has

$$A = \frac{(v_y)_0 + g/k}{(v_x)_0}$$

and

$$v_y = \left[(v_y)_0 + \frac{g}{k}\right] e^{-kt} - \frac{g}{k} \qquad (4\text{-}41)$$

On integration this becomes

$$y = -\frac{(v_y)_0 + g/k}{k} e^{-kt} - \frac{gt}{k} + C \qquad (4\text{-}42)$$

from which $\quad C = \dfrac{(v_y)_0 + g/k}{k} \quad$ since $y = 0$ when $t = 0$

Finally,

$$y = \frac{(v_y)_0 + g/k}{k}(1 - e^{-kt}) - \frac{gt}{k} \qquad (4\text{-}43)$$

and

$$x = \frac{(v_x)_0}{k}(1 - e^{-kt}) \quad \text{since } x = 0 \text{ when } t = 0 \qquad (4\text{-}44)$$

Solving for t in terms of x, one has

$$t = -\frac{1}{k} \log\left[1 - \frac{kx}{(v_x)_0}\right] \qquad (4\text{-}45)$$

which, when substituted in the expression for y, yields

$$y = \frac{(v_y)_0 + g/k}{(v_x)_0} x + \frac{g}{k^2} \log\left[1 - \frac{kx}{(v_x)_0}\right] \qquad (4\text{-}46)$$

The maximum horizontal distance traveled is obtained by setting $y = 0$ and solving for x.

Example 5: There are instances in which a differential equation serves as an auxiliary tool for carrying out mathematical operations. An example of this sort is differentiation under the integral sign, which frequently facilitates evaluation of the integral. This usually leads to a differential equation in the variable with respect to which differentiation is made. To illustrate, suppose that it is desired to evaluate the integral

$$I(y) = \int_0^\infty e^{-x^2} \cos 2xy \, dx \tag{4-47}$$

Solution: Differentiation with respect to y gives

$$\frac{dI}{dy} = -\int_0^\infty 2xe^{-x^2} \sin 2xy \, dx = [e^{-x^2} \sin 2xy]_0^\infty - \int_0^\infty 2ye^{-x^2} \cos 2xy \, dx$$
$$= -2yI \tag{4-48}$$

having integrated by parts. The solution of the differential equation

$$\frac{dI}{dy} = -2yI \tag{4-49}$$

gives
$$I = ce^{-y^2} \tag{4-50}$$

The original integral has the value $\sqrt{\pi}/2$ when $y = 0$. Thus

$$c = \frac{\sqrt{\pi}}{2} \quad \text{and} \quad I = \frac{\sqrt{\pi}}{2} e^{-y^2/2} \tag{4-51}$$

To show that

$$\int_0^\infty e^{-x^2} \, dx = \frac{\sqrt{\pi}}{2} \tag{4-52}$$

multiply the integral by another integral

$$\int_0^\infty e^{-y^2} \, dy \tag{4-53}$$

Obviously the two integrals have the same value $I(0)$. Thus, introducing polar coordinates by means of the substitution

$$x = r \cos \theta \quad y = r \sin \theta \tag{4-54}$$

whose Jacobian is equal to r, one has $dy \, dx = r \, dr \, d\theta$. (The product of the original differentials equals the absolute value of the Jacobian multiplied by the product of the new differentials.) Then

$$[I(0)]^2 = \int_0^\infty e^{-y^2} \, dy \int_0^\infty e^{-x^2} \, dx = \int_0^\infty \int_0^\infty e^{-(x^2+y^2)} \, dy \, dx$$
$$= \int_0^\infty \int_0^{\pi/2} e^{-r^2} r \, d\theta \, dr = \frac{\pi}{4} \tag{4-55}$$

It follows that $I(0) = \sqrt{\pi}/2$. Note that this is half the value of Gauss' integral

$$\int_{-\infty}^{\infty} e^{-x^2}\, dx \tag{4-56}$$

The integrand is an even function, that is, $f(x) = f(-x)$, and hence can be split into two integrals each of which is equal to $I(0)$.

Example 6: Not infrequently one is faced with the problem of solving a set of linear first-order differential equations which involve time, for example, as the independent variable. Among other methods, one that might yield a rapid solution involves the calculation of the characteristic roots of the coefficient matrix. The method will be illustrated in an example and the necessary generalities pointed out.

Consider the simple system

$$\begin{aligned}\frac{dx}{dt} &= x + y \\ \frac{dy}{dt} &= x - y\end{aligned} \tag{4-57}$$

which in matrix notation can be written as

$$\frac{d\mathbf{X}}{dt} = \mathbf{AX} \tag{4-58}$$

where

$$\mathbf{X} = \begin{pmatrix} x \\ y \end{pmatrix} \tag{4-59}$$

and the coefficient matrix is given by

$$\mathbf{A} = \begin{pmatrix} 1 & 1 \\ 1 & -1 \end{pmatrix} \tag{4-60}$$

Solution: One proceeds by assuming a solution of the form

$$\mathbf{X}^T = \mathbf{C}e^{\mathbf{A}t} \tag{4-61}$$

where $\mathbf{C}$ is the vector $\mathbf{C} = (C_1, C_2)$ and $\mathbf{X}^T = (x, y)$.

To determine $e^{\mathbf{A}t}$, subtract λ times the identity matrix from $\mathbf{A}$ and set the determinant of the resulting matrix equal to zero:

$$|\mathbf{A} - \lambda \mathbf{I}| = \begin{vmatrix} 1 - \lambda & 1 \\ 1 & -1 - \lambda \end{vmatrix} = 0 \tag{4-62}$$

This yields the quadratic equation in λ

$$\lambda^2 - 2 = 0 \tag{4-63}$$

which has two roots $\lambda_1 = \sqrt{2}$, $\lambda_2 = -\sqrt{2}$. Since the two roots are distinct, one can simply write

$$e^{\mathbf{A}t} = \frac{e^{\lambda_1 t}}{\lambda_1 - \lambda_2}(\mathbf{A} - \lambda_2 \mathbf{I}) + \frac{e^{\lambda_2 t}}{\lambda_2 - \lambda_1}(\mathbf{A} - \lambda_1 \mathbf{I}) \tag{4-64}$$

By substituting for λ_1 and λ_2, multiplying each matrix by the coefficient outside, adding, and then multiplying the result on the left by C, one has the desired solution by equating components of the vector X^T and of the vector just computed. The computation is an elementary exercise.

Remark: If the roots of A are distinct, one can write [11]

$$F(A) = \sum_{i=1}^{n} F(\lambda_i) Z_i(A) \tag{4-65}$$

where
$$Z_i(A) = \frac{\prod_{j \neq i} (A - \lambda_j I)}{\prod_{j \neq i} (\lambda_i - \lambda_j)} \qquad i, j = 1, 2, \ldots, n \tag{4-66}$$

Here, for this purpose, $F(A)$ is a polynomial in A or an exponential function in A. For example, it can be shown without difficulty that if

$$A = \begin{bmatrix} 2 & 1 \\ 1 & 2 \end{bmatrix} \tag{4-67}$$

then
$$F(A) = A^{100} = \begin{bmatrix} \dfrac{3^{100} + 1}{2} & \dfrac{3^{100} - 1}{2} \\ \dfrac{3^{100} - 1}{2} & \dfrac{3^{100} + 1}{2} \end{bmatrix} \tag{4-68}$$

4-3. Linear Difference Equations with Constant Coefficients

Difference equations, like differential equations, are frequently used in the formulation of problems and are particularly useful in queueing theory. Of the different types of difference equations occurring in practice, linear equations with constant coefficients have the widest use and are easiest to deal with.

The general nth-order linear difference equation with constant coefficients may be expressed in the form

$$p_{k+n} + a_1 p_{k+n-1} + \cdots + a_{n-1} p_{k+1} + a_n p_k = f_k \tag{4-69}$$

where $a_i (i = 1, \ldots, n)$ are constants and f_k is a function of k. This equation is nonhomogeneous unless $f_k \equiv 0$, in which case it is referred to as the homogeneous difference equation of nth order. Here two methods of solving the nonhomogeneous equation will be illustrated.

1. By finding the general solution of the homogeneous equation and adding to it a particular solution of the nonhomogeneous equation, the general solution of the nonhomogeneous equation is obtained as the sum of these two solutions. We first obtain the solution of the homogeneous part.

Let $p_k = \alpha^k$; then the above equation becomes

$$\alpha^k(\alpha^n + a_1\alpha^{n-1} + \cdots + a_{n-1}\alpha + a_n) = 0 \quad (4\text{-}70)$$

Let the n nonzero roots of this equation be $\alpha_1, \ldots, \alpha_n$ and let $\alpha_i = e^{\beta_i}$; then the general solution is given by

$$p_k = c_1 e^{\beta_1 k} + c_2 e^{\beta_2 k} + \cdots + c_n e^{\beta_n k} \quad (4\text{-}71)$$

if all the roots are distinct; otherwise, if a root of multiplicity m occurs, the exponential factor in that root, as in the solution of a differential equation, is multiplied by

$$d_1 + d_2 k + \cdots + d_m k^{m-1} \quad (4\text{-}72)$$

where the d_i are constants.

A particular solution of the nonhomogeneous equation may be obtained either by observation, by variation of parameters, or by the method of undetermined coefficients [13]. The latter method involves constructing the family of the right-hand term and expressing the particular solution as a linear combination of the family some members of which are multiplied by k if they appear in the solution of the homogeneous equation.

An illustration of this equation will be given, finding a particular solution by observation.

Example 1: Let p_1 be the probability that in a sequence of trials an event will occur on the first trial. Find p_n, the probability that the event will occur on the nth trial, given λ, the probability of its occurrence on the nth trial if it occurred on the $(n-1)$st trial, and μ, the probability of its occurrence on the nth trial if it did not occur on the $(n-1)$st trial [29].

Solution: Note that the occurrence of the event on the nth trial is preceded either by its occurrence in the $(n-1)$st trial with probability p_{n-1} or its nonoccurrence in the $(n-1)$st trial with probability $1 - p_{n-1}$. The probability that the event occurs on the $(n-1)$st trial and then on the nth trial is λp_{n-1}; the probability that it occurs on the nth trial if it failed to occur on the $(n-1)$st is $\mu(1 - p_{n-1})$. By the law of compound probabilities,

$$p_n = \lambda p_{n-1} + \mu(1 - p_{n-1}) \quad (4\text{-}73)$$

This may be written as

$$p_n + (\mu - \lambda)p_{n-1} = \mu$$

The homogeneous equation $p_n + (\mu - \lambda)p_{n-1} = 0$ has the solution

$$p_n = A(\lambda - \mu)^{n-1}$$

where A is a constant.

A particular solution of the nonhomogeneous part is

$$p_n = B$$

which, on substitution to determine B, gives

$$B = \frac{\mu}{1 + \mu - \lambda}$$

assuming $1 + \mu - \lambda \neq 0$ (the nontrivial case, for otherwise $p_n = 0$ is a particular solution).

Now A may be determined by putting p_1 (which is known) for p_n in the general solution:

$$p_n = A(\lambda - \mu)^{n-1} + \frac{\mu}{1 + \mu - \lambda}$$

which finally becomes

$$p_n = \frac{\mu}{1 + \mu - \lambda} + \left(p_1 - \frac{\mu}{1 + \mu - \lambda}\right)(\lambda - \mu)^{n-1} \quad (4\text{-}74)$$

2. The generating-function method is another way of solving the nonhomogeneous equation. In this method one determines an analytic function $P(z)$, say, inside the unit circle (particularly true when p_n are probabilities with $0 \leq p_n \leq 1$, $n = 0, 1, 2, \ldots$, and $\sum_{n=0}^{\infty} p_n = 1$). This function, called the *generating function*, is given by

$$P(z) = \sum_{n=0}^{\infty} p_n z^n \quad (4\text{-}75)$$

Then one obtains p_n by differentiating $P(z)$, n times to obtain $P^{(n)}(z)$, setting $z = 0$ and dividing by $n!$ (Laplace's method), or, since the nth derivative of $P(z)$ is given by Cauchy's formula as

$$P^{(n)}(z) = \frac{n!}{2\pi i} \int_C \frac{P(w)}{(w-z)^{n+1}} dw \quad (4\text{-}76)$$

one wants

$$p_n = \frac{P^{(n)}(0)}{n!} = \frac{1}{2\pi i} \int_C \frac{P(w)}{w^{n+1}} dw \quad (4\text{-}77)$$

where C is a simple closed contour inside the region in which $P(z)$ is analytic. The coefficients p_n may be obtained from this expression by calculating the residues (Lagrange's method).

Example 2: Find the probability of m heads in n tosses of a coin, if the probability of a head is λ and that of a tail is μ, with $\lambda + \mu = 1$.

Solution: Let $p(m,n)$ be the required probability. The difference equation for this problem can easily be set up as m heads in n trials $= m - 1$

heads in $n-1$ trials and a head in the last trial, or m heads in $n-1$ trials and a tail on the nth trial.

$$p(m,n) = \lambda p(m-1, n-1) + \mu p(m, n-1) \qquad (4\text{-}78)$$

with the initial condition
$$p(m,0) = 0 \qquad m > 0$$
$$p(0,n) = \mu^n \qquad n \geq 0$$

that is, n failures in succession. Let $P_n(z) = \sum_{m=0}^{\infty} p(m,n)z^m$ be the generating function of $p(m,n)$. Note that

$$\mu P_{n-1}(z) = \mu p(0, n-1) + \sum_{m=1}^{\infty} \mu p(m, n-1)z^m$$

$$\lambda z P_{n-1}(z) = \sum_{m=1}^{\infty} \lambda p(m-1, n-1)z^m$$

Adding these two equations and using the difference equation of the problem, the following is obtained:

$$(\lambda z + \mu)P_{n-1}(z) = P_n(z) + \mu p(0, n-1) - p(0,n)$$

having used the initial condition
$$\mu p(0, n-1) - p(0,n) = \mu^n - \mu^n = 0$$
Thus, finally, $\qquad P_n(z) = (\lambda z + \mu)P_{n-1}(z)$

This can be shown e.g. by successive substitutions to have the solution

$$P_n(z) = (\lambda z + \mu)^n P_0(z) \qquad (4\text{-}79)$$

which, on using $p(m,0) = 0$ for $m > 0$ and $p(0,0) = 1$, gives

$$P_0(z) = 1$$

When $(\lambda z + \mu)^n$ is developed in a power series of z, $p(m,n)$ is obtained as the coefficient of z^m and is given by

$$p(m,n) = \frac{n(n-1) \cdots (n-m+1)\lambda^m \mu^{n-m}}{m!} \qquad (4\text{-}80)$$

as may be verified by substitution.

Example 3: Consider the following set of linear difference equations with constant coefficients arising in the study of time-independent queues [18] studied in a later chapter:

$$\begin{aligned}\lambda p_0 - \mu p_k &= 0 & n &= 0 \\ \lambda p_n - \mu p_{n+k} - \lambda p_{n-1} &= 0 & 1 &\leq n \leq k-1 \qquad (4\text{-}81) \\ (\mu + \lambda)p_n - \mu p_{n+k} - \lambda p_{n-1} &= 0 & n &\geq k\end{aligned}$$

Define $$P(z) \equiv \sum_{n=0}^{\infty} p_n z^n \quad \text{and} \quad \rho = \frac{\lambda}{\mu} \quad (<1) \quad (4\text{-}82)$$

Solution: To obtain $P(z)$, multiply the above equations by z^n and sum over the indicated values of n and add. This gives

$$(\lambda p_0 - \mu p_k) + \sum_{n=1}^{k-1} (\lambda p_n - \mu p_{n+k} - \lambda p_{n-1}) z^n$$
$$+ \sum_{n=k}^{\infty} (\mu p_n + \lambda p_n - \mu p_{n+k} - \lambda p_{n-1}) z^n = 0$$

On simplifying, this becomes

$$\lambda \sum_{n=0}^{\infty} p_n z^n - \mu \sum_{n=0}^{\infty} p_{n+k} z^n - \lambda \sum_{n=1}^{\infty} p_{n-1} z^n + \sum_{n=k}^{\infty} \mu p_n z^n = 0$$

or $$\lambda P(z) - \frac{\mu}{z^k} \sum_{n=0}^{\infty} p_{n+k} z^{n+k} - \lambda z P(z) + \mu \sum_{n=k}^{\infty} p_n z^n = 0$$

Further simplifying yields

$$\lambda P(z) - \lambda z P(z) - \left(\frac{\mu}{z^k} - \mu\right) \sum_{n=k}^{\infty} p_n z^n = 0$$

or $$\lambda P(z) - \lambda z P(z) - \left(\frac{\mu}{z^k} - \mu\right) \left[-\sum_{n=0}^{k-1} p_n z^n + P(z) \right] = 0$$

from which one has

$$P(z) = \frac{(\mu/z^k - \mu) \sum_{n=0}^{k-1} p_n z^n}{\lambda z - \lambda + \mu/z^k - \mu} \tag{4-83}$$

Finally,

$$P(z) = \frac{\mu(1 - z^k) \sum_{n=0}^{k-1} p_n z^n}{\lambda z^{k+1} - \lambda z^k + \mu - \mu z^k} = \frac{(1 - z^k) \sum_{m=0}^{k-1} p_n z^n}{\rho z^{k+1} - (\rho + 1) z^k + 1} \tag{4-84}$$

One must somehow eliminate the unknown probabilities appearing on the right. The function $P(z)$ must converge at least inside the unit circle. Now the denominator has $k + 1$ zeros. Of these zeros, k lie on or interior to the unit circle. To show this we consider the following theorem, due to Rouché [28], as applied to the denominator. If $f(z)$ and $g(z)$ are analytic inside and on a closed contour C, and if $|g(z)| < |f(z)|$ on C, then $f(z)$ and $f(z) + g(z)$ have the same number of zeros inside C.

Thus, if $f(z) = -z^k(\rho + 1)$ and $g(z) = \rho z^{k+1} + 1$, the absolute values on the circle $|z| = 1 + \delta$ satisfy Rouché's theorem, and since $f(z)$ has k zeros inside this circle, then so does the sum of f and g, which is the denominator. One zero of the denominator is $z = 1$ and $k - 1$ zeros must coincide with those of $\sum_{n=0}^{k-1} p_n z^n$. Otherwise $P(z)$ would not converge for values of z inside the unit circle owing to the vanishing of the denominator at a zero which is not cancelled by one in the numerator. This leaves one zero of the denominator which lies outside this circle; denote it by z_0. Then one may write (equating common factors of numerator and denominator)

$$(z - 1)(z - z_0) \sum_{n=0}^{k-1} p_n z^n = A[\rho z^{k+1} - (1 + \rho)z^k + 1]$$

where A is a constant. This gives

$$P(z) = \frac{A \sum_{n=0}^{k-1} z^n}{z_0 - z} \qquad (4\text{-}85)$$

A is determined by $P(1) = 1$ to give

$$P(z) = \frac{(z_0 - 1) \sum_{n=0}^{k-1} z^n}{(z_0 - z)k} \qquad (4\text{-}86)$$

From this generating function it can be shown by differentiation that

$$p_n = \frac{1 - z_0^{-n-1}}{k} \qquad n \leq k - 1$$

and $\qquad p_n = \dfrac{z_0^{-n-1}(z_0^k - 1)}{k} \qquad n \geq k \qquad (4\text{-}87)$

Example 4: Now we determine the coefficients by Lagrange's method. Suppose that two players, I and II, play a series of games. In each game, λ is the probability that I wins, while μ is the probability that II wins the game, with $\lambda + \mu = 1$. Player I wins the series if he wins a games before II wins b games. What is the probability that I will win the series?

Solution: Let $p(m,n)$ be the probability that I will win when m games remain for him to win while II has n games left to win [29]. Then one has the following descriptive equation: (I wins m remaining games while II has n games) = (I wins the next game and has to win $m - 1$ games before II wins n games) + (he loses the next game and must win m games before II can win $n - 1$ games). Thus

$$p(m,n) = \lambda p(m - 1, n) + \mu p(m, n - 1) \quad (4\text{-}88)$$

with the initial conditions

$p(m,0) = 0 \quad m > 0 \quad$ I cannot win, II having won all his games
$p(0,n) = 1 \quad n > 0 \quad$ I wins if he has no more games to play
$p(0,0) = 0$

A particular solution of the equation is $p(m,n) = y^m z^n$, where y is determined as a function of z by substitution. This gives

$$y = \frac{\lambda}{1 - \mu z^{-1}}$$

The particular solution then becomes

$$p(m,n) = \left(\frac{\lambda}{1 - \mu z^{-1}}\right)^m z^n \quad (4\text{-}89)$$

Note that the initial conditions for the example are not satisfied. Multiplying by a function $Q(z)$ analytic in a ring about the origin, one writes

$$p(m,n) = \frac{\lambda^m}{2\pi i} \int_C \frac{w^n}{(1 - \mu w^{-1})^m} Q(w) \, dw \quad (4\text{-}90)$$

This expression is tentatively assumed to give $p(m,n)$. C is a simple closed curve in the ring. Now $Q(w)$ must be determined to satisfy the initial conditions. Thus, with $m = 0$, from the initial conditions one has

$$p(0,n) = \frac{1}{2\pi i} \int w^n Q(w) \, dw = 1 \quad n = 1, 2, \ldots$$

That is, the residue which is the coefficient of $1/w$ is unity. By putting

$$Q(w) = \frac{1}{w(w - 1)} \quad \text{with } |w| > 1$$

then no matter what n is, the coefficient of $1/w$ is always unity. Similarly, $p(m,0) = 0$ is satisfied since

$$p(m,0) = \frac{\lambda^m}{2\pi i} \int_C \frac{dw}{(1 - \mu w^{-1})^m w(w - 1)} = \frac{\lambda^m}{2\pi i} \int_C \frac{w^{m-1} \, dw}{(w - \mu)^m (w - 1)} = 0 \quad (4\text{-}91)$$

which is true by Cauchy's theorem, since $|w| > 1$. Thus this choice of $Q(w)$ solves the problem, and by expanding in series and calculating the coefficient of $1/w$ in the expression for $p(m,n)$ for this $Q(w)$, one has for the answer

$$p(m,n) = \lambda^m \left[1 + m\mu + \frac{m(m+1)}{2!} \mu^2 + \cdots + \frac{m(m+1)(m+n-2)}{(n-1)!} \mu^{n-1} \right] \quad (4\text{-}92)$$

4-4. Integral Equations

An equation involving integrals with an unknown function under the integral sign is called an *integral equation*. Sometimes a differential equation together with its boundary conditions may be transformed to an integral equation. This gives a convenient relationship between the two [13]. To solve such an equation is to determine the unknown function. Among the well-known linear integral equations (i.e., involving only linear functions of the unknown function) are the following:

1. Fredholm's equation which has the general form

$$g(x)f(x) = h(x) + \lambda \int_a^b K(x,t)f(t)\,dt \quad (4\text{-}93)$$

where λ, a, and b are constants.

2. Volterra's equation, which has the same general form as Fredholm's equation except that the upper limit b is replaced by the variable x. Such an equation is of the first kind if $g(x) = 0$ and of the second kind if $g(x) = 1$. Otherwise it is a Volterra equation of the third kind.

The following example shows how an integral equation arises in the consideration of a physical problem.

Example: Given a system with many components in which the failure distribution of some important component has been experimentally determined. Denote by $f(x)\,dx$ the failure probability during an interval of time of length dx. The system contains N representatives of this component population. It is desired to estimate the renewals needed by time t.

Solution: The initial number of renewals, i.e., renewals of the first generation at time t, are estimated by $Nf(t)$. Renewals of this first generation constitute the second-generation renewals, etc. Generally, one has the following relationship between the $(i+1)$st- and the ith-generation renewals at time t:

$$u_{i+1}(t) = \int_0^t u_i(t-x)f(x)\,dx \quad (4\text{-}94)$$

The $(i+1)$st-generation renewals at time t constitute replacement of ith-generation renewals introduced prior to t but fail at t. Now ith-generation renewals introduced at time $t-x$, that is, $u_i(t-x)$, will suffer at the end of a period x, that is, at time t, $u_i(t-x)f(x)$ failures. Hence

$(i + 1)$st-generation renewals are obtained by summing over $0 \le x < t$. If $u(t)$ denotes the total renewals at time t, then

$$u(t) = \sum_{i=1}^{\infty} u_i(t) \qquad u_1(t) = Nf(t)$$

Applying this to the previous expression, one has

$$u(t) = Nf(t) + \int_0^t u(t - x)f(x)\, dx \qquad (4\text{-}95)$$

This is Volterra's linear integral equation of the second kind.

Now the total renewals $U(t)$ by time t are obtained by integrating over $u(y)$, where y is the integration dummy variable, from zero to t, which gives

$$\begin{aligned} U(t) &= NF(t) + \int_0^t dy \int_0^y u(y - x)f(x)\, dx \\ &= NF(t) + \int_0^t dx \int_x^t u(y - x)f(x)\, dy \\ &= NF(t) + \int_0^t U(t - x)\, dF(x) \qquad (4\text{-}96) \end{aligned}$$

where
$$F(t) = \int_0^t f(x)\, dx$$

The solution of the integral equation in $u(t)$ will be obtained using the Laplace transform. One notes that the expression

$$\int_0^t u(t - x)f(x)\, dx \qquad (4\text{-}97)$$

is defined as the convolution of $u(x)$ and $f(x)$. It can be shown that the product of the Laplace transforms

$$L\{u(x)\} \equiv \int_0^{\infty} e^{-tx} u(x)\, dx \qquad (4\text{-}98)$$

and
$$L\{f(x)\} \equiv \int_0^{\infty} e^{-tx} f(x)\, dx \qquad (4\text{-}99)$$

of $u(x)$ and $f(x)$ is equal to the Laplace transform of the above convolution expression. Thus by applying Laplace transforms to the integral equation in $u(t)$ and solving for $L\{u(t)\}$ one has

$$L\{u(t)\} = \frac{NL\{f(t)\}}{1 - L\{f(t)\}} \qquad (4\text{-}100)$$

Since $f(t)$ is known, the expression on the right may be calculated, and then $u(t)$ is determined from its Laplace transform.

As an illustration, if the failure distribution is exponential, one has $f(t) = \mu e^{-\mu t}$, and $u(t)$ may be calculated as μN for $t > 0$, which yields $U(t) = \mu N t$ for the total renewals by time t.

4-5. Operators

The advantage of operator theory is that it provides algebraic models into which other types of mathematical models are transformed. Many questions relating to the original model can be examined in the new model.

When a function $f(x)$ is multiplied by a constant k, the resulting function $kf(x)$ may be regarded as a result of applying the operator k to $f(x)$. Differentiation $df(x)/dx$ of a function $f(x)$ yields a new function $f'(x)$. In this sense, the operation of taking the derivative may be considered as a transformation of functions which have a derivative to functions some of which need not have this property but are themselves derivatives. One advantage of considering d/dx as an operator D is to transform, whenever feasible, a differential equation in the former to an algebraic equation in the latter. In this manner, a solution of the differential equation can often be easily obtained.

For this purpose, the algebraic properties which these operators are required to satisfy are commutativity and associativity under addition and multiplication and the distributive law $A(B + C) = AB + AC$.

Note, for example, that d^n/dx^n may be represented by D^n, the nth power of D, since D must be applied n times. We shall illustrate the ideas with three examples, one using differential operators and two using difference operators.

Example 1: Find the general solution of the differential equation

$$\frac{d^2y}{dx^2} - 3\frac{dy}{dx} + 2y = 0 \tag{4-101}$$

Solution: In operator notation this equation becomes

$$(D^2 - 3D + 2)y = 0 \tag{4-102}$$

Now the operator expression in parentheses has the same effect as the product of the two operators $D - 2$ and $D - 1$, obtained by factoring the expression, regardless of the order in which they are applied. But

$$(D - 2)(D - 1)y = 0 \tag{4-103}$$

is satisfied if

$$(D - 1)y = 0 \tag{4-104}$$

Now one can easily show by expanding that

$$e^{-x}(D - 1)y = De^{-x}y \tag{4-105}$$

This is a special case of Heaviside's translation theorem which states that, for every rational expression $R(D)$ in the operator D, one has

$$e^{rx}R(D + r)f(x) = R(D)e^{rx}f(x) \tag{4-106}$$

Since the left-hand side above is zero, it follows that

$$De^{-x}y = 0$$

Let D^{-1}, the inverse operator which obtains the original function operated on by D, be applied to both sides of the last equation; then, since the derivative of a constant is zero, one has

$$e^{-x}y = c_1$$
or
$$y = c_1 e^x \qquad (4\text{-}107)$$

Similarly, if $D - 2$ is first applied, one has

$$y = c_2 e^{2x} \qquad (4\text{-}108)$$

The sum of these two independent solutions is also a solution which has two arbitrary constants, as required for the solution of a second-order differential equation, and hence it is the most general solution of the equation. These ideas may be similarly generalized to an nth-order linear differential equation with constant coefficients.

Example 2: Let $y_1 + y_2 + \cdots + y_n$ be a sum of numbers $y_k (k = 1, \ldots, n)$. It is desired to find an expression of the sum which depends only on n. Let E be an operator which, when applied to y_k, yields y_{k+1}, that is, $Ey_k = y_{k+1}$, but Ey_n is undefined. Let Δ be another operator such that $\Delta y_k = y_{k+1} - y_k$. E is called the *shifting operator*, and Δ is called the *difference operator*. [When applied to a function $f(x)$ one may write $Ef(x) = f(x + h)$ and $\Delta f(x) = f(x + h) - f(x)$. The three operators D, E, and Δ are then related by the formula $E = 1 + \Delta = e^{hD}$. The last operator represents the series $e^{hD} = 1 + hD + h^2D^2/2! + \cdots$. Note that $\Delta^2 y_k = \Delta(y_{k+1} - y_k) = \Delta y_{k+1} - \Delta y_k = y_{k+2} - 2y_{k+1} + y_k$, etc.]

Solution: Now one may write

$$y_1 + y_2 + y_3 + \cdots + y_n = y_1 + Ey_1 + E^2 y_1 + \cdots + E^{n-1}y_1 \qquad (4\text{-}109)$$

$$y_1 + y_2 + y_3 + \cdots + y_n$$
$$= (1 + E + E^2 + \cdots + E^{n-1})y_1 = \frac{E^n - 1}{E - 1} y_1 = \frac{(1 + \Delta)^n - 1}{\Delta} y_1$$
$$= \left[n + \frac{n(n-1)}{2!}\Delta + \frac{n(n-1)(n-2)}{3!}\Delta^2 + \cdots + \Delta^{n-1} \right] y_1 \qquad (4\text{-}110)$$

Hence, to obtain the sum, one simply calculates Δy_1, $\Delta^2 y_1$, etc.

This idea may now be applied by the reader to $\sum_{k=1}^{n} k$ and $\sum_{k=1}^{n} k^2$ to obtain sums which depend only on n.

Example 3: Solve the difference equation

$$y_{n+1} - ay_n = b_n \qquad n = 0, 1, 2, \ldots$$

Solution: This equation may be written as

$$(E - a)y_n = b_n \tag{4-111}$$

$$y_n = (E - a)^{-1}b_n = E^{-1}(1 - aE^{-1})^{-1}$$
$$= E^{-1}(1 + aE^{-1} + a^2E^{-2} + \cdots)b_n \tag{4-112}$$

Since $E^{-k}b_n = b_{n-k}$ and $b_{-1} = 0$

one has the desired solution given by

$$y_n = \sum_{k=0}^{n-1} a^k b_{n-k-1} \tag{4-113}$$

as may be verified by substitution.

Remark: The Laplace transform is another operator often used to reduce the problem of solving linear differential equations to linear algebraic equations, solving the latter and computing the inverse Laplace transform to obtain the solution of the differential equations. Again we use as an example

$$\frac{dx}{dt} = x + y$$

$$\frac{dy}{dt} = x - y$$

Here define

$$\bar{x}(\lambda) \equiv \int_0^\infty e^{-\lambda t} x(t)\, dt$$

$$\bar{y}(\lambda) \equiv \int_0^\infty e^{-\lambda t} y(t)\, dt$$

Multiply each equation by $e^{-\lambda t}$ and integrate. Use integration by parts to reduce the left-hand sides to $c_1 + \lambda\bar{x}(\lambda)$ and $c_2 + \lambda\bar{y}(\lambda)$, respectively. Solve the resulting pair of simultaneous equations in $\bar{x}$ and $\bar{y}$, then either compute or use a table to find the functions whose Laplace transforms are $\bar{x}$ and $\bar{y}$. This gives the solution of the original problem.

4-6. Dimensional Analysis [3]

In physics, the different physical variables originate largely in the operations symbolized in their dimensional formulas. Dimensional formulas and equations have a structure closely related to the operations of physical measurement. A small number of variables is selected as a function of which all other variables are expressed. Usually mass (M), length (L), and time (T) and electric charge (Q) are used as the fundamental, or primary, variables. A theorem in dimensional analysis asserts that any physical variable f is proportional to a product of powers of primary variables $P_1, P_2, \ldots, P_n$; that is, if $a_1, a_2, \ldots, a_n$ are rational numbers, then

$$f = KP_1^{a_1}P_2^{a_2} \cdots P_n^{a_n} \tag{4-114}$$

where K is a proportionality constant. Frequently K is dropped and one writes

$$[f] = [P_1^{a_1} \cdots P_n^{a_n}]$$

where the brackets indicate dimensional equivalence.

Force may be expressed in terms of the primary variables given above as

$$F = K(MLT^{-2}) \tag{4-115}$$

As an illustration, suppose that it is required to calculate the period of a simple pendulum consisting of a rigid rod of length h supporting a mass m and having an angular displacement θ.

Write the functional expression for the physical variables in which one is interested:

$$t = t(m,h,g,\theta)$$

or, dimensionally, $\quad [T] = [M^a L^b (LT^{-2})^c] \tag{4-116}$

The equations obtained by equating corresponding powers are

$$\begin{array}{llll} M: & 0 = a & \text{or} & a = 0 \\ L: & 0 = b + c & \text{or} & b = \tfrac{1}{2} \\ T: & 1 = -2c & \text{or} & c = -\tfrac{1}{2} \end{array} \tag{4-117}$$

Thus $\quad t \sim \dfrac{h^{1/2}}{g^{1/2}}$

Taking into account θ, insert an unknown function $f(\theta)$; one has

$$t = K\sqrt{\frac{h}{g}} f(\theta)$$

Experimentally, one finds that, if θ is small, $f(\theta)$ is nearly a constant or approximately 2π. Note that the period is independent of mass since the exponent of the mass is zero.

Letting $K = 1$, obtain $t = 2\pi \sqrt{h/g}$.

By similar arguments the reader will have no difficulty in verifying the well-known elementary relations of physics, $s = \tfrac{1}{2}gt^2$ and $v^2 = 2as$, where s is distance, v is velocity, a is acceleration, and g is the acceleration due to gravity.

4-7. Some Basic Statistics and Probabilistic Methods

When many factors are involved, attention is limited to a few, the purpose being prediction of new values from given ones. Some elementary methods used are as follows:

The Method of Least Squares

In the method of least squares, one assumes a polynomial or other approximation to a set of points, attempts to minimize the sum of the squares of the differences, and thus solves for the parameters. A simple case of this method will be considered.

General Discussion: If the points $(x_1, y_1), \ldots, (x_n, y_n)$ are fitted by the line $y = mx + b$, the residuals are

$$\begin{aligned} r_1 &= y_1 - (mx_1 + b) \\ r_2 &= y_2 - (mx_2 + b) \\ &\cdots\cdots\cdots\cdots\cdots \\ r_n &= y_n - (mx_n + b) \end{aligned} \qquad (4\text{-}118)$$

The squares of the residuals are

$$\begin{aligned} r_1^2 &= y_1^2 - 2my_1 x_1 - 2y_1 b + m^2 x_1^2 + 2mx_1 b + b^2 \\ &\cdots\cdots\cdots\cdots\cdots\cdots\cdots\cdots\cdots\cdots\cdots\cdots\cdots \\ r_n^2 &= y_n^2 - 2my_n x_n - 2y_n b + m^2 x^2 + 2mx_n b + b^2 \end{aligned} \qquad (4\text{-}119)$$

Adding yields

$$\sum_{i=1}^{n} r_i^2 = nb^2 + 2\left(m \sum_{i=1}^{n} x_i - \sum_{i=1}^{n} y_i\right) b \\ + \left(m^2 \sum_{i=1}^{n} x_i^2 - 2m \sum_{i=1}^{n} x_i y_i + \sum_{i=1}^{n} y_i^2\right) \qquad (4\text{-}120)$$

One must minimize this as a function of b and of m. By setting the derivative with respect to b equal to zero, one has

$$b = \frac{m \sum_{i=1}^{n} x_i - \sum_{i=1}^{n} y_i}{n} \qquad (4\text{-}121)$$

Then also setting the derivative with respect to m equal to zero yields

$$m = -\frac{b \sum_{i=1}^{n} x_i - \sum_{i=1}^{n} x_i y_i}{\sum_{i=1}^{n} x_i^2} \qquad (4\text{-}122)$$

Substituting for b gives

$$m = \frac{\sum_{i=1}^{n} x_i y_i - \left(\sum_{i=1}^{n} x_i \sum_{i=1}^{n} y_i\right)/n}{\sum_{i=1}^{n} x_i^2 - \left(\sum_{i=1}^{n} x_i\right)^2/n} \qquad (4\text{-}123)$$

With this value of m, one has

$$b = \frac{\left(\sum_{i=1}^{n} x_i\right)\left(\sum_{i=1}^{n} x_i y_i\right) - \left(\sum_{i=1}^{n} y_i\right)\left(\sum_{i=1}^{n} x_i^2\right)}{\left(\sum_{i=1}^{n} x_i\right)^2 - n\left(\sum_{i=1}^{n} x_i^2\right)} \qquad (4\text{-}124)$$

These values of m and b provide the least-square linear fit to the data. The same procedure may be used to fit the data with a polynomial of higher degree. As before, the coefficients are determined from the simultaneous equations resulting from setting the partial derivatives equal to zero in attempting to minimize the sum of the squares of the residuals with respect to the coefficients. The method is also applicable to non-polynomial least-square approximation.

It may be noted in passing that harmonic analysis [25] is another effective method of approximating a set of data by a Fourier series after it has been shown that such an approximation is justified.

Steepest Ascent Along Fitting Surfaces

Statistical methods of determining an optimum point for an unconstrained function have been investigated by G. Box and others [4, 9]. The procedure will be briefly described here. The reader may find it more convenient to read this section after familiarity with the next chapter.

Suppose that the interactions of several factors $(x_1, \ldots, x_n)$ in an experiment are related by a function $f(x_1, \ldots, x_n)$ which is unknown. Assuming that this function has an optimum such as a maximum, it is desired to determine or estimate this optimum by experimental sampling and data fitting without attempting to determine $f(x_1, \ldots, x_n)$ completely, which is usually an expensive and wasteful process.

The fitting of the data is done locally to determine a direction in which one may proceed toward the maximum; having followed a path of steepest ascent, one must stop somewhere and repeat the process.

One of the favorite attacks is the method of one factor at a time. By holding all the controlled variables except one at prescribed values and allowing the latter to vary, a maximum is obtained. It is then used as a fixed value for this variable, and another variable is allowed to vary, etc.

A more reliable procedure leads to a maximum by following a path of steepest ascent in steps, assuming approximations by planes or quadratic surfaces in small regions as may be justified by series expansions. The plan is to predict the best-fitting plane in a small portion of the experimental region, using a least-squares procedure. A steepest-ascent path is indicated by changes in variables equal to the direction cosines which are proportional to the coefficients. Experiments are then performed along this path until a decrease in value is noted. Repetition of the procedure around this point can confirm whether a maximum has been reached or whether a new path of ascent should be determined. If the approximating surface indicates flatness, it does not imply that a maximum has been reached. The surface may contain a ridge which is stationary, it may be slowly rising, or a saddle point may even have been reached. Further investigation is needed, depending on the type of solution in the practical problem.

It may be found that the linear and quadratic terms in the series expansion (see Chap. 5) may be used as an approximation, and a least-squares method to determine the coefficients is then applied. Because of the increase in the number of coefficients, larger samples for their determination are required, since at least N observations are necessary to fit an approximating function containing N constants. Reduction to canonical form by a translation and orthogonal rotation facilitates the determination of the steepest-ascent path in terms of the latent roots of the associated characteristic equation.

Information Theory [27]

An appreciation of the use of information theory to provide solutions of some problems can best be developed by means of an example. Without too much detail, one defines the entropy content (the negative of this quantity is defined as the information content) of a set of n mutually exclusive events by

$$-\sum_{i=1}^{n} p_i \log p_i \qquad (4\text{-}125)$$

where p_i is the probability of occurrence of the ith event. The logarithm arises from requiring additivity in the sense that the sum of the entropies obtained from two events is equal to the entropy from the events taken together. Symbolically, this yields the functional equation

$$f(x) + f(y) = f(xy) \qquad (4\text{-}126)$$

for which one has $f(x) = \log x$ as the solution.

The definition of entropy will now be used to find an a priori upper bound to N, where N is the number of coins from among which a single counterfeit may be isolated on an equal-arm balance in a given number of weighings n, and determine whether the counterfeit is heavier or lighter than a good coin [20, 23].

It may be assumed equally probable that any of the coins is the counterfeit and hence is heavier or lighter than any other coin. Since there are N coins, this gives, in all, $2N$ possibilities. Since any of these possibilities is equally probable, one has

$$p_i = \frac{1}{2N} \quad \text{for } i = 1, \ldots, 2N$$

The total entropy is then easily calculated as $\log 2N$. Note that the entropy would be less if it were not true that all coins are equally probable. In order to isolate the counterfeit coin in n weighings, it is necessary that the weighings have at least total entropy equal to $\log 2N$. If the entropy of the weighings is less than $\log 2N$, generally the counterfeit coin cannot be identified.

Now from a single weighing one has three equally probable results: the left arm of the balance is lighter, it is heavier, or it is equal to the right arm. Each possibility is equally probable and since there are, in all, three possibilities, the entropy of a weighing is given by

$$-\sum_{i=1}^{3} w_i \log w_i = \log 3 \quad \text{where } w_i = 1/3 \quad (4\text{-}127)$$

Note that i ranges over the number of possible outcomes. The entropy in n weighings is given as the sum of the entropies of each weighing and is

$$\log 3 + \cdots + \log 3 = \log 3^n \quad (4\text{-}128)$$

Thus to identify the counterfeit coin the entropy from the weighings cannot be less than $\log 2N$ or, putting it differently, $\log 2N$ must be less than $\log 3^n$. Suppose that the two are equal. Then

$$\log 3^n = \log 2N$$

or

$$N = \frac{3^n}{2} \quad (4\text{-}129)$$

Since this is not an integer, one takes the greatest integer smaller than this number, which gives

$$N = \frac{3^n - 1}{2} \quad (4\text{-}130)$$

It will now be shown that the entropy of the weighings is insufficient to identify a counterfeit coin from among $(3^n - 1)/2$ coins. Note first that N is of the form $3M + 1$, where M is an integer. It is not difficult to see that the maximum gain of entropy is obtained if the first move is to divide the coins into three equal collections. A collection of M coins is placed in each pan, leaving aside $M + 1$ coins. If the pans balance, the entropy to be obtained in the remaining $n - 1$ weighings must at least equal

$$\log 2(M + 1) = \log (3^{n-1} + 1) \quad (4\text{-}131)$$

having used the fact that $M = (N - 1)/3$ and that $N = (3^n - 1)/2$.

Since $\log(3^{n-1} + 1)$ is greater than the maximum entropy obtained in $n - 1$ weighings, that is, $\log 3^{n-1}$, it is clear that the problem is not solvable for

$$N = \frac{3^n - 1}{2} \quad (4\text{-}132)$$

Thus the number of coins cannot exceed

$$N = \frac{3^n - 3}{2} \quad (4\text{-}133)$$

It can be shown that the problem is actually solvable for this number of coins. This simple example effectively demonstrates how, by proceeding from abstract considerations, one obtains quantitative formulas as the solution of a problem.

REFERENCES

1. *American Mathematical Monthly*, vol. 28, p. 144, 1921.
2. Belgodère, P.: "Aide-mémoire de mathématiques générales" and "Aide-mémoire de calcul différentiel et intégral," Centre de Documentation Universitaire, Tournier et Constans, Paris.
3. Bridgeman, P.: "Dimensional Analysis," rev. ed., Yale University Press, New Haven, Conn., 1931.
4. Box, G. E. P.: The Exploration and Exploitation of Response Surfaces, *Biometrics*, vol. 10, 1954.
5. Box, G. E. P., and J. S. Hunter: A Confidence Region for the Solution of Simultaneous Equations with an Application to Experimental Design, *Biometrika*, vol. 41, pp. 190–200, 1954.
6. Box, G. E. P., and J. S. Hunter: "Multifactor Designs," Institute of Statistics Mimeograph Series, Raleigh, N.C., 1954.
7. Box, G. E. P., and K. B. Wilson: On the Experimental Attainment of Optimum Conditions, *Journal of the Royal Statistical Society*, Series B, p. 13, 1951.
8. Brown, R. H.: A Stochastic Analysis of Lanchester's Theory of Combat, *Operations Research Office Technical Memorandum* ORO T-323, 1955.
9. Davies, O. L.: "Design and Analysis of Industrial Experiments," chap. 11, Oliver & Boyd, Ltd., Edinburgh and London, 1954.
10. DeBaun, R. M.: Block Effects in the Determination of Optimum Conditions, unpublished manuscript.
11. Frazer, R. A., W. J. Duncan, and A. R. Collar: "Elementary Matrices," Cambridge University Press, London, 1946.
12. Hildebrand, F. B.: "Advanced Calculus for Engineers," Prentice-Hall, Inc., Englewood Cliffs, N.J., 1949.
13. Hildebrand, F. B.: "Methods of Applied Mathematics," Prentice-Hall, Inc., Englewood Cliffs, N.J., 1952.
14. Hildebrand, F. B.: "Introduction to Numerical Analysis," McGraw-Hill Book Company, Inc., New York, 1956.
15. Hotelling, H.: Experimental Determination of the Maximum of a Function, *Annals of Mathematical Statistics*, vol. 12, pp. 20–45, 1941.
16. Hunter, J. S.: Searching for Optimum Conditions, *Transactions of the New York Academy of Science*, series 11, 17, 1954.

METHODS USED IN MODEL FORMULATION

17. Isbell, J., and W. Marlow: Attrition Games, *Naval Research Logistics Quarterly,* vol. 3, pp. 71–94, 1956.
18. Jackson, R. R. P., and D. G. Nickols: Some Equilibrium Results for the Queuing Process $E_k/M/1$, *Journal of the Royal Statistical Society, Series B*, vol. 18, pp. 275–279, 1956.
19. Kamke, E.: "Differentialgleichungen Reeller Funktionen," Chelsea Publishing Company, New York, 1947.
20. Kellogg, P. J., and D. J. Kellogg: Entropy of Information and the Odd Ball Problem, *Journal of Applied Physics*, vol. 45, p. 1438, 1953.
21. Lanchester, F.: "Aircraft in Warfare, the Dawn of the Fourth Arm," Constable & Co., Ltd., London, 1916.
22. Lotka, A.: "Elements of Mathematical Biology," Dover Publications, New York, 1956.
23. Mitchell, F. H., and R. N. Whitehurst: Further Remarks on the Odd Ball Problem as an Example in Information Theory, *Journal of Applied Physics*, vol. 27, pp. 778–779, 1955.
24. Quastler, H. (ed.): "Information Theory in Biology," University of Illinois Press, Urbana, Ill., 1953.
25. Reddick, H. W., and F. H. Miller: "Advanced Mathematics for Engineers," 2d ed., John Wiley & Sons, Inc., New York, 1947.
26. Shannon, C. E.: A Mathematical Theory of Communication, *Bell System Technical Journal*, vol. 27, pp. 379–423, 623–656, 1948.
27. Shannon, C. E., and W. Weaver: "Mathematical Theory of Communication," University of Illinois Press, Urbana, Ill., 1949.
28. Titchmarsh, E. C.: "The Theory of Functions," Oxford University Press, London, 1939.
29. Uspensky, J. V.: "Introduction to Mathematical Probability," McGraw-Hill Book Company, Inc., New York, 1937.
30. Wiener, N.: "Cybernetics," John Wiley & Sons, Inc., New York, 1948.
31. Woodward, P. M.: "Probability and Information Theory with Applications to Radar," McGraw-Hill Book Company, Inc., New York, 1953.
32. Volterra, V.: "Leçon sur la théorie mathématique de la lutte pour la vie," Gauthier-Villars et Cie, Editeurs, Paris, 1931.
33. Whittaker, E. T., and G. N. Watson: "A Course of Modern Analysis," 4th ed., Cambridge University Press, London, 1952.

PART 2

OPTIMIZATION, PROGRAMMING, AND GAME THEORY

CHAPTER 5

OPTIMIZATION

5-1. Introduction

It was previously remarked that a type of problem occurring frequently in operations research is that of maximizing the use of resources or minimizing the effort exerted or the cost in attaining these objectives.

The simple assumption, often made, that most people dealing with operational problems have a desire to obtain the maximum utility from the efforts exerted proves to be both correct and useful. Many operational problems can be mathematically formulated with the objective of obtaining an accurate description and subsequently manipulating the resulting model to obtain an optimum (a maximum or a minimum). The advantage here is twofold. (1) The essential variables of the problem are brought together in one model accounting for the constraints and the utility function to be optimized. (2) The problem is now given a familiar structure which can be analyzed for solutions, their existence, uniqueness, and construction. Comparisons in the solution, based on varying the given conditions, are also possible.

A large number of important problems arising in operations receive this type of formulation. To solve such problems requires a good deal of understanding of the underlying concepts, many of which we give in this chapter. We shall not justify each section from the standpoint of utility.

The reader will find several applications of these ideas in illustrative examples both in this and in the next part of the book.

It is important to point out that in this chapter we first study the constraint sets in order to lay the groundwork for optimization subject to constraints. The latter constitutes a type of optimization of greater significance and frequency of occurrence than optimization without constraints. The constraints, when given as inequalities, offer a large number of solution points. This makes it possible to introduce a function which assumes a maximum or a minimum value at one or more of these points.

Optimization is discussed first not because it has application precedence over probabilistic and statistical methods, but rather to keep the reader aware of the historical transition from deterministic to nondeterministic models. An attempt is made to provide an interesting structure for the subject of optimization which should furnish ample motivation for attempting to discover why certain parts of the subject are more difficult to deal with.

Because of space limitations, very few proofs will be given; the aim is essentially to state the general problems and indicate methods of solution. Wherever possible, useful applications are presented.

NO OPTIMIZATION: THE CONSTRAINT SETS

In this portion of the chapter solutions to the constraint sets are studied to derive properties required later in optimization of functions subject to such constraints.

5-2. Solutions of Single Equations

In this section the basic problem is one of finding the roots, or acceptable approximations thereto, of a given equation. The two basic theorems in this area will first be given.

A polynomial of degree n in x is defined by

$$f(x) = a_0 x^n + a_1 x^{n-1} + \cdots + a_n \tag{5-1}$$

$a_0 \neq 0$, and the coefficients a_i ($i = 1, \ldots, n$) are real numbers. An algebraic equation is given by $f(x) = 0$, where $f(x)$ is a polynomial.

If the equation is algebraic and of degree n, then the following theorem, due to Gauss, ensures the existence of at least one root.

Theorem 5-1: Every equation of the form

$$f(x) = a_0 x^n + a_1 x^{n-1} + \cdots + a_n = 0 \qquad a_0 \neq 0 \tag{5-2}$$

has at least one complex root.

This theorem is assumed implicitly in most college algebra courses and apparently cannot be proved rigorously without rather advanced mathematical methods. It is often called "The fundamental theorem of algebra."

The other basic theorem is much less deep and depends on the observation that the nth-degree polynomial $f(x)$ can be divided by $x - r$, where r is any constant, yielding a quotient $Q(x)$ and a remainder R:

$$f(x) = (x - r)Q(x) + R \tag{5-3}$$

If r_1 is a root of the equation $f(x) = 0$ (such a root exists by Theorem 5-1), then setting $x = r_1$ yields

$$0 = 0 + R$$

and so (since $R = 0$) one has simply

$$f(x) = (x - r_1)Q(x) \tag{5-4}$$

In other words, if r_1 is a root of $f(x) = 0$, then $x - r_1$ is a factor of $f(x)$. The quotient $Q(x)$ is a polynomial of degree $n - 1$ to which the same argument can be applied. One can then find successive factors $x - r_1$, $x - r_2$, ..., of $f(x)$ corresponding to the roots $r_1, r_2, \ldots$, of the equation $f(x) = 0$. An nth-degree polynomial cannot have more than n first-degree factors, and so the process terminates after n steps. Thus one has the following:

Theorem 5-2: If $f(x)$ is a polynomial of degree n,

$$f(x) = a_0 x^n + a_1 x^{n-1} + \cdots + a_n \qquad a_0 \neq 0$$

then there exists one and only one set of constants $r_1, \ldots, r_n$ (not necessarily distinct) such that

$$f(x) = a_0(x - r_1)(x - r_2) \cdots (x - r_n) \tag{5-5}$$

and $r_1, \ldots, r_n$ are precisely the roots of the equation $f(x) = 0$.

The reader will recall the general formula

$$x = \frac{-b \pm \sqrt{b^2 - 4ac}}{2a} \tag{5-6}$$

for the solutions of the quadratic equation

$$f(x) = ax^2 + bx + c = 0 \qquad a \neq 0 \tag{5-7}$$

Similar (but more complicated) formulas exist for the solution of third-degree and fourth-degree equations and can be found in standard algebra texts. The straightforward procedure, employing radicals, breaks down for equations of degree greater than four. (The impossibility of general solutions of equations of higher degree by radicals was one of the important first results of group theory.)

Several numerical techniques are available for solving fifth- and higher-degree algebraic equations. Among these may be mentioned the methods of Horner and Graeffe. Graphical methods can also be useful on occasion. In addition, there are methods of approximation, such as that discovered by Newton, which can handle equations involving transcendental (i.e., nonalgebraic) functions. An example of equations of this type is

$$\sin^2 x + e^x = 1$$

For an arbitrary function $f(x)$ there may be no real value x for which $f(x) = 0$. A simple example is $f(x) = e^x$. It is for this reason that theorems such as those given above must specify certain properties of the equation (e.g., belonging to the class of algebraic equations).

Newton's Approximation Method

If either from a graph or by trial calculations it is found that $f(x) = 0$ has a real root r in the neighborhood of $x = a_1$ such that $f(a_1)$ and $f''(a_1)$

(the number of primes indicates the order of the derivative of f) have the same sign and if $f''(a_1)$ does not change sign in the interval $a_1 \leq x \leq r$, then a_1 may be used to calculate a point $x = a_2$ which is a closer approximation to the root r under study. In general, approximations to any desired accuracy may be obtained from the formula

$$a_n = a_{n-1} - \frac{f(a_{n-1})}{f'(a_{n-1})} \qquad n = 2, 3, 4, \ldots \qquad (5\text{-}8)$$

where $f'(a_{n-1})$ is the derivative of $f(x)$ with respect to x evaluated at a_{n-1}. Geometrically, a_n is the x coordinate of the point in which the tangent to the curve $y = f(x)$ at $[a_{n-1}, f(a_{n-1})]$ meets the x axis. See Fig. 5-1.

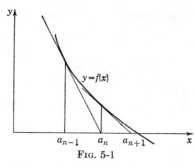

Fig. 5-1

An iterative method is applied to equations which can be put in the form

$$x = f(x) \qquad (5\text{-}9)$$
$$\text{with} \quad |f'(x)| < 1$$

Once a_1 has been chosen, say by graphical means, the method is applied successively as follows:

$$a_2 = f(a_1)$$
$$a_3 = f(a_2)$$
$$\cdots \cdots$$
$$a_n = f(a_{n-1})$$
$$(5\text{-}10)$$

If $f'(x) > 0$, then $a_1, \ldots, a_n$ approach the solution in the same direction (through a ladder), as in Fig. 5-2. If $f'(x) < 0$, $a_1, a_3, \ldots, a_{2n-1}$

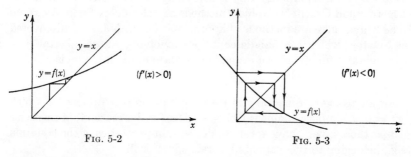

Fig. 5-2   Fig. 5-3

approach the solution in one direction, and $a_2, \ldots, a_{2n}$ approach it in the opposite direction, so that a spiral results, as in Fig. 5-3. If $|f'(x)| > 1$, the inverse function can be used instead.

The process of iteration can be extended to several equations with

OPTIMIZATION

several unknowns of the form

$$x_i = f_i(x_1, \ldots, x_n) \quad i = 1, \ldots, n \tag{5-11}$$

If a first approximation $(x_1^0, \ldots, x_n^0)$ has been found, then a better approximation is

$$\begin{aligned} x_1^1 &= f_1(x_1^0, \ldots, x_n^0) \\ &\cdots\cdots\cdots\cdots\cdots \\ x_n^1 &= f_n(x_1^0, \ldots, x_n^0) \end{aligned} \tag{5-12}$$

Successively better approximations can be obtained by repeating this procedure. The method can be shown to converge if the partial derivatives of the functions f_i are all less than $1/n$, where n is the number of unknowns appearing in the equations. As an illustration, consider $x + e^{-x} = 0$. Write $a_n = -e^{-a_{n-1}}$, $n = 2, 3, \ldots$, and, starting with $a_1 = 0$, obtain $a_2 = -1$, etc.

5-3. A Set of Simultaneous Linear Equations: Some Methods of Solution [23]

The general problem can be stated in the following manner: Find the set of values of x_j, $j = 1, 2, \ldots, n$, which satisfy the system of linear equalities

$$\sum_{j=1}^{n} a_{ij}x_j = c_i \quad i = 1, \ldots, m \tag{5-13}$$

Two matrices may be considered:

1. The matrix of coefficients $[a_{ij}]$ which is an m by n matrix called the *coefficient matrix*.
2. The m by $n + 1$ matrix formed by taking the matrix $[a_{ij}]$ together with the column vector of constants $[c_i]$. This is called the *augmented matrix*.

The rank of a matrix is the order of the largest square array in the matrix whose determinant does not vanish.

Existence

The following theorem provides information as to the existence of a solution of the above system of equations.

Theorem 5-3: A set of linear equations possesses a solution if and only if the rank r of the augmented matrix is equal to the rank of the coefficient matrix.

If $r = n$, the only solution to the homogeneous set of equations (obtained by setting $c_i = 0$) is the trivial solution $x_j = 0, j = 1, \ldots, n$. If $r < n$ (thus also when $m < n$), an infinite set of solutions exists, depending on a number of arbitrary variables. The number of arbitrary variables is called the *defect* (or *degree of freedom*) $d = n - r$, that is, the order minus the rank.

In the case $m < n$, c_i not all zero, the set of equations can be reduced (for example, by the Gauss-Jordan elimination method), to obtain solutions in terms of $n - r$ variables which are assigned arbitrary values. The remaining variables are determined from these. This method proceeds as follows: Select an equation in which the coefficient of x_1 is not zero. The equation, having been divided by this coefficient, is multiplied by the coefficients of x_1 appearing in another equation and then is subtracted from that equation. This is done for all the remaining equations. As before, the variable x_2 is eliminated from all but one of the equations in which the coefficient of x_1 is zero by using one of these equations. In this manner the process is continued r times, where either $r = m$ or $r < m$ and the left side of the remaining $m - r$ equations is zero. If the right side of any of these equations is not zero then the equations are inconsistent and have no solution. If this is not the case one now has r instead of m equations, in which, therefore, r of the variables may be expressed in terms of $n - r$ arbitrary variables.

Method of Solution by Cramer's Rule when $m = n$

The Gauss-Jordan method also takes care of this case. When the determinant of the coefficient matrix is not zero, i.e., when $r = n$, the set of equations has a unique solution which may be obtained directly, using Cramer's rule: The expression for any x_j is the ratio of two determinants, the denominator being the determinant of the matrix of coefficients and the numerator being the determinant of the matrix obtained by replacing the column of the coefficients of x_j in the coefficient matrix by the column of the right-hand members.

Among others, a well-known method due to Crout [23], which is essentially the same as the Gauss-Jordan method, permits the calculation of approximate solutions to a set of simultaneous linear equations, minimizing the number of operations required on a desk calculator. There are several iterative procedures for obtaining solutions to a set of simultaneous equations. To list a few in passing [14], there are (1) the method of Withmeyer-Hertwig-Cesari, (2) the least-squares method of Rosser-Hestenes (viewed by experienced computors as one of the best), and (3) gradient-type methods, illustrated below.

When the determinant of the coefficient matrix does not vanish in the nonhomogeneous case, the following iterative methods can be used for computing A^{-1}, the inverse of the coefficient matrix $A \equiv [a_{ij}]$:[1] Hertwig's

---

[1] Recall that $A^{-1} \equiv \left[\dfrac{(-1)^{i+j} A_{ji}}{\text{determinant } A} \right]$, where A_{ji} is the determinant of the submatrix obtained after eliminating the ith row and jth column. It occupies the jth row and the ith column.

OPTIMIZATION

linear convergence method, Schulz's quadratic convergence method, Bodewig's [64] nth-degree convergence method, or the Monte Carlo method (but the last is not recommended). Details of these methods may be obtained from the references.

5-4. A Set of Simultaneous General Equations $f_i(x_1, \ldots, x_n) = 0$, $i = 1, \ldots, m$

The following is an elementary illustration of the simultaneous solution of equations and of root extraction arising in a physical problem.

Example: In the movement of projectiles, there are an infinity of parabolas (see Fig. 5-4) tangent to the line AB whose equation is given by $x + y = h$. Let T be the point of contact of one of these trajectories with AB. Show that the time t required to traverse the distance OT is an absolute constant; i.e., depends only on g, the acceleration due to gravity, and h, and not on the initial velocity v_0 and the angle α.

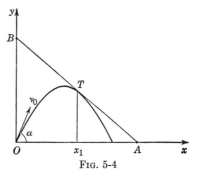

Fig. 5-4

Proof: The parabola has the equation

$$y = x \tan \alpha - \frac{gx^2}{2v_0^2 \cos^2 \alpha} \quad (5\text{-}14)$$

obtained by eliminating t between the following equations of motion of a projectile,

$$x = v_0 t \cos \alpha \qquad y = v_0 t \sin \alpha - \tfrac{1}{2}gt^2 \quad (5\text{-}15)$$

From the equation of AB one has

$$y = x - h$$

which is then substituted in the first equation, and the resulting quadratic is solved for the abscissa x_1 of the point of tangency T. This yields a double root for which the discriminant must vanish:

$$v_0 \cos \alpha = \frac{2gh}{1 + \tan \alpha}$$

Using this, one has

$$x_1 = \frac{1}{g}(1 + \tan \alpha)v_0^2 \cos^2 \alpha = \frac{2h}{1 + \tan \alpha} \quad (5\text{-}16)$$

The time required to go from 0 to T may be obtained by dividing the horizontal distance x_1 by the velocity component in that direction:

$$t = \frac{x_1}{v_0 \cos \alpha} = \sqrt{\frac{2h}{g}} \tag{5-17}$$

Newton's Method

The perturbations of Newton's method used to solve a single equation may be extended to the solution of a finite number of equations:

$$f_i(x_1, \ldots, x_n) = 0 \qquad i = 1, \ldots, m$$

The method requires that one start with some initial approximation $(x_1^0, \ldots, x_n^0)$ and then solve the system of linear equations

$$f_i(\mathbf{x}^k) + \sum_{j=1}^{n} \frac{\partial f_i(\mathbf{x}^k)}{\partial x_j}(x_j^{k+1} - x_j^k) = 0 \qquad i = 1, \ldots, m \tag{5-18}$$

to obtain the $(k+1)$st approximation defined by

$$\mathbf{x}^{k+1} = (x_1^{k+1}, \ldots, x_n^{k+1}) \tag{5-19}$$

Note that this procedure may be needed in order to solve sets of equations resulting from the use of Lagrange's multiplier method, discussed later.

Cauchy's Gradient (Steepest-descent or Saddle-point) Method

Given a system of simultaneous equations to be solved, one ordinarily starts by reducing them to a single equation by means of elimination and then solving the latter whenever possible. It is important to note the following:

1. In general, the elimination may not be possible.
2. Even if elimination is possible, the resulting equation is generally very complicated, although the original equations may be simple.

For these two reasons, it is clearly very desirable to have a general method which can produce a direct solution for a system of simultaneous equations.

The Gradient of a Function. Consider the surface defined by $z = f(x,y)$. By assigning a value c to z and allowing x and y to vary, a contour of the surface in the xy plane is obtained. The vector defined by

$$\nabla f \equiv \left(\frac{\partial f}{\partial x}, \frac{\partial f}{\partial y}\right)$$

whose components (directional numbers) are the first partial derivatives of f with respect to x and y at any point, respectively, points in the direction of the normal to this contour and is called the *gradient of f*. Note that the gradient points in the direction of maximum increase in f. This may be shown by using the distance s along the contour as a parameter, where $ds^2 = dx^2 + dy^2$, and obtaining from the equation of the contour x and y as functions of s. Thus

$$\frac{df}{ds} = \frac{\partial f}{\partial x}\frac{dx}{ds} + \frac{\partial f}{\partial y}\frac{dy}{ds} \tag{5-20}$$

But this is the scalar product of ∇f and the tangent vector to the contour at any point. (The scalar or dot product of two vectors is the sum of the products of corresponding components of the two vectors. It is also equal to the product of the magnitudes of the two vectors multiplied by the cosine of the angle between them.) Here $dx/ds = \cos\alpha$ and $dy/ds = \sin\alpha$, where α is the angle which the tangent line makes with the x axis.

Now the *directional derivative* $\frac{df}{ds}$, when evaluated at a given point (x_0, y_0) on a contour, gives the rate of change of $f(x,y)$ with respect to the distance s traversed by a point moving on that contour. It has a value for each of the contours passing through the given point. At this given point, the directional derivative is a function of the angle α. That α which yields the maximum value of the directional derivative may be determined by equating to zero the derivative of (5-20) with respect to α, solving the resulting equation in α, and using that value of α which satisfies the second derivative test for a maximum. There is a unique value α_0 between 0 and 2π which yields the maximum. This gives

$$\tan\alpha_0 = \frac{f_y(x_0,y_0)}{f_x(x_0,y_0)}$$

Actually, the point (x_0, y_0) may be replaced by an arbitrary point (x,y), then, by assigning different values to α_0, one obtains a set of curves. At every point of each of these curves the directional derivative is maximum in a direction perpendicular to the derivative to the curve. This may be verified by comparing their slopes. By considering a right-angle triangle with $\tan\alpha_0$ as given above, one may determine $\cos\alpha_0$ and $\sin\alpha_0$ and substitute these values in the expression for $\frac{df}{ds}$. The value of the latter in the direction α_0 is $\sqrt{f_x^2 + f_y^2}$, which is the magnitude of the gradient vector. Thus the directional derivative has the largest value when it lies along the gradient vector.

The directional derivative in the direction α_0 is normal to the contour. This follows from the fact that when the directional derivative at (x_0,y_0) is equated to zero, a value of α, say α_1, is obtained, which gives the direction of the tangent to the contour, hence its slope. This leads to

$$-\tan\alpha_1 = \frac{f_x}{f_y} = \frac{1}{\tan\alpha_0}$$

Therefore, on combining the extremes of this relation, one has

$$\alpha_0 = \frac{\pi}{2} + \alpha_1$$

Hence, the gradient is normal to the contour. The method presented below utilizes the foregoing results by moving along the gradient toward a solution.

The Gradient Method. The following method, due to Cauchy, is taken directly from one of his papers [6]. To simplify the notation, it will be assumed that one is dealing with a function

$$z = f(x,y) \tag{5-21}$$

of only two variables which is nonnegative and well behaved in the region of the (x,y) values under discussion. Because the function is nonnegative, a reasonable way to find a solution of the equation

$$z = 0 \tag{5-22}$$

is to provide a procedure which, beginning with any particular (x_0,y_0), yields successive points (x_1,y_1), (x_2,y_2), ... , at which z takes on steadily decreasing values. To this end, note that, for a point $(x_0 + \Delta x, y_0 + \Delta y)$ near the initial point (x_0,y_0), the "first-order approximation" to $f(x_0 + \Delta x, y_0 + \Delta y)$ is

$$z = f(x_0 + \Delta x, y_0 + \Delta y) \approx z_0 + X_0 \Delta x + Y_0 \Delta y \tag{5-23}$$

where $z_0 = f(x_0,y_0)$, $X_0 = \partial z/\partial x$ evaluated at (x_0,y_0), and $Y_0 = \partial z/\partial y$ evaluated at (x_0,y_0). Geometrically, the approximation involves replacing the part of the surface $z = f(x,y)$ near (x_0,y_0,z_0) by the tangent plane through (x_0,y_0,z_0).

Suppose that, in particular, one chooses a positive quantity θ and selects $\Delta x = -\theta X_0$, $\Delta y = -\theta Y_0$. Then the approximation for z becomes

$$z = f(x_0 - \theta X_0, y_0 - \theta Y_0) \approx z_0 - \theta(X_0{}^2 + y_0{}^2) \tag{5-24}$$

which strongly suggests that, if θ is not too large (so that the first-order approximation remains a good one), then z is less than z_0. This is indeed true although no rigorous proof will be given here. An even stronger statement is, in fact, true: As θ increases from 0 through positive values, the point $(x_0 - \theta X_0, y_0 - \theta Y_0)$ moves away from (x_0,y_0) along a certain ray [the value of θ is proportional to the distance of the moving point from (x_0,y_0)], and it can be shown that this is precisely that ray emanating from (x_0,y_0) along which z decreases most rapidly [near (x_0,y_0)].

The last approximation suggests that in order to decrease z rapidly one should choose θ large (i.e., go far out along the ray), but if one does so, then the approximation itself is no longer valid. What one can do, however, is to choose a θ which corresponds to a (relative) minimum value of z along the ray; such a θ can be found by solving the equation $dz/d\theta = 0$

(which is believed to be simpler than the original equation $z = 0$). Take $(x_0 + \theta X_0, y_0 + \theta Y_0)$ as (x_1, y_1) and repeat the process, etc.

For a trivial illustration (in which only one application of the procedure is necessary), suppose that

$$z = x^2 + y^2 - 2xy$$

and that one begins with $(x_0, y_0) = (1,0)$. Then $X = 2$ and $Y = -2$, so that the "moving point" is $(1 - 2\theta, 2\theta)$ and the value of z at this point is

$$z = (1 - 2\theta)^2 + (2\theta)^2 - 2(1 - 2\theta)2\theta = 16\theta^2 - 8\theta + 1$$

and

$$\frac{dz}{d\theta} = 32\theta - 8$$

and so the equation $dz/d\theta = 0$ yields $\theta = \frac{1}{4}$ and $(x_1, y_1) = (\frac{1}{2}, \frac{1}{2})$. But $z = 0$ at $(\frac{1}{2}, \frac{1}{2})$, and so the equation $z = 0$ has already been solved and one need not repeat the process.

Geometrically, the process led from $(x_0, y_0) = (1,0)$ along the ray perpendicular to the line $x = y$ which gives all solutions of $z = 0$ [since actually $z = (x - y)^2$] and stopped "motion" along the ray when the solution line was reached. For small values of z_0 the right side of the approximation involving θ may be set equal to zero and this choice used.

The restriction to functions of two variables was completely unnecessary; if

$$z = f(x_1, \ldots, x_n)$$

is nonnegative and well behaved, then the same method (and essentially the same geometric interpretation) applies to the equation $z = 0$. Furthermore, if the functions $z_1, \ldots, z_r$ are well behaved (but not necessarily nonnegative), then the method applies to the equation

$$(z_1)^2 + \cdots + (z_r)^2 = 0$$

and thus to the (equivalent) system of *simultaneous* equations

$$z_1 = 0, \ldots, z_r = 0$$

Finally, even if the single well-behaved function z also takes on negative values, the method aids in searching for the minimum of z (though not in finding a solution of $z = 0$).

5-5. The Use of Inequalities

As will be seen later, it may be desired to study the optimum of a function subject to constraints. In that case its domain is usually defined by one or several inequalities which may or may not be strict. It is even possible that some of the constraints are actually equations. For example, the domain may be composed of a portion of the xy plane interior to a closed curve in the plane. If this curve is represented by a unique equation $g(x,y) = 0$, the interior area and the boundary are represented by $g \leq 0$. This same idea applies to portions of ordinary space of n dimensions limited by a surface [22].

Again, for example, if the problem is concerned with the interior of a square with the center at the origin and with sides parallel to the axes, such a region is defined by the inequalities

$$-a \leq x \leq a \qquad -a \leq y \leq a$$

A spherical cap is defined on the surface of the sphere $x^2 + y^2 + z^2 = 1$ by the equation of the sphere together with the inequality

$$z > a \qquad -1 < a < 1$$

If all the inequalities defining the domain to be studied are strict, the corresponding points are strictly interior to the domain.

Elementary Inequality Relations

The following properties of simple inequalities using real numbers may be used to derive the remaining ones. Note that $a > b$ means that $a - b > 0$.

1. If a is a real number, then either (a) $a > 0$ or (b) $-a > 0$ or (c) $a = 0$, and these are mutually exclusive.
2. If $a > 0$ and $b > 0$, then $a + b > 0$.
3. If $a > 0$ and $b > 0$, then $ab > 0$.

These yield the following relations: If $a < b$, $c < d$, then $a + c < b + d$. If, in addition, a, $c > 0$, then $ac < bd$. An inequality preserves its sense if a constant is added to or subtracted from both sides. Multiplication by a negative quantity reverses the sense of the inequality. If $ab > 0$, then either $a > 0$ and $b > 0$ or $a < 0$ and $b < 0$. On the other hand, $ab < 0$ implies that either $a < 0$ and $b > 0$ or $a > 0$ and $b < 0$. If $ab = 0$, then one or the other or both of a and b are zero.

Examples: Find all real x which satisfy the following:

1. $2x + 3 > 7$. Subtract 3 from both sides and multiply by $\frac{1}{2}$ to obtain $x > 2$.

2. $6 > -2x + 4 > -10$. Subtract 4 in the double inequality to obtain $2 > -2x > -14$; then multiply by $-\frac{1}{2}$. This gives $-1 < x < 7$ as the final answer.

3. $x^2 + 7 < 3x + 5$. Subtract $3x + 5$ from both sides and factor to obtain $(x - 2)(x - 1) < 0$. Hence, either $(x - 2) < 0$ and $(x - 1) > 0$, that is, $x < 2$ and $x > 1$ (or simply $1 < x < 2$), or $x - 2 > 0$ and $x - 1 < 0$, that is, $x > 2$ and $x < 1$, which is impossible. Thus, $1 < x < 2$ is the set of values of x which satisfies the given problem.

Inequalities and Absolute Values

Inequalities and absolute values frequently occur together in analysis, particularly in the study of limits. Here the definition of "absolute value" and some of its properties which involve inequalities are given, together with illustrations. The definition of the absolute value of a is

$$|a| = a \quad \text{if } a \geq 0 \quad \text{and} \quad |a| = -a \quad \text{if } a < 0$$

One immediately has the fact that $|a| \geq 0$ and $|a| = 0$ if and only if $a = 0$. The absolute value of a product is the product of the absolute values, that of a sum is less than or equal to the sum of the absolute values, and the absolute value of a difference is greater than or equal to the absolute value of the difference of the absolute values. Two important properties are as follows: If $\epsilon > 0$ and $|a| < \epsilon$, then $-\epsilon < a < \epsilon$, and conversely. Also $|a| > \epsilon$ is true if and only if $a < -\epsilon$ or $a > \epsilon$.

Examples: Find all real x which satisfy the following:

1. $|12 - 7x| > 2$. One has either $12 - 7x < -2$, that is, $x > 2$, or $12 - 7x > 2$, that is, $x < 10/7$. Thus all $x < 10/7$ and all $x > 2$ satisfy the problem.

2. $|x - 2| < \epsilon$. This gives $-\epsilon + 2 < x < \epsilon + 2$.

3. $|x^2 - |x| + 2| = 4$. This gives either (a) $x^2 - |x| + 2 = 4$, which yields $|x| = x^2 - 2$ which in turn gives either (i) $x = x^2 - 2$, that is, $x^2 - x - 2 = 0$, $(x - 2)(x + 1) = 0$ which is satisfied if $x = 2$ or $x = -1$ or both, or (ii) $x = -x^2 + 2$, that is, $x^2 + x - 2 = 0$, $(x + 2)(x - 1) = 0$ which is satisfied if $x = -2$ or $x = 1$ or both; or (b) $x^2 - |x| + 2 = -4$ which yields $|x| = x^2 + 6$ which in turn gives either (i) $x = x^2 + 6$, that is, $x^2 - x + 6 = 0$ with no real roots, or (ii) $x = -x^2 - 6$, that is, $x^2 + x + 6 = 0$ with no real roots. Of the values $x = \pm 1, x = \pm 2$, the latter pair satisfies the problem.

4. $|x^2 - 2| = 7$ and $|x - 1| = 2$. The first relation gives either (a) $x^2 - 2 = 7$, that is, $x = \pm 3$, or (b) $x^2 - 2 = -7$, that is, $x^2 = -5$ with no real roots. The second relation gives either (a) $x = 3$ or (b) $x = -1$. Thus $x = 3$ satisfies both relations.

Some Useful Inequality Relations

The occurrence of inequalities in calculations is an essential part of solving problems, particularly when it is desired to prove a property such as the convergence of a series majorized term by term, by another series, or to provide upper and lower estimates.

There are a large number of well-known useful inequalities, a few of which will be given here. The derivation of such relations is usually either entirely algebraic or it uses properties of convex functions and properties of maxima and minima of a function often subject to constraints. To illustrate an algebraic proof, suppose that it is desired to prove that if a_i, $i = 1, \ldots, n$ are real numbers then

$$\frac{\left(\sum_{i=1}^{n} a_i\right)^2}{\sum_{i=1}^{n} a_i^2} \leq n \tag{5-25}$$

Consider the expression

$$\sum_{i=1}^{n} (a_i - \lambda b_i)^2 \tag{5-26}$$

which is a quadratic nonnegative expression in λ and where b_i ($i = 1, \ldots, n$) are arbitrary real numbers. On expanding, this expression becomes

$$\sum_{i=1}^{n} a_i^2 - 2\lambda \sum_{i=1}^{n} a_i b_i + \lambda^2 \sum_{i=1}^{n} b_i^2 \tag{5-27}$$

Since the expression is nonnegative for all λ, the discriminant of the quadratic must be nonpositive, i.e., must satisfy the relation (called the Cauchy-Schwarz inequality)

$$\left(\sum_{i=1}^{n} a_i b_i\right)^2 - \sum_{i=1}^{n} a_i^2 \sum_{i=1}^{n} b_i^2 \leq 0 \tag{5-28}$$

On putting $b_i = 1$ ($i = 1, \ldots, n$) one has the desired proof.

A generalization of the Cauchy-Schwarz inequality is given by Hölder's inequality

$$\left|\sum_{i=1}^{n} a_i b_i\right| \leq \left(\sum_{i=1}^{n} |a_i|^p\right)^{1/p} \left(\sum_{i=1}^{n} |b_i|^q\right)^{1/q} \tag{5-29}$$

where $1/p + 1/q = 1$ and $p > 0$, $q > 0$.

From this one also has

$$\frac{1}{n}\left|\sum_{i=1}^{n} a_i b_i\right| \leq \left(\frac{1}{n}\sum_{i=1}^{n} |a_i|^p\right)^{1/p} \left(\frac{1}{n}\sum_{i=1}^{n} |b_i|^q\right)^{1/q} \tag{5-30}$$

Minkowski's inequality gives

$$\left(\sum_{i=1}^{n} |a_i + b_i|^p\right)^{1/p} \leq \left(\sum_{i=1}^{n} |a_i|^p\right)^{1/p} + \left(\sum_{i=1}^{n} |b_i|^p\right)^{1/p} \tag{5-31}$$

where $p \geq 1$. The opposite inequality holds for $p \leq 1$. Note that equality holds if $p = 1$ or if $a_i = cb_i$ ($i = 1, \ldots, n$), where c is a constant. Both Hölder's (hence also Cauchy-Schwarz's) and Minkowski's inequalities hold if the summation is replaced by an integral, a_i by a function $f(x)$, and b_i by a function $g(x)$, where $|f(x)|^p$ and $|g(x)|^q$ are integrable in the case of Hölder's inequality and $|f(x)|^p$ and $|g(x)|^p$ are integrable in the case of Minkowski's inequality.

Jensen's inequality is given by

$$\left(\sum_{i=1}^{n} |a_i|^r\right)^{1/r} > \left(\sum_{i=1}^{n} |a_i|^s\right)^{1/s} \quad \text{if } 0 < r < s \quad (5\text{-}32)$$

Other useful inequalities are

$(1 + x)^p > 1 + px$ if $x > -1$, $x \neq 0$, and $p > 1$
$x^p - 1 > p(x - 1)$ if $x > 1$ and p is a real number > 1
$x^p - 1 < p(x - 1)$ if $x > 1$ and $0 < p < 1$
$x^a y^b < ax + by$ if $x \neq y$ and $a + b = 1$, $a > 0$, $b > 0$
$x \leq \dfrac{1 - p^x}{1 - p} \leq xp^{x-1}$ if $0 \leq x \leq 1$ and $0 < p < 1$

In the last inequality note that p may be an expression such as $1 - ae^{-by/x}$, where $a, b > 0$. Thus the inequality may be applied to show that

$$\lim_{x \to 0} \frac{1 - (1 - ae^{-by/x})^x}{x} = 0$$

which is the derivative of the numerator at the origin.

The following relation holds among the harmonic, geometric, and arithmetic means of the real numbers a_i ($i = 1, \ldots, n$) (from left to right):

$$\frac{n}{1/a_1 + \cdots + 1/a_n} < \sqrt[n]{a_1 \cdots a_n} < \frac{a_1 + \cdots + a_n}{n} \quad (5\text{-}33)$$

Each of these means is bounded above by the largest of the a_i and bounded below by the smallest of the a_i. When these three means are defined for an integrable function the above relation becomes

$$\frac{b - a}{\int_a^b dx/f(x)} \leq e^{1/(b-a) \int_a^b \log f(x)\, dx} \leq \frac{1}{b - a} \int_a^b f(x)\, dx \quad (5\text{-}34)$$

The geometric mean requires that $f(x)$ be positive.

Jensen's theorem asserts that for a convex function $f(x)$ and for positive a_i ($i = 1, \ldots, n$)

$$f\left(\frac{\sum_{i=1}^{n} a_i x_i}{\sum_{i=1}^{n} a_i}\right) \leq \frac{\sum_{i=1}^{n} a_i f(x_i)}{\sum_{i=1}^{n} a_i} \tag{5-35}$$

where x_i ($i = 1, \ldots, n$) are n arbitrary values of x. The definition of a convex function, also given in a later section, is obtained by letting $n = 2$, $a_1 = a_2 = 1$, and $x_1 < x_2$.

This theorem has many interesting consequences and applications, both in the case of sums and in the case of integrals. The theorem itself applies to integrals over an interval (a,b) if the sum is replaced by an integral, the x_i by a bounded function $g(x)$, and the a_i by a nonnegative function $a(x)$ with a positive integral over (a,b).

A general theorem for harmonic, geometric, and arithmetic means gives for positive a_i and positive b_i ($i = 1, \ldots, n$)

$$\frac{\sum_{i=1}^{n} b_i}{\sum_{i=1}^{n} b_i/a_i} \leq \exp\left(\frac{\sum_{i=1}^{n} b_i \log a_i}{\sum_{i=1}^{n} b_i}\right) \leq \frac{\sum_{i=1}^{n} a_i b_i}{\sum_{i=1}^{n} b_i} \tag{5-36}$$

One also has

$$\exp\left[\frac{\sum_{i=1}^{n} (b_i/a_i) \log a_i}{\sum_{i=1}^{n} b_i/a_i}\right] < \frac{\sum_{i=1}^{n} b_i}{\sum_{i=1}^{n} b_i/a_i}$$

and

$$\frac{\sum_{i=1}^{n} a_i b_i}{\sum_{i=1}^{n} b_i} < \exp\left(\frac{\sum_{i=1}^{n} a_i b_i \log a_i}{\sum_{i=1}^{n} a_i b_i}\right)$$

These three results have an immediate generalization to the continuous case. For further details the interested reader should consult Ref. [65].

5-6. A Set of Simultaneous Linear Inequalities

This section will treat briefly such systems as

$$\sum_{j=1}^{n} a_{ij} x_j \begin{cases} \geq \\ = \end{cases} b_i \qquad i = 1, \ldots, m \tag{5-37}$$

where both inequalities and equalities can occur. A part of this discussion is postponed to Chap. 6 on linear programming.

Geometrically, each inequality generally defines a half space with an $(n - 1)$-dimensional hyperplane for a boundary in n-dimensional affine space which is the set of ordered points $(c_1, \ldots, c_n)$, where the c_i are real numbers. A set is convex in affine space if it contains the segment joining any two of its points. The above inequalities intersect in a convex set. This set has vertices, edges, etc., similar to a polyhedron which may be unbounded. If all the b_i are zero, the system is said to be homogeneous and the boundary hyperplanes pass through the origin. When the b_i are not all zero, the system is called inhomogeneous. An inhomogeneous system of inequalities may be changed to a homogeneous one by a change of variables and by adjoining an additional strict inequality to the system, as will be shown later. In that case all the boundary hyperplanes can be made to pass through the origin of $(n + 1)$-dimensional space. Since the above inhomogeneous system may be reduced to a homogeneous one, it suffices to study the problem of solvability of a homogeneous system. Thus, without loss of generality, assume that all the b_i are zero. The resulting homogeneous system may then be represented in matrix notation by $\mathbf{Ax} \left\{ \begin{array}{c} \geq \\ = \end{array} \right. \mathbf{0}$, where the coefficient matrix $\mathbf{A}$ is defined by

$$\mathbf{A} = \begin{bmatrix} a_{11} a_{12} & \cdots & a_{1n} \\ a_{21} a_{22} & \cdots & a_{2n} \\ \cdots & \cdots & \cdots \\ a_{m1} a_{m2} & \cdots & a_{mn} \end{bmatrix}$$

and the column vectors $\mathbf{x}$ and $\mathbf{0}$ by

$$\mathbf{x} = \begin{bmatrix} x_1 \\ \cdots \\ x_n \end{bmatrix} \quad \mathbf{0} = \begin{bmatrix} 0 \\ \cdots \\ 0 \end{bmatrix}$$

To determine whether the homogeneous set of inequalities has a solution $\mathbf{x} \neq 0$, it is necessary only to know the signs of the minors of order r, where r is the rank of $\mathbf{A}$. The first step is to select any r linearly independent columns of $\mathbf{A}$. In this set, consider every system of $r + 1$ rows and let $d_1, d_2, \ldots, d_{r+1}$ be the rth-order determinants (d_i is the determinant of the matrix obtained by deleting the ith row) of this $r + 1$ by r matrix $\mathbf{B}$. Consider the quantities

$$d_1, -d_2, d_3, -d_4, \ldots, (-1)^r d_{r+1}$$

The system has no solution if $(-1)^k d_{k+1}$ is nonzero for $k = 1, \ldots, r + 1$ and has the same sign for all k such that the kth row of $\mathbf{B}$ corresponds to an inequality [49].

The following useful theorem due to Farkas gives the conditions for a homogeneous inequality to be satisfied by a solution of a homogeneous system of inequalities: In order that the linear inequality $b'x \geq 0$ hold for x satisfying a system of homogeneous linear inequalities $Ax \geq 0$, it is necessary and sufficient that $b = Ax^0$ for some $x^0 \geq 0$. Here b is a column vector with components b_i ($i = 1, \ldots, n$), b' is the corresponding row vector, and x^0 is a column vector with components $x_i^0 \geq 0$ ($i = 1, \ldots, n$).

The rest of this section is devoted to studying the existence of solutions of different linear systems and obtaining the duality theorem of linear programming. Now consider the systems

$$Ax = 0$$
$$Bx \geq 0 \qquad \text{(i)}$$
$$Cx = 0$$

and

$$\lambda'A + \mu'B + \nu'C = 0$$
$$\lambda' \geq 0 \qquad \mu' \geq 0 \qquad \nu' \neq 0 \qquad \text{(ii)}$$

where A, B, and C are rectangular matrices and x, λ', μ', and ν' are arbitrary row vectors.

Transposition Theorem 5-4: One and only one of the systems (i) and (ii) is solvable. This is the most general duality theorem for finite systems of linear inequalities [51].

An inhomogeneous system of linear inequalities and equations can be replaced by a homogeneous system if y_j/z is substituted for x_j and the additional inequality $z > 0$ is adjoined. When applied to the inhomogeneous system

$$Ax \geq b \qquad \text{(IH)}$$

for instance, this process yields the homogeneous system

$$[0, \ldots, 0, 1] \begin{bmatrix} y \\ z \end{bmatrix} > 0 \qquad \text{(H)}$$

$$[A \quad -b] \begin{bmatrix} y \\ z \end{bmatrix} \geq 0$$

Any solution of (IH) yields a solution of (H) with $z = 1$; conversely, any solution of (H) can be normalized so that $z = 1$ and then yields a solution of (IH) [20].

The most important application of these results is now presented.

Example: Let A be a matrix, b and c vectors, and M a number. First consider the system

$$Ax \geq b$$
$$x \geq 0 \qquad \text{(iii)}$$
$$c'x \leq M$$

Using the technique explained above, one passes from (iii) to the homogeneous system

$$[0, \ldots, 0, 1] \begin{bmatrix} \mathbf{y} \\ \mathbf{z} \end{bmatrix} > 0$$

$$\begin{bmatrix} \mathbf{A} & -\mathbf{b} \\ \mathbf{I} & 0 \\ -\mathbf{c}' & M \end{bmatrix} \begin{bmatrix} \mathbf{y} \\ \mathbf{z} \end{bmatrix} \geq 0 \qquad \text{(iv)}$$

The transposition theorem asserts that this last system has a solution $\begin{bmatrix} \mathbf{y} \\ \mathbf{z} \end{bmatrix}$ if and only if the system

$$\begin{array}{l} \lambda'[0, \ldots, 0, 1] + \mathbf{\mu}' \begin{bmatrix} \mathbf{A} & -\mathbf{b} \\ \mathbf{I} & 0 \\ -\mathbf{c}' & M \end{bmatrix} = 0 \\ \lambda' \geq 0 \quad \lambda' \neq 0 \\ \mathbf{\mu}' \geq 0 \end{array} \qquad \text{(v)}$$

has no solutions λ', $\mathbf{\mu}'$. To rewrite the last system in a more convenient form, let s be the sum of the components of λ' and note that the conditions $\lambda' \geq 0 (\lambda' \neq 0)$ imply $s > 0$. Also, make the partition

$$\mathbf{\mu}' = (\mathbf{u}', \mathbf{v}', t)$$

where t is a number, and note that $\mathbf{\mu}' \geq 0$ is equivalent to $\mathbf{u}' \geq 0$, $\mathbf{v}' \geq 0$, $t \geq 0$. Thus, one has

$$\begin{array}{l} \mathbf{u}'\mathbf{A} + \mathbf{v}' = t\mathbf{c}' \\ \mathbf{u}' \geq 0 \\ \mathbf{u}'\mathbf{b} = tM + s \\ t \geq 0 \\ s > 0 \\ \mathbf{v}' \geq 0 \end{array} \qquad \text{(vi)}$$

which in turn is equivalent to

$$\begin{array}{l} \mathbf{u}'\mathbf{A} \leq t\mathbf{c}' \\ \mathbf{u}' \geq 0 \\ \mathbf{u}'\mathbf{b} > M \\ t \geq 0 \end{array} \qquad \text{(vii)}$$

Thus one and only one of (iii) and (vii) has solutions.
Next consider the system

$$\begin{array}{l} \mathbf{A}\mathbf{x} \geq \mathbf{b} \\ \mathbf{x} \geq 0 \\ \mathbf{c}'\mathbf{x} < M \end{array} \qquad \text{(viii)}$$

Arguing as above, one finds that one and only one of (viii) and

$$\begin{aligned} \mathbf{u'A} &\leq t\mathbf{c'} \\ \mathbf{u'} &\geq 0 \\ \mathbf{u'b} &\geq tM \\ t &\geq 0 \\ \mathbf{u'b} &> 0 \quad \text{if } t = 0 \end{aligned} \qquad \text{(ix)}$$

has solutions.

These results lead to the well-known *duality theorem of linear programming*, whose significance will be discussed later.

Theorem 5-5: If one of the two problems

Minimize $\mathbf{c'x}$ subject to the constraints $\mathbf{Ax} \geq \mathbf{b}$, $\mathbf{x} \geq 0$ (Min)
Maximize $\mathbf{u'b}$ subject to the constraints $\mathbf{u'A} \leq \mathbf{c'}$, $\mathbf{u'} \geq 0$ (Max)

has a solution, then so does the other. If solutions exist, then the "maximum" and "minimum" are equal.

Proof: To prove this, suppose first that (Min) has a solution, and take M above to be the minimum. Then (iii) has a solution, and so (vii) does not; in particular, (vii) has no solution with $t = 1$. Also, (viii) has no solution, and so (ix) does have a solution. This solution cannot have $t = 0$, however, for in that case $\mathbf{u'A} \leq 0$ [from (ix)] and $\mathbf{x} \geq 0$ [from (iii)] would imply $\mathbf{u'Ax} \leq 0$, and also $\mathbf{Ax} \geq \mathbf{b}$ [from (iii)] and $\mathbf{u'} \geq 0$ [from (ix)] would imply $\mathbf{u'Ax} \geq \mathbf{u'b}$; the two "implied" statements would yield $\mathbf{u'b} \leq 0$, contradicting $\mathbf{u'b} > 0$ [from (ix)]. Thus the solutions of (ix) have $t > 0$ and can be normalized to have $t = 1$. We have proved that (ix) has solutions with $t = 1$ but that (vii) does not; this shows that M is precisely the maximum desired in (Max), which therefore has a solution.

Conversely, suppose that (Max) has a solution; now take M to be the maximum. Then (ix) has a solution (indeed, a solution with $t = 1$), and so (viii) has no solution. Also, (vii) has no solution with $t = 1$, and thus no solution with $t > 0$ (since such a solution could be normalized to yield $t = 1$). But (vii) has no solution with $t = 0$, either, for if such a solution, $\mathbf{u'_1}, t = 0$, existed, then one would have $\mathbf{u'_1} \geq 0$, $\mathbf{u'_1 A} \leq 0$ by (ix) and so if $\mathbf{u'}$ is a solution of (Max) then $\mathbf{u'} + \mathbf{u'_1}$ would obey the constraints of (Max) and would be such that

$$(\mathbf{u'} + \mathbf{u'_1})\mathbf{b} = M + \mathbf{u'_1 b} > 2M$$

contradicting the fact that M is the maximum. Thus (vii) has no solution at all, and so (iii) has a solution. The fact that (iii) has a solution but (viii) does not shows that M is the desired minimum for (Min), which therefore has a solution. This completes the proof of the theorem.

Some methods of obtaining solutions to a system of linear inequalities will now be given with illustrations.

Elimination Method

This method consists of successively eliminating the variables from all the inequalities in the system [52, 55]

$$\sum_{j=1}^{n} a_{ij}x_j \geq b_i \qquad i = 1, \ldots, m \qquad (5\text{-}38)$$

Let the first variable to be eliminated be x_k. Solve each of the m inequalities for the term containing x_k to obtain

$$a_{ik}x_k \geq b_i - \sum_{j \neq k} a_{ij}x_j \equiv F_i(x_j) \qquad i = 1, \ldots, m \qquad (5\text{-}39)$$

This relation may be divided into three classes:

$$\begin{array}{ll} \text{Class A:} & \text{If } a_{ik} \geq 0, \text{ then } x_k \geq F_a(x_j) \quad j \neq k \\ \text{Class B:} & \text{If } a_{ik} \leq 0, \text{ then } x_k \leq F_b(x_j) \quad j \neq k \\ \text{Class C:} & \text{If } a_{ik} = 0, \text{ then } 0 \geq F_c(x_j) \quad j \neq k \end{array} \qquad (5\text{-}40)$$

The switch in the subscripts of F enables a distinction among the three different classes.

The inequalities of Class C have x_k already absent. To eliminate the x_k from Classes A and B, combine each inequality of Type A with each of Type B and obtain

$$F_b(x_j) - F_a(x_j) \geq 0$$

in addition to
$$F_c(x_j) \leq 0$$

Repetition of this process will eventually lead to a system of inequalities for one unknown, say x_n, which defines the permissible range of values for this unknown. By fixing a definite value to x_n (in the permissible range) and repeating this process, one is led to a range of values in which another x_i, say x_{n-1}, must lie, etc. In this way a solution may be obtained.

The chief disadvantage of the elimination method is that the number of inequalities may increase enormously after each step in the operation. It is difficult to estimate the number of elementary operations required for a solution by this method. For example, in the case of n linear equations in n unknowns, the number of elementary operations (essentially multiplications) is of the order of n^3. In the case of k_n inequalities, the process requires approximately nk_n operations for the first elimination, $(n-1)k_{n-1}$ at the second, etc. Thus, in all, approximately $nk_n + (n-1)k_{n-1} + \cdots$ operations are needed. Note that $k_{n-1} \leq k_n^2/4$ but k_{n-1} may exceed k_n.

To illustrate this rather obvious method, consider the following example:

$$\begin{aligned} 2x_1 - 5x_2 + x_4 &\geq 3 \\ x_1 - 2x_2 + 3x_3 - x_4 &\geq 6 \\ -x_1 - 2x_2 - 2x_3 + 2x_4 &\geq -5 \\ x_2 - x_3 + 3x_4 &\geq -1 \\ 4x_2 - x_3 &\geq 0 \end{aligned} \quad (5\text{-}41)$$

By applying the existence statement given above to this system with x_j replaced by y_i and adjoining to it the inequality $z > 0$, it can easily be shown to have a solution. Now eliminating x_1,

$$\begin{array}{lll} & & \text{Class} \\ (a) & x_1 \geq 3/2 + 5/2 x_2 - 1/2 x_4 & A \\ (b) & x_1 \geq 6 + 2x_2 - 3x_3 + x_4 & A \\ (c) & x_1 \leq 5 - 2x_2 - 2x_3 + 2x_4 & B \\ (d) & 0 \leq 1 + x_2 - x_3 - 3x_4 & C \\ (e) & 0 \leq 4x_2 - x_3 & C \end{array} \quad (5\text{-}42)$$

Combine (a) and (c) and (b) and (c) to obtain

$$\begin{aligned} 7/2 - 9/2 x_2 - 2x_3 + 5/2 x_4 &\geq 0 \\ -1 - 4x_2 + x_3 + x_4 &\geq 0 \\ x_2 - x_3 - 3x_4 &\geq -1 \\ 4x_2 - x_3 &\geq 0 \end{aligned} \quad (5\text{-}43)$$

Eliminating x_2, one has (always ordering the inequalities according to the classes)

$$\begin{array}{ll} & \text{Class} \\ x_2 \geq -1 + x_3 + 3x_4 & A \\ x_2 \geq 1/4 x_3 & A \\ x_2 \leq 2/9 (7/2 - 2x_3 + 5/2 x_4) & B \\ x_2 \leq 1/4 (-1 + x_3 + x_4) & B \end{array} \quad (5\text{-}44)$$

Combine each inequality of Class A with every inequality of Class B to obtain

$$\begin{aligned} 16/9 - 13/9 x_3 - 22/9 x_4 &\geq 0 \\ 7/9 - 25/36 x_3 + 5/9 x_4 &\geq 0 \\ 3/4 - 3/4 x_3 - 11/4 x_4 &\geq 0 \\ -1/4 + 1/4 x_4 &\geq 0 \end{aligned} \quad (5\text{-}45)$$

Eliminate x_4:

$$\begin{array}{ll} & \text{Class} \\ x_4 \geq 9/5(-7/9 + 25/36 x_3) = -7/5 + 5/4 x_3 & A \\ x_4 \geq 1 & A \\ x_4 \leq 9/22 (16/9 - 13/9 x_3) = 16/22 - 13/22 x_3 & B \\ x_4 \leq 4/11 (3/4 - 3/4 x_3) = 3/11 - 3/11 x_3 & B \end{array} \quad (5\text{-}46)$$

The possible combinations are

$$234/110 - 81/44 x_3 \geq 0$$
$$-3/11 - 13/22 x_3 \geq 0$$
$$92/55 - 67/44 x_3 \geq 0 \qquad (5\text{-}47)$$
$$-8/11 - 3/11 x_3 \geq 0$$

yielding $x_3 \leq -8/3$ which satisfies all these conditions. For convenience, let $x_3 = -3$. When this value is substituted in the sixth step (5-46) and the common region of values of x_4 is chosen, one has $1 \leq x_4 \leq 12/11$. By setting $x_3 = -3$ and $x_4 = 1$ in the fourth step (5-44), $x_2 = -3/4$ is obtained. Finally, with these values of x_2, x_3, and x_4 substituted in the second step (5-42), one has $x_1 = 29/2$. Thus the point $(29/2, -3/4, -3, 1)$ is a solution of the problem, as can be readily verified by substitution.

Relaxation Method

Assume that one has a set of inequalities

$$y_i \equiv \sum_{j=1}^{n} a_{ij} x_j + b_i \geq 0 \quad i = 1, 2, \ldots, m \qquad (5\text{-}48)$$

each of which is nontrivial in the sense that some $a_{ij} \neq 0$ for each i. Then, by dividing both sides of the ith inequality by $\sum_{j=1}^{n} a_{ij}^2 > 0$, a system obeying the useful normalization condition

$$\sum_{j=1}^{n} a_{ij}^2 = 1 \quad i = 1, 2, \ldots, m \qquad (5\text{-}49)$$

is obtained. Finding a solution to the given system of inequalities proceeds as follows: Select any initial set of values for the x_j, say $(x_1^0, \ldots, x_n^0)$, and substitute them obtaining corresponding values $(y_1^0, \ldots, y_m^0)$. If all y_i^0 are nonnegative, then the set of x_i^0 constitutes a solution. If not, choose the algebraically largest of the negative y_i^0 (call it y_k^0) and form new values of the x_j by using

$$x_j^1 = x_j^0 - a_{kj} y_k^0 \quad j = 1, \ldots, n \qquad (5\text{-}50)$$

The resulting values

$$y_1^1, \ldots, y_m^1$$

will then include $y_k^1 = 0$; that is, the x_j^1 have been so chosen that the kth inequality (the one most badly violated by the x_i^0) is now satisfied and indeed satisfied as an equality. The process is now repeated, and in most cases a solution will be reached in finitely many steps. It is theoretically possible, however, that the process fails to terminate, and such cases occasionally occur in practice. In Ref. [1] it is shown that even then the process converges to a solution.

Geometrically, the procedure can be described as follows: Each equation

$$y_i = \sum_{j=1}^{n} a_{ij}x_j + b_i = 0 \qquad i = 1, \ldots, m \qquad (5\text{-}51)$$

defines a "hyperplane" in the n-dimensional space of points $x = (x_1, \ldots, x_n)$ just as a linear equation in two unknowns defines a line in two-dimensional space and a linear equation in three unknowns defines a plane in three-dimensional space. Each hyperplane $y_i = 0$ divides the n-dimensional space into two half spaces; the inequality $y_i \geq 0$ ($i = 1, \ldots, m$) defines that half space for each hyperplane in which a solution is found. Beginning with the initial point

$$x^0 = (x_1^0, \ldots, x_n^0)$$

if this is not already a solution of the system, then it lies on the "wrong side" of one or more of the hyperplanes $y_i = 0$. If x^0 lies farthest on the wrong side of the particular hyperplane $y_k = 0$, then the foot of the perpendicular from x^0 to this hyperplane is certainly a better approximation to a solution of the system than is x^0. The equations defining x_j^1 ($j = 1, \ldots, m$) give precisely the foot x^1 of this perpendicular; the simple form of these equations is due to the normalization.

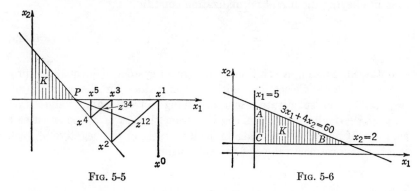

Fig. 5-5   Fig. 5-6

To illustrate a case where an infinite sequence of iterations is required and to suggest a way to circumvent this, suppose that the region K (see Fig. 5-5) is a section of the set of solutions of the original set of inequalities. Let x^0 be the first point chosen (the first guess), and let x^1 be the result of solving for x_j^1. Successively, the method yields the points x^2, x^3, x^4, x^5, ..., which approach P (the nearest point in the region K). However, the point P can also be reached by finding the mid-point z^{12} of x^1 and x^2, the mid-point z^{34} of x^3 and x^4, and then P as the intersection of the line through z^{12} and z^{34} with the line through x^1 and x^3. Consider the following problem in two-space:

$$\begin{aligned} y_1 &= x_1 - 5 & \geq 0 \\ y_2 &= x_2 - 2 & \geq 0 \\ y_3 &= -3x_1 - 4x_2 + 60 &\geq 0 \end{aligned} \quad (5\text{-}52)$$

Normalizing the coefficients of x_1 and x_2 yields

$$\begin{aligned} y_1 &= x_1 - 5 & \geq 0 \\ y_2 &= x_2 - 2 & \geq 0 \\ y_3 &= -.6x_1 - .8x_2 + 12 &\geq 0 \end{aligned} \quad (5\text{-}53)$$

Geometrically, the inequalities form a closed triangle any point of which is a solution to the system. Thus if the initial choice x^0 were to be made in the region K the given system of inequalities would then be satisfied by x^0. If x^0 were chosen in the region of either angle A or angle B an infinite process would result from the relaxation method; the "midpoint method" previously suggested would give a point A (or B) as a solution. If x^0 were initially chosen in any other region, not more than two applications of the iteration on x_j would be needed either to approach a solution or to place one on the sides of an acute angle.

In particular, let x^0 be $(25,5)$. Then

$$x_1^0 = 25 \qquad x_2^0 = 5$$

so that $\qquad y_1^0 = 20 \qquad y_2^0 = 3 \qquad y_3^0 = -7$

Thus $\qquad\qquad\qquad y_k^0 = y_3^0 = -7$

An iteration yields

$$x_1^1 = 25 - (-.6)(-7) = 20.8$$
$$x_2^1 = 5 - (-.8)(-7) = -.6$$

so that $\qquad y_1^1 = 15.8 \qquad y_2^1 = -2.6 \qquad y_3^1 = 0$

Here $\qquad\qquad\qquad y_k^1 = y_2^1 = -2.6$

and a second iteration gives

$$x_1^2 = 20.8 - 0(-2.6) = 20.8$$
$$x_2^2 = -.6 - 1(-2.6) = 2.0$$

Thus $\qquad y_1^2 = 15.8 \qquad y_2^2 = 0 \qquad y_3^2 = -2.08$

Again, $\qquad y_k^2 = y_3^2 = -2.08$

yielding $\qquad x_1^3 = 20.8 - (-.6)(-2.08) = 19.552$
$$x_2^3 = 2.0 - (-.8)(-2.08) = .336$$

and $\qquad y_1^3 = 14.552 \qquad y_2^3 = -1.664 \qquad y_3^3 = 0$

For the final iteration

$$y_k^3 = y_2^3 = -1.664$$
$$x_1^4 = 19.552 - 0(-1.664) = 19.552$$
$$x_2^4 = 0.336 - 1(-1.664) = 2.000$$

Having determined four values of x, a solution $(5\frac{2}{3}, 2)$ is obtained by the mid-point method.

Such special devices for obtaining a solution when convergence is slow might be too complicated in large systems of inequalities containing many variables, since intuition and judgment may not be of much use. It is often better to change the initial guess radically in the hope of arriving in a region of rapid convergence.

Exponential Method [55]

To obtain a point $(x_1, \ldots, x_n)$ which satisfies the system of inequalities

$$y_i = \sum_{j=1}^{n} a_{ij}x_j + b_i \geq 0 \qquad i = 1, \ldots, m \qquad (5\text{-}54)$$

form
$$f = \sum_{i=1}^{m} e^{-\lambda_i y_i} \qquad (5\text{-}55)$$

where $\lambda_i \geq 0$ $(i = 1, \ldots, m)$ are constants.

Remark: The point $x \equiv (x_1, \ldots, x_n)$ satisfies the system if and only if the corresponding point $y \equiv (y_1, \ldots, y_m)$ satisfies the conditions $y_i \geq 0$ $(i = 1, \ldots, m)$. (This condition asserts that y_i must lie in the nonnegative orthant of the y space, i.e., the part of the space where all the variables assume nonnegative values.) Thus, the function f is used to adjust the point x in such a way that y moves toward the positive orthant in the y space.

Let y_i^0 $(i = 1, \ldots, m)$ be the value of the system of inequalities obtained from an initial choice $x^0 \equiv (x_1^0, \ldots, x_n^0)$. If $y_i^0 \geq 0$ for every i, then x^0 is a solution. If not, compute

$$f_j^0 = \left(\frac{\partial f}{\partial x_j}\right)_{x_j=x_j^0} = \sum_{i=1}^{m} -\lambda_i a_{ij} e^{-\lambda_i y_i^0} \qquad j = 1, \ldots, n \qquad (5\text{-}56)$$

Since f increases (decreases) most rapidly in the direction of the gradient whose components here are $f_j^0(-f_j^0)$, the second choice is obtained as follows (see Cauchy's method):

$$x_j^1 = x_j^0 - \theta f_j^0 \qquad \theta > 0 \qquad j = 1, \ldots, n \qquad (5\text{-}57)$$

Note that in general
$$f^1 < f^0 \qquad (5\text{-}58)$$

where f^0 and f^1 are the values of f computed from y^0 and y^1 corresponding to x^0 and x^1.

Also note from the expression defining it that f decreases as y moves into the positive orthant. Thus, y^1 is closer to the positive orthant than y^0 and x^1 is closer to the set defined by the inequality system than x^0. Repetition of the process will eventually produce convergence to a solution.

Special cases of the exponential method are as follows:

1. The optimum gradient method (Cauchy's method). Note from the definitions of f and f^1 that f^1 may be considered as a function of θ only. If θ minimizes f^1, then use this value of θ to determine x_j^1.

2. The constant gradient method. In the above process let θ be a constant throughout.

3. The infinitesimal gradient method. Instead of a constant θ, use a sequence of θ's which may all be different (if desired) to determine each x_j^1 $(j = 1, \ldots, n)$.

Another method of solving a set of simultaneous linear inequalities is the double description method [47, 54] mentioned only in passing.

The Simplex Process

The inequalities either have a nonvoid intersection in the positive orthant of the space or can be made to have one by a change of variables. By adjoining a linear function to be maximized or minimized subject to the set of inequalities and finding a solution of the resulting linear-programming problem, the latter will serve as a solution of the inequalities. (The simplex process will be discussed at length in another chapter.)

5-7. A Set of Simultaneous General Inequalities

The system of general inequalities $g_i(x_1, \ldots, x_n) \leq 0, i = 1, \ldots, m$, may be reduced to a system of equations and nonnegative constraints in single variables when the latter, called *slack variables*, are added to each inequality.

A possible procedure for solving the resulting set of equalities may be carried out by starting with a point from the feasible region of the slack variables; after substitution into the system, the equalities are solved by methods discussed previously.

OPTIMIZATION: NO CONSTRAINTS

5-8. Definitions

The word *extremum* or the phrase *extreme value* is used in mathematics to refer to a value of a function (perhaps subject to constraints) which is either a maximum or a minimum. The word *optimum* is used for the particular type of extremum desired in the problem at hand. For instance, if $f(x,y)$ (perhaps with x and y subject to constraints) is a "cost function," then the type of extremum one may seek is a minimum; the optimum is the minimum value of $f(x,y)$, and it is desired to find a pair (x_0,y_0) at which this minimum is attained. But if $f(x,y)$ were an "output function," then the optimum would be a maximum. Note that the term *extreme point* is used to refer to a point in the domain of definition of the function which yields an extremum of the function. It is a well-known fact that a continuous function attains its maxima and minima in its domain of definition.

A point $P_0 \equiv (x_1^0, \ldots, x_n^0)$ of a function $f(x_1, \ldots, x_n)$ defined in a domain D is an absolute maximum in D if the following inequality holds for any point of the domain D:

$$f(x_1, \ldots, x_n) \leq f(x_1^0, \ldots, x_n^0)$$

It is an absolute minimum if for every point in D

$$f(x_1, \ldots, x_n) \geq f(x_1^0, \ldots, x_n^0)$$

A maximum (minimum) is said to be strict only if the strict inequalities above hold.

P_0 is a maximum (minimum) in the large if there are no points P in D for which $f(P) = f(P_0)$.

The function $f(x_1, \ldots, x_n)$ has a relative or a local maximum (minimum) if $\epsilon > 0$ can be found such that f has an absolute maximum or minimum at P_0 in a subset of points of D which satisfy

$$|x_1 - x_1^0| < \epsilon, \ldots, |x_n - x_n^0| < \epsilon$$

For example, $\sin^2 x \cos x$ in the domain $-\pi/2 < x < \pi/2$ has a relative minimum at $x = 0$ which is strict since $\sin^2 x \cos x > 0$ for $\pi/2 > |x| \neq 0$. Note that $x = 0$ is not an absolute minimum when the domain of definition is extended to include the entire real axis.

5-9. Suboptimization

Suboptimization is a case of optimization for one phase of an operation, without taking into consideration every factor which has a bearing on the problem, whether in an obvious or in a subtle way. Thus, in optimizing performance in a given naval operation, one does not have to consider the entire set of objectives of the Navy. The suboptimization approach is useful, particularly when neither the formulation of the problem nor the available techniques enable one to obtain a reasonable answer. Although a true optimum is not obtained, it at least provides a rational technique for approaching the optimum. In suboptimization one must frequently assume that studying the operation in this restricted manner will produce the desired improvements without causing injury to the greater activity involving several objectives. This type of analysis may be done on any size of operation, ignoring all or a part of the facts relating it to

OPTIMIZATION

the activity to which it belongs. This procedure is often necessary because of economic and practical considerations, the finiteness of time, and the difficulty which one encounters in attempting to produce quick and coherent answers if one considers the problem in its broadest aspects. Suboptimizing frequently comprises reducing the number of objectives and including some that are nonquantifiable. There are, of course, the errors of concentrating on the simple factors and objectives while omitting the subtle ones. In most practical cases suboptimization is the only resort to solving a problem. Suboptimization is also recommended in the early study of an optimization problem to estimate weights for the effect, on the final answer, of the factors used [25, 26, 41].

5-10. Optimization of Nonconstrained Functions

A Differentiable Function

The extremum of a differentiable function of a single variable may be obtained directly by setting the first derivative equal to zero (necessary condition). The second derivatives provide a test (sufficient condition) to determine whether the extremum so obtained is a maximum or a minimum. Depending on the specific problem, either may be the desired optimum.

The following is a simple example illustrating the use of differentiation to determine the maximum of a function.

Example: A publisher sells a book at a profit of \$2.00 a copy and loses \$5.00 for each unsold copy. Assuming that the demands are unknown but can be estimated from previous data by a distribution $f(y)\,dy$, with a maximum sale of y_0 copies, it is desired to calculate the number of books to be printed in order to maximize the profit.

Solution: If the publisher prints x copies of which y copies are sold, then the profit is given by

$$p = \begin{matrix} 2y - 5(x - y) = 7y - 5x & \text{if } y < x \\ 2x & \text{if } y \geq x \end{matrix} \qquad (5\text{-}59)$$

The expected profit is determined in a straightforward manner, as

$$E[p] = \int_0^x (7y - 5x)f(y)\,dy + \int_x^{y_0} 2xf(y)\,dy \qquad (5\text{-}60)$$

with
$$\int_0^{y_0} f(y)\,dy = 1$$

This gives
$$\int_x^{y_0} f(y)\,dy = 1 - \int_0^x f(y)\,dy$$

$E[p]$ is maximized by equating to zero the derivative with respect to x.

Thus

$$\frac{dE}{dx} = \frac{d}{dx}\left[\int_0^x (7y - 5x)f(y)\,dy + 2x - 2x\int_0^x f(y)\,dy\right]$$

$$= 7xf(x) - 7xf(x) - 7\int_0^x f(y)\,dy + 2$$

$$= 0 \tag{5-61}$$

or

$$\int_0^x f(y)\,dy = 2/7$$

This shows that x must be chosen so that the integral has the value $2/7$.

For an extremum of a function of several variables, set the first partial derivatives equal to zero and solve simultaneously in the variables. This is the necessary condition. The following discussion is to provide background for a well-known sufficiency test.

Sufficiency Tests. Taylor's formula gives for the series expansion of a function of two variables $f(x,y)$ in the neighborhood of a point (x_0,y_0) the following:

$$f(x_0 + h, y_0 + k) = f(x_0,y_0) + \sum_{i=1}^{n} \frac{1}{i!}\left(h\frac{\partial}{\partial x} + k\frac{\partial}{\partial y}\right)^i f(x_0,y_0)$$

$$+ \frac{1}{(n+1)!}\left(h\frac{\partial}{\partial x} + k\frac{\partial}{\partial y}\right)^{n+1} f(x_0 + \theta h, y_0 + \theta k) \qquad 0 < \theta < 1 \tag{5-62}$$

If (x_0,y_0) is an extreme point and for small h and k, $f(x_0 + h, y_0 + k) - f(x_0,y_0)$ is positive or negative, depending on whether the terms involving the second partial derivatives are positive or negative, since for small values of h and k the latter is the dominant expression. The two terms with the first partial derivatives as coefficients vanish as a necessary condition for an extremum.

A similar consideration of the simpler case of a function of a single variable $f(x)$ leads to the well-known sufficient condition which requires that $f''(x) < 0$ for a maximum and $f''(x) > 0$ for a minimum. More generally, if all the derivatives up to but not including the nth-order derivative vanish at a point x_0, then if n is even $f(x)$ has a minimum (maximum) at x_0 if $f^{(n)}(x_0) > 0(<0)$, and if n is odd $f(x)$ increases (decreases) at x_0 if $f^{(n)}(x_0) > 0$ (<0).

In the case of several variables these conditions generalize to a quadratic form whose coefficients are the second partial derivatives. A quadratic form in n variables is given by

$$\sum_{i=1}^{n}\sum_{j=1}^{n} a_{ij}\lambda_i\lambda_j \qquad \text{where } a_{ij} = a_{ji} \tag{5-63}$$

The matrix $[a_{ij}]$ is called the *matrix of the quadratic form*. Note that $a_{ij} \equiv f_{x_i x_j}$ is given by the second partial derivatives of f. A quadratic form is positive definite if it is nonnegative for all real values of the variables and is zero if each of the variables is zero. It is positive semidefinite if it can also vanish for nonzero values of the variables.

Testing for a Minimum. Let $|D_m|$ be the determinant of the matrix $[D_m]$ obtained by deleting all elements which do not simultaneously lie in the first m rows and columns of the matrix whose elements are the second partial derivatives $f_{x_i x_j}$ evaluated at the stationary point (a point at which the first derivatives vanish) under study. It can be shown that the matrix $[a_{ij}]$ is positive definite if and only if each of the n determinants $|D_m|$ is positive [23]. The matrix is positive definite if the quadratic form (5-63) is.

The above condition on the determinants of the matrix is necessary and sufficient for the quadratic form to be positive definite which, in turn, is a sufficient condition for a minimum.

Note that to maximize a functon f is the same as to minimize $-f$. Thus, the above argument can also be used as a sufficiency test for a maximum after appropriate adjustment.

In order to determine whether a stationary point of a function $f(x,y)$ is a relative minimum, the following inequalities resulting from the condition on the determinants $|D_m|$ must be simultaneously satisfied:

$$f_{xx} > 0 \qquad f_{yy} > 0 \qquad f_{xx}f_{yy} - (f_{xy})^2 > 0 \qquad (5\text{-}64)$$

The subscripts indicate the variables with respect to which the partial derivatives are to be taken and their order.

For a relative maximum these conditions become

$$f_{xx} < 0 \qquad f_{yy} < 0 \qquad f_{xx}f_{yy} - (f_{xy})^2 > 0 \qquad (5\text{-}65)$$

In the case of a function of three variables, $f(x,y,z)$, the condition for a minimum gives

$$f_{xx} > 0 \qquad f_{xx}f_{yy} - (f_{xy})^2 > 0 \qquad \begin{vmatrix} f_{xx} f_{xy} f_{xz} \\ f_{yx} f_{yy} f_{yz} \\ f_{zx} f_{zy} f_{zz} \end{vmatrix} > 0 \qquad (5\text{-}66)$$

with $f_{xy} = f_{yx}$, $f_{xz} = f_{zx}$, and $f_{yz} = f_{zy}$.

Remark: There are many instances in which an optimization problem with constraints is simpler than the corresponding problem without constraints, in the sense that the constrained-variable problem may have a solution, although the unconstrained problem may not.

On the other hand, an example of a function of several variables which has no minimum in all variables at once, but which has one in $n-1$

variables if one of them is set equal to zero, is

$$f(x_1, \ldots, x_n) = (2x_1 + x_2)^2 + (2x_1 + x_3)^2 + \cdots \\ + (2x_1 + x_n)^2 - x_1^2 \quad (5\text{-}67)$$

It is positive in the remaining variables if not all are zero. It is zero if they are all zero. It is negative if

$$x_1 = \epsilon > 0, \; x_2 = x_3 = \cdots = x_{n-1} = x_n = -2\epsilon$$

Thus it does not have a minimum for the values

$$x_1 = x_2 = \cdots = x_n = 0$$

Monotone Functions and Convex Functions

Two well-known types of functions which are related and which occur frequently are monotone functions and convex functions.

A function $f(x)$ is monotone increasing on an interval if for any two values x_1 and x_2 in the interval with $x_1 < x_2$ the relation $f(x_1) \leq f(x_2)$ holds. It is strictly increasing if only the inequality holds. It is monotone decreasing if $\geq$ holds. If the derivative of a function is nonnegative (nonpositive) in an interval, then the function is monotone increasing (decreasing) in the interval.

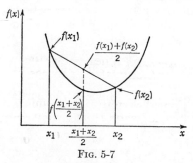

Fig. 5-7

Convex functions occur in an important way in optimization theory. Here we give conditions for determining when a function is concave or convex. These conditions are related to the foregoing discussion of quadratic forms as applied to determining an extremum of an unconstrained function.

A function of a single variable $f(x)$ is convex (also called concave upward) on an interval if, for any two points x_1, x_2 on the interval with $x_1 < x_2$ (see Fig. 5-7),

$$f\left(\frac{x_1 + x_2}{2}\right) \leq \frac{f(x_1) + f(x_2)}{2} \quad (5\text{-}68)$$

Geometrically, this says that the value of the function at a point which is the average of x_1 and x_2 is less than its average value at the two points. This condition may also be stated as

$$f[(1 - \theta)x_1 + \theta x_2] \leq (1 - \theta)f(x_1) + \theta f(x_2) \quad \text{for } 0 \leq \theta \leq 1 \quad (5\text{-}69)$$

It is strictly convex only if the inequality holds. A twice-differentiable function $f(x)$ on an open interval (not including the end points of the interval) is convex if and only if $d^2f/dx^2 \geq 0$ on the interval. This definition may be generalized to a function of several variables.

If D is a convex set of the space of points $(x_1, \ldots, x_n)$, a real-valued function defined in D is convex in D if for x and y in D the following holds:

$$f[(1 - \theta)x + \theta y] \leq (1 - \theta)f(x) + \theta f(y) \qquad \text{for } 0 \leq \theta \leq 1 \quad (5\text{-}70)$$

It is strictly convex if $<$ holds for $0 < \theta < 1$ and x and y are distinct points in D. If $f(x_1, \ldots, x_n)$ is a twice-differentiable function in an open convex set D, it is convex in D if and only if the quadratic form

$$\sum_{i=1}^{n} \sum_{j=1}^{n} \frac{\partial^2 f}{\partial x_i \, \partial x_j} \lambda_i \lambda_j \qquad (5\text{-}71)$$

is positive semidefinite for every point in D. Also $f(x_1, \ldots, x_n)$ is convex in D if and only if it is convex on every straight segment in D. It is strictly convex if the quadratic form is positive definite [11].

A function $f(x_1, \ldots, x_n)$ is concave (strictly concave) if $-f(x_1, \ldots, x_n)$ is convex (strictly convex).

The above condition involving the quadratic form is identical with the condition previously given to test for a minimum of a function of several variables, except that here the determinants $|D_m|$ are only required to be nonnegative. The conditions on the $|D_m|$ generally provide a set of inequalities in the variables from which the region or regions in which the function is convex may be determined. Let it be required to determine the region of concavity of the function

$$f(x,y) = \frac{1}{2\pi\sigma^2} \exp\left[-\frac{1}{2\sigma^2}(x^2 + y^2) \right] \qquad (5\text{-}72)$$

The second partial derivatives are given by

$$\begin{aligned} \frac{\partial^2 f}{\partial x^2} &= \frac{1}{2\pi\sigma^4}\left(\frac{x^2}{\sigma^2} - 1\right)\exp\left[-\frac{(x^2+y^2)}{2\sigma^2}\right] \\ \frac{\partial^2 f}{\partial y^2} &= \frac{1}{2\pi\sigma^4}\left(\frac{y^2}{\sigma^2} - 1\right)\exp\left[-\frac{(x^2+y^2)}{2\sigma^2}\right] \\ \frac{\partial^2 f}{\partial x \, \partial y} &= \frac{1}{2\pi\sigma^4}\frac{xy}{\sigma^2}\exp\left[-\frac{(x^2+y^2)}{2\sigma^2}\right] \end{aligned} \qquad (5\text{-}73)$$

and for concavity the following conditions must be simultaneously satisfied:

$$\frac{y^2}{\sigma^2} - 1 \leq 0$$
$$\left(\frac{x^2}{\sigma^2} - 1\right)\left(\frac{y^2}{\sigma^2} - 1\right) - \left(\frac{xy}{\sigma^2}\right)^2 \geq 0 \tag{5-74}$$

which, on simplifying, become

$$y^2 \leq \sigma^2 \quad \text{or} \quad -\sigma \leq y \leq \sigma$$
$$x^2 + y^2 \leq \sigma^2 \quad \text{a circle of radius } \sigma \tag{5-75}$$

Obviously every y which satisfies the second condition also satisfies the first condition, but not conversely. Hence the region of concavity is defined by the region with the circle $x^2 + y^2 = \sigma^2$ as boundary.

When the extremum of a function subject to constraints does not lie in the region defined by the constraints, it is useful to know if the function is concave (convex); this then makes it possible to determine the extremum on the boundary of the constraint region.

An Integral Containing a "Variable" Function. Calculus of Variations: Euler's Equation [8, 23, 61]

One step removed from the previous discussion is the problem of finding a function, instead of a variable, which yields an extremum to an integral. Suppose that one wishes to determine a function $y(x)$ which yields an extremum to

$$I = \int_{x_1}^{x_2} f(x,y,y') \, dx \tag{5-76}$$

Two limitations are imposed on the integrand: (1) $f(x,y,y')$ has continuous second partial derivatives in x,y,y' and (2) satisfies the "end" conditions

$$y(x_1) = y_1 \qquad y(x_2) = y_2 \tag{5-77}$$

A necessary condition for $y(x)$ to maximize (minimize) I is that f satisfies Euler's equation

$$\frac{d}{dx}\left(\frac{\partial f}{\partial y'}\right) - \frac{\partial f}{\partial y} = 0 \tag{5-78}$$

which may be written as

$$f_{y'y'} \frac{d^2 y}{dx^2} + f_{y'y} \frac{dy}{dx} + (f_{y'x} - f_y) = 0 \tag{5-79}$$

Example: Let (x_1,y_1) and (x_2,y_2) be given points in the (x,y) plane. Find the curve $y = y(x)$ between these points which, when rotated around the x axis, yields a surface of revolution of minimum area.

Solution: One must minimize an expression familiar from calculus,

$$2\pi \int_{x_1}^{x_2} y(1 + y'^2)^{1/2}\, dx \tag{5-80}$$

subject to the end-condition constraints. Here

$$f(x,y,y') = y(1 + y'^2)^{1/2}$$

and Euler's equation, after simplification, becomes

$$yy'' = 1 + y'^2 \tag{5-81}$$

a differential equation whose general solution is

$$y(x) = c_1 \cosh\left(\frac{x}{c_1} + c_2\right) \tag{5-82}$$

A particular solution is obtained by choosing c_1 and c_2 so that the end-condition constraints are satisfied, yielding the desired curve.

If it is desired to find $u(x,y)$, $v(x,y)$ which maximize (minimize)

$$\iint_R f(x,y,u,v,u_x,u_y,v_x,v_y)\, dx\, dy \tag{5-83}$$

where x and y are the independent variables and R is a simply connected two-dimensional region (such as a circle) of the xy plane, the necessary condition is given by the pair of second-order partial differential equations

$$\frac{\partial}{\partial x}\left(\frac{\partial f}{\partial u_x}\right) + \frac{\partial}{\partial y}\left(\frac{\partial f}{\partial u_y}\right) - \frac{\partial f}{\partial u} = 0 \tag{5-84}$$

and

$$\frac{\partial}{\partial x}\left(\frac{\partial f}{\partial v_x}\right) + \frac{\partial}{\partial y}\left(\frac{\partial f}{\partial v_y}\right) - \frac{\partial f}{\partial v} = 0 \tag{5-85}$$

The first term of the first equation, for example, is given by

$$f_{u_x u_x}\frac{\partial^2 u}{\partial x^2} + f_{u_x u_y}\frac{\partial^2 u}{\partial x\, \partial y} + f_{uu_x}\frac{\partial u}{\partial x} + f_{u_x v_x}\frac{\partial^2 v}{\partial x^2} + f_{u_x v_y}\frac{\partial^2 v}{\partial x\, \partial y}$$
$$+ f_{vu_x}\frac{\partial v}{\partial x} + f_{xu_x} \tag{5-86}$$

Remark: We give a brief sketch of the proof of Euler's condition. One perturbs the function $y(x)$ by replacing it by $\bar{y}(x) + \epsilon\eta(x)$. Here $\bar{y}(x)$ is assumed to yield a relative minimum and $\eta(x)$ vanishes at the end points but is otherwise arbitrary and twice differentiable; ϵ is a real number which is chosen small so that all the curves $y(x)$ are contained in a small neighborhood of $\bar{y}(x)$. Thus, for small values of ϵ, the integral when evaluated at $\epsilon = 0$ has the least value. Consequently, for a minimum value of the new integral, one differentiates under the integral sign with respect to ϵ and sets the resulting integral equal to zero at $\epsilon = 0$. Then one integrates by parts the component of the derivative taken with respect to y' multiplied by the derivative of y' with respect to ϵ and uses the fact that $\eta(x)$ vanishes at the end points. The resulting integrand consists of $\eta(x)$ multiplied by the left side of Euler's equation. The integral vanishes if this part vanishes, since $\eta(x)$ is arbitrary. This results in Euler's equation.

OPTIMIZATION SUBJECT TO CONSTRAINTS

In this part of the chapter the function to be optimized is subjected to a set of constraints. The resulting problem is considerably more interesting than that which has already been discussed. The difficulty occurs mainly when the maximum cannot be obtained by ordinary differentiation because it does not lie in the region defined by the constraint set. For example, if it is required to find values of x and y which maximize the function

$$f_1(x) + f_2(y)$$

where $\quad\quad f_1(x) = 2x \quad\quad f_2(y) = y$

subject to the constraints

$$x + y = 1 \quad\quad x, y \geq 0$$

then $f_1'(x) = 2$ and $f_2'(y) = 1$ everywhere. However, the maximum is attained at the boundary point (1,0), as can be readily verified by the geometry of the problem. If the function to be optimized is continuous, then it has a maximum either in the region or on the boundary. A maximum on the boundary is determined by gradient methods and by exploring along the boundary, usually after having decided by ordinary methods that it is not in the region. The problem of finding the boundary point which yields a maximum is simpler if the function is concave or convex.

5-11. An Optimum on the Boundary

As has been previously mentioned, the methods of optimization without constraints do not apply to the study of an extremum at a boundary point of the domain considered. Those techniques apply to cases where the domain is indefinite, such as finding an extremum in the entire space. Even though it is not specified, the domain of optimization is not necessarily infinite. For example, it may be desired to determine an extremum on a closed surface such as a sphere.

If the optimum lies in the domain, then the problem can be considered the same as if the inequalities were not present. The latter do not interfere with the condition that the relative extremum of the function be in the domain, and one may proceed as if the domain were infinite. On the other hand, if the extreme point is on the boundary, then the inequalities on which this point lies may be replaced by equalities (the others are neglected); on differentiation, using Lagrange's method, these equalities will then furnish the conditions which must be satisfied by the optimized function when one moves to the interior of the domain from the optimum point. It will be seen that for such displacements the increase in f should be positive for a minimum and negative for a maximum.

In the case of a function $f(x)$ of a single variable defined on an interval $a \leq x \leq b$, the necessary condition for a minimum (maximum) at a is that $f'(a_+) \geq 0$ (≤ 0) or at b that $f'(b_-) \leq 0$ (≥ 0). The derivatives at the end points must be right- and left-hand derivatives, respectively, since the function is not defined outside the interval. Note that, if $f(x)$ is monotone increasing or decreasing, the minimum and the maximum value of $f(x)$ are the appropriate ones of $f(a)$, $f(b)$.

5-12. The General Programming Problem

The general programming problem may be stated as follows: Find the extremum values of the function $f(x_1, \ldots, x_n)$ subject to the $m + n$ constraints $x_j \geq 0$ ($j = 1, \ldots, n$) and $g_i(x_1, \ldots, x_n) \leq b_i$ ($i = 1, \ldots, m$), where the b_i are constants [34]. Often the b_i are transposed to the left and considered as a part of the g_i. Then one may write $g_i \leq 0$, instead of the above. There is no general algorithm for solving this type of problem. However, there are existence theorems. Note that an interesting simple special case of this problem is the linear-programming problem where f and g are linear expressions in the variables. The remainder of this chapter is essentially concerned with studying different aspects of the general problem. The question may arise as to why one considers a single function to be optimized. Note that, in general, one cannot simultaneously optimize several functions. The complexity is increased when a set of constraints is introduced. If it is possible to decide on the relative importance of the functions, then each function can be assigned an appropriate "weight" and the techniques presented are then applied to optimize the resulting weighted average. The prescription of these weights, although difficult in practice, is a nonmathematical problem.

5-13. Equality Constraints: Lagrange Multiplier Techniques

In order to develop methods for determining an optimum on the boundary, and in harmony with current techniques usually introduced in calculus, Lagrange's multiplier method will be studied first. The method provides a necessary condition for an optimum when the constraints are given as equations. This is a special case of the more general problem with inequality constraints studied in the next section.

Suppose that it is desired to find an extremum of a differentiable function $f(x,y)$ whose variables are subject to a constraint

$$g(x,y) = 0$$

where g is also differentiable. If such an extremum occurs at a point (x_0,y_0) at which at least one of the partial derivatives g_x, g_y (say g_y) does not vanish, then one can proceed as follows: Near (x_0,y_0) the equation of the curve $g(x,y) = 0$ can be written in the form

$$y = h(x)$$

Since g vanishes all along the curve, one has

$$\frac{dg\,[x,h(x)]}{dx} = g_x + g_y \frac{dh}{dx} = 0 \qquad \text{at } (x_0,y_0) \tag{5-87}$$

and since (x_0,y_0) yields the constrained extreme value one also has

$$\frac{df\,[x,h(x)]}{dx} = f_x + f_y \frac{dh}{dx} = 0 \qquad \text{at } (x_0,y_0) \tag{5-88}$$

Since $g_y \neq 0$ at (x_0,y_0), one can define a quantity λ by

$$f_y(x_0,y_0) + \lambda g_y(x_0,y_0) = 0$$

If Eq. (5-87) is multiplied by λ and the result then added to Eq. (5-88), one obtains

$$f_x(x_0,y_0) + \lambda g_x(x_0,y_0) = 0$$

Thus the equations

$$f_x + \lambda g_x = 0$$
$$f_y + \lambda g_y = 0$$

hold at (x_0,y_0). These equations can be written as

$$\frac{\partial F}{\partial x} = 0$$
$$\frac{\partial F}{\partial y} = 0$$

if one sets

$$F(x,y,\lambda) = f(x,y) + \lambda g(x,y) \tag{5-89}$$

and the original constraint $g(x,y) = 0$ is just

$$\frac{\partial F}{\partial \lambda} = 0$$

In other words, the necessary conditions for an *unconstrained* extremum of F (namely, the vanishing of the three partial derivatives of F) are also necessary conditions for a *constrained* extremum of $f(x,y)$ (under the assumption that both g_x and g_y do not vanish at the point in question).

These arguments generalize readily. Suppose that one wishes to find an extremum of a differentiable function $f(x_1, \ldots, x_n)$ whose variables are subject to the m constraints

$$g_i(x_1, \ldots, x_n) = 0 \quad i = 1, \ldots, m; \, m \leq n$$

where the g_i are also differentiable. Form the *Lagrangian function*

$$F(x_1, \ldots, x_n; \lambda_1, \ldots, \lambda_m) = f(x_1, \ldots, x_n) + \sum_{i=1}^{m} \lambda_i g_i(x_1, \ldots, x_n)$$

involving the *Lagrange multipliers* $\lambda_1, \ldots, \lambda_m$. Then the necessary conditions for an unconstrained extremum of F (namely, the vanishing of F's first partial derivatives) are also necessary conditions for a constrained extremum of $f(x_1, \ldots, x_n)$, provided that the matrix of partial derivatives $\partial g_i/\partial x_j$ has rank m at the point in question. These necessary conditions are a system of $m + n$ equations

$$\frac{\partial F}{\partial x_j} = \frac{\partial f}{\partial x_j} + \sum_{i=1}^{m} \lambda_i \frac{\partial g_i}{\partial x_j} = 0 \quad j = 1, \ldots, n \quad (5\text{-}90)$$

$$\frac{\partial F}{\partial \lambda_i} = g_i = 0$$

which one can then (at least in theory) solve for the $m + n$ unknowns $x_1, \ldots, x_n; \lambda_1, \ldots, \lambda_m$. One is really interested only in obtaining $x_1, \ldots, x_n$.

Example 1: Find the dimensions of a rectangular parallelepiped with largest volume whose sides are parallel to the coordinate planes, to be inscribed in the ellipsoid

$$g(x,y,z) = \frac{x^2}{a^2} + \frac{y^2}{b^2} + \frac{z^2}{c^2} - 1 = 0 \quad (5\text{-}91)$$

Solution: If the dimensions are x, y, and z, then the volume is given by

$$f(x,y,z) = xyz \quad (5\text{-}92)$$

Form the Lagrangian function

$$F(x,y,z,\lambda) = f(x,y,z) + \lambda g(x,y,z)$$

Differentiate with respect to each variable and set the results equal to zero. Solve for λ, x, y, z in the four resulting equations

$$F_x = yz + \frac{2\lambda x}{a^2} = 0$$
$$F_y = xz + \frac{2\lambda y}{b^2} = 0$$
$$F_z = xy + \frac{2\lambda z}{c^2} = 0 \quad (5\text{-}93)$$
$$F_\lambda = \frac{x^2}{a^2} + \frac{y^2}{b^2} + \frac{z^2}{c^2} - 1 = 0$$

Multiplying the first three equations by x, y, z, respectively, adding, and then making use of the last equation yield

$$3f(x,y,z) + 2\lambda = 0$$

Thus, $\quad \lambda = -\tfrac{3}{2} f(x,y,z)$

With this value of λ substituted in the first three equations, respectively, one has

$$x = \frac{a}{\sqrt{3}} \quad y = \frac{b}{\sqrt{3}} \quad z = \frac{c}{\sqrt{3}} \quad (5\text{-}94)$$

for the answer. Note that these results hold also for the special case of the sphere obtained by putting $a = b = c = 1$.

As an elementary illustration, this problem can be formulated for a modern auditorium with a hemisphere for an outer structure, for example, if for ventilation reasons it is desired to have a parallelepiped for the inside. This method can also be applied to problems concerned with the inner structure of spheres used to house equipment for high-atmosphere experiments.

Example 2: Minimize $f(x_1, \ldots, x_n)$ subject to the constraint

$$\sum_{i=1}^{n} c_i x_i = C$$

Solution: By equating to zero the partial derivatives of the Lagrangian and solving for λ, one has:

$$\lambda = -\frac{1}{c_i} \frac{\partial f}{\partial x_i} \quad i = 1, \ldots, n$$

Suppose that $(x_1^0, \ldots, x_n^0)$ is an estimate of the x_i. This estimate determines for each of the above n equations a value of λ, which we denote by λ_i, where in general $\lambda_i \neq \lambda_j$ for $i \neq j$. The object is to find an iterative procedure for estimating the x_i which would successively improve the values of the λ_i until they all equal the same constant λ. One would then have an $(x_1^0, \ldots, x_n^0)$, which solves the problem.

To do this, note that each λ_i is a function of $(x_1, \ldots, x_n)$ and hence, using up to the linear term of series expansion, one has

$$\lambda_i = \lambda_i^0 + \sum_{j=1}^{n} \frac{\partial \lambda_i}{\partial x_j}\bigg|^0 (x_j - x_j^0) \qquad i = 1, \ldots, n$$

where $\lambda_i^0 (i = 1, \ldots, n)$ and $\dfrac{\partial \lambda_i}{\partial x_j}\bigg|^0$ are computed at $(x_1^0, \ldots, x_n^0)$

Since it is ultimately desired to make all the λ_i equal by successive choices of $(x_1, \ldots, x_n)$, one replaces λ_i in the last set of equations by λ, then solves the linear equations for x_j. These expressions then contain λ as an unknown. Finally, by substituting the x_j in $\sum_{i=1}^{n} c_i x_i = C$ and solving for λ, one obtains a new estimate $(x_1^1, \ldots, x_n^1)$. Note that in general, $\lambda_i \neq \lambda_j$, $i \neq j$ for the new estimate. However, the procedure can now be repeated with the new point replacing the previous one. One has convergence if the initial choice is in the neighborhood of the optimum. The process is also applicable when the constraint is nonlinear, in which case the problem of solving for λ to determine the new value is slightly more complicated. In an actual operation the existing solution may be used for the initial choice.

When there are several constraints, the iterative procedure proceeds as previously described in the solution of a set of simultaneous general equations.

There are instances in which the inequality constraints are changed to equality constraints by replacing the variable by the squares of a new variable and introducing a slack variable as the square of a third variable. The resulting problem can then be solved by Lagrange's method. For example, to calculate the minimum of $f(x) = (x + 2)^2$, where $0 \leq x \leq 1$, one replaces x by t^2 and the constraint on x by $t^2 + u^2 = 1$. It has been demonstrated that this procedure can prove very useful in practice [30].

5-14. Inequality Constraints

This section deals with a necessary condition (on the gradients) for a maximum on the boundary.

Theorem 5-6: A necessary condition for a function to have a maximum at a point (x_0, y_0) on the boundary C of the region R defined by $g(x,y) \leq 0$, where g is differentiable, is that the gradient of f at (x_0, y_0) be a nonnegative multiple of the gradient of g at (x_0, y_0).

Proof: For simplicity, assume that the region defined by g is as in Fig. 5-8. Since, at (x_0,y_0), $f(x_0,y_0) \geq f(x,y)$ for all points (x,y) lying in R and on C, it follows that

$$\nabla f \cdot \mathbf{V} \leq 0 \tag{5-95}$$

for any vector $\mathbf{V}$ lying on the same side of the tangent line $\mathbf{t}$ as the region R. Note that ∇f, when pointing into the region R along any vector $\mathbf{V}$, points in a direction of decreasing values of f, since the maximum is reached at (x_0,y_0).

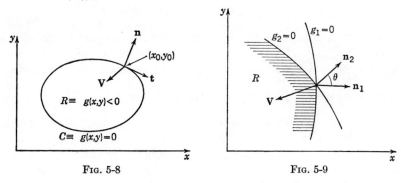

Fig. 5-8   Fig. 5-9

Now if a coordinate system with (x_0,y_0) as origin and with $\mathbf{t}$ and $\mathbf{n}$, the normal line to g at (x_0,y_0), as axes is adopted, then obviously one may write

$$\mathbf{V} = a\mathbf{t} - b\mathbf{n} \tag{5-96}$$

where a is arbitrary and b is nonnegative. Using this expression for $\mathbf{V}$ in the previous inequality, one has

$$\nabla f \cdot a\mathbf{t} - \nabla f \cdot b\mathbf{n} \leq 0 \tag{5-97}$$

Since this inequality must hold for an arbitrary vector $\mathbf{V}$, it follows, if the inequality is to be always satisfied for arbitrary a, that the coefficient of a should vanish:

$$\nabla f \cdot \mathbf{t} = 0 \tag{5-98}$$

Thus ∇f is perpendicular to $\mathbf{t}$ and hence must lie along the normal $\mathbf{n}$. But ∇g also lies along the normal $\mathbf{n}$; therefore ∇f must be a positive multiple of ∇g.

The foregoing theorem will now be proved in case the boundary is that of a region R defined by $g_1(x,y) \leq 0$ and $g_2(x,y) \leq 0$, and the point (x_0,y_0) which yields a maximum to the function $f(x,y)$ coincides with the intersection of $g_1(x,y) = 0$ and $g_2(x,y) = 0$. Here ∇f is shown to be a nonnegative linear combination of $\nabla g_i (i = 1, 2)$. Let $\mathbf{n}_1$ and $\mathbf{n}_2$ be the normals to g_1 and g_2 at (x_0,y_0). It is clear that $\mathbf{n}_1$ and $\mathbf{n}_2$ are linearly independent; that is, $\mathbf{n}_1 \cdot \mathbf{n}_2 \neq 1$. This is true since at a "corner" the angle between the tangents to g_1 and g_2 is nonzero, which is then also true of the normals (see Fig. 5-9).

Let $\mathbf{V}$ be any vector directed from (x_0, y_0) into R. Then as before, with $\mathbf{n}_1$ and $\mathbf{n}_2$ as axes with origin at (x_0, y_0), one has

$$\mathbf{V} = a\mathbf{n}_1 + b\mathbf{n}_2 \qquad \text{with } a,b \leq 0 \qquad (5\text{-}99)$$

This follows from the fact that $\mathbf{V}$ is contained in the common region where both $\mathbf{n}_1$ and $\mathbf{n}_2$ are negative.

Now
$$\nabla f = c\mathbf{n}_1 + d\mathbf{n}_2 \qquad (5\text{-}100)$$

and it remains to show that $c, d \geq 0$, for since ∇g_1 and ∇g_2 lie along $\mathbf{n}_1$ and $\mathbf{n}_2$, respectively, it would follow that ∇f is a nonnegative linear combination of the latter.

Also as before,

$$\nabla f \cdot \mathbf{V} = ac + bd + (ad + bc)(\mathbf{n}_1 \cdot \mathbf{n}_2) \leq 0 \qquad (5\text{-}101)$$

having used the fact that $\mathbf{n}_i \cdot \mathbf{n}_i = 1$ ($i = 1, 2$), $\mathbf{n}_i$ being unit vectors. Since $\mathbf{V}$ is arbitrary, let $\mathbf{V}_1$ be such that $a = 0$ and $b < 0$; then the above inequality becomes

$$bd + bc(\mathbf{n}_1 \cdot \mathbf{n}_2) \leq 0$$

or, because $b < 0$,

$$d + c(\mathbf{n}_1 \cdot \mathbf{n}_2) \geq 0$$

Again let $\mathbf{V}_2$ be such that $b = 0$, $a < 0$, which leads to the inequality

$$c + d(\mathbf{n}_1 \cdot \mathbf{n}_2) \geq 0 \qquad (5\text{-}102)$$

On multiplying the latter of the two inequalities by $(\mathbf{n}_1 \cdot \mathbf{n}_2)$ and subtracting from the former, one has

$$d \geq d(\mathbf{n}_1 \cdot \mathbf{n}_2)^2 \qquad (5\text{-}103)$$

which implies that $d \geq 0$, since $(\mathbf{n}_1 \cdot \mathbf{n}_2)^2 = \cos^2 \theta < 1$. θ is the angle between $\mathbf{n}_1$ and $\mathbf{n}_2$. Similarly, $c \geq 0$ can be shown to hold, and the proof is complete.

The above argument can be easily generalized to n dimensions. If it is desired to minimize f subject to $g_i \leq 0$, then the necessary condition requires that ∇f be a nonpositive linear combination of the constraints. If, in addition, the constraints $x_j \geq 0$ ($j = 1, \ldots, n$) are introduced, then they can be adjoined to the $g_i \leq 0$ by writing $-x_j \leq 0$. The general theorem for maxima on the boundary with $x_j \geq 0$ may now be stated and proved in condensed notation. We give a different method of proof.

Theorem 5-7: A necessary condition that the boundary point

$$\mathbf{x}^0 \equiv (x_1^0, \ldots, x_n^0)$$

yield a maximum to the general programming problem with differentiable functions is that there exist

$$\lambda_i \geq 0 \ (i = 1, \ldots, m) \quad \text{and} \quad \mu_j \geq 0 \ (j = 1, \ldots, n)$$

such that $\quad \nabla f(\mathbf{x}^0) = \sum_{i=1}^{m} \lambda_i \nabla g_i(\mathbf{x}^0) - \mathbf{\mu} \quad \mathbf{\mu} \equiv (\mu_1, \ldots, \mu_n) \quad$ (5-104)

where if $x_j{}^0 > 0$, then $\mu_j = 0$ and, if $g_i(\mathbf{x}^0) < 0$, then $\lambda_i = 0$. [Recall that $\nabla \equiv (\partial/\partial x_1, \ldots, \partial/\partial x_n)$.]

Proof: The proof given below uses the duality theorem of linear programming derived early in this chapter.

It is clear that $\mathbf{x}^0$ may lie only on the boundary of some of the constraints $g_i \leq 0$. Let $M(\mathbf{x}^0)$ be the set of subscript i such that $g_i(\mathbf{x}^0) = 0$. Let the vector $(\mathbf{x} - \mathbf{x}^0)$ be directed from $\mathbf{x}^0$ to $\mathbf{x} \equiv (x_1, \ldots, x_n)$ with $x_i \geq 0 \ (i = 1, \ldots, n)$. It follows that, as before,

$$\nabla f(x^0) \cdot (\mathbf{x} - \mathbf{x}^0) \leq 0 \quad \text{and} \quad \nabla g_i(\mathbf{x}^0) \cdot (\mathbf{x} - \mathbf{x}^0) \leq 0 \quad (5\text{-}105)$$

for all subscripts i in the set $M(\mathbf{x}^0)$ and all $\mathbf{x} \geq 0$.

It is clear that the g_i decrease as one moves from the boundary into the constraint region. The first inequality and the second set of inequalities are linear in the nonnegative vector $\mathbf{x}$. Thus the problem of finding $\mathbf{x}^0$ which maximizes f subject to the constraints $g_i = 0$, where i assumes values in $M(\mathbf{x}^0)$, becomes one of finding $\mathbf{x}^0 \geq 0$ which maximizes $\nabla f(\mathbf{x}^0) \cdot \mathbf{x}$ subject to the linear constraints

$$\nabla g_i(\mathbf{x}^0) \cdot \mathbf{x} \leq \nabla g_i(\mathbf{x}^0) \cdot \mathbf{x}^0 \quad (5\text{-}106)$$

The dual of this linear-programming problem requires that $\lambda_i \geq 0$ be found which minimize

$$\sum_{i \in M(\mathbf{x}^0)} \lambda_i [\nabla g_i(\mathbf{x}^0) \cdot \mathbf{x}^0] \quad (5\text{-}107)$$

subject to the linear inequality constraints

$$\sum_{i \in M(\mathbf{x}^0)} \lambda_i [\nabla g_i(\mathbf{x}^0)] \geq \nabla f(\mathbf{x}^0) \quad (5\text{-}108)$$

The remaining λ_i for i not in $M(\mathbf{x}^0)$ are all zero and correspond to those $g_i(\mathbf{x})$ such that $g_i(\mathbf{x}^0) < 0$. Thus from the last set of inequalities one may write

$$\nabla f(\mathbf{x}^0) = \boldsymbol{\lambda} \nabla G - \mathbf{\mu} \quad (5\text{-}109)$$

where $\boldsymbol{\lambda} \equiv (\lambda_1, \ldots, \lambda_n)$, ∇G is the matrix of first partial derivatives of g_i for all i evaluated at $\mathbf{x}^0$, and $\mathbf{\mu} \equiv (\mu_1, \ldots, \mu_n)$ consists of nonnegative slack variables which reduce the relation to equalities. On taking the scalar product with the vector $\mathbf{x}^0$ one has

$$\mathbf{\mu} \cdot \mathbf{x}^0 = (\lambda \, \nabla G) \cdot \mathbf{x}^0 - \nabla f(\mathbf{x}^0) \cdot \mathbf{x}^0 = 0 \qquad (5\text{-}110)$$

since the duality theorem requires that the two expressions in the middle should be equal. From the equation $\mathbf{\mu} \cdot \mathbf{x}^0 = 0$ and from the fact that both $\mathbf{\mu}$ and $\mathbf{x}^0$ are nonnegative, it follows that, when $x_j > 0$, the corresponding μ_j must be zero, and the proof is complete.

The assumption throughout this section has been based upon the fact that $\mathbf{x}^0$ is a local maximum. If the region defined by the constraints is convex, then a local maximum of a concave function defined on this region is a maximum in the large (global maximum). It is clear that the minimum value of such a function is attained on the boundary. State the above theorem when minimizing subject to the constraints.

Example 1: Maximize

$$f(x,y) = \int_0^x e^{-t^2/2} \, dt + \int_0^y e^{-t^2/2} \, dt \qquad (5\text{-}111)$$

subject to the constraints

$$x \geq 0 \qquad y \geq 0 \qquad g(x,y) \equiv x + y \leq c, \, c > 0$$

Solution: Note that $f \geq 0$ and has a minimum at $(0,0)$. However, the quadratic-form test

$$\frac{\partial^2 f}{\partial x^2} = -x e^{-x^2/2} \leq 0$$

$$\frac{\partial^2 f}{\partial x^2} \frac{\partial^2 f}{\partial y^2} - \frac{\partial^2 f}{\partial y \, \partial x} \equiv xy e^{-(x^2+y^2)/2} \geq 0 \qquad (5\text{-}112)$$

shows that the function is concave in the region under study, and since it has no maximum inside the region, it attains this maximum on the boundary. Since f is concave, it is easy to show that the maximum lies on the line $x + y = c$. Hence the gradient of f is a multiple of the gradient of g. This gives

$$[e^{-(x^2/2)}, e^{-(y^2/2)}] = \lambda[1,1] \qquad (5\text{-}113)$$

which is true for $x = y$. Since the maximum is on $x + y = c$, it follows that $(c/2, c/2)$ is the desired answer obtained by putting $x = y$.

Example 2: Formulate and solve a nonlinear-programming problem in which the nonlinearity in the profit function is due to a drift toward saturation (diminishing returns) [58].

Solution: A method of solution is given for the special case where a linear factor dominates the profit function for small production rates and a negative exponential describes the higher rates. In a system with several components, a profit function showing saturation may be represented by

$$z = \sum_{i=1}^{n} a_i x_i \exp\left(-\sum_{i=1}^{n} \alpha_i x_i\right) \quad \text{with } x_i \geq 0, \alpha_i \geq 0 \ (i = 1, \ldots, n)$$
(5-114)

For small values of the components x_i (for example, the product in the single-component system), this equation is approximated by the linear function

$$z = \sum_{i=1}^{n} a_i x_i \tag{5-115}$$

which is a particular solution of

$$\sum_{i=1}^{n} x_i \frac{\partial z}{\partial x_i} = z \tag{5-116}$$

as can be easily verified.

It is not difficult to show that Eq. (5-114) is a solution of the more general equation

$$\sum_{i=1}^{n} x_i \frac{\partial z}{\partial x_i} = \left(1 - \sum_{i=1}^{n} \alpha_i x_i\right) z \quad \text{with } \alpha_i \geq 0 \ (i = 1, \ldots, n) \tag{5-117}$$

by substituting $x = uv$ in the latter expression and solving the resulting two equations obtained from the condition that the product uv must be a solution.

To interpret this model, note that if one x_i is allowed to increase, keeping the remaining ones fixed, Eq. (5-117) reduces to

$$\frac{dz}{z} = -\alpha_i \, dx_i \tag{5-118}$$

which says that the fractional drop in profit, at high rates of production, is proportional to the added rate of production.

To illustrate the ideas we maximize

$$z = (4x + 3y)e^{-.002(x+y)} \tag{5-119}$$

subject to the constraints

$$0 \leq x \leq A$$
$$0 \leq y \leq A \tag{5-120}$$

Then the necessary condition for a maximum in the interior is the vanishing of both $\partial z/\partial x$ and $\partial z/\partial y$, which gives, respectively,

$$4x + 3y = 2000 \qquad (5\text{-}121)$$
$$4x + 3y = 1500 \qquad (5\text{-}122)$$

As in Fig. 5-10, the last two equations represent two parallel lines with no common point.

Note that, on the line $4x + 3y = 1500$, z has the value

$$1500e^{.006(x-1500)}$$

which increases as x increases from 0 to 375. Similarly, on the line

$$4x + 3y = 2000$$

Fig. 5-10

z increases as x increases from 0 to 500.

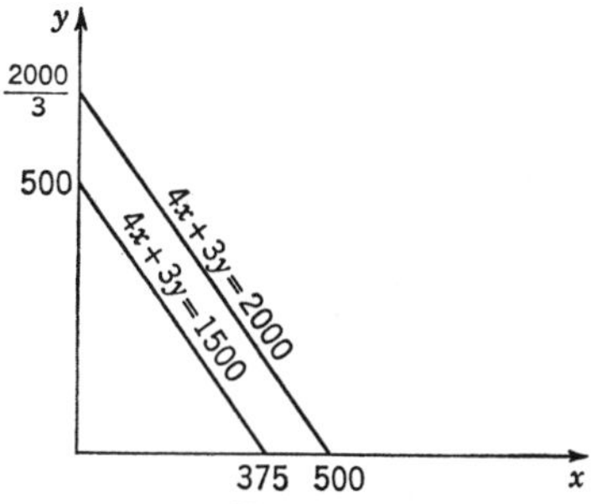

It is not difficult to verify that the maximum of z is given as follows:

Values of A	The point which maximizes z
$0 \leq A \leq 200\%$	(A,A)
$200\% < A \leq 375$	$\left(A, \dfrac{1500 - 4A}{3}\right)$
$375 \leq A \leq 500$	$(A,0)$
$500 \leq A$	$(500,0)$

A simple example with integer solution follows. (This example cannot be readily solved by the foregoing theory, owing to its discreteness, although solution may be possible by modifying the approach. For this type of problem we have no general theory to give. Each such problem receives special treatment.) It is desired to advance a vehicle 1,000 miles from an original position with a minimum expenditure of fuel. The vehicle has maximum fuel capacity of 500 units, which it uses at the uniform rate of one unit per mile. The vehicle must stock its own storage points from its one tank by making round trips between storage points. Determine the storage points along the route which minimize the fuel consumption for the entire trip. Determine the number of round trips required between each pair of storage points. Determine the minimum amount of fuel required at the start.

If s_i is the amount of fuel stored at the ith storage point ($i = 0, 1, \ldots, n$), d_i is the distance between the $(i-1)$st and the ith storage points, and k_i is the number of round trips the vehicle makes between these two points, then

$$s_{i-1} = s_i + 2k_i d_i + d_i \qquad i = 1, \ldots, n \qquad (5\text{-}123)$$

Thus, the amount of fuel stored at the $(i-1)$st point is equal to the amount stored at the ith point plus the amount consumed along the route in making k_i round trips and a single forward trip.

It is not difficult to see that a minimum use of fuel is made if for each trip the vehicle proceeds loaded at full capacity. For in that case, a smaller number of trips would be made and, hence, less fuel is consumed in travel. Also, in order to have no fuel left at the end, it is necessary that the vehicle be loaded with 500 units of fuel at the 500-mile point. From these two facts it follows that the ith storage point should be located so that the vehicle makes k_{i+1} round trips between this point and the $(i+1)$st storage point and one last forward trip, always fully loaded and ultimately leaving no fuel behind at the ith storage point. Working backward, the last statement is valid back to the first storage point but is not possible for the starting point since the position of that point is predetermined. Hence, the vehicle will make its last forward trip between the starting point and the first storage point with a load $c \leq 500$.

Thus one has

$$\begin{aligned} s_i &= k_i(500 - 2d_i) + 500 - d_i & i = 2, \ldots, n \\ s_1 &= k_1(500 - 2d_1) + c - d_1 \end{aligned}$$ (5-124)

It is desired to minimize s_0 given by Eq. (5-123):

$$s_0 = s_1 + 2k_1 d_1 + d_1$$

Now, using for s_1 the value given in (5-124), one has

$$s_0 = 500 k_1 + c \quad (5\text{-}125)$$

Since the vehicle can travel the last 500 miles without need for stored fuel, in order to minimize s_0, it suffices to put $s_n = 500$ and to place the storage points along the first 500 miles of the route. Thus, one has

$$\sum_{i=1}^{n} d_i = 500 \quad (5\text{-}126)$$

Now from the second equation of the problem one has

$$d_i = \frac{500 k_i + 500 - s_i}{2k_i + 1} \quad i = 2, \ldots, n \quad (5\text{-}127)$$

Replacing i by $i + 1$ in the first equation and substituting for s_i in the expression for d_i, one has

$$d_i = \frac{500 k_i + 500 - s_{i+1} - 2k_{i+1} d_{i+1} - d_{i+1}}{2k_i + 1}$$

$$i = 2, \ldots, n \qquad k_{n+1} \equiv 0 \quad (5\text{-}128)$$

OPTIMIZATION

Finally, replacing s_{i+1} by its equal from the second equation and simplifying yield

$$d_i = \frac{500(k_i - k_{i+1})}{2k_i + 1} \qquad i = 2, \ldots, n \qquad k_{n+1} \equiv 0 \qquad (5\text{-}129)$$

Similarly, one has

$$d_1 = \frac{500(k_1 - k_2) + c - 500}{2k_1 + 1}$$

Since $d_i > 0$, it follows that $k_i > k_{i+1}$. Thus it is desired to minimize

$$s_0 = 500k_1 + c \qquad (5\text{-}130)$$

subject to the constraint

$$\sum_{i=1}^{n} d_i = \left(500 \sum_{i=1}^{n} \frac{k_i - k_{i+1}}{2k_i + 1}\right) + \frac{c - 500}{2k_1 + 1} = 500$$

This can be written as

$$\sum_{i=1}^{n} \frac{k_i - k_{i+1}}{2k_i + 1} - \frac{1 - c/500}{2k_1 + 1} = 1 \qquad k_{n+1} \equiv 0 \qquad (5\text{-}131)$$

The second term on the left is less than the least value which any term in the sum can assume, i.e.,

$$\frac{1}{2k_1 + 1}$$

This relation simply stated requires a choice of k_i for which the sum exceeds unity with a minimum k_1.

It will now be shown that the minimum value of k_1 is obtained by taking $k_i - k_{i+1} = 1$ ($i = 1, \ldots, n$) and $k_n = 1$. Suppose that for $i = i_0$, $k_{i_0} - k_{i_0+1} = m$, an integer > 1. Then for the i_0th term, one has, by working backwards,

$$\frac{m}{2(k_{i_0+1} + m) + 1}$$

where the denominator jumps from $2k_{i_0+1} + 1$ to $2(k_{i_0+1} + m) + 1$.

Now, by taking $k_{i_0} - k_{i_0+1} = 1$, this jump may be replaced by a gradually increasing sum of terms

$$\frac{1}{2(k_{i_0+1} + 1) + 1} + \frac{1}{2(k_{i_0+1} + 2) + 1} + \cdots + \frac{1}{2(k_{i_0+1} + m) + 1}$$

This sum is greater than the previous expression which is m times the smallest term in the sum. It follows that a minimum k_1 is attained by using unit differences. These expressions with unit differences will determine n, the number of storage points. When the number of storage points is decreased, then the number of round trips between two adjacent points must be increased and the above argument shows that this is inefficient. Again, if the number of storage points is increased one automatically increases k_1 which is also inefficient. Furthermore, by taking $k_n = 1$, one obtains a more rapidly increasing sum from the monotonically increasing denominators. Because of the monotone property of the k_i, one chooses n such that $k_n = 1$, $k_{n-1} = 2$, ... , in such a manner as to produce k_2 which brings the sum in the constraint to less than unity. In this case, $k_2 = 6$. The corresponding d_i are then calculated, working backward from the 500-mile point. The distance which remains is taken as d_1 with a corresponding $k_1 = 7$ round trips. This choice of k_1 minimizes s_0, and the corresponding choice of d_i satisfies the distance constraint.

Example 3: Find nonnegative values of x_j $(j = 1, \ldots, n)$ which maximize

$$F(x_1, \ldots, x_n) = Q \sum_{j=1}^{n} (x_j + s_j) \sum_{j=1}^{n} \frac{p_j x_j}{x_j + s_j} - \sum_{j=1}^{n} x_j \quad (5\text{-}132)$$

subject to the conditions

$$s_j > 0 \ (j = 1, \ldots, n) \ n > 1 \qquad 0 < Q < 1 \qquad 0 \le p_j \le 1$$

and

$$\sum_{j=1}^{n} p_j = 1$$

Solution: The solution yielding the only positive maxima of F is given, without proof, by [28]

$$\begin{aligned} x_i &= 0 & i &= 1, \ldots, t-1 \\ x_i &= \lambda_t \sqrt{p_i s_i} - s_i & i &= t, \ldots, n \end{aligned} \quad (5\text{-}133)$$

where t and λ_t are furnished by successively testing

$$\rho_v > \frac{1}{\lambda_v^2} = \frac{q + \sum_{j=1}^{v-1} p_j}{\sum_{j=1}^{v-1} s_j} \quad (5\text{-}134)$$

Here

$$q = \frac{1 - Q}{Q}$$

and

$$\rho_j = \frac{p_j}{s_j}$$

the subscripts having been initially assigned so that $\rho_1 \leq \rho_2 \leq \cdots \leq \rho_n$. The smallest v for which this is satisfied is t.

Remarks:
1. There is always a nonnegative maximum of F; if

$$\max_{1 \leq j \leq n} \frac{p_j}{s_j} > \frac{1}{Q\Sigma s_i} \qquad (5\text{-}135)$$

this maximum is positive.
2. The maximal F never occurs when all the x_i are positive.
3. If some of the s_j are zero, a solution need not exist.

This problem applies to horse-race bets as follows: If the win probability of the jth horse in a race is p_j, the amount wagered on him by a better is x_j, the amount wagered by all others is s_j, and if the total sum wagered is multiplied by $Q(0 < Q < 1)$ prior to distribution to the winning betters, then the problem is the one posed above.

A Special Case

Suppose that both the effort x exerted in a task and its effect $f(x)$ can be measured and that the effect of over-all effort exerted in more than one task is the sum of the separate effects of the effort exerted on each one [31]. This assumes that the tasks have no interaction among them and hence are separate. Thus if the efforts are denoted by $x_i \geq 0$ $(i = 1, \ldots, n)$ satisfying the relation

$$\sum_{i=1}^{n} x_i = X \qquad (5\text{-}136)$$

where X is the total effort, and if the total effect or return is given as

$$f(x_1, \ldots, x_n) = \sum_{i=1}^{n} f_i(x_i) \qquad (5\text{-}137)$$

then the problem is one of finding a point $(x_1^0, \ldots, x_n^0)$ which maximizes f and satisfies the constraints. Generally the solution is on the boundary, and its determination is facilitated if all the f_i (and consequently f) are known to be convex or concave.

This simple problem may, with appropriate interpretations, be applied to such situations as the determination of response in a sales campaign; search for a lost object by distributing effort to several areas, each having a separate return function; and the distribution of destructive effort between distant enemy targets with different strategic values.

Without loss of generality, let $X = 1$; then it is possible to characterize the solution vectors $(x_1^0, \ldots, x_n^0)$ as follows:

Theorem 5-8: A necessary condition that $(x_1^0, \ldots, x_n^0)$ be a solution of the above problem is that there exists a real number λ such that

$$\begin{aligned} f_i'(x_i^0) &= \lambda & \text{if } x_i^0 > 0 \\ f_i'(x_i^0) &\leq \lambda & \text{if } x_i^0 = 0 \end{aligned} \quad (5\text{-}138)$$

This theorem is an immediate corollary to Theorem 5-7.

This condition is also sufficient if the functions are concave:

$$f_i''(x_i) \leq 0 \quad \text{for } 0 \leq x_i \leq 1$$

This theorem dates back to Josiah Willard Gibbs who used it around 1876 to solve a problem on the equilibrium of heterogeneous substances. In the case where functions are concave, a sufficient condition is also provided by the following theorem:

Theorem 5-9: Having obtained λ as in the previous theorem, a sufficient condition for $(x_1^0, \ldots, x_n^0)$ to be a solution of the problem is that

$$\begin{aligned} f_i'(0) &> \lambda & \text{if and only if } x_i^0 > 0 \\ f_i'(0) &\leq \lambda & \text{if and only if } x_i^0 = 0 \end{aligned} \quad (5\text{-}139)$$

If the f_i are convex, i.e., if

$$f''(x_i) \geq 0 \quad \text{for } 0 \leq x_i \leq 1$$

and a minimum is required, then the condition of the first theorem becomes

$$\begin{aligned} f_i'(x_i^0) &= \lambda & \text{if } x_i^0 > 0 \\ f_i'(x_i^0) &\geq \lambda & \text{if } x_i^0 = 0 \end{aligned} \quad (5\text{-}140)$$

which is true regardless of convexity but is also sufficient with it, and the condition of the second theorem becomes

$$\begin{aligned} f_i'(0) &< \lambda & \text{if and only if } x_i^0 > 0 \\ f_i'(0) &\geq \lambda & \text{if and only if } x_i^0 = 0 \end{aligned} \quad (5\text{-}141)$$

Example: Find $\mathbf{x}^0 = (x_1^0, \ldots, x_n^0)$ which minimizes

$$\sum_{i=1}^{n} a_i e^{-b_i x_i} \quad (5\text{-}142)$$

with
$$a_i, b_i > 0$$

and subject to the constraints

$$x_i \geq 0 \qquad \sum_{i=1}^{n} x_i = 1$$

OPTIMIZATION

Solution: From the above discussion [10]

$$-a_i b_i e^{-b_i x_i^0} \begin{cases} = \lambda & \text{if } x_i^0 > 0 \\ \geq \lambda & \text{if } x_i^0 = 0 \end{cases} \quad \lambda \leq 0 \quad (5\text{-}143)$$

and because of the strict convexity of the $f_i(x_i)$

$$\begin{aligned} -a_i b_i &< \lambda \quad \text{if and only if } x_i > 0 \\ -a_i b_i &\geq \lambda \quad \text{if and only if } x_i^0 = 0 \end{aligned} \quad (5\text{-}144)$$

In order to determine nonzero x_i^0 one has, from the above,

$$x_i^0 = \frac{1}{b_i} \log - \frac{a_i b_i}{\lambda} \quad \text{for } -a_i b_i < \lambda \quad (5\text{-}145)$$

This is now used together with the constraint

$$\sum_{i=1}^{n} x_i^0 = 1$$

in order to determine λ.
Define

$$f(\lambda) \equiv \sum \frac{1}{b_i} \log - \frac{a_i b_i}{\lambda} \quad (5\text{-}146)$$

where i ranges over all values for which the inequality $-a_i b_i < \lambda$ is satisfied. Then $f(\lambda)$ is a continuous strictly increasing function of λ which is equal to 1 for some value of λ. The process of determining λ is a simple one if

$$a_i b_i > a_j b_j \quad \text{for } i < j$$

With this assumption, let $\lambda_1 = -a_2 b_2$ and calculate the sum

$$f(\lambda_1) = \frac{1}{b_1} \log \frac{a_1 b_1}{a_2 b_2}$$

Then let $\lambda_2 = -a_3 b_3$, from which

$$f(\lambda_2) = \frac{1}{b_1} \log \frac{a_1 b_1}{a_3 b_3} + \frac{1}{b_2} \log \frac{a_2 b_2}{a_3 b_3}$$

Let i_0 be the last subscript for which the calculated sum is less than unity. Then one must find λ_0(the cutoff point) such that

$$-a_{i_0+1} b_{i_0+1} \geq \lambda_0 > -a_{i_0} b_{i_0}$$

for which

$$f(\lambda_0) = 1$$

In that case,

$$f(\lambda_0) - f(-a_{i_0}b_{i_0}) = 1 - f(-a_{i_0}b_{i_0})$$
$$= \sum_{i=1}^{i_0} \frac{1}{b_i} \log \frac{-a_i b_i}{\lambda_0} - \sum_{i=1}^{i_0} \frac{1}{b_i} \log \frac{-a_i b_i}{-a_{i_0}b_{i_0}}$$
$$= \log \frac{-a_{i_0}b_{i_0}}{\lambda_0} \sum_{i=1}^{i_0} \frac{1}{b_i} \tag{5-147}$$

Now $f(-a_{i_0}b_{i_0})$ has been calculated. This gives

$$\lambda_0 = -a_{i_0}b_{i_0} \exp\left[\frac{f(-a_{i_0}b_{i_0}) - 1}{\sum_{i=1}^{i_0} 1/b_i}\right] \tag{5-148}$$

Finally $x_i^0 > 0$ may be calculated from

$$-a_i b_i e^{-b_i x_i^0} = \lambda_0$$

Remarks: A function, which, by a linear transformation of the variables, can be reduced to the sum of functions each in a single variable, may be approximated to any desired degree of accuracy by piecewise linear functions. This is also true of convex and concave functions.

Note that a vector $\mathbf{x}^0 = (x_1^0, \ldots, x_n^0)$ which yields a maximum to

$$\prod_{i=1}^{n} f_i(x_i)$$

where $f_i(x_i) > 0$ for all i subject to constraints, also yields a maximum to

$$\sum_{i=1}^{n} \log f_i(x_i)$$

subject to the same constraints. If the total-effort constraint is replaced by

$$\sum_{i=1}^{n} x_i \leq X$$

then the solution can be obtained "stepwise" by setting

$$F_1(X) = \max_{0 \leq x_1 \leq X} f_1(x_1)$$
$$F_k(X) = \max_{0 \leq x_k \leq X} [F_{k+1}(X - x_k) + f_k(x_k)] \tag{5-149}$$

$F_n(X)$ is the desired maximum, as can be proved by mathematical induction on n. The basic idea of the process is roughly that of adding the variables x_k to the problem "one at a time." This is an illustration of the "principle of optimality" used in dynamic programming, a brief discussion of which will appear later.

In a RAND Corporation memorandum O. Gross determined a necessary and sufficient condition that $\mathbf{x} = (x_1, \ldots, x_n)$, with the x_i assuming nonnegative integer values satisfying the condition

$$\sum_{i=1}^{n} x_i = m \tag{5-150}$$

yields a minimum to

$$\sum_{i=1}^{n} f_i(x_i) \tag{5-151}$$

where the f_i are convex. This condition is

$$\min_{i} [f_i(x_i + 1) - f_i(x_i)] \geq \max_{i \in s^+(x)} [f_i(x_i) - f_i(x_i - 1)] \tag{5-152}$$

where $s^+(x)$ is the set of indices i for which $x_i > 0$.

The vector $\mathbf{x}^m = (x_1^m, \ldots, x_n^m)$ which yields a minimum subject to the constraints is defined as follows: Define

$$x_j^0 = 0 \qquad j = 1, \ldots, n$$

For $k > 0$, let $i(k)$ be an index providing

$$\min_{i} [f_i(x_i^k + 1) - f_i(x_i^k)]$$

and set

$$\begin{aligned} x_i^{k+1} &= x_i^k + 1 \quad \text{for } i \text{ in } i(k) \\ x_i^{k+1} &= x_i^k \quad \text{for } i \text{ not in } i(k) \end{aligned} \tag{5-153}$$

This gives the desired solution.

Equivalence with Saddle-value Problem

A function $F(x,\lambda)$ (where x,λ may be subject to constraints) is said to have a *saddle point* at a particular pair (x^0,λ^0) if

$$F(x,\lambda^0) \leq F(x^0,\lambda^0) \leq F(x^0,\lambda)$$

for all x,λ admitted by the constraints. (It is, of course, assumed that x^0, λ^0 satisfy the constraints [32, 34].)

To see why the term "saddle point" is used, note that a saddle rises away from its center in either direction parallel to the horse's spine and falls away from its center in either direction perpendicular to the spine.

If the saddle surface is given by the equation

$$z = F(x,\lambda)$$

(where the λ axis is horizontal and parallel to the spine, the x axis is horizontal and perpendicular to the spine, and the positive z axis is directed vertically upward), then F will indeed have a saddle point (as defined above) at the pair (x^0,λ^0) yielding its center.

Example: The function $F(x,\lambda) = \lambda^2 - x^2$ has a saddle point at $(0,0)$. Now note that the definition given above also applies when $\mathbf{x}$, $\boldsymbol{\lambda}$ are *vectors*, i.e., are convenient abbreviations for sets of variables

$$\mathbf{x} = (x_1, \ldots, x_n)$$
and
$$\boldsymbol{\lambda} = (\lambda_1, \ldots, \lambda_m)$$

Theorem 5-10: If $F(\mathbf{x},\boldsymbol{\lambda})$ is a differentiable function of an n-vector $\mathbf{x}$ with components $x_i \geq 0$ and an m-vector $\boldsymbol{\lambda}$ with components $\lambda_j \geq 0$, then a necessary condition that nonnegative vectors $\mathbf{x}^0$ and $\boldsymbol{\lambda}^0$ exist, such that

$$F(\mathbf{x},\boldsymbol{\lambda}^0) \leq F(\mathbf{x}^0,\boldsymbol{\lambda}^0) \leq F(\mathbf{x}^0,\boldsymbol{\lambda}) \qquad \text{for all } \mathbf{x}, \boldsymbol{\lambda} \geq 0$$

is that

$$\mathbf{F}_x^0 \equiv \left(\frac{\partial F}{\partial x_i}\right)^0 \leq 0 \qquad \mathbf{F}_x^0 \cdot \mathbf{x}^0 = 0 \qquad \mathbf{x}^0 \geq 0 \qquad \text{that is, } x_i^0 \geq 0 \text{ for all } i \tag{5-154}$$

$$\mathbf{F}_\lambda^0 \equiv \left(\frac{\partial F}{\partial \lambda_j}\right)^0 \geq 0 \qquad \mathbf{F}_\lambda^0 \cdot \boldsymbol{\lambda}^0 = 0 \qquad \boldsymbol{\lambda}^0 \geq 0 \qquad \text{that is, } \lambda_j^0 \geq 0 \text{ for all } j \tag{5-155}$$

For sufficiency, in addition to these conditions the following must be satisfied:

$$F(\mathbf{x},\boldsymbol{\lambda}^0) \leq F(\mathbf{x}^0,\boldsymbol{\lambda}^0) + \mathbf{F}_x^0 \cdot (\mathbf{x} - \mathbf{x}^0) \tag{5-156}$$
$$F(\mathbf{x}^0,\boldsymbol{\lambda}) \geq F(\mathbf{x}^0,\boldsymbol{\lambda}^0) + \mathbf{F}_\lambda^0 \cdot (\boldsymbol{\lambda} - \boldsymbol{\lambda}^0) \tag{5-157}$$

for all $\mathbf{x}, \boldsymbol{\lambda} \geq 0$.

Theorem 5-11: In order that $\mathbf{x}^0$ be a solution of the general programming problem, it is *necessary* that $\mathbf{x}^0$ and some $\boldsymbol{\lambda}^0$ satisfy conditions (5-154) and (5-155) and *sufficient* that $\mathbf{x}^0$ and $\boldsymbol{\lambda}^0$ satisfy conditions (5-154), (5-155), and (5-156), where $F(\mathbf{x},\boldsymbol{\lambda}) \equiv f(\mathbf{x}) + \boldsymbol{\lambda} \cdot \mathbf{g}(\mathbf{x})$.

Theorem 5-12: Suppose that all $g_i(\mathbf{x}) \geq 0$ and $f(\mathbf{x})$ are concave and differentiable for $\mathbf{x} \geq 0$. Then $\mathbf{x}^0$ is a solution of the programming problem if and only if $\mathbf{x}^0$ and some $\boldsymbol{\lambda}^0$ give a solution of the saddle-value problem for $F(\mathbf{x},\boldsymbol{\lambda}) = f(\mathbf{x}) + \boldsymbol{\lambda} \cdot \mathbf{g}(\mathbf{x})$.

Remark: When the foregoing is applied to the linear-programming problem [the $g_i(\mathbf{x})$ and $f(\mathbf{x})$ above are all linear], the well-known duality relation of linear programming is produced.

5-15. Dynamic Programming

The subject of dynamic programming has been expertly developed and investigated by Richard Bellman [2]. Because of the large number of publications currently available on the subject, the brief discussion given here is only a sketchy introduction. The literature should be consulted for further details.

Suppose that it is desired to divide an amount of effort x into two parts y and $x - y$, with returns from y given by $g(y)$ and from $x - y$ given by $h(x - y)$. After a part of the original amount x has been expended, there remains an amount of effort equal to $ay + b(x - y)$, where $0 < a$, $b < 1$. The latter amount is again divided into two parts with the same return functions. In this manner the process is continued in several stages. How should this allocation of effort be made at each stage in order that the total return is maximized for a finite number of stages? Formally, the problem is described by taking

$$R_1(x,y) = g(y) + h(x - y)$$

as the total return from the first stage,

$$R_2(x,y_1,y_2) = g(y_1) + h(x_1 - y_1) + g(y_2) + h(x_2 - y_2)$$

where $x_1 = x$, $x_2 = ay_1 + b(x_1 - y_1)$, and $0 \leq y_1 \leq x_1$, as the total return from two stages, and

$$R_N(x,y_1, \ldots ,y_N) = \sum_{i=1}^{N} [g(y_i) + h(x_i - y_i)] \qquad (5\text{-}158)$$

as the total return from successive allocation $y_1, y_2, \ldots, y_N$, where

$$x_1 = x \qquad x_{i+1} = ay_i + b(x_i - y_i) \ (i = 1, \ldots, N) \quad \text{and}$$
$$0 \leq y_i \leq x_i \ (i = 1, \ldots, N) \qquad (5\text{-}159)$$

Any choice of y_i $(i = 1, \ldots, N)$ is called a *policy*, and a choice which maximizes R_N is called an *optimum policy*. For an optimum policy, R_N depends only on the original effort x and on the number of stages N.

In order to derive the functional equation of this problem, let $f_N(x)$ be the total return from any N stages using an optimum policy and starting with an initial amount x.

Note that, for the first stage, x is divided into y and $x - y$, and then $g(y) + h(x - y)$ is computed. The remainder $ay + b(x - y)$ is used optimumly for the remaining $N - 1$ stages no matter what the return $f_{N-1}[ay + b(x - y)]$ of y is. Therefore

$$R_N(x,y) = g(x) + h(x - y) + f_{N-1}[ay + b(x - y)]$$

Thus by definition one has

$$f_N(x) = \max_{0 \le y \le x} R_N(x,y) = \max_{0 \le y \le x} \{g(y) + h(x - y) + f_{N-1}[ay + b(x - y)]\} \quad (5\text{-}160)$$

which is the basic functional equation. For a large number of stages, f_N and f_{N-1} are approximately the same and may be denoted by a function f which is independent of N. This gives the simpler equation

$$f(x) = \max_{0 \le y \le x} \{g(y) + h(x - y) + f[ay + b(x - y)]\} \quad (5\text{-}161)$$

Existence and uniqueness theorems are available for this equation.

With the requirement that $f(0) = 0$ this may be solved by successive approximations as follows: Let

$$y = \begin{cases} 0 & \text{if } \dfrac{g(x)}{1-a} \ge \dfrac{h(x)}{(1-b)x} \\ x & \text{if } \dfrac{g(x)}{1-a} < \dfrac{h(x)}{(1-b)x} \end{cases}$$

Otherwise choose

$$\frac{g(y)}{(1-a)y} = \frac{h(x-y)}{(1-b)(x-y)}$$

Let $f_0(x) = f_g(x)$ be the return calculated by recurrence, using one or the other of these policies. Successive approximations may now be computed using the functional equation in N stages. This process leads to improved approximations at each step since $f_i \le f_j$ for $i \le j$.

By making assumptions on the properties of g and h, it is possible to prescribe methods for computing

$$y_N = y_N(x) \quad \text{and} \quad f_N(x)$$

For example, if both g and h are strictly convex in x an optimum policy requires that $y = 0$ or that $y = x$. Several applications to bottleneck and other types of problems are given in Bellman's papers.

5-16. An Integral Subject to Constraints

Suppose that it is desired to find a $y(x)$ yielding an extreme value of

$$\int_{x_1}^{x_2} f(x,y,y') \, dx \quad (5\text{-}162)$$

where, in addition to the "end" conditions

$$y(x_1) = y_1 \qquad y(x_2) = y_2 \quad (5\text{-}163)$$

there is an additional constraint

$$\int_{x_1}^{x_2} g(x,y,y') \, dx = C \quad (C \text{ a given constant}) \quad (5\text{-}164)$$

One introduces the Lagrange multiplier λ and the Lagrangian function

$$F(x,y,y';\lambda) = f(x,y,y') + \lambda g(x,y,y')$$

and uses the Euler-equation technique to find an extreme value for $\int_{x_1}^{x_2} F \, dx$ subject to the end conditions. The resulting partial differential equation is

$$\left[\frac{d}{dx}\left(\frac{\partial f}{\partial y'}\right) - \frac{\partial f}{\partial y}\right] + \lambda \left[\frac{d}{dx}\left(\frac{\partial g}{\partial y'}\right) - \frac{\partial g}{\partial y}\right] = 0 \quad (5\text{-}165)$$

which is to be solved together with Eqs. (5-163) and (5-164). The same method applies if there are several constraints like (5-164); each constraint leads to an additional Lagrange multiplier.

For a two-dimensional problem of the same type, suppose that one wishes to find $u(x,y)$, $v(x,y)$ which yield an extreme value of

$$\iint_R f(x,y,u,u_x,u_y,v,v_x,v_y) \, dx \, dy \quad (5\text{-}166)$$

Here subscripts on u, v indicate partial derivatives and R is a given region in the (x,y) plane. Corresponding to the end conditions in the one-dimensional case, one has "boundary conditions" u and v specified on the boundary of R. If, in addition, there is a constraint

$$g(u,v) = 0 \quad (5\text{-}167)$$

then by combining the techniques of this part of the chapter one is led to the partial differential equation

$$g_v \left[\frac{d}{dx}\left(\frac{\partial f}{\partial u_x}\right) - \frac{\partial f}{\partial u}\right] + g_u \left[\frac{d}{dx}\left(\frac{\partial f}{\partial v_x}\right) - \frac{\partial f}{\partial v}\right] = 0 \quad (5\text{-}168)$$

which is to be solved together with the boundary conditions and (5-167).

Note that the necessary condition gives the result as a solution to a system of partial differential equations. Sufficient conditions are very difficult to derive and will not be discussed here. Sometimes physical arguments are used to show that the result obtained is the desired one.

The following example will illustrate the ideas discussed above.

Example 1: For a set of continuous variables, with joint density function f, the entropy measure of information is defined by $-E[\log f]$, the expected value of $\log f$ (see Chap. 8 on probability theory).

Theorem 5-13: The density function with finite mean and variance, in one variable, with maximum entropy, is normal.

Proof: The entropy from the above definition is

$$-E[\log f] = -\int_{-\infty}^{\infty} f(x) \log f(x)\, dx \qquad (5\text{-}169)$$

This is to be maximized, subject to

$$\int_{-\infty}^{\infty} f(x)\, dx = 1 \qquad \int_{-\infty}^{\infty} xf(x)\, dx = 0 \qquad \int_{-\infty}^{\infty} x^2 f(x)\, dx = \sigma^2 \qquad (5\text{-}170)$$

The first condition follows from the property of a density function; the second requires the mean to be at the origin. This can be done by means of a translation. The last condition assigns the standard deviation. From the preceding discussion the problem becomes one of maximizing

$$\int_{-\infty}^{\infty} f(x)[\lambda_1 x^2 + \lambda_2 x + \lambda_3 - \log f(x)]\, dx \qquad (5\text{-}171)$$

Applying Euler's equation to this problem with $y \equiv f(x)$, one has

$$-1 - \log f(x) + \lambda_1 x^2 + \lambda_2 x + \lambda_3 = 0$$

or
$$f(x) = \exp(\lambda_1 x^2 + \lambda_2 x + \lambda_3 - 1) \qquad (5\text{-}172)$$

The problem is to determine $\lambda_1, \lambda_2, \lambda_3$, and $f(x)$ such that this condition, together with the three conditions of the problem, is satisfied. The last expression for $f(x)$ may now be substituted in the constraints. Thus multiplying the first condition by λ_2 and the second by $2\lambda_1$, adding, and integrating yield

$$e^{\lambda_3 - 1}[e^{\lambda_2 x + \lambda_1 x^2}]_{-\infty}^{\infty} = \lambda_2 \qquad (5\text{-}173)$$

In order for this to exist one must have $\lambda_1 < 0$, $\lambda_2 = 0$. Instead of λ_1, $-|\lambda_1|$ will be used. With these values and using the first condition, one has

$$e^{\lambda_3 - 1} \int_{-\infty}^{\infty} e^{-|\lambda_1| x^2}\, dx = e^{\lambda_3 - 1} \sqrt{\frac{\pi}{|\lambda_1|}} = 1 \qquad (5\text{-}174)$$

which gives
$$e^{\lambda_3 - 1} = \sqrt{\frac{|\lambda_1|}{\pi}}$$

A slight variation of this integral has been encountered and evaluated previously.

Again

$$e^{\lambda_3 - 1} \int_{-\infty}^{\infty} x^2 e^{|\lambda_1| x^2}\, dx = \sigma^2 = e^{\lambda_3 - 1} \frac{1}{2|\lambda_1|} \sqrt{\frac{\pi}{|\lambda_1|}} \qquad (5\text{-}175)$$

which gives
$$|\lambda_1| = \frac{1}{2\sigma^2}$$

or, since λ_1 is negative,

$$\lambda_1 = -\frac{1}{2\sigma^2}$$

having used for e^{λ_2-1} its value obtained in the previous calculation. Thus finally

$$f(x) = \frac{1}{\sqrt{2\pi}\sigma} e^{-(1/2\sigma^2)x^2} \qquad (5\text{-}176)$$

The entropy is then given by

$$\log \sqrt{2\pi e \sigma^2} \qquad (5\text{-}177)$$

This theorem may be generalized to obtain the n-dimensional normal distribution.

Example 2: The following example [43, 44] is of considerable interest in itself and also involves some theoretical points not yet mentioned. Find the most profitable schedule of production of a commodity over a given period of time $(0,T)$ so as to meet the following requirements:

1. The initial inventory h_0 is given.
2. The sales schedule is given, $S(t)$ being the cumulative sales from 0 to t (dS/dt need not be continuous).
3. Inventory can never be negative.
4. The terminal inventory is 0.
5. The cost of production is given: Let $f(x)$ be the cost, per unit time, of producing x units of product per unit time; $f(x)/x$ is then the average cost, and $f'(x)$ is the marginal cost which is increasing.
6. The cost of storage is α per unit product and per unit time.

The known production schedule $X(t) = h_0 +$ cumulative output up to time t. This ensures continuity for $X(t)$ but not for $X'(t)$. No restrictions on $X(t)$ other than the above are imposed. The assumptions are that the sales function is any piecewise continuous function and the marginal cost function is monotone increasing and differentiable. The problem then becomes

1. $X(0) = h_0$.
2. $X(T) = S(T)$.
3. $X(t) \geq S(t)$ for $0 \leq t \leq T$.
4. $X(t)$ is continuous and nondecreasing and has a piecewise continuous derivative.
5. $C = \int_0^T [\alpha(X - S) + f(X')]\, dt$, where the first term of the integrand is storage cost per unit at time t and the second term is the production cost per unit time at time t.

Determine $X(t)$ which minimizes the total cost subject to conditions 1 to 4.

Solution: The solution is derived from the conditions for the minimization of an integral of the form

$$\int_{x_0}^{x_1} f(x,y,y')\, dx$$

which are as follows:

1. Euler's equation $\frac{\partial f}{\partial y} - \frac{d}{dx}\frac{\partial f}{\partial y'} = 0$ must be satisfied as a necessary condition. In this case it is given by $X''f''(X') = \alpha$, which may also be written as

$$\frac{d}{dt}[f'(X') - \alpha t] = 0 \qquad (5\text{-}178)$$

Thus $X(t)$ must be chosen with a rate of production X' determined in such a manner that the marginal cost of production equals αt plus a constant; the constant is to satisfy the given conditions.

2. The corner condition must be satisfied. Since production is a discrete process, production curves with corners can occur. At these corners X' does not exist, from which it follows that Euler's equation does not apply. At any corner point (x_0, y_0) of the solution curve, modification of Euler's equation gives

$$f(x_0, y_0, y'_{+0}) - f(x_0, y_0, y'_{-0}) = (y'_{+0} - y'_{-0})\frac{\partial f(x_0, y_0, y'_{-0})}{\partial y'} \qquad (5\text{-}179)$$

which for this problem is

$$f(X'_{+0}) - f(X'_{-0}) = (X'_{+0} - X'_{-0})f'(X'_{-0}) \qquad (5\text{-}180)$$

where the left-hand derivative X'_{-0} and the right-hand derivative X'_{+0} exist.

3. There is one more complication. Recall how necessary conditions for an extremum of a function at an interior point (viz., vanishing of the partial derivatives) must be changed to deal with a boundary point (viz., derivatives ≥ 0 or ≤ 0). Similarly, the constraint $X(t) \geq S(t)$ determines a "region of functions" within which one seeks a function yielding a minimum for the integral C. Euler's equation [which is $X''f''(X') - \alpha = 0$] is the necessary condition for the extremum at an "interior function," but it must be changed to an inequality

$$X''f''(X') - \alpha \geq 0 \qquad (5\text{-}181)$$

"along the boundary," i.e., along any interval of t values for which $X(t) = S(t)$. On such an interval, then, the above condition may also be written as $S''f''(S') \geq \alpha$.

To summarize, then, after $X(t)$ is determined it is made (1) tangent to S at two points or (2) tangent to S and passing through one of the extreme points $(0, h_0)$ and $[T, S(T)]$ or (3) to pass through both extreme points. The last condition ensures that the solution curve coincides with $S(t)$ for those values of t on which it cannot be determined as above.

Example 3: By analogy with the "special case" in Sec. 5-14, if given that

$$f(t) \geq 0 \qquad (5\text{-}182)$$

OPTIMIZATION

integrable in the sense of Lebesgue, and it is required to find a measurable function $x(t)$ which maximizes

$$\int f(t)x(t)\,dt \tag{5-183}$$

subject to the conditions

$$\int x(t)\,dt = 1 \qquad 0 \leq x(t) \leq A < \infty \quad (A \text{ is a constant}) \tag{5-184}$$

then the following theorem may be applied:

Theorem 5-14: If $x_0(t)$ maximizes (5-183), then there exists a real number λ such that, almost everywhere,

$$f'(t) \begin{cases} \geq \lambda & \text{if } x_0(t) = A \\ = \lambda & \text{if } 0 < x_0(t) < A \\ \leq \lambda & \text{if } x_0(t) = 0 \end{cases} \tag{5-185}$$

We wish to remark in ending this chapter that it is possible to provide solutions to optimization problems by purely geometric methods. Here we point out the use of the conceptual approach to a problem. We give a sketch of Steiner's famous proof of the isoperimetric problem in the plane, which, of course, can be solved using the calculus of variations. This problem states that among all simple closed curves C, the circle encloses the largest area [9]. (See also Steiner's collected works, Berlin 1881–1882.)

Steiner first shows that a solution to the problem exists by proving that for a given perimeter, of all even-sided polygons, a regular even-sided polygon has the maximum area. He then shows that all the vertices of this regular polygon lie on a circle. By allowing the number of vertices of the polygon to increase indefinitely, the polygon tends to a circle in the limit. Thus the problem has a solution.

Now, to see how one obtains a circle from an arbitrary simple closed curve C, he first shows that the region enclosed by C is convex. If it is concave anywhere, it is possible to reflect the portion of the perimeter where it is concave so that one has convexity and, in addition, a larger area. Then C is divided into two arcs of equal length, and a straight line is passed through the division points. That half which contains the larger area is taken. It is then shown that in order to contain the maximum area it must be a semicircle (by reflection the other half may be obtained). This is demonstrated by selecting a point on the half perimeter under consideration and joining it to the end points which have a line connecting them. This gives a triangle. The angle subtended at the point is then increased or decreased, using the point as a hinge, as is required to produce a right angle. This gives another (a right-angle) triangle whose area can easily be seen to be greater than that of the first triangle. This increases the area under the curve. Since the point chosen was arbitrary, and since a right triangle is subtended at any point of the circumference of a circle by a diameter, one can obtain a right triangle at any point of the perimeter. This yields a semicircle. By reflection one has the entire circle. Note that if the circumference of the circle is c then the area A of any figure whose perimeter is C satisfies the inequality $A \leq \pi \left(\dfrac{c}{2\pi}\right)^2$.

REFERENCES

1. Agmon, S.: The Relaxation Method for Linear Inequalities, *Canadian Journal of Mathematics*, vol. 6, pp. 382–392, 1954.
2. Bellman, R.: Some Applications of the Theory of Dynamic Programming—A Review, *Journal of the Operations Research Society of America*, vol. 2, no. 3, August, 1954, as well as various other publications on dynamic programming.
3. Bliss, G. A.: "Lectures on the Calculus of Variation," University of Chicago Press, Chicago, 1946.
4. Blumenthal, L. M.: Two Existence Theorems for Systems of Linear Inequalities, *Pacific Journal of Mathematics*, vol. 2, pp. 523–530, 1952.
5. Booth, A. D.: "Numerical Methods," Academic Press, Inc., New York, 1955.
6. Cauchy, A. L.: Méthode générale pour la resolution des systèmes d'équations simultanée, *Comptes rendus de l'académie des sciences de Paris*, vol. 25, pp. 536–538, 1847.
7. Charnes, A., and C. E. Lemke: Minimization of Non-linear Separable Convex Functionals, *Naval Research Logistic Quarterly*, vol. 1, no. 4, December, 1954.
8. Courant, R., and D. Hilbert: "Methods of Mathematical Physics," vol. 1, Interscience Publishers, Inc., New York, 1953.
9. Courant, R., and H. Robbins: "What Is Mathematics?" Oxford University Press, London, 1941.
10. Danskin, J.: Various mimeographed notes, Operations Evaluations Group, Washington, D.C., 1953.
11. Fenchel, W.: Convex Cones, Sets and Functions (Lecture notes by D. W. Blackett), *Naval Research Logistics Project Report*, Princeton University, Department of Mathematics, 1953.
12. Fite, W. L.: Maximization of Return from Limited Resources, *Journal of the Society for Industrial and Applied Mathematics*, vol. 1, no. 2, December, 1953.
13. Fleming, W. J.: Relative Maxima in Variational Problems with Inequality Constraints, *U.S. Air Force Project RAND Report RM* 1529, May, 1955.
14. Forsythe, G. E.: Solving Linear Algebraic Equations Can Be Interesting, *Bulletin of the American Mathematical Society*, no. 2, pp. 299–329, July, 1953.
15. Forsythe, G. E.: "Theory of Selected Methods of Finite Matrix Inversion and Decomposition," National Bureau of Standards, Los Angeles, Calif.
16. Fox, L.: "An Introduction to the Calculus of Variations," Oxford University Press, London, 1950.
17. Gaddum, J. W.: A Theorem on Convex Cones with Applications to Linear Inequalities, *Proceedings of the American Mathematical Society*, vol. 3, no. 6, pp. 957–960, December, 1952.
18. Gale, D.: Mathematics and Economic Models, *American Scientist*, pp. 33–44, January, 1956.
19. Gale, D.: The Basic Theorems of Real Linear Equations, Linear Programming and Game Theory, *Naval Research Logistics Quarterly*, vol. 3, no. 3, p. 193, September, 1956.
20. Goldman, A. J., and A. W. Tucker: Theory of Linear Programming, in H. W. Kuhn and A. W. Tucker (eds.), "Linear Inequalities and Related Systems," Princeton University Press, Princeton, N.J., 1956.

21. Green, J. W.: Recent Applications of Convex Functions, *American Mathematical Monthly*, September, 1954.
22. Hadamard, J.: "Leçons sur le calcul des variations," Hermann & Cie, Paris, 1910.
23. Hildebrand, F. B.: "Methods of Applied Mathematics," Prentice-Hall, Inc., Englewood Cliffs, N.J., 1952.
24. Hildebrand, F. B.: "Introduction to Numerical Analysis," McGraw-Hill Book Company, Inc., New York, 1956.
25. Hitch, C.: Sub-optimization in Operations Research: *Journal of the Operations Research Society of America*, vol. 1, no. 3, p. 87, May, 1953.
26. Hitch, C., and R. McKean: Sub-optimization in Operations Problems, in J. F. McCloskey and F. N. Trefethen (eds.), "Operations Research for Management," Johns Hopkins Press, Baltimore, 1954.
27. Hoffman, A. J.: On Approximate Solutions of Systems of Linear Inequalities, *Journal of Research of the National Bureau of Standards*, vol. 49, pp. 263–265, 1952.
28. Isaacs, R.: Optimal Horse Race Bets, *American Mathematical Monthly*, pp. 310–315, May, 1953.
29. Kimball, W. S.: "Calculus of Variations," Butterworth & Co. (Publishers) Ltd., London, 1952.
30. Klein, B.: Direct Use of Extremal Principles in Solving Certain Optimizing Problems Involving Inequalities, *Journal of the Operations Research Society of America*, vol. 3, no. 2, pp. 168–175, May, 1955.
31. Koopman, B. O.: The Optimum Distribution of Effort, *Journal of the Operations Research Society of America*, vol. 1, 1953.
32. Kose, T.: Solutions of Saddle Value Problems by Differential Equations, *Econometrica*, vol. 24, no. 1, January, 1956.
33. Kuhn, H. W.: Solvability and Consistency for Linear Equations and Inequalities, *American Mathematical Monthly*, pp. 217–232, April, 1956.
34. Kuhn, H. W., and A. W. Tucker: Non-linear Programming, in J. Neyman (ed.), "Second Berkeley Symposium on Mathematical Statistics and Probability," pp. 481–492, University of California Press, Berkeley, Calif., 1951.
35. Kuhn, H. W., and A. W. Tucker (eds.): Linear Inequalities and Related Systems, *Annals of Mathematical Studies*, no. 38, Princeton University Press, Princeton, N.J., 1956.
36. Kuhn, H. W., and A. W. Tucker (eds.): Contributions to the Theory of Games, vol. 1, *Annals of Mathematical Studies*, no. 24, Princeton University Press, Princeton, N.J., 1950.
37. Kuhn, H. W., and A. W. Tucker (eds.): Contributions to the Theory of Games, vol. 2, *Annals of Mathematical Studies*, no. 28, Princeton University Press, Princeton, N.J., 1953.
38. Landau, E.: "Gründlagen der Analysis," Chelsea Publishing Company, New York, 1948.
39. Marcus, M.: On the Optimum Gradient Method for Systems of Linear Equations, *Proceedings of the American Mathematical Society*, vol., 7, no. 1, p. 77, February, 1956.
40. Margenau, H., and G. Murphy: "The Mathematics of Physics and Chemistry," D. Van Nostrand Company, Inc., Princeton, N.J., 1943.
41. Marshall, A. W.: A Mathematical Note on Sub-optimization, *Journal of the Operations Research Society of America*, vol. 1, no. 3, p. 100, May, 1953.
42. Miles, E. J.: Notes on College Algebra, Yale University, New Haven, Conn., 1951.
43. Modigliani, F., and F. Hohn: Production Planning, Overtime and the Nature of the Expectation and Planning Horizon, *Econometrica*, vol. 23, p. 46, 1955.

44. Morin, F.: Note on An Inventory Problem, *Econometrica*, vol. 23, p. 447, 1955.
45. Morse, M.: "The Calculus of Variations in the Large," American Mathematical Society Colloquium Publications, vol. 18, American Mathematical Society, New York, 1934.
46. Morse, P. M., and H. Feshbach: "Methods of Theoretical Physics," McGraw-Hill Book Company, Inc., New York, 1953.
47. Motzkin, T. S., H. Raiffa, G. L. Thompson, and R. M. Thrall; The Double Description Method, *Annals of Mathematical Studies*, vol. 28, pp. 51–73, Princeton University Press, Princeton, N.J., 1956.
48. Motzkin, T. S.: Beitrage zur Theorie der Linearen Ungleichungen, Inaugural Dissertation, University of Basel, Jerusalem, 1936.
49. Motzkin, T. S.: The Theory of Linear Inequalities, Doctoral Dissertation, University of Basel, 1933; also *U.S. Air Force Project RAND Publication* T-22, March, 1952.
50. Motzkin, T. S., and I. J. Schoenberg: The Relaxation Method for Linear Inequalities, *Canadian Journal of Mathematics*, vol. 6, pp. 393–404, 1954.
51. Motzkin, T. S.: Two Consequences of the Transposition Theorem on Linear Inequalities, *Econometrica*, vol. 19, p. 184, 1951.
52. O'Brien, G. G.: Mimeographed notes on Linear Programming, Melpar, Inc., Alexandria, Va., 1953.
53. Orden, A.: Solution of Systems of Linear Inequalities on a Digital Computer, *Proceedings of the Association for Computing Machinery*, pp. 91–95, May, 1952.
54. Pollack, S.: The Double Description Method on the SEAC, *National Bureau of Standards Report* 2961, Dec. 9, 1954.
55. Project SCOOP, Symposium on Linear Inequalities and Programming. Planning Research Division, Director of Management and Analysis Service, Comptroller, Hq., U.S. Air Force, and National Bureau of Standards, Washington, D.C., June 14–16, 1951.
56. Rietz, H. L., and A. R. Crathorne: "Introductory College Algebra," Henry Holt and Company, Inc., New York, 1933.
57. Schlauch, H. M.: Mixed Systems of Linear Equations and Inequalities, *American Mathematical Monthly*, vol. 39, p. 218, 1932.
58. Slade, J. J.: Some Observations on Formal Models for Programming, *Transactions of the American Society of Mechanical Engineers*, vol. 78, pp. 47–53, January, 1956.
59. Tucker, A. W.: Linear and Nonlinear Programming, *Operations Research*, vol. 5, no. 2, p. 244, April, 1957.
60. Vazsonyi, A.: Optimizing a Function of Additively Separated Variables Subject to a Simple Restriction, *Proceedings of the Second Symposium on Linear Programming*, vol. 2, p. 453, sponsored by Directorate of Management Analysis, DCS Comptroller, Hq., U.S. Air Force, and National Bureau of Standards, Washington, D.C., Jan. 27–29, 1955.
61. Weinstock, R. P.: "Calculus of Variations," McGraw-Hill Book Company, Inc., New York, 1952.
62. Willers, A.: "Practical Analysis," Dover Publications, New York, 1948.
63. Camp, G. D.: Inequality-constrained Stationary-value Problem, *Journal of the Operations Research Society of America*, vol. 3, no. 4, p. 548, November, 1955.
64. Bodewig, E.: "Matrix Calculus," North-Holl and Publishing Co., Amsterdam, Interscience Publishers, Inc., New York, 1956.
65. Hardy, G. H., J. E. Littlewood, and G. Polya: "Inequalities," Cambridge University Press, New York, 1934.

CHAPTER 6

LINEAR AND QUADRATIC PROGRAMMING

6-1. Introduction

Programming is the planning of economic activities for the sake of optimization. The latter may involve, for example, minimizing costs or maximizing returns. When linear input-output relationships are assumed, together with a linear objective—e.g. profit or cost—function, optimization involves solving a linear-programming problem. The assumptions of linearity for the constraint sets are frequently justifiable in practice. In fact, the successful application of linear programming depends on such reasonable assumptions.

It often happens that when the objective function is assumed to be quadratic, a better description is obtained. In this chapter solutions of problems with linear constraints and both linear and quadratic objective functions are studied and illustrated and applications given.

Here is a simple illustration of linear programming: A thrifty patient of limited means was advised by his doctor to increase the consumption of liver and frankfurters in his diet. (What a horrible diet!) In each meal he must get no less than 200 calories from this combination and no more than 14 units of fat. When he consulted his diet book he found the following information: There are 150 calories in a pound of liver and 200 calories in a pound of frankfurters. However, there are 14 units of fat in a pound of liver and 4 units in a pound of frankfurters.

So he reasoned as follows: I must eat an amount x_1 of liver and an amount x_2 of frankfurters, paying the least amount of money (the price is 30 cents per pound of liver and 50 cents per pound of frankfurters) and meeting the requirements set by the doctor. There are $150x_1$ calories in x_1 pounds of liver and $200x_2$ in a pound of frankfurters; the total amount of calories obtained, that is, $150x_1 + 200x_2$, must be no less than 200, and so one writes

$$150x_1 + 200x_2 \geq 200$$

Similarly, the total fat should be no greater than 14 units:

$$14x_1 + 4x_2 \leq 14$$

and the total cost $30x_1 + 50x_2$ must be a minimum. He is faced with minimizing the total cost subject to the medical constraints (hence the term *constrained optimization*). He must eat something but just enough to meet the requirements and at minimum cost.

This problem can be solved in several ways. For a more complicated problem with many more dietary items which the patient must eat and with further restrictions such as the amount of vitamins obtained, one uses the well-known simplex process for its solution. For the simple

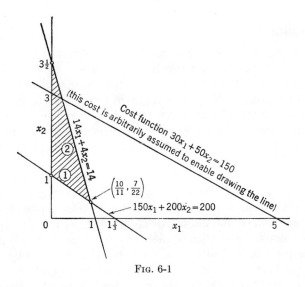

Fig. 6-1

problem given above, one can plot a picture and find the desired solution. Now one must find the food quantities x_1 and x_2 to satisfy the constraining inequalities and yield the minimum cost.

To plot the inequalities one simply plots a straight line, using only the equality sign, as in Fig. 6-1. Then one notices that the first inequality is satisfied by all the points to the right of the line and the second inequality by the points to the left of the line. Thus the points common to both inequalities must satisfy both at the same time. Now the cost expression cannot be drawn in any fixed position. But one gives it any value he desires at first. Then he notices that the position it should have in order to have the smallest value, i.e., to be nearest to being zero, is the point $(10/11, 7/22)$. Consequently, this value, which is roughly one pound of liver and a third of a pound of frankfurters, satisfies the requirements and has a minimum cost. The shaded region in Fig. 6-1 has the points common to both inequalities. The cost function can be moved back and forth in a parallel direction. For a solution it must touch the shaded region. In fact, it must touch it at a point where it is nearest to the origin O. This point is $x_1 = {}^{10}\!/_{11}$, $x_2 = {}^{7}\!/_{22}$, as indicated by the arrow.

6-2. Historical Background

George Stigler, in 1945, studied the problem of determining adequate diet at minimum cost with 77 different foods in terms of nine nutrients contained in them. He found that, for a minimum of $39.93 for the year 1939, optimum diet consisted of wheat flour, cabbage, and dried navy beans. With 1944 prices, navy beans were eliminated in the solution and pork liver added, at a cost of $59.88 for the year.

Hitchcock in 1941 and Koopmans in 1947 independently formulated the transportation problem which is given below. The Air Force's Project SCOOP conducted organized research in the field of linear programming. The outstanding contribution of this project was the simplex method developed by George Dantzig [22] and its application to the solution of large linear-programming problems on the UNIVAC and the SEAC. Sufficient interest was stimulated in the field so that activity in research and applications has become widespread. The useful simplicity of these ideas has undoubtedly contributed to the outcome.

6-3. Statement of the Problem

The objective of each problem must be stated. In the example given in Sec. 6-1, it was minimizing costs. In another situation, one may be interested in maximizing returns. Both these objectives are of frequent occurrence, and they may be stated first verbally, to clarify the ideas, and then mathematically.

Verbal Statement

Assume that prescribed amounts of inputs of materials or efforts are available. Given the amount of output per unit input and the return of a unit of output, the problem is how to use a combination of inputs to produce outputs with maximum returns. It is assumed that the total quantities of each resource used and the total return are linear homogeneous functions of the various commodities produced.

Mathematical Statement

This problem may be stated mathematically as follows: Find values $(x_1, \ldots, x_n)$ which maximize the linear form

$$x_1 c_1 + \cdots + x_n c_n \qquad (6\text{-}1)$$

subject to the conditions

$$x_j \geq 0 \; (j = 1, \ldots, n) \qquad (6\text{-}2)$$

and
$$\sum_{j=1}^{n} a_{ij}x_j \leq b_i \quad i = 1, \ldots, m \quad (6\text{-}3)$$

where the a_{ij}, b_i, c_j are constants.

In matrix notation the linear-programming problem may be stated as follows (vector inequalities are equivalent to component inequalities):

Find $\mathbf{x}$ to maximize $\mathbf{c'x}$ subject to $\mathbf{x} \geq 0$ and $\mathbf{Ax} \leq \mathbf{b}$, where, if

$$\mathbf{c} = \begin{bmatrix} c_1 \\ \cdots \\ c_n \end{bmatrix} \quad \text{then} \quad \mathbf{c'} = [c_1, \ldots, c_n] \quad (6\text{-}4)$$

is its transpose, and

$$\mathbf{b} = \begin{bmatrix} b_1 \\ \cdots \\ b_n \end{bmatrix} \quad (6\text{-}5)$$

$$\mathbf{A} = \begin{bmatrix} a_{11}a_{12} & \cdots & a_{1n} \\ a_{21}a_{22} & \cdots & a_{2n} \\ \cdots & \cdots & \cdots \\ a_{m1}a_{m2} & \cdots & a_{mn} \end{bmatrix} \quad (6\text{-}6)$$

Note that maximizing $\mathbf{c'x}$ is equivalent to minimizing $-\mathbf{c'x}$, and vice versa.

6-4. Applications

By examining F. V. Rhode's paper [8], the interested reader may obtain from this bibliography the desired references for the applications given here. Only some of the more interesting applications will be explained. While there is only one abstract linear-programming problem, there are several types of general formulations of the problem to which the different applications may be considered related. No effort will be made to categorize all these applications. However, of the several main types of problems, a discussion of diet, transportation, scheduling and inventory, and smoothing problems will be given. Several other applications are also mentioned. The reader may wish to return to this section after reading the rest of this chapter.

Diet Problem

Given the information in Table 6-1 concerning the foods under consideration (slightly more general than the introductory example)—for example, there are four units of vitamin A per unit of butter, and one unit of butter costs 9 cents—it is desired to determine a minimum-cost diet which provides at least five units of vitamin A and four units of vitamin B. Mathematically, the problem becomes:

LINEAR AND QUADRATIC PROGRAMMING 169

Find values of x_1, x_2, x_3 which minimize the linear form

$$2x_1 + 9x_2 + x_3 \qquad (6\text{-}7)$$

subject to the constraints

$$x_i \geq 0$$
$$x_1 + 4x_2 + 2x_3 \geq 5 \qquad (6\text{-}8)$$
$$3x_1 + x_2 + 2x_3 \geq 4$$

Thus the x_i, $i = 1, 2, 3$, are the (nonnegative) quantities of each food to be eaten to yield minimum cost and satisfy the vitamin conditions.

TABLE 6-1

Vitamins	Item			
	x_1 Bread	x_2 Butter	x_3 Milk	
A	1	4	2	≥ 5
B	3	1	2	≥ 4
Cost	2	9	1	= min
Requirements	Vitamin A ≥ 5 Vitamin B ≥ 4			

Transportation Problem [4, 45]

Suppose that a homogeneous product is to be shipped from m origins to n destinations, each origin furnishing a stated amount of the item and each destination requiring a stated amount, so that the total supply and total demand are equal. Let a_i be the amount available for shipment from origin i $(i = 1, \ldots, m)$ and let b_j be the amount required at destination j $(j = 1, \ldots, n)$; these are known quantities. Then

$$\sum_{i=1}^{m} a_i = \sum_{j=1}^{n} b_j \qquad (6\text{-}9)$$

with $a_i, b_j \geq 0$ for all i and j.

Let x_{ij} be the unknown quantity to be shipped from the ith origin to the jth destination. Then

$$\sum_{j=1}^{n} x_{ij} = a_i \qquad i = 1, \ldots, m \qquad (6\text{-}10)$$

$$\sum_{i=1}^{m} x_{ij} = b_j \qquad j = 1, \ldots, n \qquad (6\text{-}11)$$

$$x_{ij} \geq 0 \qquad \text{for all } i \text{ and } j \qquad (6\text{-}12)$$

Let c_{ij} be the cost of shipping a unit amount of the item from the ith origin to the jth destination. These values are also given. The problem is to find x_{ij} satisfying the above constraints which will minimize

$$\sum_{j=1}^{n} \sum_{i=1}^{m} c_{ij} x_{ij} \qquad (6\text{-}13)$$

Balancing of Production and Inventories [53]

A company desires to schedule production according to a reliable sales forecast. The plant supervisor's interest is to stabilize his labor force and let inventories supply the slack, to which the treasurer objects since investing capital in inventories is not a very profitable enterprise. It is necessary to study the problem with a view to minimizing inventory costs and meeting sales forecasts. If x_i, y_i are the number of units produced in regular shifts in the ith period and the number of units produced in overtime in this period, respectively; s_i, the sales forecast for the period; m_i, n_i, the maximum number produced in a regular shift and in an overtime shift in the ith period, respectively; P, the marginal cost of production in overtime; R, the inventory charge per unit per period; k, the number of periods; I_i, the inventory at the end of the ith period; and C_i, the accumulated inventory charge at the end of the ith period, then linear programming gives

$$I_i = I_{i-1} + x_i + y_i - s_i \qquad (6\text{-}14)$$

$$C_k = P \sum_{i=1}^{k} y_i + R \sum_{i=1}^{k} I_i \qquad (6\text{-}15)$$

Substituting from Eq. (6-14) into (6-15) and simplifying yield

$$C_k = kRx_1 + (k-1)Rx_2 + \cdots + Rx_k + (P + kR)y_1 \\ + [P + (k-1)R]y_2 + \cdots + (P + R)y_k + L \qquad (6\text{-}16)$$

where L is a constant involving sales forecasts.

It is desired to find x_i and y_i which minimize (6-16) subject to the constraints

$$\begin{aligned} x_i &\leq m_i \\ y_i &\leq n_i \\ I_i &\geq 0 \end{aligned} \qquad (6\text{-}17)$$

It is possible to cut costs to 44 per cent of the previous schedule cost by this method and to recommend spreading vacations over an extended period of three months instead of closing the plant down for two weeks, as had been previously done.

Smooth Patterns of Production [47]

Let the constants $r_i > 0$ $(i = 1, \ldots, n)$ be the shipping requirements for various months. Let the variables $x_i \geq 0$ $(i = 1, \ldots, n)$ represent production in the various months. Let

$$R_t = \sum_{i=1}^{t} r_i \qquad (6\text{-}18)$$

and

$$X_t = \sum_{i=1}^{t} x_i \qquad t = 1, \ldots, n \qquad (6\text{-}19)$$

The shipping requirement is given as

$$X_t \geq R_t \qquad t = 1, \ldots, n \qquad (6\text{-}20)$$

Let s_t be the part of total production at the end of month t that has not been shipped and thus must be stored. Then

$$X_t - R_t = s_t \qquad t = 1, \ldots, n \qquad (6\text{-}21)$$

The problem is then to find $x_i \geq 0$, $s_i \geq 0$ $(i = 1, \ldots, n)$ satisfying the last expression which minimizes the cost function

$$\Sigma s_t + \lambda \Sigma (x_t - x_{t-1})_+ \qquad (6\text{-}22)$$

where $(x_t - x_{t-1})_+ = x_t - x_{t-1}$ when the latter quantity is positive and equals 0 otherwise and where one unit of increased production costs the same as λ units of storage.

The examples just given are essentially the four main different forms of the linear-programming problem of interest here. Further applications are given below which illustrate one or another of these forms.

Personnel-assignment Problem [72, 90]

Given m persons, n jobs, and the expected productivity c_{ij} of the ith person on the jth job, find an assignment of persons

$$x_{ij} \geq 0 \qquad \text{for all } i \text{ and } j$$

to the n jobs such that the average productivity of the persons assigned is a maximum, subject to

$$\sum_{j=1}^{n} x_{ij} \leq a_i$$
$$\sum_{i=1}^{m} x_{ij} \leq b_j \qquad (6\text{-}23)$$

where a_i is the number of persons in personnel category i and b_j is the number of jobs in personnel category j.

The Caterer Problem [50]

A caterer knows that, in connection with the meals he has arranged to serve during the next n days, he will need $r_j \geq 0$ fresh napkins on the jth day, $j = 1, 2, \ldots, n$. Laundering normally takes p days; that is, a soiled napkin sent for laundering immediately after use on the jth day is returned in time to be used again on the $(j + p)$th day. However, the laundry also has a higher-cost service which returns the napkins in $q < p$ days (p and q are integers). Having no usable napkins on hand or in the laundry, the caterer will meet his immediate needs by purchasing napkins at a cents each. Laundering costs b and c cents per napkin for the normal and high-cost service, respectively. How does he arrange matters to meet his needs and minimize his outlay for the n days?

Let x_j represent the napkins purchased for use on the jth day; the remaining requirements, if any, are supplied by laundered napkins. Of the r_j napkins which have been used on that day plus any other soiled napkins on hand, let y_j be the number sent for laundering under normal service and z_j the number under the rapid service. Finally, since soiled napkins need not be sent to the laundry immediately after use, let s_j be the stock of soiled napkins on hand after $y_j + z_j$ have been shipped to the laundry; they will be available for laundering on the next day, together with those from the r_{j+1} used on that day. Consequently,

$$y_j + z_j + s_j - s_{j-1} = r_j \tag{6-24}$$

That is, the napkins used on the jth day are equal to the number sent to the laundry plus the change in the stock of soiled napkins.

The returns from the laundry are equal to the amounts shipped for normal service p days earlier plus the amounts shipped for rapid service q days earlier. Together with purchases, these provide for the needs on the same day. This gives

$$x_j + y_{j-p} + z_{j-q} = r_j \tag{6-25}$$

The total cost to be minimized, subject to these constraints, is

$$\sum_j (ax_j + by_j + cz_j) \tag{6-26}$$

where $a > b$, $c > b$; $x_j, y_j, z_j, s_j \geq 0$.

Linear Programming in Bid Evaluation [33]

Consider a problem with n depots and m separate bidders. Each of the m bidders wishes to produce an amount not exceeding a_i ($i = 1, \ldots, m$). The demands at n depots are b_j ($j = 1, \ldots, n$). It costs an amount c_{ij}

to deliver a unit from the ith bidder to the jth depot. If x_{ij} denotes the quantity purchased from the ith manufacturer for shipment to the jth destination, then the problem is to minimize

$$\sum_{i,j} c_{ij} x_{ij} \qquad (6\text{-}27)$$

subject to the constraints

$$\begin{aligned} x_{ij} &\geq 0 \qquad \text{for all } i \text{ and } j \\ \sum_j x_{ij} &\leq a_i \qquad \text{for each } i \\ \sum_i x_{ij} &= b_j \qquad \text{for each } j \end{aligned} \qquad (6\text{-}28)$$

The problem is solvable by the simplex process to obtain the contract x_{ij}.

When time is introduced in the problem, $x_{ij}{}^k$ denotes the amount shipped in the kth month, $b_j{}^k$ is the requirement at the depot in the month k, and a_i is the total amount the ith bidder wishes to produce. The problem becomes one of minimizing

$$\sum_{i,j} c_{ij} \left(\sum_k x_{ij}{}^k \right) \qquad (6\text{-}29)$$

subject to
$$x_{ij}{}^k \geq 0$$

$$\begin{aligned} \sum_{i=1}^{m} x_{ij}{}^k &= b_j{}^k \qquad \text{for all } j \text{ and } k \\ \sum_{j,k=1}^{n,r} x_{ij}{}^k &\leq a_i \qquad \text{for each } i \end{aligned} \qquad (6\text{-}30)$$

$$\sum_{j=1}^{n} \frac{x_{ij}{}^1}{P_1} = \sum_{j=1}^{n} \frac{x_{ij}{}^2}{P_2} = \cdots = \sum_{j=1}^{n} \frac{x_{ij}{}^r}{P_r} \qquad (6\text{-}31)$$

for each i, assuming r months, where $P_k = T_k/T$ and where T_k is the requirement for all depots in month k and T is the total requirement for all depots.

Optimum Estimation of Executive Compensation by Linear Programming [21]

The object is to determine a consistent plan of executive compensation in an industrial concern. Salary, job ranking, and the amounts of each factor required on the ranked job level are taken into consideration by the constraints of linear programming.

An Application of Linear Programming to Efficiency in Operation of a System of Dams [57]

The problem is to determine variations in water storage of six Missouri River dams which generate power so as to maximize the energy obtained from the entire system. The physical limitations of storage appear as inequalities.

The Trim Problem [75]

A firm manufactures rolls of newsprint to meet customers' specifications with regard to width and diameter. In cutting these customer rolls from larger rolls of paper, trim losses are incurred. An application of linear programming is made to reduce total trim loss incurred by six of the paper machines. Four types of solutions are determined which cover all the roll sizes which are likely to be encountered in practice.

An Application of Linear Programming to the Policy of a Firm [27]

Consider a firm which has access to certain factors of production whose supply, for one reason or another, cannot be increased in the time period in view. These resources limit the opportunities open to the firm. They may be utilized in various ways or not at all, and, depending on what is done with them, the resources, expenses, and profits of the firm will vary. The problem facing management is to find the productive program which will make the profits of the firm as great as possible, subject to the limitation that this program must not require more than the total available supply of any resource. The production problem becomes one of choosing which productive forces to use and the level at which to use each of them. Applications of linear programming to this problem made it possible to state in advance how many different processes had to be used in order to maximize profit. This enabled the selection of an optimum program.

Determination of the Optimum Operations Level in a Textile Manufacturing Factory [74]

This problem deals with the optimum operating level of a factory in which many brands of textile goods are manufactured. Linear programming is used to resolve the conflicting requirements among market prices, manufacturing costs, and maintenance expenses of the factory.

Other Applications

Linear programming has also been applied to problems of gasoline blending, structural design, scheduling of traffic signals, economic analysis of industrial relationships, scheduling of a military tanker fleet, minimizing the number of carriers to meet a fixed schedule, the least ballast shipping required to meet a specified shipping program, the distribution of a product by several properties, scheduling by means of dynamic linear programming, commercial transportation, communications, the economics of the coal industry, cost cutting in business, fabrication scheduling, profit scheduling, aircraft routing, computation of maximum flows in networks, steel-production planning, stocks and flows, the balancing of assembly lines, etc. These applications, although large in number, are inadequate for classifying the types of operations to which linear programming promises applicability. It is evident that transportation problems and scheduling problems are amenable to the techniques of linear programming.

6-5. Geometry of Linear-programming Problem

Geometrically, a solution of the linear-programming problem is a point of the convex set defined by the constraints which also maximizes (minimizes) the objective function. This point is generally a vertex of the

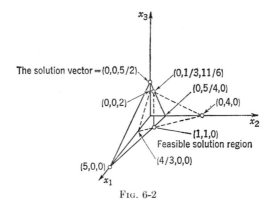

Fig. 6-2

boundary. For simplicity, assume that constraints and objective function are normalized. Then this is the vertex $(x_1^0, x_2^0, \ldots, x_n^0)$ for which the objective function assumes its maximum (or minimum) distance from the origin. H. Weyl [92] has shown that the maximum (minimum) is attained on the boundary and generally at a vertex of the polyhedron.

The diet problem of Sec. 6-4 may be solved geometrically. This technique is possible only for $n \leq 3$ and where m is not very large where n is the number of variables and m is the number of inequalities. Figure 6-2 gives the boundaries of the half spaces defined by the inequalities. The condition $x_i \geq 0$ limits the problem to the positive orthant where all the variables assume nonnegative values. The intersection vertices are also labeled. The objective function is evaluated at each of the intersection vertices of the feasible region, and the vertex yielding the minimum is adopted as the solution of the entire problem.

6-6. The Dual of the Problem [34]

An interesting theorem which occurs in linear programming is that of the presence of a dual problem which is related to the original problem both in formulation and in the existence of a solution. The dual of the original problem (the primal) may be stated as follows: Find $y_i \geq 0$ $(i = 1, \ldots, m)$ such that

$$\sum_{i=1}^{m} a_{ij} y_i \geq c_j \quad j = 1, \ldots, n \qquad (6\text{-}32)$$

$$\sum_{i=1}^{m} b_i y_i = \min \qquad (6\text{-}33)$$

where a_{ij}, b_i, and c_j are the constants of the primal.

A duality theorem previously given, due to Gale, Kuhn, and Tucker, asserts that if either a problem or its dual has a solution then the minimum (maximum) value of the dual is equal to the maximum (minimum) value of the original problem.

Remark: As an application of the dual, it is suggested that the reader dualize the diet problem.

An Interpretation [42] *of an Original Problem*

With a prescribed unit value for each output and a prescribed upper limit for the quantity of each input, how much of each output should be produced in order to maximize the value of the total output?

$$\begin{array}{c} \text{Output } j \geq 0 \\ \sum_j \left(\frac{\text{input } i}{\text{output } j}\right) \text{output } j \leq \text{input } i \\ \sum_j \left(\frac{\text{value}}{\text{output } j}\right) \text{output } j = \text{value} \end{array} \qquad (6\text{-}34)$$

An Interpretation of the Dual

With a prescribed quantity of each input and a prescribed lower limit of unit value for each output, what unit values should be assigned to each input in order to minimize the value of the total input?

$$\frac{\text{value}}{\text{input } i} \geq 0$$

$$\sum_i \left(\frac{\text{input } i}{\text{output } j}\right) \frac{\text{value}}{\text{input } i} \geq \frac{\text{value}}{\text{output } j} \quad (6\text{-}35)$$

$$\sum_i (\text{input } i) \frac{\text{value}}{\text{input } i} = \text{value}$$

Note that "value" differs in meaning in the two problems. The cost coefficients in the dual are referred to as "shadow prices."

Remark: Solving a linear-programming problem is equivalent to solving a set of simultaneous inequalities which include the constraints of the primal and the dual, with the additional inequality which requires the objective function of the primal (or the dual), which is to be minimized, to be greater than or equal to the objective function of the dual (or of the primal) which is to be maximized. This will be better understood on reading Chap. 7, where the relationship between a linear program and a game is illustrated.

6-7. The Simplex Technique

Geometric Description and the Number of Iterations [78]

Briefly, the *simplex process* is one of several iterative procedures used in solving a linear-programming problem. Each iteration of this process consists of translating the hyperplane corresponding to the objective function parallel to itself, each time evaluating its distance from the origin at the intersected vertex of the convex set defined by the inequalities. The process is repeated in such a way that the translated hyperplane in each step yields improved results leading to the optimum distance. The simplex process also involves a useful criterion which eliminates some of the vertices of the convex set as possible trial points.

The number of iterations involved is decided for each problem separately. Although experience has shown that the iterations of the simplex process are approximately equal to twice the total number of inequalities, in general, no definite proof has been given of this fact for all dimensions. However, it is worth noting that the following upper bounds for the number of vertices of a polyhedron in n dimensional space are available:

Let F_i ($i = 0, 1, \ldots, n - 1$) denote the number of ith dimensional faces of an n-dimensional polyhedron. A crude estimate which does not consider the fact that a polyhedron is convex but includes all intersections of the planes defined by the inequalities of the problem is given by the combinatorial formula

$$F_0 \le \frac{F_{n-1}!}{(F_{n-1}-n)!n!} \qquad (6\text{-}36)$$

where n is the dimension of the space, F_0 is the number of vertices, and F_{n-1} is the number of $(n-1)$-dimensional hyperplanes.

A finer upper bound which is dominated by the above gives

$$F_0 \le \frac{2}{n}\frac{F_{n-1}!}{(F_{n-1}-n+1)!(n-1)!} \le \frac{F_{n-1}!}{(F_{n-1}-n)!n!} \qquad (6\text{-}37)$$

When $n = 3$, then it has been shown that

$$F_0 \le 2F_2 - 4$$

which, in fact, shows that the number of iterations is no more than twice the number of inequalities minus 4.

When $n = 4$, the result is given by

$$F_0 \le \frac{F_3}{2}(F_3 - 3) \qquad (6\text{-}38)$$

When n is even, the general result is given by

$$F_0 \le \frac{2}{n-2}\left\{\sum_{p=1}^{(n-2)/2}\binom{F_{n-1}}{n-2p} - F_{n-1} - 2^n + \left[2 + \frac{n(n+3)}{2}\right]\right\} \qquad (6\text{-}39)$$

When n is odd, the general result is given by

$$F_0 \le \frac{2}{n-2}\left\{\sum_{p=1}^{(n-3)/2}\binom{F_{n-1}}{n-2p} + F_{n-1} - 2^n + \frac{(n-1)n}{2}\right\}$$

$$F_0 \le \frac{2}{n-2}\left\{\sum_{p=1}^{(n-3)/2}\binom{F_{n-1}}{n-2p} - \frac{n-2}{2}F_{n-1} - 2^n + (n-1)n\right\} \qquad (6\text{-}40)$$

Algebraic Procedure [4, 22]

Let

$$\sum_{j=1}^{n} a_{ij}x_j \ge b_i \qquad i = 1, \ldots, m \qquad (6\text{-}41)$$

be a set of linear inequalities. The linear inequality can be replaced by a linear equality of nonnegative variables by writing instead

$$\sum_{j=1}^{n} a_{ij}(x_{j1} - x_{j2}) - y_i = b_i \qquad y_i \geq 0 \qquad x_{j1} \geq 0$$

$$x_{j2} \geq 0 \qquad i = 1, \ldots, m \qquad (6\text{-}42)$$

Note that no assumptions are made on x_j being ≥ 0; but the x_{j1} and x_{j2} are nonnegative.

The basic problem then, without loss of generality, becomes one of finding $(x_1, \ldots, x_n)$ which maximize the linear form

$$x_1 c_1 + \cdots + x_n c_n \qquad (6\text{-}43)$$

subject to the conditions that

$$x_j \geq 0 \qquad j = 1, \ldots, n$$

and

$$\sum_{j=1}^{n} a_{ij} x_j = b_i \qquad i = 1, \ldots, m \qquad (6\text{-}44)$$

where a_{ij}, b_i, c_j are constants $(i = 1, \ldots, m; j = 1, \ldots, n)$.

Discussion: Consider the augmented matrix

$$\begin{bmatrix} a_{11} a_{12} & \cdots & a_{1n} b_1 \\ a_{21} a_{22} & \cdots & a_{2n} b_2 \\ \cdots & \cdots & \cdots \\ a_{m1} a_{m2} & \cdots & a_{mn} b_m \end{bmatrix} = [\mathbf{P}_1, \ldots, \mathbf{P}_n; \mathbf{P}_0] \qquad (6\text{-}45)$$

Each column may be viewed as representing the coordinates of a point in Euclidean R_m space. Let $\mathbf{P}_j$ denote the jth column with the exception of $\mathbf{P}_0$, which is to denote the last column.

The problem now is to find $x_j \geq 0$ which satisfy

$$\begin{aligned} x_1 \mathbf{P}_1 + x_2 \mathbf{P}_2 + \cdots + x_n \mathbf{P}_n &= \mathbf{P}_0 \\ x_1 c_1 + x_2 c_2 + \cdots + x_n c_n &= \max \end{aligned} \qquad (6\text{-}46)$$

Definition: A set x_j which satisfies the constraints without necessarily yielding the maximum to the objective function will be termed a *feasible solution*. One which also maximizes will be called a *maximum feasible solution*.

Remark: The simplex method consists of constructing a feasible solution first and then a maximum feasible solution.

Assumption: The assumption is made that every subset of m points from the set $[\mathbf{P}_1, \ldots, \mathbf{P}_n; \mathbf{P}_0]$ is linearly independent; i.e., the determinant of the submatrix defined by the subset does not vanish. This is called the *nondegenerate case*. A method of handling the degenerate problem using the epsilon technique is given later.

The following two theorems have proofs independent of the above assumption [22].

Theorem 6-1: If one feasible solution exists, then there exists a feasible solution (called a *basic feasible* solution) with, at most, m points P_i with positive weights x_i and $n - m$ or more points P_i with $x_i = 0$.

Theorem 6-2: If the values of the cost function for the class of feasible solutions have a finite upper bound, then there exists a maximum feasible solution which is a basic feasible solution.

The simplex method will be applied to the diet problem given above. The long method of calculation is used here to clarify the ideas. A shorter "tableau" form is used in the quadratic-programming example.

The constraint set for the diet problem is

$$x_1 + 4x_2 + 2x_3 \geq 5$$
$$3x_1 + x_2 + 2x_3 \geq 4 \quad (6\text{-}47)$$
$$x_i \geq 0 \quad i = 1, 2, 3$$

and the form to be minimized is

$$2x_1 + 9x_2 + x_3 \quad (6\text{-}48)$$

Step 1: Change to equalities by introducing slack variables which are nonnegative variables used to reduce the inequalities to equalities.

$$\begin{aligned} x_1 + 4x_2 + 2x_3 - x_4 &= 5 & x_4 &\geq 0 \\ 3x_1 + x_2 + 2x_3 - x_5 &= 4 & x_5 &\geq 0 \\ -2x_1 - 9x_2 - x_3 &= \max & & \end{aligned} \Bigg\} \text{slack variables} \quad (6\text{-}49)$$

Step 2: Write out the matrix

$$[P_1, P_2, P_3, P_4, P_5; P_0] = \begin{bmatrix} 1 & 4 & 2 & -1 & 0 & | & 5 \\ 3 & 1 & 2 & 0 & -1 & | & 4 \end{bmatrix} \quad (6\text{-}50)$$

and write $c_1 = -2 \quad c_2 = -9 \quad c_3 = -1 \quad c_4 = 0 \quad c_5 = 0$

Step 3: Start by selecting a basis, which is a set of vectors which are linearly independent (i.e., the determinant of their matrix does not vanish), and every other vector can be expressed as a linear combination of the basis vectors. A linear combination of vectors P_1 and P_2, for example, may be given as $aP_1 + bP_2$, where a and b are real numbers. In this case, two vectors are needed to form a basis, as can be seen from the matrix.

Remark: Using a set of artificial vectors as a basis avoids an initial choice of nonfeasible vectors, i.e., a set of vectors for which the corresponding variables are negative. Artificial vectors are of the form $(1,0)$ and $(0,1)$.

Suppose that the initial choice is P_1 and P_2 as a basis. Express P_0 as a linear combination of them. Obtain $P_0 = P_1 + P_2$ in this case. In this problem the cost function will be negative for positive x_i. Hence, if a

LINEAR AND QUADRATIC PROGRAMMING

positive cost is obtained for P_0 as a combination of the basis vectors chosen, then at least one basis vector must be changed. Also express the remaining vectors as a linear combination of P_1 and P_2, obtaining

	P_1	P_2
P_1	1	0
P_2	0	1
P_3	$6/11$	$4/11$
P_4	$1/11$	$-3/11$
P_5	$-4/11$	$1/11$

(6-51)

Let β_{ij} be the coefficient in the above array whose first subscript is the same as that of the vector on top and the second subscript the same as that of the vector to its left. For example, to obtain P_4 as a linear combination of P_1 and P_2, write $P_4 = aP_1 + bP_2$. Note that $\beta_{14} = a$, $\beta_{24} = b$. To show how a and b are obtained, the relation among the vectors may be written as

$$\begin{matrix} P_4 & & P_1 & & P_2 \\ \begin{bmatrix} -1 \\ 0 \end{bmatrix} & = a \begin{bmatrix} 1 \\ 3 \end{bmatrix} & + b \begin{bmatrix} 4 \\ 1 \end{bmatrix} & = \begin{bmatrix} a \\ 3a \end{bmatrix} + \begin{bmatrix} 4b \\ b \end{bmatrix} & = \begin{bmatrix} a + 4b \\ 3a + b \end{bmatrix} \end{matrix}$$

(6-52)

or by equating sides: $-1 = a + 4b$, $0 = 3a + b$. On solving these two simultaneous equations in a and b one has $a = 1/11$, $b = -3/11$. Similarly, all other vectors are expressed as linear combinations of the basis P_1 and P_2.

Step 4: Consider $z_j = \beta_{1j}c_1 + \beta_{2j}c_2$ ($j = 1, \ldots, 5$).

Remark: In general, if the basis vectors have subscripts p, q, r, etc., write

$$z_j = \beta_{pj}c_p + \beta_{qj}c_q + \beta_{rj}c_r + \cdots \quad (6\text{-}53)$$

$z_1 = c_1 \qquad\qquad\qquad = -2 \qquad$ Compare with $c_1 = -2$
$z_2 = c_2 \qquad\qquad\qquad = -9 \qquad$ Compare with $c_2 = -9$
$z_3 = \quad 6/11\, c_1 + 4/11\, c_2 = -48/11 \quad$ Compare with $c_3 = -1$
$z_4 = \quad 1/11\, c_1 - 3/11\, c_2 = 25/11 \quad$ Compare with $c_4 = 0$
$z_5 = -4/11\, c_1 + 1/11\, c_2 = -1/11 \quad$ Compare with $c_5 = 0$

Compare z_j with c_j as indicated.

If $z_j \geq c_j$ for all j, the process is finished, that is, $P_0 = P_1 + P_2$, and the cost would be (when maximizing, negative) given as follows: If $P_0 = aP_1 + bP_2$, then the cost is $ac_1 + bc_2$. In this case one has

$$-2 - 9 = -11 \text{ units}$$

i.e., 11 units for the minimum. If one were directly minimizing, then the criterion is $z_j \leq c_j$.

Remark: It is clear that $z_j < c_j$ for some j. Then consider max $(c_j - z_j)$. In this case $c_3 - z_3 = $ max. Hence for a next choice of basis one uses $\mathbf{P}_3$ along with either $\mathbf{P}_1$ or $\mathbf{P}_2$. One decides which of $\mathbf{P}_1$ or $\mathbf{P}_2$ to use by the following method (If $c_j \geq z_j$ for all j and $\beta_{ij} \leq 0$ for all i, the maximum feasible solution is infinite.):

Step 5:
$$\mathbf{P}_0 = \mathbf{P}_1 + \mathbf{P}_2$$

expressing $\mathbf{P}_0$ as a linear combination of $\mathbf{P}_1$ and $\mathbf{P}_2$.

$$\theta \mathbf{P}_3 = \theta \tfrac{6}{11} \mathbf{P}_1 + \theta \tfrac{4}{11} \mathbf{P}_2$$

expressing $\mathbf{P}_3$ as a linear combination of $\mathbf{P}_1$ and $\mathbf{P}_2$ and multiplying by θ. Subtracting the second equation from the first, one has

$$\mathbf{P}_0 = \theta \mathbf{P}_3 + (1 - \tfrac{6}{11}\theta)\mathbf{P}_1 + (1 - \tfrac{4}{11}\theta)\mathbf{P}_2$$

Choose $\quad \theta = \min \left(\dfrac{1}{6/11}, \dfrac{1}{4/11} \right) = \dfrac{11}{6}$

Hence one obtains

$$\mathbf{P}_0 = \tfrac{11}{6} \mathbf{P}_3 + \tfrac{2}{6} \mathbf{P}_2$$

and the new basis will consist of $\mathbf{P}_2$ and $\mathbf{P}_3$. The cost in this case is given by

$$\tfrac{11}{6}c_3 + \tfrac{2}{6}c_2 = \tfrac{11}{6}(-1) + \tfrac{2}{6}(-9) = -\tfrac{29}{6}$$

which is clearly an improvement on the previous cost since one is maximizing.

Remark: In general, if $\mathbf{P}_0 = \alpha_1 \mathbf{P}_1 + \cdots + \alpha_q \mathbf{P}_q$, $\mathbf{P}_1, \ldots, \mathbf{P}_q$ being the basis vectors, and if $c_j - z_j = \max\limits_{j}$ yields $\mathbf{P}_j$ for the new vector and

$$\mathbf{P}_j = \beta_{1j}\mathbf{P}_1 + \cdots + \beta_{qj}\mathbf{P}_q \tag{6-54}$$

then choose $\quad \theta = \min\limits_{i} \left(\dfrac{\alpha_i}{\beta_{ij}} \right) \quad \beta_{ij} > 0 \tag{6-55}$

In this manner one of the vectors in

$$\mathbf{P}_0 = \theta \mathbf{P}_j + (\alpha_1 - \theta\beta_{ij})\mathbf{P}_1 + \cdots + (\alpha_q - \theta\beta_{qj})\mathbf{P}_q \tag{6-56}$$

is eliminated.

Step 6: Again express the remaining vectors as a linear combination of $\mathbf{P}_2$ and $\mathbf{P}_3$:

	$\mathbf{P}_2$	$\mathbf{P}_3$
$\mathbf{P}_1$	$-\tfrac{2}{3}$	$\tfrac{11}{6}$
$\mathbf{P}_2$	1	0
$\mathbf{P}_3$	0	1
$\mathbf{P}_4$	$-\tfrac{1}{3}$	$\tfrac{1}{6}$
$\mathbf{P}_5$	$\tfrac{1}{3}$	$-\tfrac{2}{3}$

(6-57)

Consider now $z_j = \beta_{2j}c_2 + \beta_{3j}c_3$.

$$z_1 = (-\tfrac{2}{3})(-9) + 1\tfrac{1}{6}(-1) = 2\tfrac{5}{6} \quad \text{Compare with } c_1 = -2$$
$$z_2 = c_2 = -9 \quad \text{Compare with } c_2 = -9$$
$$z_3 = c_3 = -1 \quad \text{Compare with } c_3 = -1$$
$$z_4 = (-\tfrac{1}{3})(-9) + (\tfrac{1}{6})(-1) = 1\tfrac{7}{6} \quad \text{Compare with } c_4 = 0$$
$$z_5 = (\tfrac{1}{3})(-9) - (\tfrac{2}{3})(-1) = -\tfrac{7}{3} \quad \text{Compare with } c_5 = 0$$

Here one has $c_5 - z_5 = \max$. Hence $\mathbf{P}_5$ is to replace either $\mathbf{P}_2$ or $\mathbf{P}_3$ in the next choice of basis. To decide which, write

$$\mathbf{P}_0 = 1\tfrac{1}{6}\mathbf{P}_3 + \tfrac{2}{6}\mathbf{P}_2$$
$$\theta\mathbf{P}_5 = -\theta\tfrac{2}{3}\mathbf{P}_3 + \theta\tfrac{1}{3}\mathbf{P}_2$$

Subtracting the second from the first, one has

$$\mathbf{P}_0 = \theta\mathbf{P}_5 + (1\tfrac{1}{6} + \tfrac{2}{3}\theta)\mathbf{P}_3 + (\tfrac{2}{6} - \tfrac{1}{3}\theta)\mathbf{P}_2$$

Thus $\theta = 1$ since one considers only those values of β_{ij} that are greater than zero. This choice gives

$$\mathbf{P}_0 = \mathbf{P}_5 + 1\tfrac{5}{6}\mathbf{P}_3$$

and the cost is given by

$$1(0) + 1\tfrac{5}{6}(-1) = -\tfrac{5}{2}$$

which is greater than the preceding costs.

Step 7: Again, express the remaining vectors as a linear combination of $\mathbf{P}_3$ and $\mathbf{P}_5$:

	$\mathbf{P}_3$	$\mathbf{P}_5$
$\mathbf{P}_1$	½	-2
$\mathbf{P}_2$	2	3
$\mathbf{P}_3$	1	0
$\mathbf{P}_4$	$-\tfrac{1}{2}$	-1
$\mathbf{P}_5$	0	1

(6-58)

Consider once more $z_j = \beta_{3j}c_3 + \beta_{5j}c_5$.

$$z_1 = -\tfrac{1}{2} \quad \text{Compare with } c_1 = -2$$
$$z_2 = -2 \quad \text{Compare with } c_2 = -9$$
$$z_3 = -1 \quad \text{Compare with } c_3 = -1$$
$$z_4 = \tfrac{1}{2} \quad \text{Compare with } c_4 = 0$$
$$z_5 = 0 \quad \text{Compare with } c_5 = 0$$

It is clear that all $z_j \geq c_j$, and the solution is complete. In other words, since $\mathbf{P}_0 = 1\tfrac{5}{6}\mathbf{P}_3 + \mathbf{P}_5$, one uses $1\tfrac{5}{6}$ of the quantities denoted under milk.

The total cost in that case will be $5/2$ (having changed the sign to obtain the minimum). The vector $\mathbf{P}_5$ which has zero cost contributes nothing. There is no item corresponding to it.

Note that the simplex process guarantees no integer values for the solution vector. There is no a priori reason why any vertex of the polyhedron of feasible solutions should be a lattice point, i.e., have integer coordinates. Approximating the coordinate values with integers does not necessarily yield a feasible vector since the vertex yielding a solution may be nearest to a lattice point outside the region of feasible solutions and the nearest feasible lattice point is nowhere near to justify any approximation. Rapid progress is being made to provide methods of solving such problems.

Solution of the Dual

If the basis leading to a maximum solution of the primal has been obtained by the simplex process, then the solution of the dual is obtained as follows:

Let $\mathbf{B}$ be the matrix whose column vectors constitute the basis vectors leading to the solution with inverse $\mathbf{B}^{-1}$ and let $\mathbf{C}^*$ be the cost vector corresponding to the solution basis. Then the solution of the dual is $\mathbf{y} = \mathbf{C}^*\mathbf{B}^{-1}$, where the components of the vector $\mathbf{y}$ comprise the solution $(y_1^0, \ldots, y_m^0)$.

Proof: Note that $z_j = \mathbf{C}^*(\mathbf{B}^{-1}\mathbf{A})$. Since $z_j - c_j \geq 0$, then $\mathbf{C}^*(\mathbf{B}^{-1}\mathbf{A}) - \mathbf{c} \geq 0$, which becomes (on substituting $\mathbf{y} = \mathbf{C}^*\mathbf{B}^{-1}$) $\mathbf{y}\mathbf{A} - \mathbf{c} \geq 0$ or $\mathbf{y}\mathbf{A} \geq \mathbf{c}$, which is the dual problem.

Given a linear-programming problem with three constraints and n variables, it is possible to obtain the simplex solution by forming the dual and solving geometrically. To show this, suppose for simplicity that (y_1, y_2, y_3) are the values constituting the solution. Then $\mathbf{P}_1$, $\mathbf{P}_2$, and $\mathbf{P}_3$ are the basis vectors in the simplex solution of the dual; when this submatrix is taken from the matrix of the problem, inverted, and multiplied by the vector whose components are the cost coefficients corresponding to $\mathbf{P}_1$, $\mathbf{P}_2$, $\mathbf{P}_3$, the solution (x_1, x_2, x_3) of the primal is obtained, that is,

$$(x_1, x_2, x_3) = (\text{cost vector})(\mathbf{B}^{-1})$$

Once the values of x_1, x_2, and x_3 are known, the corresponding vectors from the matrix constitute the basis vectors leading to the solution of the primal. A similar argument applies for the simpler case of two constraints in the primal.

As will be seen later, a linear-programming problem may be reduced to a game problem and solved by methods used in solving games.

Degeneracy [19]

Degeneracy occurs when the same value of θ is obtained for more than one value of i. This can happen if the vectors chosen or those qualifying for a basis are not linearly independent. This difficulty is represented graphically by the fact that more than n vectors pass through the same point which, in turn, is described by the fact that more than m of the original $m + n$ variables are zero. In this case the value of the objective function remains the same, and the new iteration does not improve the situation. This condition might hold for a number of iterations. There is a possibility that one of the preceding bases, in the set which yields the same value of the objective function, might reappear. In this case the simplex procedure is said to have cycled [15, 46], and on successive iterations the same set of bases reappear. Thus the simplex procedure will not terminate. To resolve this difficulty the epsilon technique is applied.

The procedure to handle degeneracy is as follows:

Subtract from the right-hand side of

$$\sum_{j=1}^{n} a_{ij}x_j \leq b_i \qquad i = 1, \ldots, m$$

the polynomial
$$\epsilon^{m-i} + \sum_{j=1}^{n} a_{ij}\epsilon^{m+n-j} \qquad (6\text{-}59)$$

where $\epsilon > 0$ is arbitrarily small and is set equal to zero at the termination of the process. (The perturbation may be more simply effected by only adding ϵ^i.) In any case, the value of θ for the choice of a new vector using this technique would be unique, as can be easily verified. Since the coefficients appear at the same time as coefficients of the nonbasic variables, it is unnecessary to write down the polynomials in epsilon. The operations can be carried out exactly as before and without introducing the polynomials in epsilon but following a simple rule which has the same effect as if these polynomials had been recorded. To do this, suppose that $\mathbf{P}_j$ is the vector to go into the new basis and suppose that the θ criterion yields a tie between $\mathbf{P}_1$ and $\mathbf{P}_2$ of the form $\alpha_1/\beta_{1j} = \alpha_2/\beta_{2j}$, for example; then form the ratios β_{1k}/β_{1j}, β_{2k}/β_{2j} for another vector $\mathbf{P}_k$. If these two ratios are different, select the smaller; then the vector to be eliminated has the same row subscript as the first subscript of the smaller ratio. If equal, compare a similar ratio for another vector, etc., until one qualifies.

The Simplex Process Applied to a Transportation Problem [71]

When the simplex process is applied to a transportation problem with integral coefficients, integer solutions are obtained. This is illustrated in the following example. Let

$$a_1 = \sum_{j=1}^{5} x_{1j} = 24 \qquad b_1 = \sum_{i=1}^{4} x_{i1} = 10$$

$$a_2 = \sum_{j=1}^{5} x_{2j} = 18 \qquad b_2 = \sum_{i=1}^{4} x_{i2} = 20$$

$$a_3 = \sum_{j=1}^{5} x_{3j} = 20 \qquad b_3 = \sum_{i=1}^{4} x_{i3} = 10 \qquad (6\text{-}60)$$

$$a_4 = \sum_{j=1}^{5} x_{4j} = 16 \qquad b_4 = \sum_{i=1}^{4} x_{i4} = 18$$

$$b_5 = \sum_{i=1}^{4} x_{i5} = 20$$

and let the cost matrix be

$$\begin{bmatrix} 4 & 9 & 8 & 10 & 12 \\ 6 & 10 & 3 & 2 & 3 \\ 3 & 2 & 7 & 10 & 3 \\ 3 & 5 & 5 & 4 & 8 \end{bmatrix} \qquad (6\text{-}61)$$

Thus, the cost to ship a unit from the second origin to the first destination is 6, from the fourth origin to the fifth destination is 8, etc. One is faced with the problem of choosing the quantities x_{ij} to be shipped (i.e., choosing the values in the shipping matrix) subject to the constraints, and such that the cost function which now can be written as

$$4x_{11} + 9x_{12} + 8x_{13} + 10x_{14} + 12x_{15} + 6x_{21} + 10x_{22} + 3x_{23} + 2x_{24}$$
$$+ 3x_{25} + 3x_{31} + 2x_{32} + 7x_{33} + 10x_{34} + 3x_{35} + 3x_{41} + 5x_{42}$$
$$+ 5x_{43} + 4x_{44} + 8x_{45} \qquad (6\text{-}62)$$

is a minimum.

The constraints constitute a set of $4 + 5 = 9$ equations in 20 unknowns. Of the equations, $m + n - 1$ are independent (recall that $\sum_{i=1}^{4} a_i = \sum_{j=1}^{5} b_j$). The minimizing solution, therefore, requires, at most, $m + n - 1$ routes of positive shipments.

Here $m + n - 1 = 4 + 5 - 1 = 8$.

To find a feasible solution containing eight or fewer nonzero shipping values is simple. It is done as follows:

Form a table with the proper number of origins and the proper number of destinations. Place the given values of a_i in an additional column on the right and b_j as an additional row on the bottom, as illustrated in Table 6-2.

TABLE 6-2

		Destinations					Totals
		1	2	3	4	5	
Origins	1						$24 = a_1$
	2						$18 = a_2$
	3						$20 = a_3$
	4						$16 = a_4$
	Totals	$10 = b_1$	$20 = b_2$	$10 = b_3$	$18 = b_4$	$20 = b_5$	78

The spaces must be assigned numbers which add up to the totals indicated. To do this, put in space (1,1) the minimum of (a_1, b_1), that is, the minimum of the two values 24 and 10; here the value is 10. Move one space in the direction of the maximum of (a_1, b_1), that is, to position (1,2), the first row and second column. Record in this position the minimum of $(a_1 - 10, b_2)$ with $a_1 = 24$, $b_2 = 20$, which is 14. Next, move one space in the direction of the maximum of $(a_1 - 10, b_2)$, namely, to position (2,2), and record in that position the minimum of

$$(b_2 - 14, a_2) = (20 - 14, 18) = 6$$

Continuing this process will give a feasible solution. It is clear that, at most, $(m + n - 1) = 8$ nonzero shipping values will be required.

While the above method will always give a feasible solution, and hence a starting point for the simplex routine, a better feasible solution can generally be obtained by a little more thought, often greatly reducing the work which follows. This means choosing more than the positions with the lowest shipping costs for the positive shipments; it means choosing the positions whose shipping costs seem, by inspection, to give the "best bargain." To illustrate, consider the first column of the cost matrix; the third and fourth positions are the ones with lowest cost in the column, yet the best choice seems to be the first position. This follows from the fact that, in the third row, there is a better choice than 3, that is, 2, and in the fourth row there is a choice almost as good as 3, that is, 4. However, in the first row if one fails to use the first element which is 4 one must choose between costs exceeding this number by from 4 to 8. With this in mind Table 6-2 will now be filled with a feasible solution, as given in

Table 6-3. The vacant positions represent zero shipments. Note that fewer than $m + n - 1$ nonzero values have been used. Note also that the cost of this choice is

$$40 + 80 + 40 + 54 + 40 + 56 + 16 = 326$$

In order to improve this feasible solution, form a new table containing the same number of elements as the cost matrix and using the given cost

TABLE 6-3. FIRST SHIPPING TABLE

	j\i	Destinations					
		1	2	3	4	5	Totals
Origins	1	10		10	4		24
	2					18	18
	3		20				20
	4				14	2	16
	Totals	10	20	10	18	20	78

values in each position containing a nonzero shipment. (See the circled values in Table 6-4.) The elements in this "pseudo-cost" table will be labeled $\bar{C}_{ij}$. Hence $C_{ij} = \bar{C}_{ij}$ for all values which have the same subscripts as those of a nonzero shipping value. As for the remaining values, form

TABLE 6-4. FIRST PSEUDO-COST TABLE

④	4	⑧	⑩	14
−7	−7	−3	−1	③
2	②	6	8	12
−2	−2	2	④	⑧

an auxiliary table with spaces for values U_i $(i = 1, \ldots, m)$ and V_j $(j = 1, \ldots, n)$. To see how this is done, choose any circled value $\bar{C}_{ij}$ and arbitrarily assign real values to U_i and V_j so that

$$U_i + V_j = \bar{C}_{ij} = C_{ij}$$

Find a circled value having one subscript the same as C_{ij}, say C_{ik}, and choose V_k such that the previously chosen U_i plus V_k equals C_{ik}. If no such circled value exists, choose at random a new C_{kp} from the circled values, and assign to U_k and V_p values such that $U_k + V_p = C_{kp}$. Continue this until the auxiliary Table 6-5 is complete.

TABLE 6-5. FIRST AUXILIARY TABLE

	U_i	V_j
1	4	0
2	-7	0
3	2	4
4	-2	6
5		10

To see how this table is developed, choose position (1,1) and arbitrarily let $U_1 = C_{11} = 4$. Since $U_1 + V_1 = C_{11}$, it follows that $V_1 = 0$. Position (1,3) is also occupied by a nonzero shipping value. Since U_1 and V_3 correspond to that position, one must have $U_1 + V_3 = C_{13}$ or $4 + V_3 = 8$, yielding $V_3 = 4$. Similarly, $U_1 + V_4 = C_{14}$ which gives $V_4 = 6$. However, in that column one also has a nonzero shipping value in position (4,4). Hence $U_4 + V_4 = C_{44}$. Since $V_4 = 6$ it follows that $U_4 = -2$, etc. With the auxiliary table completed in this manner, all $\bar{C}_{ij} = U_i + V_j$ may now be computed and then compared with the original costs C_{ij}. If all $\bar{C}_{ij} \leq C_{ij}$, the shipping allocation yields the minimum cost. Otherwise, consider only those positions for which the difference $\bar{C}_{ij} - C_{ij}$ is positive and choose the position for which this difference is maximum. In this case $\bar{C}_{ij} > C_{ij}$ only in positions (1,5) and (3,5), and the latter gives the maximum difference. Rewrite the old shipping table with a θ_1 placed in position (3,5). In order to obtain the total a_3, θ_1 must be subtracted from the nonzero shipping value in position (3,2). However, it must also be placed somewhere in the second column to yield b_2. Hence it is placed in the (4,2) position and subtracted from the value in the (4,5) position. The totals are now correctly obtained everywhere. When there are two possible nonzero values to which θ_1 may be added, as is the case in the fourth row, the value with the smaller total cost is used. This gives Table 6-6.

Among the expressions in which θ_1 appears, select those for which $x_{ij} > 0$. Then assign to θ_1 the smallest of these x_{ij} and introduce this value in the table. This gives a new shipping table. In the present example $\theta_1 = 2$, which leaves the shipping table with eight nonzero shipping values as required by the condition that $m + n - 1$ of the constraints be independent, and one has shipping Table 6-7.

190 OPTIMIZATION, PROGRAMMING, AND GAME THEORY

TABLE 6-6. SHIPPING TABLE

Origins

i \ j	1	2	3	4	5	Totals
1	10		10	4		24
2					18	18
3		$20 - \theta_1$			θ_1	20
4		θ_1		14	$2 - \theta_1$	16
Totals	10	20	10	18	20	78

TABLE 6-7. SECOND SHIPPING TABLE

Origins

i \ j	1	2	3	4	5	Totals
1	10		10	4		24
2					18	18
3		18			2	20
4		2		14		16
Totals	10	20	10	18	20	78

The process is now repeated until all $\bar{C}_{ij} \leq C_{ij}$. When this occurs, the solution yields the minimum shipping cost. The steps are given in Tables 6-8 to 6-17 without further elaboration.

The new cost is

$$40 + 80 + 40 + 54 + 36 + 6 + 10 + 56 = 322$$

TABLE 6-8. SECOND PSEUDO-COST TABLE

④	11	⑧	⑩	12
-5	2	-1	1	③
-5	②	-1	1	③
-2	⑤	2	④	6

TABLE 6-9. SECOND AUXILIARY TABLE

U_i	V_i
4	0
-5	7
-5	4
-2	6
	8

The only $\bar{C}_{ij} > C_{ij}$ is $\bar{C}_{12} = 11$, where $C_{12} = 9$.

TABLE 6-10. SHIPPING TABLE

i \ j	Destinations					
	1	2	3	4	5	Totals
1	10	θ_2	10	$4 - \theta_2$		24
2					18	18
3		18			2	20
4		$2 - \theta_2$		$14 + \theta_2$		16
Totals	10	20	10	18	20	78

Origins

The smallest $\theta_2 > 0$ is $\theta_2 = 2$, and the table is given as Table 6-11.

TABLE 6-11. THIRD SHIPPING TABLE

i \ j	Destinations					
	1	2	3	4	5	Totals
1	10	2	10	2		24
2					18	18
3		18			2	20
4				16		16
Totals	10	20	10	18	20	78

Origins

The cost is now

$$40 + 18 + 80 + 20 + 54 + 36 + 6 + 64 = 318$$

which is an improvement on the previous cost.

TABLE 6-12. THIRD PSEUDO-COST TABLE

④	⑨	⑧	⑩	10
−3	2	1	3	③
−3	②	1	3	③
−2	3	2	④	4

TABLE 6-13. THIRD AUXILIARY TABLE

U_i	V_i
4	0
−3	5
−3	4
−2	6
	6

Here $\bar{C}_{24} > C_{24}$.

Table 6-14. Shipping Table

	Destinations					
i \ j	1	2	3	4	5	Totals
1	10	$2 + \theta_3$	10	$2 - \theta_3$		24
2				θ_3	$18 - \theta_3$	18
3		$18 - \theta_3$			$2 + \theta_3$	20
4				16		16
Totals	10	20	10	18	20	78

(Origins on the left)

The minimum $\theta_3 > 0$ is $\theta_3 = 2$, giving Table 6-15.

Table 6-15. Final Shipping Table

	Destinations					
i \ j	1	2	3	4	5	Totals
1	10	4	10			24
2				2	16	18
3		16			4	20
4				16		16
Totals	10	20	10	18	20	78

(Origins on the left)

The cost is now

$$40 + 36 + 80 + 4 + 48 + 32 + 12 + 64 = 316$$

a still lower cost than before.

Table 6-16. Final Pseudo-cost Table

④	⑨	⑧	9	10
-3	2	1	②	③
-3	②	1	2	③
-1	4	3	④	5

Table 6-17. Final Auxiliary Table

U_i	V_i
4	0
-3	5
-3	4
-1	5
	6

Every $\bar{C}_{ij} \leq C_{ij}$ so that the last shipping table gives the optimum solution with cost 316.

If the supply exceeds the demand in the formulation of the problem, then

$$\sum_{i=1}^{m} a_i > \sum_{j=1}^{n} b_j \qquad (6\text{-}63)$$

and if the demand exceeds the supply the above inequality is reversed. In the former case, one can add an extra column to the shipping and cost matrices with the costs being zero (indicating storage). In the latter case, an extra row may be added to the shipping and to the cost matrices with costs M in each space. This M will be considered a very high cost so that the row will be used only to indicate the destinations that are left without their demands fulfilled.

It is a characteristic feature of transportation problems that they need not have unique solutions, and there may exist a whole family of shipping programs each of which costs no more than any program not contained in this family. This lack of uniqueness is a consequence of the assumption that the specific transportation cost for a shipping route is independent of the amount shipped along this route. In many fields of application, this assumption is not realistic, because congestion of the shipping routes must be expected to increase the transportation cost. If the assumption is made that the specific shipping cost for a route increases linearly with the amount shipped over this route, the optimum shipping program is then unique [77].

The Dual Simplex Method [55]

In this method one is concerned with obtaining an "optimum (e.g., maximum) solution" in the sense that a basis B is determined such that

$$\begin{aligned} B\mathbf{x} &= \mathbf{b} \\ \mathbf{x} &= B^{-1}\mathbf{b} \\ z_j - c_j &\geq 0 \end{aligned} \qquad (6\text{-}64)$$

where the vector $\mathbf{x}$ is not restricted to be nonnegative. The computational procedure is to transform the components of $\mathbf{x}$ in such a manner as to allow the components of $\mathbf{x}$ to become nonnegative while always keeping $z_j - c_j \geq 0$. This is a finite iterative procedure, the first feasible solution of which will also be an optimum solution.

6-8. Parametric Linear Programming [35, 36]

The following two sections illustrate ideas pertinent to Le Chatelier's principle which has the following statement [83]: "If the external conditions of a thermodynamic system are altered, the equilibrium of the system will tend to move in such a direction as to oppose the change in external conditions." An extension of Le Chatelier's principle is as follows: In a linear-programming problem, for any small change in the cost coefficients c_i the change in x_i will be smaller every time a new constraint is added to the system.

In a linear-programming problem it may be desirable to study the behavior of solutions when the cost coefficients are parametrized. The problem is then to find the set of x_j $(j = 1, 2, \ldots, n)$ which minimizes the linear form $\sum_{j=1}^{n} \lambda_j x_j$ and satisfies the constraints $x_j \geq 0$ $(j = 1, \ldots, n)$ and $\sum_{j=1}^{n} a_{ij} x_j \leq b_i$ $(i = 1, \ldots, m)$, where λ_j are parameters and a_{ij} and b_i are constants. Using the simplex process, it is possible to generate systematically the intervals of parameter values in the case of a single parameter, with their corresponding solutions. A generalization to several parameters is possible, but there is no systematic way of generating all the regions in parameter space. In the single-parameter case suppose that $z_j - c_j$ in the simplex algorithm is expressed as $\alpha_j + \lambda \beta_j$. Then the region of values of λ corresponding to a selected solution basis is obtained from

$$\max_{\beta_j < 0} \frac{-\alpha_i}{\beta_j} \leq \lambda \leq \min_{\beta_j \geq 0} \frac{-\alpha_i}{\beta_j} \tag{6-65}$$

Note that when c_j is replaced by a single parameter λ in the primal it is also replaced in the dual. Hence it suffices to dualize a given problem and study the behavior of solutions of the latter when a parameter is introduced.

Brief geometric considerations described below show that (1) the values of λ corresponding to a solution are either a point or a closed interval; (2) the number of intervals is finite; (3) two intervals meet in, at most, one point, and (4) the collection of intervals forms a connected set which is either a whole line or a half line. Proceeding to the argument, it is evident that the convex set defined by the constraints of the dual has a finite number of vertices. In fact, as mentioned previously, their number cannot exceed $\binom{m+n}{n}$. In both the primal and the dual let some c_j be replaced by a parameter λ. A minimum of the dual problem must, of course, be at a vertex of the boundary. Suppose that this vertex is not included in the parametrized hyperplane. Then by allowing λ to vary continuously on the real numbers the boundary of the parametrized half space is translated parallel to itself, increasing or decreasing the size of the region of feasible solutions. As long as a convex region of feasible solutions is defined and the parametrized hyperplane does not meet the vertex yielding the minimum, one has a solution corresponding to the entire set of values of λ up to and including that value of λ bringing the hyperplane to this vertex. This set of values of λ is a half line.

Since each vertex is determined by a set of hyperplanes, the optimum vertex is changed only when the parametrized plane passes through an optimum vertex determined by a set of hyperplanes not including the parametrized hyperplane. The values of λ at which this shift in solutions occurs are the end points of the intervals. Two neighboring intervals clearly correspond to two different solutions (degeneracy excluded).

It can easily be shown that the solution of a two-person zero-sum game with the coefficients of an entire row of its payoff matrix replaced by a parameter can be obtained from this problem.

Geometric considerations (of the dual) analogous to the one-parameter case show that in the general case the space of parameters is entirely or partly tessellated by a finite set of convex hyperpolyhedra. No two cells correspond to the same solution yielding a minimum. Note that fixing some of the parameters corresponds to projections of the tessellated figure in parameter n-space onto an $(n - k)$-dimensional hyperplane, where k is the number of parameters whose values are fixed. It is now clear that small perturbations in the coefficients c_j do not necessarily entail changing the solution vectors. The region in parameter space corresponding to a solution vector is convex, and always either the entire space or more than a half space is tessellated with these convex regions. We now show that these regions are convex.

Suppose that every c_j is replaced by a parameter λ_j. Then one has the following:

Theorem 6-3: A solution $(x_1^0, \ldots, x_n^0)$ of the original problem which yields a minimum corresponding to $(\lambda_1^0, \ldots, \lambda_n^0)$ yields a minimum to a convex region in parameter space containing $(\lambda_1^0, \ldots, \lambda_n^0)$.

Proof: If it also yields a minimum corresponding to $(\lambda_1^1, \ldots, \lambda_n^1)$, then one has

$$\lambda_1^0 x_1^0 + \cdots + \lambda_n^0 x_n^0 \leq \lambda_1^0 x_1 + \cdots + \lambda_n^0 x_n$$

for all $x_i \geq 0$, and (6-66)

$$\lambda_1^1 x_1^0 + \cdots + \lambda_n^1 x_n^0 \leq \lambda_1^1 x_1 + \cdots + \lambda_n^1 x_n$$

for all $x_i \geq 0$.

If $a + b = 1, a \geq 0$ and $b \geq 0$, then one must show that $a(\lambda_1^0, \ldots, \lambda_n^0) + b(\lambda_1^1, \ldots, \lambda_n^1)$ has $(x_1^0, \ldots, x_n^0)$ for a minimum solution. This follows immediately by multiplying the first inequality by a and the second by b and adding the two inequalities.

Thus in a problem where the a_{ij} and the b_i are fixed but the c_j are allowed to fluctuate, it is possible to generate the entire set of solutions and their corresponding convex sets in parameter space. We have recently completed a method of solving the more general problem of parametrizing all the coefficients.

Remark: Additional situations of interest arise where the coefficients are not exactly known but must be described statistically. Problems of this sort are referred to as programming in the face of uncertainty.

6-9. Bounds on the Value of the Objective Function

A linear-programming problem with a large number of inequalities and unknowns generally requires a considerable amount of expense and time for its solution. In fact, a solution may not be obtainable at all if a digital computer is not available. Furthermore, the precise value of (for example) the minimum of the objective function may often not be required. That is to say, a lower and an upper bound on its value may be sufficient to answer many questions as effectively as the complete solution and may be obtainable with much less effort.

For example, a knowledge of the *order of magnitude* of the minimum value of the cost function may permit a decision not to undertake a complete solution because the minimum cost will be prohibitively large. Conversely, an approximate solution of the type suggested may indicate that the situation presently prevailing may be very far from optimum and a detailed solution is therefore justified.

Alternatively, when computing facilities are not available, the approximate solution suggested may be all that is feasible in a reasonable time. A justification and a method follow:

It is clear that increasing the number of constraints in a linear-programming problem generally entails an increase in the minimum value of the objective function. This follows from the fact that the convex region of the larger set of inequalities is contained in the convex region of the smaller set and thus the minimum of the smaller set is attained at a greater distance from the origin.

The approximation problem is one of solving linear-programming problems with the same objective function relative to a subset of the constraints.

From the above statement one has the following relation:

$$\min P \geq \max (\min P_i) \tag{6-67}$$

where the quantity on the left is the minimum value of the cost function for the entire problem and the quantity in parentheses on the right is a minimum for each of the several partition problems solved separately with the original objective function.

One also has

$$\min (\max P_i) \geq \max P \qquad (6\text{-}68)$$

An algorithm for carrying out the approximation process may be described as follows: When minimizing, solve the linear-programming problems corresponding to subsets (partitions) of the constraints with the objective function of the original problem. For a lower bound to the minimum of the original problem, take the maximum of the minima obtained.

Note that, to obtain the minimum value in the case of single partitions, the trial vertices may be obtained from the appropriate intersection points of the constraint inequalities with the coordinate axes. The objective function is then evaluated at each vertex and the minimum of these values selected. When this is done for every constraint individually taken with the objective function, the maximum of the several minima provides a simple lower bound to the minimum of the entire problem. This result may be further refined by taking the constraints (some or all, depending on the value obtained) two at a time and solving the corresponding simple linear-programming problem, etc. The method may also be applied to the dual, yielding a further refinement of the problem of placing bounds on the values of the objective function.

To illustrate with a simple example, consider the diet problem studied previously. Let P_1 be the inequality whose intersection vertices with the coordinate axes are $(5,0,0)$, $(0,5/4,0)$, and $(0,0,5/2)$ and let P_2 be the inequality with intersections $(4/3,0,0)$, $(0,4,0)$, and $(0,0,2)$.

The minimum value of the cost function $2x_1 + 9x_2 + x_3$ at the intersection points of P_1 is attained at $(0,0,5/2)$ and is equal to $5/2$, that is, $\min P_1 = 5/2$; similarly, $\min P_2 = 2$. Finally, the maximum of these two values which must be less than or equal to the minimum value of the entire problem is $5/2$, which is actually the minimum value.

6-10. Quadratic Programming

We now consider the slightly more general problem of optimizing a quadratic function subject to linear constraints. This provides another useful model. The quadratic-programming problem considered here may be presented in the following vector notation. Minimize the quadratic functions of n variables

$$f(\mathbf{x}) = \mathbf{p}\mathbf{x} + \mathbf{x}'\mathbf{C}\mathbf{x}$$

subject to $\mathbf{A}\mathbf{x} \leq \mathbf{b}$, $\mathbf{x} \geq 0$. Here, $\mathbf{x}$ is a column vector with n conponents, $\mathbf{x}'$ is its transpose, $\mathbf{p}$ is a row vector with n components, $\mathbf{b}$ is a column vector with m components, and $\mathbf{C}$ and $\mathbf{A}$ are n by n and m by n matrices, respectively. The notation will be made clearer by studying a specific example. Note that $f(\mathbf{x})$ is a quadratic form. It is assumed here that

$f(\mathbf{x})$ is convex which requires that the coefficient matrix $\mathbf{C}$ be positive semidefinite.

A method of solving a quadratic-programming problem, due to P. Wolfe and M. Frank [94], will now be illustrated. Minimize

$$f(\mathbf{x}) = x_1^2 + 3x_2^2 - 4x_1 - 6x_2 \tag{6-69}$$

subject to
$$x_1 \geq 0 \qquad x_2 \geq 0$$
and
$$x_1 + 2x_2 \leq 4 \tag{6-70}$$

It is easy to see that the minimum must lie on the boundary. Geometrically, the contours of the function $f(x_1, x_2)$ = a constant may be drawn for different values of the constant and the maximum noted. It can be observed that the maximum actually lies on the boundary line $x_1 + 2x_2 = 4$. It may be immediately obtained by Lagrange's method. However, the problem will be solved by the simplex process to show that a systematic approach to solving complex problems of this type is available.

Using the above notation one has

$$\mathbf{x} = \begin{bmatrix} x_1 \\ x_2 \end{bmatrix} \geq 0 \qquad \mathbf{x}' = [x_1, x_2] \tag{6-71}$$

$$\mathbf{A} = [1, 2] \qquad \mathbf{b} = [4] \qquad \mathbf{p} = [-4, -6] \qquad \mathbf{C} = \begin{bmatrix} 1 & 0 \\ 0 & 3 \end{bmatrix} \tag{6-72}$$

Form the matrix equation

$$\mathbf{Bw} = \begin{bmatrix} \mathbf{b} \\ -\mathbf{p}' \end{bmatrix} \equiv \mathbf{d} \tag{6-73}$$

where $\mathbf{p}'$ is the transpose of $\mathbf{p}$,

$$\mathbf{B} = \begin{bmatrix} \mathbf{A} & 0 & \mathbf{I} & 0 \\ 2\mathbf{C} & \mathbf{A}' & 0 & -\mathbf{I} \end{bmatrix} \tag{6-74}$$

the vector of variables

$$\mathbf{w}' = [\mathbf{x}', \mathbf{u}, \mathbf{y}', \mathbf{v}] \tag{6-75}$$

is the transpose of $\mathbf{w}$, $\mathbf{x}'$ and $\mathbf{y}'$ are the transposes of $\mathbf{x}$ and the slack vector $\mathbf{y}$, and where $\mathbf{x}, \mathbf{y}, \mathbf{u}, \mathbf{v} \geq 0$ are vectors to be determined such that

$$\mathbf{Ax} + \mathbf{y} = \mathbf{b} \qquad \Delta f(\mathbf{x}) = -\mathbf{uA} + \mathbf{v} \qquad \mathbf{vx} + \mathbf{uy} = 0 \tag{6-76}$$

The last two conditions follow from the conditions for an optimum on the boundary studied in the previous chapter.

By combining the above conditions, the problem becomes one of finding $\mathbf{w}$ which minimizes (to zero) the convex quadratic function

$$g(\mathbf{w}) \equiv \mathbf{vx} + \mathbf{uy} = \mathbf{ub} + \mathbf{px} + 2\mathbf{x}'\mathbf{Cx} \tag{6-77}$$

subject to the linear constraints defined by

$$\begin{matrix} & \mathbf{B}_1 & \mathbf{B}_2 & \mathbf{B}_3 & \mathbf{B}_4 & \mathbf{B}_5 & \mathbf{B}_6 & \mathbf{w} & & \mathbf{d} \end{matrix}$$
$$\begin{bmatrix} 1 & 2 & 0 & 1 & 0 & 0 \\ 2 & 0 & 1 & 0 & -1 & 0 \\ 0 & 6 & 2 & 0 & 0 & -1 \end{bmatrix} \begin{bmatrix} x_1 \\ x_2 \\ u \\ y \\ v_1 \\ v_2 \end{bmatrix} = \begin{bmatrix} 4 \\ 4 \\ 6 \end{bmatrix} \quad (6\text{-}78)$$

together with the nonnegativity constraints on the coefficients of $\mathbf{w}$, which can be written as $\mathbf{w} \geq 0$.

Phase 1: An initial feasible solution is reached just as in the simplex method. Let the initial basis be $\mathbf{B}_3$, $\mathbf{B}_4$, $\mathbf{B}_6$; then the following matrix comprises the coefficients of the vectors $\mathbf{d}$ and $\mathbf{B}_i$ expressed as linear combinations of the initial basis $\mathbf{B}_3$, $\mathbf{B}_4$, $\mathbf{B}_6$.

	$\mathbf{w}_1^*$	1	2	3	4	5	6
$\mathbf{B}_3$	4	2	0	1	0	−1	0
$\mathbf{B}_4$	4	1	2	0	1	0	0
$\mathbf{B}_6$	2	4	−6	0	0	−2	1

(6-79)

An asterisk is used over $\mathbf{w}_1$ to indicate that it is written in contracted form. From $\mathbf{w}_1^*$, by using its coefficients in the positions indicated by the subscripts of the basis vectors and zeros elsewhere, its transpose may now be written as

$$\mathbf{w}_1' = [\underbrace{0\ \ 0}_{\mathbf{x}'}\ \underbrace{4\ \ 4}_{\mathbf{u}\ \mathbf{y}'}\ \underbrace{0\ \ 2}_{\mathbf{v}}] \quad (6\text{-}80)$$

For this to yield a solution to the problem, $g(\mathbf{w})$ must be zero. However, substitution shows that

$$g(\mathbf{w}) = \mathbf{v}\mathbf{x} + \mathbf{u}\mathbf{y} = [0,2]\begin{bmatrix} 0 \\ 0 \end{bmatrix} + 4 \cdot 4 = 16 > 0 \quad (6\text{-}81)$$

Therefore the minimum is not yet obtained. Note that, if $g(\mathbf{w})$ had vanished, $\mathbf{x} = \begin{bmatrix} 0 \\ 0 \end{bmatrix}$ would solve the original problem. Also note that if M is the minimum value sought then

$$f(\mathbf{x}) - M \leq g(\mathbf{w}) = 16$$

Thus, since in this case $\mathbf{x} = \begin{bmatrix} 0 \\ 0 \end{bmatrix}$, one has $f(x) = 0$ and hence $M \geq -16$.

In general, obtain $\nabla g(\mathbf{w}) = [\mathbf{v}, \mathbf{y}', \mathbf{u}, \mathbf{x}']$. This is the gradient vector obtained by taking the partials of g with respect to the components of w

in the order in which they appear in w. Thus

$$\nabla g(w) = \left[\frac{\partial g}{\partial x_1}, \frac{\partial g}{\partial x_2}, \frac{\partial g}{\partial u}, \frac{\partial g}{\partial y}, \frac{\partial g}{\partial v_1}, \frac{\partial g}{\partial v_2}\right] \quad (6\text{-}82)$$

gives the desired vector on expanding $g(\mathbf{w}) = \mathbf{vx} + \mathbf{uy} = v_1 x_1 + v_2 x_2 + \mathbf{uy}$. This vector is now evaluated at $\mathbf{w} = \mathbf{w}_1$ to yield $\nabla g(\mathbf{w}_1) = [0,2,4,4,0,0]$. Now $\nabla g(\mathbf{w}_1)$ is used as the cost-coefficient vector of a linear program to be minimized subject to the linear constraints given above. The simplex-method test $z_j - c_j$ is applied in order to obtain a new basis. If this basis does not provide the desired minimum, at least one step of the programming problem must be done.

| | | | c_1 | c_2 | c_3 | c_4 | c_5 | c_6 |
| | | | $\mathbf{c} = \nabla g(\mathbf{w}_1) = 0$ | 2 | 4 | 4 | 0 | 0 |
	$\mathbf{c}^*$	$\mathbf{w}_1^*$						
$\mathbf{B}_3$	4	4	2	0	1	0	-1	0
$\mathbf{B}_4$	4	4	1	2	0	1	0	0
$\mathbf{B}_6$	0	2	4	-6	0	0	-2	1
			$z_j = 12$	8	4	4	-4	0

(6-83)

Here $\mathbf{c}^*$ gives the costs corresponding to the basis vectors, and z_j is the product of the $\mathbf{c}^*$ and the corresponding column of the matrix above z_j.

Now, the $c_j - z_j$ for $j = 1, \ldots, 6$ are given by $-12, -6, 0, 0, 4, 0$. (Recall that one must have $c_j - z_j \geq 0$ for a minimum which is not true here. Hence the vector corresponding to the most negative of this difference will be introduced into the basis.) Since -12 is most negative, $\mathbf{B}_1$ is introduced into the basis. To find which vector is to be eliminated, one computes

$$\theta = \min \frac{\text{components of basic solution } w_1}{\text{positive components of first column}}$$
$$= \min (4/2, 4/1, 2/4) = 1/2$$

Thus, $\mathbf{B}_6$ is eliminated from the basis and $\mathbf{B}_1$ is used instead.

When using the tableau form illustrated below, a basis change in the simplex method may be carried out in the following manner:

Assume that the basis vector $\mathbf{B}_i$ is to be replaced by the vector $\mathbf{B}_j$. Then take the following steps:

1. Divide the row to the left of which $\mathbf{B}_i$ appears by the "pivot element" which is at the intersection of this row with the column expressing

B_j as a linear combination of the basis. This yields unity in the pivot position.

2. Subtract from each other original row such a multiple of the "pivot row" that 0 appears over the pivot element, and insert. This gives the new matrix with B_6 replaced by B_1.

After this transformation one obtains w_2^* which is used to calculate $\nabla g(w_2)$. This is the new cost vector **c**. The new **c*** is obtained from **c** and not as a result of the transformation. Also note that w_2^* turns out to be the vector **d** expressed as a linear combination of the new basis.

	c*	w_2^*		c_1	c_2	c_3	c_4	c_5	c_6
			$\mathbf{c} = \nabla g(w_2) =$ 0	0	7/2	3	1/2	0
B_3	7/2	3		0	3	1	0	0	$-1/2$
B_4	3	7/2		0	7/2	0	1	1/2	$-1/4$
B_1	0	1/2		1	$-3/2$	0	0	$-1/2$	1/4
			$z_j =$ 0	21	7/2	3	3/2	$-5/2$
			$c_j - z_j =$ 0	-21	0	0	-1	5/2

(6-84)

This method makes the tableau form considerably more useful in applying the simplex process.

It is easily checked that this matrix is the expression of B in terms of $[B_3, B_4, B_1]$ by showing that each B_i, when expressed as a linear combination of the new basis, has the indicated coefficients. Then one obtains w_2' from w_2^*. It is given as

$$w_2' = [\underbrace{1/2}_{x} \ \underbrace{0}_{u} \ \underbrace{3}_{} \ \underbrace{7/2}_{y} \ \underbrace{0 \ 0}_{v}] \tag{6-85}$$

from which one has

$$g(w_2) = vx + uy = 3(7/2) = 21/2 \tag{6-86}$$

Note that the value of $g(w)$ has improved. If this were not the case, then one would use the interpolation subroutine; one finds that point **w** on the segment $w_1 w_2$ which minimizes $g(w)$, where

$$g(w) = g[w_1 + \lambda(w_2 - w_1)] \quad 0 \leq \lambda \leq 1 \tag{6-87}$$

Now $g(w)$ is quadratic in λ, and the problem is then to find λ which minimizes $g(w)$. This would give the new **w** for which ∇g is calculated, yielding the cost coefficients. Note that here c_i are obtained from evaluating

$\nabla g(\mathbf{w}_2)$ which is given by

$$\nabla g(\mathbf{w}_2) = [\underbrace{0 \quad 0}_{\mathbf{v}} \quad \underbrace{\tfrac{7}{2}}_{\mathbf{y}} \quad \underbrace{3}_{\mathbf{u}} \quad \underbrace{\tfrac{1}{2} \quad 0}_{\mathbf{x}}] \qquad (6\text{-}88)$$

This also gives $\mathbf{c}^*$ which is then used to compute the z_j's as before.

From the previous tableau it can be seen that $\mathbf{B}_2$ will replace either $\mathbf{B}_3$ or $\mathbf{B}_4$. Assume that it is $\mathbf{B}_3$. The new tableau is

	$\mathbf{w}_3^*$						
$\mathbf{B}_2$	1	0	1	$\tfrac{1}{3}$	0	0	$-\tfrac{1}{6}$
$\mathbf{B}_4$	0	0	0	$-\tfrac{7}{6}$	1	$\tfrac{1}{2}$	$\tfrac{1}{3}$
$\mathbf{B}_1$	2	1	0	$\tfrac{1}{2}$	0	$-\tfrac{1}{2}$	0

(6-89)

Remark: It is advisable not to compute the entire tableau unless one has computed $g(\mathbf{w})$ first.

Now,

$$\mathbf{w}_3' = [\underbrace{2 \quad 1}_{\mathbf{x}} \quad \underbrace{0 \quad 0}_{\mathbf{u}} \quad \underbrace{0}_{\mathbf{y}} \quad \underbrace{0}_{\mathbf{v}}] \qquad (6\text{-}90)$$

which yields $\qquad g(\mathbf{w}_3) = \mathbf{vx} + \mathbf{uy} = 0 \qquad (6\text{-}91)$

And the process has terminated in a solution.

The solution is provided by the components of the vector $\mathbf{x}$ obtained from $\mathbf{w}_3$, that is, $x_1 = 2$, $x_2 = 1$ for which

$$f(\mathbf{x}) = 4 + 3 - 8 - 6 = -7$$

This elegant method of solving a quadratic problem has the virtue of combining the essential ideas of an optimum on the boundary, in order to linearize the problem, together with the systematic approach of the simplex process and obtaining a convergent method yielding a solution.

There are instances in which one may wish to optimize, subject to linear equality and inequality constraints, a linear function

$$\sum_{j=1}^{n} r_j x_j \qquad (6\text{-}92)$$

where r_j ($j = 1, \ldots, n$) are random variables with means μ_j and covariances σ_{jk} ($\sigma_{jj} = \sigma_j^2$ is the variance) and where $x_j \geq 0$ ($j = 1, \ldots, n$). The expected value of this payoff function is

$$E = \sum_{j=1}^{n} \mu_j x_j \qquad (6\text{-}93)$$

and its variance is

$$V = \sum_j \sum_k \sigma_{jk} x_j x_k \quad (6\text{-}94)$$

Depending on which of the last two expressions one wishes to deal with, the problem may be pursued as one requiring a minimum V for a given E and subject to the constraints, or a maximum E for a given V subject to the constraints. Such combinations are called *efficient combinations*. It may be desired to find all $(x_1, \ldots, x_n)$ called efficient, which yield such efficient E, V combinations.

Note that V is a quadratic payoff function which can then be optimized subject to the linear constraints. It can be shown [60] that, if $V(x)$ is a strictly convex function over the convex set defined by the linear constraints (including the nonnegativity constraints), then it attains a minimum value at one and only one point $\underline{x}$. The set of efficient (E,V) combinations and the corresponding x can be described by a single-valued continuous function

$$\left. \begin{array}{r} V(E) \\ x(E) \end{array} \right\} \begin{array}{l} \underline{E} \leq E \leq \bar{E} \\ \text{or} \\ \underline{E} \leq E \leq \infty \end{array} \quad (6\text{-}95)$$

where $\underline{E} \equiv E(\underline{x})$ and $\bar{E} \equiv E(\bar{x})$, where $\bar{x}$ maximizes $E(x)$ for a given V.

REFERENCES

Remark: References 1 to 11 are works of interest on the subject.

1. Charnes, A., W. W. Cooper, and A. Henderson: "An Introduction to Linear Programming," John Wiley & Sons, Inc., New York, 1953.
2. Eisemann, K.: Linear Programming, *Quarterly of Applied Mathematics*, vol. 13, pp. 209–232, 1955.
3. Gass, S. I.: "Linear Programming: Methods and Applications," McGraw-Hill Book Company, Inc., New York, 1958.
4. Koopmans, T. C. (ed.): "Activity Analysis of Production and Allocation," John Wiley & Sons, Inc., New York, 1951.
5. Kuhn, H. W., and A. W. Tucker (eds.): Contributions to the Theory of Games, vol. 1, *Annals of Mathematical Studies*, No. 24, Princeton University Press, Princeton, N.J., 1950.
6. Kuhn, H. W., and A. W. Tucker (eds.): Contributions to the Theory of Games, vol. 2, *Annals of Mathematical Studies*, No. 28, Princeton University Press, Princeton, N.J., 1953.
7. Kuhn, H. W., and A. W. Tucker (eds.): Papers on Linear Inequalities and Related Systems, *Annals of Mathematical Studies*, No. 38, Princeton University Press, Princeton, N.J., 1956.
8. Rhode, F. V.: Bibliography on Linear Programming, *Operations Research*, vol. 5, pp. 45–62, February, 1957.
9. Project SCOOP, *Symposium on Linear Inequalities and Programming*, Planning Research Division, Director of Management and Analysis Service, Comptroller,

Hq., U.S. Air Force, and National Bureau of Standards, Washington, D.C., June 14–16, 1951.
10. Second Symposium on Linear Programming, sponsored by Directorate of Management Analysis, DCS Comptroller, Hq., U.S. Air Force, and National Bureau of Standards, Washington, D.C., Jan. 27–29, 1955.
11. Vajda, S.: "Theory of Games and Linear Programming," John Wiley & Sons, Inc., New York, 1956.
12. Antosiewicz, H. A., and A. J. Hoffman: A Remark on the Smoothing Problem, *Management Science*, vol. 1, pp. 92–95, 1954.
13. Barankin, E. W.: Some Investigations in Linear Programming, Project SCOOP, *Symposium on Linear Inequalities and Programming*, pp. 68–73, Planning Research Division, Director of Management and Analysis Service, Comptroller, Hq., U.S. Air Force, and National Bureau of Standards, Washington, D.C., June 14–16, 1951.
14. Beale, E. M. L.: An Alternative Method for Linear Programming, *Proceedings of the Cambridge Philosophical Society*, vol. 50, pp. 512–523, 1954.
15. Beale, E. M. L.: Cycling in the Dual Simplex Algorithm, *Naval Research Logistics Quarterly*, vol. 2, pp. 269–275, 1954.
16. Bearman, J. E.: Cutting Costs with Linear Programming, *Proceedings of the Second Annual Computer Applications Symposium*, Illinois Institute of Technology, Armour Research Foundation, Chicago, Oct. 24–25, 1955.
17. Bowman, E. H.: Production Scheduling by the Transportation Method of Linear Programming, *Operations Research*, vol. 4, pp. 100–103, 1956.
18. Brown, G. W., and T. C. Koopmans: Computational Suggestions for Maximizing a Linear Function Subject to Linear Inequalities, chap. XXV, in "Activity Analysis of Production and Allocation," John Wiley & Sons, Inc., New York, 1951.
19. Charnes, A.: Optimality and Degeneracy in Linear Programming, *Econometrica*, vol. 20, no. 2, pp. 160–170, 1952.
20. Charnes, A., and C. E. Lemke: Computational Problems of Linear Programming, *Proceedings of the Association for Computing Machinery*, pp. 97–98, May, 1952.
21. Charnes, A., and W. W. Cooper: Optimal Estimation of Executive Compensation by Linear Programming, *Management Science*, vol. 1, p. 138, January, 1955.
22. Dantzig, G. B.: Maximization of a Linear Function of Variables Subject to Linear Inequalities, chap. XXI, in "Activity Analysis of Production and Allocation," John Wiley & Sons, Inc., New York, 1951.
23. Dantzig, G. B.: A Note on a Dynamic Leontief Model with Substitution, *Econometrica*, vol. 21, p. 179, 1953. Abstract.
24. Dantzig, G. B., D. R. Fulkerson, and S. Johnson: Solution of a Large Scale Traveling-salesman Problem, *Journal of the Operations Research Society of America*, vol. 2, pp. 393–410, 1954.
25. Dantzig, G. B., and S. Johnson: A Production Smoothing Problem, *Proceedings of the Second Symposium in Linear Programming*, pp. 151–176, sponsored by Directorate of Management Analysis, DCS Comptroller, Hq., U.S. Air Force, and National Bureau of Standards, Washington, D.C., Jan. 27–29, 1955.
26. Davie, J. W.: Use of Linear Programming in Selective Blending Studies, *Econometrica*, vol. 23, no. 336, 1955. Abstract.
27. Dorfman, R.: "Application of Linear Programming to the Theory of the Firm," University of California Press, Berkeley, Calif., 1951.
28. Dorfman, R.: Mathematical, or "Linear" Programming: A Nonmathematical Exposition, *American Economic Review*, December, 1953.
29. Fenchel, W.: Convex Cones, Sets and Functions (Lecture notes by D. W.

Blackett), *Naval Research Logistics Project Report*, Princeton University, Department of Mathematics, 1953.
30. Flood, M. M.: On the Hitchcock Distribution Problem, *Pacific Journal of Mathematics*, vol. 3, pp. 369–386, 1953.
31. Freeman, R. J., and J. G. Hocking: Discrete Linear Programming, *BRL Memorandum Report*, 924, Ballistic Research Laboratory, Aberdeen Proving Ground, Md., September, 1955.
32. Gaddum, J. W., A. J. Hoffman, and D. Sokolowsky: On the Solution of the Caterer Problem, *Naval Research Logistics Quarterly*, vol. 1, pp. 223–229, 1954.
33. Gainen, L., D. P. Honig, and E. D. Stanley: Linear Programming in Bid Evaluations, *Naval Research Logistics Quarterly*, vol. 1, pp. 48–54, 1954.
34. Gale, D., H. W. Kuhn, and A. W. Tucker: Linear Programming and the Theory of Games, in "Activity Analysis of Production and Allocation," pp. 317–329, John Wiley & Sons, Inc., New York, 1951.
35. Gass, S. I., and T. L. Saaty: The Parametric Objective Function, Part I, *Journal of the Operations Research Society of America*, vol. 2, pp. 316–319, 1954; Part II, *ibid.*, vol. 3, pp. 395–401, 1955.
36. Gass, S. I., and T. L. Saaty: The Computational Algorithm for the Parametric Objective Function, *Naval Research Logistics Quarterly*, vol. 2, pp. 39–45, 1955.
37. Gass, S. I.: A First Feasible Solution to the Linear Programming Problem, *Proceedings of the Second Symposium in Linear Programming*, pp. 495–500, sponsored by Directorate of Management Analysis, DCS Comptroller, Hq., U.S. Air Force, and National Bureau of Standards, Washington, D.C., Jan. 27–29, 1955.
38. Gleyzal, A.: An Algorithm for Solving the Transportation Problem (to appear in *Journal of Research of the National Bureau of Standards*).
39. Goldman, A. J.: Optimal Rays for Linear Programs, Papers on Linear Inequalities and Related Systems, *Annals of Mathematical Studies*, vol. 38, pp. 613–614, Princeton University Press, Princeton, N.J. Abstract. Also in Ref. 10. Abstract.
40. Goldstein, L.: Problem of Contract Awards, Project SCOOP, *Symposium on Linear Inequalities and Programming*, pp. 147–154, Planning Research Division, Director of Management and Analysis Service, Comptroller, Hq., U.S. Air Force, and National Bureau of Standards, Washington, D.C., June 14–16, 1951.
41. Halsbury, Earl of: From Plato to the Linear Program, *Journal of the Operations Research Society of America*, vol. 3, pp. 239–254, 1955.
42. Harrison, J. O., Jr.: Linear Programming and Operations Research, in J. F. McCloskey and F. N. Trefethen (eds.): "Operations Research for Management," pp. 217–237, Johns Hopkins Press, Baltimore, 1954.
43. Heller, I.: Least Ballast Shipping Required to Meet a Specified Shipping Program, Project SCOOP, *Symposium on Linear Inequalities and Programming*, pp. 164–171, Planning Research Division, Director of Management and Analysis Service, Comptroller, Hq., U.S. Air Force, and National Bureau of Standards, Washington, D.C., June 14–16, 1951.
44. Heller, I.: On the Traveling-salesman's Problem, *Proceedings of the Second Symposium in Linear Programming*, pp. 643–665, sponsored by Directorate of Management Analysis, DCS Comptroller, Hq., U.S. Air Force, and National Bureau of Standards, Washington, D.C., Jan. 27–29, 1955.
45. Hitchcock, F. L.: The Distribution of a Product from Several Sources to Numerous Localities, *Journal of Mathematical Physics*, vol. 20, pp. 224–230, 1941.
46. Hoffman, A. J.: Cycling in the Simplex Algorithm, *National Bureau of Standards Report*, Dec. 16, 1953 (unpublished).

47. Hoffman, A. J., and W. W. Jacobs: Smooth Patterns of Production, *Management Science*, vol. 1, pp. 86–91, 1954.
48. Hoffman, A. J., M. Mannos, D. Sokolowsky, and N. A. Wiegmann: Computational Experience in Solving Linear Programs, *Journal of the Society for Industrial and Applied Mathematics*, vol. 1. no. 1, pp. 17–33, 1953.
49. Index of Publications, The RAND Corporation, Santa Monica, Calif.
50. Jacobs, W. W.: The Caterer Problem, *Naval Research Logistics Quarterly*, vol. 1, pp. 154–165, 1954.
51. Jackson, J. R.: On the Existence Problem of Linear Programming, *Pacific Journal of Mathematics*, vol. 4, pp. 29–36, 1954.
52. Joseph, J. A.: The Application of Linear Programming to Weapon Selection and Target Analysis, *OA Technical Memorandum* 42, Jan. 5, 1954.
53. Kingsberry, S.: Application of Inventory Control, presented at Operations Evaluations Group Decennial Conference on Operations Research.
54. Kuhn, H. W.: The Hungarian Method for the Assignment Problem, *Naval Research Logistics Quarterly*, vol. 2, nos. 1 and 2, pp. 83–97, 1955.
55. Lemke, C. E.: The Dual Method of Solving the Linear Programming Problem, *Naval Research Logistics Quarterly*, vol. 1, pp. 36–47, 1954.
56. Leontief, W. W.: Computational Problems Arising in Connection with Economic Analysis of Industrial Relationships, *Proceedings of a Symposium on Large-scale Digital Calculating Machinery*, pp. 169–175, Harvard University Press, Cambridge, Mass., 1948.
57. Mannos, M.: An Application of Linear Programming to Efficiency in Operation of a System of Dams, *Econometrica*, vol. 23, p. 335, 1955. Abstract.
58. McLynn, J. M., and C. M. Tompkins: Application of a Duality Theorem in the Calculation of Some Linear Maximizing and Minimizing Problems, *Office of Naval Research Logistics Papers* IV.
59. Magee, J. I.: "Studies in Operations Research. I: Application of Linear Programming to Production Scheduling," Arthur D. Little, Inc., Cambridge, Mass.
60. Markowitz, H.: The Optimization of a Quadratic Function Subject to Linear Constraints, *Naval Research Logistics Quarterly*, vol. 3, pp. 111–133, 1956; also, Portfolio Selection, *Journal of Finance*, vol. 7, p. 77, 1952.
61. Manne, A. S.: A Linear Programming Model of the U.S. Petroleum Refining Industry, *Econometrica*, vol. 23, p. 337, 1955. Abstract.
62. McCloskey, J. T., and F. N. Trefethen (eds.): "Operations Research for Management," Johns Hopkins Press, Baltimore, 1954.
63. Mayberry, J. P.: A Geometrical Interpretation of the Simplex Method, Project SCOOP, *Symposium on Linear Inequalities and Programming*, pp. 56–65, Planning Research Division, Director of Management and Analysis Service, Comptroller, Hq., U.S. Air Force, and National Bureau of Standards, Washington, D.C., June 14–16, 1951.
64. Motzkin, T. S.: New Techniques for Linear Inequalities and Optimization, and Remarks on the History of Linear Inequalities, Project SCOOP, *Symposium on Linear Inequalities and Programming*, Planning Research Division, Director of Management and Analysis Service, Comptroller, Hq., U.S. Air Force, and National Bureau of Standards, Washington, D.C., June 14–16, 1951.
65. Motzkin, T. S.: The Assignment Problem, *Proceedings of the Sixth Symposium in Applied Mathematics*, 1956.
66. Natrella, J. V.: New Applications of Linear Programming, *Computers and Automation*, vol. 4, no. 11, 22, item 48, 1955. Abstract.
67. Nelson, R. T.: An Engineering Analysis of the Scheduling Problem—Initial Results, Management Sciences Research Project, University of California, Los Angeles.

68. Nelson, R. T.: Setup and Transportation as Elements of a Linear Scheduling Model, Management Sciences Research Project, University of California, Los Angeles.
69. Nelson, R. T.: Job Shop Scheduling: An Application of Linear Programming, Management Sciences Research Project, University of California, Los Angeles.
70. Von Neumann, J.: A Numerical Method to Determine Optimum Strategy, *Naval Research Logistics Quarterly*, vol. 1, pp. 109–115, 1954.
71. O'Brien, G. G.: Mimeographed notes on linear programming, Melpar, Inc., Alexandria, Va., 1953.
72. Orden, A., and D. F. Votaw: Personnel Assignment Problem, Project SCOOP, *Symposium on Linear Inequalities and Programming*, pp. 155–163, Planning Research Division, Director of Management and Analysis Service, Comptroller, Hq., U.S. Air Force, and National Bureau of Standards, Washington, D.C., June 14–16, 1951.
73. Orden, A.: Application of the Simplex Method to a Variety of Matrix Problems, Project SCOOP, *Symposium on Linear Inequalities and Programming*, pp. 28–50, Planning Research Division, Director of Management and Analysis Service, Comptroller, Hq., U.S. Air Force, and National Bureau of Standards, Washington, D.C., June 14–16, 1951.
74. Osawa, Y.: On the Determination of the Optimal Operating Level in a Textile Manufacturing Factory, *Industrial Administration Research Memorandum*, Osaka Prefectural Institute for Industrial Management, vol. 1, 1955.
75. Paull, A. E., and J. R. Walter: The Trim Problem: An Application of Linear Programming to the Manufacture of Newsprint Paper, *Econometrica*, vol. 23, p. 336, 1955. Abstract.
76. Pollack, S.: The Double Description Method on the SEAC, *National Bureau of Standards Report*, Dec. 9, 1954.
77. Prager, W.: On the Role of Congestion in Transportation Problems, *Zeitschrift für angewandte Mathematik und Mechanik*, vol. 35, pp. 264–268, 1955.
78. Saaty, T. L.: The Number of Vertices of a Polyhedron, *American Mathematical Monthly*, vol. 62, pp. 326–331, May, 1955.
79. Saaty, T. L.: Approximation to the Value of the Objective Function in Linear Programming by the Method of Partitions, *Operations Research*, vol. 4, pp. 352–353, 1956.
80. Salveson, M. E.: On a Quantitative Method in Production Planning and Scheduling, *Econometrica*, vol. 22, no. 1, October, 1952.
81. Salveson, M. E.: A Computational Technique for the Fabrication Scheduling Problem, Management Sciences Research Project, University of California, Los Angeles, 1953.
82. Salveson, M. E.: An Introduction to Mathematical Methods in Production Control, Management Sciences Research Project, University of California, Los Angeles, 1953.
83. Samuelson, P. A.: The Le Chatelier Principle in Linear Programming, *RAND Report RM*-210, Aug. 4, 1949.
84. Schell, E. D.: Distribution of a Product by Several Properties, *Proceedings of the Second Symposium in Linear Programming*, pp. 615–618, sponsored by Directorate of Management Analysis, DCS Comptroller, Hq., U.S. Air Force, and National Bureau of Standards, Washington, D.C., Jan. 27–29, 1955.
85. Smith, L. W., Jr.: Current Status of the Industrial Use of Linear Programming, *Management Science*, vol. 2, no. 2, 1956.
86. Smith, L. W., Jr.: Experience in Scheduling by Means of Dynamic Linear Programming, *Journal of the Operations Research Society of America*, vol. 3, p. 357, item E3, 1955. Abstract.

87. Suzuki, G.: A Transportation Simplex Algorithm for Machine Computation Based on the Generalized Simplex Method, *David Taylor Model Basin Report* 959, Washington, D.C., June, 1955.
88. Symonds, G. H.: "Application of Linear Programming to the Solution of Refinery Problems," Esso Standard Oil Co., New York, 1955.
89. Tompkins, C. B.: Projection Methods in Calculation of Some Linear Problems, *Office of Naval Research Logistics Papers* IV.
90. Votaw, D. R., Jr.: Methods of Solving Some Personnel Classification Problems, *Psychometrika*, vol. 17, pp. 255–266, 1952.
91. Waugh, F. V., and G. L. Burrows: A Short Cut to Linear Programming, *Econometrica*, vol. 23, pp. 18–29, 1955.
92. Weyl, H.: The Elementary Theory of Convex Polyhedra, Contributions to the Theory of Games, *Annals of Mathematical Studies*, No. 24, pp. 3–18, Princeton University Press, Princeton, N.J., 1950.
93. Wolfe, P.: Reduction of Systems of Linear Relations, *Proceedings of the Second Symposium in Linear Programming*, pp. 449–451, sponsored by Directorate of Management Analysis, DCS Comptroller, Hq., U.S. Air Force, and National Bureau of Standards, Washington, D.C., Jan. 27–29, 1955. Abstract.
94. Wolfe, P., and M. Frank: An Algorithm for Quadratic Programming, *Naval Research Logistics Quarterly*, vol. 3, nos. 1 and 2, pp. 95–110, March and June, 1956.

CHAPTER 7

THE THEORY OF GAMES

7-1. Introduction

A *game* is essentially a set of rules describing the formal structure of a competitive situation. These rules specify (1) the alternatives among which the "players" must choose at each stage of "play," (2) the information available to each player when making such a choice, and (3) the "payoff" to each player after any particular contest. A *strategy* for a player is a set of directions for playing the game from beginning to end, which is "complete" in the sense that it includes instructions on what to do in every situation that might possibly arise during play. Note that a player is assumed ignorant of the strategy used by his opponent.

The game is finite if the number of strategies available to each player is finite; otherwise it is an infinite game.

An n-person game is one in which there are n opposing interests (thus bridge is a two-person game). The study of coalition formation in n-person games with $n > 2$ is one of the most interesting and controversial branches of game theory.

Attention will be confined to the particular class of games for which a fully satisfactory theory exists. These are two-person games which are "zero-sum" in the sense that all payoffs are from one player to the other; no money enters from "outside" nor does any leave the "system" consisting of the two players. (A simple example of a nonzero-sum game is this: Players X and Y will receive \$10 from an outside source if they can decide in advance how to divide it.)

7-2. Finite Games: Zero-sum Games

Suppose that the number of strategies is finite. The competition may be represented by the money or any other transferable utility which each player hopes to gain by using his strategies against his opponents. Thus, corresponding to each strategy of a player, there is an amount (a payoff) to be paid or received corresponding to each of the opponent's possible strategies. In the case of two-person zero-sum games, a convenient way of representing these quantities is by means of a payoff matrix. A nonzero-sum game may be reduced to a zero-sum game with an additional player to be responsible for the leakage of payoff into and out of the system.

Every participant of a game must be regarded as desiring to maximize his returns and minimize his losses. These desires are obviously conflicting since some player must lose if another is to win. Thus, the solution of a game consists of specifying the strategies for each player which will maximize the return to the winner and simultaneously minimize the deprivation of the loser.

A large number of examples of finite zero-sum games is given by J. D. Williams, author of "The Compleat Strategyst," in an interesting exposition of the subject.

In this example of a finite zero-sum game, the following is the payoff matrix corresponding to two strategies for each of the players in some game.

	Y_1	Y_2
X_1	10	-7
X_2	13	15

Note that in this matrix the element a_{ij} is the payoff which player Y makes to player X when Y uses his jth strategy against X's ith strategy. When the negatives of these coefficients are considered, there results a payoff matrix to player Y. It is assumed that the strategies, whatever they may be, are listed as shown above and denoted by X_1, X_2, for player X and by Y_1, Y_2 for player Y. Even though the specific game and its strategies are not known, it is possible to make recommendations to the players as to which strategies to use in order to minimize the losses of Y and maximize the winnings of X, simply by considering the payoff matrix. Of course, depending on the amount of payoff expected from each strategy "expected payoff," strategies for X will be preferred to others because of a greater expected return. However, in the face of an intelligent adversary, the player must play conservatively in such a manner as not to leave himself vulnerable. If, for example, Y played Y_2, then X by playing X_2 can always assure himself of the largest payoff, which is 15 in the above matrix. His expectation is therefore 15. However, Y realizes this and tries to minimize the loss by playing Y_1. In this case, X will play X_2 in order to obtain a payoff of 13. If he plays X_1, he will get only 10.

Saddle Point

It can be observed in the above matrix that Y_1 minimizes Y's loss and X_2 maximizes X's return with a payoff to X of 13 units. Thus the value of the game is 13, and X and Y play the pure strategies X_2 and Y_1, respectively. When a pure strategy is the best choice for each player to use, the matrix is said to have a saddle point which is at the intersection of the two pure strategies. The value of the game is the saddle value. By analogy with the discussion of saddle points previously given, a saddle value is that value which is both the maximum of the row minima and the minimum of the column maxima. In the above example the saddle value is 13 and is at the intersection of the strategies X_2 and Y_1.

Mixed Strategy

If no saddle point is found in a game there is no single safest strategy for each player. In that case a mixture of strategies is used. The opponent cannot discover the strategy if, instead of a player using a single strategy, he chooses a probability distribution over the set of strategies, a situation which combines optimization and probability. A mixed strategy for X is a vector

$$\mathbf{x} = \begin{bmatrix} x_1 \\ \cdots \\ x_n \end{bmatrix}$$

where x_i, the probability of selecting the ith strategy, satisfy

$$x_i \geq 0 \quad i = 1, \ldots, n$$

$$\sum_{i=1}^{n} x_i = 1$$

A mixed strategy for Y is a vector $\mathbf{y}$, which is similarly defined. Let $\mathbf{A}$ be the payoff matrix and let $\mathbf{x}'$ be the transpose of $\mathbf{x}$. Then the payoff to X from strategy $\mathbf{x}$ is easily shown to be $\mathbf{x}'\mathbf{Ay}$. Von Neumann's minmax theorem for finite games states that $\max_\mathbf{x} \min_\mathbf{y} \mathbf{x}'\mathbf{Ay} = \min_\mathbf{y} \max_\mathbf{x} \mathbf{x}'\mathbf{Ay} = v$.

The term on the left states that no matter what mixed strategy is chosen by X he expects at least $\min_\mathbf{y} \mathbf{x}'\mathbf{Ay}$ (the minimum taken over all Y's mixed strategies). Thus X can choose the mixed strategy $\mathbf{x}$ which maximizes this return. On the other hand, the second expression indicates that the largest payment which Y, for using mixed strategy $\mathbf{y}$, must make to X is $\max_\mathbf{x} \mathbf{x}'\mathbf{Ay}$. Player Y can then choose the strategy $\mathbf{y}$ which minimizes his loss. The theorem asserts that these two expressions have the same value v, called the value of the game.

To solve a finite game is to find vectors $\mathbf{x}$ and $\mathbf{y}$ (optimum strategies) which satisfy the minmax theorem and consequently obtain the expected payoff v, the value of the game. Note that every finite game has a solution. It can be shown that each pure strategy of either player which enters into his optimum strategy yields the value of the game when played against the opponent's optimum strategy. Also, each player has either one or an infinite number of optimum mixed strategies.

7-3. Methods of Solving Finite Two-person Zero-sum Games

There are no easy methods for solving finite games. A few of the methods will be illustrated.

The Method of Fictitious Play

This method, due to G. Brown and J. Robinson [5, 20], uses an infinite sequence of steps to solve a game. Because of its simplicity, it is presented first. Consider the zero-sum two-person game in which each player independently selects an integer from the set of integers 1, 2, 3. The player with the smaller number wins one point unless his number is less than his opponent's by one unit, in which case he loses two points. When the numbers are equal, there is no score. The payoff matrix for player X is given by

X \ Y	Y_1	Y_2	Y_3
X_1	0	-2	1
X_2	2	0	-2
X_3	-1	2	0

Payments are made by player Y to player X. The game is to be played several times. By analyzing the game in advance, it will be possible to find the optimum mixture of strategies. This game is symmetric. In a symmetric game each person has the same opportunities, that is, $a_{ij} = -a_{ji}$, where a_{ij} is the coefficient in the ith row and jth column of the payoff matrix. The value of such a game is zero, and both players have the same optimum strategies. Suppose that X, the maximizing player, chooses the strategy X_1 and that Y chooses Y_1; Y assumes that X will play X_1 again in the second round. As can be seen from the payoff matrix, he plays Y_2 in order to pay the least; in fact, in this case he wins 2. X now knows the outcome of Y's reasoning; thus for his second move he chooses X_3 which brings him the largest return against Y's Y_2 strategy. Thus, at the end of the second play X has played X_1 once and X_3 once, whereas Y has played Y_1 and Y_2.

Thus, from these choices Y expects to pay:

Play	X's strategies	Y expects to win		
		Y_1	Y_2	Y_3
1	X_1	0	2	-1
2	X_3	1	-2	0
Totals		1	0	-1

THE THEORY OF GAMES 213

Thus Y would have done better by playing Y_1 on both plays. Hence he plays Y_1 in the third play. On the other hand, X expects to win:

Play	Y's strategies	X expects to win		
		X_1	X_2	X_3
1	Y_1	0	2	−1
2	Y_2	−2	0	2
Totals		−2	2	1

And X decides that X_2 should be used in the third play since it brings him greater returns, etc.

TABLE 7-1

Play	X chooses	Y's total expectation			v_i Negative of row maxima	Y chooses	X's total expectation			$\bar{v}_i$ Row maxima
		Y_1	Y_2	Y_3			X_1	X_2	X_3	
1	X_1	0	2	−1	−2	Y_1	0	2	−1	2
2	X_3	1	0	−1	−1	Y_2	−2	2	1	2
3	X_2	−1	0	1	−1	Y_1	−2	4	0	4
4	X_2	−3	0	3	−3	Y_3	−1	2	0	2
5	X_2	−5	0	5	−5	Y_3	0	0	0	0
6	X_1	−5	2	4	−4	Y_3	1	−2	0	1
7	X_1	−5	4	3	−4	Y_3	2	−4	0	2
8	X_1	−5	6	2	−6	Y_2	0	−4	2	2
9	X_3	−4	4	2	−4	Y_2	−2	−4	4	4
10	X_3	−3	2	2	−2	Y_2	−4	−4	6	6
11	X_3	−2	0	2	−2	Y_3	−3	−6	6	6
12	X_3	−1	−2	2	−2	Y_3	−2	−8	6	6
13	X_3	0	−4	2	−2	Y_3	−1	−10	6	6
14	X_3	1	−6	2	−2	Y_3	0	−12	6	6
15	X_3	2	−8	2	−2	Y_3	1	−14	6	6
16	X_3	3	−10	2	−3	Y_1	1	−12	5	5
17	X_3	4	−12	2	−4	Y_1	1	−10	4	4
18	X_3	5	−14	2	−5	Y_1	1	−8	3	3
19	X_3	6	−16	2	−6	Y_1	1	−6	2	2
20	X_3	7	−18	2	−7	Y_1	1	−4	1	1
21	X_1	7	−16	1	−7	Y_1	1	−2	0	1
22	X_1	7	−14	0	−7	Y_1	1	0	−1	1
23	X_1	7	−12	−1	−7	Y_1	1	2	−2	2
24	X_2	5	−12	1	−5	Y_1	1	4	−3	4
25	X_2	3	−12	3	−3	Y_1	1	6	−4	6
26	X_2	1	−12	5	−5	Y_3	2	4	−4	4
27	X_2	−1	−12	7	−7	Y_3	3	2	−4	3
28	X_1	−1	−10	6	−6	Y_3	4	0	−4	4
29	X_1	−1	−8	5	−5	Y_3	5	−2	−4	5
30	X_1	−1	−6	4	−4	Y_3	6	−4	−4	6

Note that, if any two of the coefficients in the totals have the same value, one of the corresponding strategies is randomly chosen. By carrying the totals from one row to the next, 30 plays of the game give Table 7-1 ($\underline{v}_i$ and $\bar{v}_i$ are discussed below).

Note that X plays his strategies in the ratio $10/30:7/30:13/30$ obtained by recording the relative frequencies in which X_1, X_2, and X_3 appear. Similarly, this ratio for Y is given by $12/30:4/30:14/30$. It can be shown that if the limits of these calculated ratios exist then they give the desired solution. Note that they may actually oscillate when the solution is not unique. In this case every convergent subsequence defines an optimum strategy. In this game the limiting ratio will be demonstrated, by another method, to be $2/5:1/5:2/5$. In 75 plays of the game it was found that $X = (\cdot 41, \cdot 23, \cdot 36)$ and $Y = (\cdot 32, \cdot 28, \cdot 45)$. The value of the game is (approximately) estimated by selecting the finest combination of the inequalities $\underline{v}_i/i \leq v \leq \bar{v}_i/i$, $i = 1, 2, \ldots, 30$, where $\bar{v}_i$ is the maximum in the ith row of X's expectation table and $\underline{v}_i$ is the negative of the maximum in the ith row of Y's expectation table. When these inequalities are listed for the above example, it can be seen that the smallest $\bar{v}_i/i$ is zero and the largest $\underline{v}_i/i$ is $-3/25$. Therefore, v lies between these two values. The actual value of the game is, of course, zero, since the game is symmetric.

It can be shown that v is the greatest lower bound of $\bar{v}_i/i$ and the least upper bound of $\underline{v}_i/i$, $i = 1, 2, \ldots, n$. Hence, v can be estimated to any desired accuracy by increasing the number of steps. For further discussion, see Ref. [12].

Another Method

Let a zero-sum 2 by 2 game be represented as

X \ Y	Y_1	Y_2
X_1	a_{11}	a_{12}
X_2	a_{21}	a_{22}

To solve this game one begins by looking for a saddle-point solution. If there is none, then to obtain X's optimum mixed strategies the second column of the payoff matrix is subtracted from the first. The resulting column is

$$\begin{bmatrix} a_{11} - a_{12} \\ a_{21} - a_{22} \end{bmatrix}$$

Then X's optimum mixed strategy is $\mathbf{x} \equiv \begin{bmatrix} x_1 \\ x_2 \end{bmatrix}$, where

$$x_1 = \frac{|a_{21} - a_{22}|}{|a_{11} - a_{12}| + |a_{21} - a_{22}|} \quad \text{and} \quad x_2 = \frac{|a_{11} - a_{12}|}{|a_{11} - a_{12}| + |a_{21} - a_{22}|}$$

Note the use of absolute values.

For Y's mixed strategy the second row of the payoff matrix is subtracted from the first. This gives $\mathbf{y} \equiv \begin{bmatrix} y_1 \\ y_2 \end{bmatrix}$, where

$$y_1 = \frac{|a_{12} - a_{22}|}{|a_{11} - a_{21}| + |a_{12} - a_{22}|} \quad \text{and} \quad y_2 = \frac{|a_{11} - a_{21}|}{|a_{11} - a_{21}| + |a_{12} - a_{22}|}$$

Note that adding the same constant to each coefficient of the payoff matrix does not change $\mathbf{x}$ or $\mathbf{y}$. Thus, one may assume that all the coefficients are positive. The value of the game is given by multiplying $\mathbf{x}'$, the transpose of $\mathbf{x}$, by any column which is the payoff of a strategy of Y appearing in the solution for Y's strategies. In this case, either column yields the same answer because both appear in the solution. Thus

Also,
$$v = x_1 a_{11} + x_2 a_{21}$$
$$v = x_1 a_{12} + x_2 a_{22}$$

The two values of v can easily be shown to be the same. Note that the coefficients are assumed to have been transformed by adding a constant to each. This increases the value of the game by this constant amount.

How to Solve a Finite m by n Game. (See Ref. [22] for further details.)

1. If there is a saddle point, then use the pure strategy whose intersection is the saddle point. The value of the game is the saddle value.

2. Eliminate a row dominated by any other row and a column dominating any other column. Each coefficient of a dominated row is less than or equal to the corresponding coefficient of a row dominating it. Each coefficient of a dominating column is greater than or equal to the corresponding coefficient of a column which it dominates.

Remark: An optimum mixed strategy of an m by n game with $m \leq n$ contains, at most, m nonzero components.

3. Try the largest (that is, m by m) square submatrices to obtain the solution. How to handle any of these may be explained as follows:

First note the effect of the following operations on a matrix:

   a. Interchanging any two rows or any two columns does not affect the solution.

   b. Adding a constant to each element of the payoff matrix increases the value of the game by this constant and does not change the solution. Similarly, multiplying every element by a constant multiplies the value of the game by this constant.

c. The dominance principle may be applied not only when one row dominates another but also when a convex linear combination of k rows does. (If each row R_i of the collection is multiplied by a positive constant a_i with

$$0 \le a_i \le 1 \quad \sum_{i=1}^{k} a_i = 1$$

then the expression

$$\sum_{i=1}^{k} a_i R_i$$

is their convex linear combination.)

A similar statement applies to column dominance. These operations on the payoff matrix can be used to introduce zeros and eliminate fractions which will facilitate calculations.

To obtain the optimum strategy **x** for player X, subtract the last column from every other column, thus leaving an m by $m-1$ matrix. Now x_i is obtained by calculating the absolute value of the determinant of the $(m-1)$ by $(m-1)$ matrix resulting from omitting the row corresponding to X's ith strategy and dividing its value by the sum of all the calculated determinants. Similarly the optimum strategy **y** is obtained. (See the example below.) The optimum strategy for X is tested against each pure strategy of Y appearing in his optimum strategy to see if the same value of v is obtained. This provides a test for a solution. In this manner all m by m submatrices are examined for a solution. If none is obtained, all $(m-1)$ by $(m-1)$ submatrices are examined, and so on.

Example: The game presented as an illustration of the fictitious play method will now be solved by the technique discussed above. It has no saddle point. The game is

$X \backslash Y$	Y_1	Y_2	Y_3
X_1	0	-2	1
X_2	2	0	-2
X_3	-1	2	0

Solution: Subtracting the third column from each of the first two columns yields

$$\begin{bmatrix} -1 & -3 \\ 4 & 2 \\ -1 & 2 \end{bmatrix}$$

Thus $x_1 = 10/25$, $x_2 = 5/25$, $x_3 = 10/25$, which gives for X

$$\mathbf{x} = \begin{bmatrix} 2/5 \\ 1/5 \\ 2/5 \end{bmatrix}$$

if the above instructions are used.

Similarly, subtracting the third row from the first two rows and computing the determinants yield

$$\mathbf{y} = \begin{bmatrix} 2/5 \\ 1/5 \\ 2/5 \end{bmatrix}$$

Since all Y's strategies are active, the value of the game may be obtained by multiplying $\mathbf{x}$ by any column of the payoff matrix, e.g., the first, yielding

$$v = [2/5,\ 1/5,\ 2/5] \begin{bmatrix} 0 \\ 2 \\ -1 \end{bmatrix} = 0 + 2/5 - 2/5 = 0$$

Note that, in general, the solution may not be unique and thus all 2 by 2 submatrices may be examined for other solutions. In this problem the solution is unique.

The Kernel Method

An alternative method for solving finite games, the kernel method, due to H. Kuhn [16], will now be described.

A largest square nonsingular submatrix is multiplied by the corresponding component of the vector $\mathbf{x}$ and equated to the vector $\mathbf{v}$ which has identical components yet to be determined. To this system of simultaneous equations is adjoined the equation in which the sum of the x_i's under consideration is equal to unity. This system determines the x_i's and v. The transpose of the submatrix is then also multiplied by the corresponding y's from the vector Y, etc., as before. The v's determined in both cases are the same. To determine whether optimum strategies have been obtained, one multiplies each vector of the original matrix by the solution vector of x's with zeros in the appropriate positions, to determine whether v is the minmax value; if not, one proceeds to another square submatrix of highest order and so on to square matrices of lower order until a solution is obtained.

The Geometric Method

This method is restrictive in that it applies to m by 3 or 3 by m games [19] and is only mentioned in passing.

The Simplex Process

This method is applied to an example in the next section.

7-4. Relation to Linear Programming

Reduction of a Linear-programming Problem to a Game Problem

Below is an illustration of the fact that a linear-programming problem can be reduced to a game problem with a skew-symmetric payoff matrix [8]. Dantzig's method is used.

Example: Consider the following set of inequalities representing a programming problem:

$$3x_1 - x_2 + x_3 \geq 1$$
$$x_2 + 5x_3 \geq -3 \qquad (7\text{-}1)$$
$$2x_1 + 3x_2 + 4x_3 \geq \underline{M} \qquad (7\text{-}2)$$

$\underline{M}$ is the minimum of this linear form, subject to the constraints, and

$$x_i \geq 0 \qquad i = 1, 2, 3$$

This problem has the dual form

$$3y_1 \qquad\quad \leq 2$$
$$-y_1 + y_2 \leq 3 \qquad (7\text{-}3)$$
$$y_1 + 5y_2 \leq 4$$
$$y_1 - 3y_2 \leq \bar{M} \qquad (7\text{-}4)$$

$\bar{M}$ is the maximum of this linear form subject to the constraints, and

$$y_i \geq 0 \qquad i = 1, 2, 3$$

Now in (7-1) multiply the first inequality by y_1, and the second inequality by y_2, and add, obtaining

$$y_1 - 3y_2 \leq (3x_1 - x_2 + x_3)y_1 + (x_2 + 5x_3)y_2$$
$$= (3y_1)x_1 + (-y_1 + y_2)x_2 + (y_1 + 5y_2)x_3 \qquad (7\text{-}5)$$

In (7-3) multiply the first inequality by x_1, the second by x_2, the third by x_3, and add, obtaining

$$(3y_1)x_1 + (-y_1 + y_2)x_2 + (y_1 + 5y_2)x_3 \leq 2x_1 + 3x_2 + 4x_3 \qquad (7\text{-}6)$$

It is seen that $\bar{M} \leq \underline{M}$. In fact, it is known that $\bar{M} = \underline{M}$. The reverse inequality to (7-5), making use of (7-3), is given by

$$-(2x_1 + 3x_2 + 4x_3) + (y_1 - 3y_2) \geq 0 \qquad (7\text{-}7)$$

A simultaneous solution of (7-1), (7-3), and (7-7), because of relations (7-6), will be an optimizing solution.

To put (7-1), (7-3), and (7-7) in homogeneous form, set $c_j = c_j z$ and $b_i = b_i z$, where the c_j and the b_i are the constant coefficients on the right in (7-3) and (7-1), respectively.

It is desired to solve this system for $x_i \geq 0$, $y_j \geq 0$ and $z = 1$. To transform the system into a game problem, one seeks a solution to the homogeneous system, with the additional condition that

$$\sum_{i=1}^{3} x_i + \sum_{j=1}^{2} y_j + z = 1 \tag{7-8}$$

and with the restriction that $z > 0$. By dividing by z, a solution to the original system will then be obtained.

The following game is equivalent to (7-1), (7-3), and (7-7) in homogeneous form:

$$\begin{aligned} 3x_1 - x_2 + x_3 - z &\geq M \\ x_2 + 5x_3 + 3z &\geq M \\ -3y_1 + 2z &\geq M \\ y_1 - y_2 + 3z &\geq M \\ -y_1 - 5y_2 + 4z &\geq M \\ -(2x_1 + 3x_2 + 4x_3) + (y_1 - 3y_2) &\geq M \end{aligned} \tag{7-9}$$

where M is a maximum. A solution of (7-9) with $M = 0$ is equivalent to (7-1), (7-3), and (7-7) in homogeneous form. The payoff matrix associated with (7-9) is skew-symmetric and is given by

$$\begin{array}{cccccc} y_1 & y_2 & x_1 & x_2 & x_3 & z \end{array}$$
$$\begin{bmatrix} 0 & 0 & 3 & -1 & 1 & -1 \\ 0 & 0 & 0 & 1 & 5 & 3 \\ -3 & 0 & 0 & 0 & 0 & 2 \\ 1 & -1 & 0 & 0 & 0 & 3 \\ -1 & -5 & 0 & 0 & 0 & 4 \\ 1 & -3 & -2 & -3 & -4 & 0 \end{bmatrix}$$

This matrix is skew-symmetric since it has the form

$$\begin{bmatrix} 0 & \mathbf{A} & -\mathbf{b} \\ -\mathbf{A}' & 0 & \mathbf{c} \\ \mathbf{b}' & -\mathbf{c}' & 0 \end{bmatrix} \tag{7-10}$$

where $\mathbf{A}$ is the programming-problem matrix given by

$$\mathbf{A} = \begin{bmatrix} 3 & -1 & 1 \\ 0 & 1 & 5 \end{bmatrix} \tag{7-11}$$

and $\mathbf{A}'$ is the transpose of $\mathbf{A}$ and $\mathbf{b}'$ and $\mathbf{c}'$ are the transposes of the column vectors **b** and **c**, the coefficients on the right of (7-1) and on the left of (7-2), respectively.

Remark 1: The value of a game with a skew-symmetric payoff matrix is always zero. If a solution of (7-1) and (7-2) exists, then a solution of (7-9) with $M = 0$ can be obtained. Thus, an optimum mixed strategy exists for (7-10) with $z > 0$.

Remark 2: If a solution of (7-10) exists with $z > 0$ [a solution of (7-10) always exists but not necessarily with $z > 0$], then a solution of (7-1), (7-3), and (7-7) is obtained.

Reduction of a Game Problem to a Linear-programming Problem and Its Solution [8]

A finite zero-sum two-person game problem can be reduced to a linear-programming problem. The example illustrating the method of fictitious play will be used.

The minimum expected payoff for player X, if he uses a mixed strategy x_1, x_2, x_3, is given by

$$M = \min_j \sum_{i=1}^{3} a_{ij} x_j$$

where

$$\sum_{i=1}^{3} x_i = 1$$

and $x_i \geq 0$ ($i = 1, 2, 3$). The optimum mixed strategy is given by determining x_i which maximize the value of M, above.

The game problem can be written in the form

$$\begin{aligned} 2x_2 - x_3 &\geq M \\ -2x_1 + 2x_3 &\geq M \\ x_1 - 2x_2 &\geq M \\ x_1 + x_2 + x_3 &= 1 \end{aligned}$$

where $x_i \geq 0$ ($i = 1, 2, 3$).

By introducing slack variables, this system reduces to

$$\begin{aligned} 2x_2 - x_3 - x_4 &= M \\ -2x_1 + 2x_3 - x_5 &= M \\ x_1 - 2x_2 - x_6 &= M \\ x_1 + x_2 + x_3 &= 1 \end{aligned}$$

with $x_i \geq 0$ ($i = 1, \ldots, 6$).

This system of equations is reduced to a linear-programming problem by subtracting the first equation from the second and third equations. The resulting system is equivalent to the linear-programming problem of maximizing a linear form of nonnegative variables subject to the restrictions

THE THEORY OF GAMES

$$2x_2 - x_3 - x_4 = M = \max$$
$$-2x_1 - 2x_2 + 3x_3 + x_4 - x_5 = 0$$
$$x_1 - 4x_2 + x_3 + x_4 - x_6 = 0$$
$$x_1 + x_2 + x_3 = 1$$

and $x_i \geq 0$ $(i = 1, \ldots, 6)$.

It is clear that a solution of this problem exists and that the maximum in this case is the same as the maximum of the game problem above. The simplex process will be used to solve this problem. Thus, one writes

$$[P_1, P_2, \ldots, P_6; P_0] = \begin{bmatrix} -2 & -2 & 3 & 1 & -1 & 0 & 0 \\ 1 & -4 & 1 & 1 & 0 & -1 & 0 \\ 1 & 1 & 1 & 0 & 0 & 0 & 1 \end{bmatrix}$$

From $2x_2 - x_3 - x_4 = M$, one has $c_1 = 0$, $c_2 = 2$, $c_3 = -1$, $c_4 = -1$, $c_5 = 0$, $c_6 = 0$.

Let P_1, P_2, P_3 be chosen as the initial basis. It is clear that this choice is reasonable since one would expect all, or some, of these vectors to appear in the answer.

Thus, one has

$$P_0 = \tfrac{2}{5}P_1 + \tfrac{1}{5}P_2 + \tfrac{2}{5}P_3$$

and

	P_1	P_2	P_3
P_1	1	0	0
P_2	0	1	0
P_3	0	0	1
P_4	0	$-\tfrac{1}{5}$	$\tfrac{1}{5}$
P_5	$\tfrac{1}{5}$	0	$-\tfrac{1}{5}$
P_6	$-\tfrac{1}{5}$	$\tfrac{1}{5}$	0

Hence,

$z_1 = \quad c_1 \qquad\qquad\qquad = 0$
  Compare with $c_1 = 0$
$z_2 = \quad c_2 \qquad\qquad\qquad = 2$
  Compare with $c_2 = 2$
$z_3 = \qquad\qquad\quad c_3 = -1$
  Compare with $c_3 = -1$
$z_4 = 0 \times 0 + (-\tfrac{1}{5})2 + \tfrac{1}{5} \times (-1) = -\tfrac{3}{5}$
  Compare with $c_4 = -1$
$z_5 = \tfrac{1}{5} \times 0 + 0 \times 2 + (-\tfrac{1}{5})(-1) = \tfrac{2}{5}$
  Compare with $c_5 = 0$
$z_6 = (-\tfrac{1}{5}) \times 0 + \tfrac{1}{5} \times 2 + 0 \times (-1) = \tfrac{2}{5}$
  Compare with $c_6 = 0$

All $z_j \geq c_j$ and one has a maximum feasible solution.

The optimum strategy for both players is a mixed one. The numbers are chosen in the same proportion as are the coefficients appearing in the linear expression of P_0 in terms of the basis vectors; i.e., one chooses

x_1 with probability $2/5$
x_2 with probability $1/5$
x_3 with probability $2/5$

7-5. Infinite Zero-sum Games

In infinite zero-sum games, the two players X and Y choose their strategies by distribution functions $F(x)$ and $G(y)$, respectively, on the interval $[0,1]$. The payoff of Y to X corresponding to the choice of strategies is given by $M(x,y)$ (a continuum of strategies).

The expectation of X for a given strategy y of Y is given by

$$\int_0^1 M(x,y) \, dF(x)$$

But y is chosen using $G(y)$; hence the total expectation is given by

$$E(F,G) = \int_0^1 \left[\int_0^1 M(x,y) \, dF(x) \right] dG(y)$$

The minmax theorem for continuous games is as follows: If M is continuous in both variables, then the game has a solution; that is, $\max_F \min_G E(F,G) = \min_G \max_F E(F,G) = v$, and optimum strategies F^* and G^* exist such that

$$\max_F E(F,G^*) = \min_G E(F^*,G) = v$$

The actual job of solving a continuous game may be a difficult matter. Methods of solution can be prescribed for games where $M(x,y)$, in addition to being continuous, is

1. Separable, i.e., of the form $M(x,y) = \sum_{j=1}^{n} \sum_{i=1}^{m} a_{ij} r_i(x) s_j(y)$
2. Convex in Y's strategies for each of X's strategies, i.e., for every $0 \leq a \leq 1$ and pair of strategies y_1 and y_2

$$M(x, ay_1 + (1-a)y_2) \leq aM(x_1 y_1) + (1-a)M(x,y_2)$$

In addition to McKinsey's book on methods of solving infinite games, Refs. [3, 6, 7, 13] are four excellent works.

Example 1: An example of an infinite strategy game is a pistol duel in which two opponents, who start out of range, move toward each other. Each has one shot to fire. The probability of a hit increases as the range between the opponents decreases. The problem is then to determine the best instant (or distribution of instants, if more than one shot were allowed) in which a duelist should fire.

Example 2: Another example is given by encounters on a line segment of unit lengths. If the distance between the contestants is less than a prescribed amount a, the second player pays the first player a unit amount. Otherwise, he pays nothing and receives nothing. It is desired to determine optimum strategies for the players and the value of the game. This interesting simple problem is left as an exercise to the reader [3].

Example 3: Another example is given by a "Blotto game" in which a commander (Colonel Blotto) and his opponent must divide their forces between several battlefields on which simultaneous battles (not all of the same importance) are to be fought. A Blotto game with the following conditions will be considered:

1. There are three battles, B_1, B_2, B_3, with respective "payoffs" a_1, a_2, a_3 to the winner.
2. At each battle the larger force wins.
3. The total forces of Blotto and his opponent are equal in size.

A pure strategy for Blotto is a triple $\mathbf{x} = (x_1, x_2, x_3)$ with $x_i \geq 0$ and $\sum_{i=1}^{3} x_i = 1$; x_i is the fraction of his force sent to B_i. With a similar interpretation, a pure strategy for the opponent is a triple (y_1, y_2, y_3) with $y_j \geq 0$ and $\sum_{j=1}^{3} y_j = 1$.

Suppose, first, that one of a_1, a_2, a_3 exceeds the sum of the other two (i.e., one of the battles is more important than the other two taken together). In this case it is fairly clear that there is a unique optimum strategy for both sides, which consists of sending the entire force to the extremely important battle.

The case just considered is precisely the case in which the numbers a_1, a_2, a_3 cannot be the lengths of the sides of a triangle. One now considers the complementary case, in which none of a_1, a_2, a_3 exceeds the sum of the other two. In this case one can form a triangle T with sides of lengths a_1, a_2, a_3. This T is used to give a geometrical description of a certain *mixed* strategy. That this strategy is optimum will not be proved here; for such a proof see Ref. [11].

The strategy is described as follows: Inscribe a circle in T and erect a hemisphere H on this circle. Choose a point P' on H "at random," in the sense that the probability of choosing P' in any region on H is proportional to the area of the region. Project P' straight down into a point P inside (or possibly on the boundary of) T. By joining P to the vertices of T, one divides T into three triangles T_i ($i = 1, 2, 3$), where T_i has a_i

as base. Let A_i be the area of T_i. Then choose x so that

$$x_1:x_2:x_3 = A_1:A_2:A_3$$

This is a mixed strategy because P' (and thus, the final x) were chosen in a definite probabilistic way.

Remark: In showing that this mixed strategy (which is denoted by F^*) is optimum, note that the game is symmetric and, therefore, has value 0. Thus it suffices to show that F^* yields an expected payoff ≥ 0 against any y. (This shows that F^* is optimum for Blotto; by symmetry it is also optimum for his opponent.)

Example 4: Solve the game with payoff function

$$\sum_{i=1}^{n} p_i x_i e^{-k_i y_i / x_i}$$

where the strategies of X and Y are n-tuples

$$\mathbf{x} = (x_1, \ldots, x_n)$$
$$\mathbf{y} = (y_1, \ldots, y_n)$$

subject to $x_i \geq 0$, $y_i \geq 0$, $\sum_{i=1}^{n} x_i = \sum_{i=1}^{n} y_i = 1$ and where p_i and k_i are constants with $k_i > 0$.

Note that $\partial(\text{payoff})/\partial x_i > 0$ if $x_i > 0$. Thus, in any optimum $\mathbf{x}$, all $x_i = 0$ or 1. Since $\Sigma x_i = 1$, exactly one $x_i = 1$; all the rest $= 0$. Player I will, of course, choose $x_i = 1$ for that i which maximizes the resulting payoff

$$p_i e^{-k_i y_i}$$

The problem is then to find

$$\min_{y_i} \max_{i} p_i e^{-k_i y_i}$$

subject to

$$\sum_{i=1}^{n} y_i = 1$$

Suppose that $(y_1^0, \ldots, y_n^0)$ is the $\mathbf{y}$ strategy yielding the minimum. Let I_0 be the set of i's for which $p_i e^{-k_i y_i}$ is maximum with value λ. Let I_1 be the complement of this set, that is, $I_1 = I - I_0$. Thus, if i belongs to I_1, then $p_i e^{-k_i y_i^0} < \lambda$. If $y_i^0 > 0$ for this i, then decreasing y_i^0 slightly still yields

$$p_i e^{-k_i y_i^0} < \lambda$$

The amount taken from y_i^0 may be used to increase all the remaining y's whose subscripts are in I_0. This decreases the maximum value. Since

$(y_1^0, \ldots, y_n^0)$ yields the smallest maximum, a contradiction is obtained. Thus $y_i^0 = 0$ for i belonging to I_1. This proves the following:

Theorem 7-1: There is a real number λ such that

$$p_i e^{-k_i y_i^0} \begin{cases} = \lambda & \text{if } y_i^0 > 0 \\ \leq \lambda & \text{if } y_i^0 = 0 \end{cases}$$

The solution which follows from the theorem is given by

$$y_i^0 = \frac{1}{k_i} \log \frac{p_i}{\lambda} \quad \text{if } p_i > \lambda$$

and $\qquad y_i^0 = 0 \qquad \text{if } p_i \leq \lambda$

where λ is determined from the condition $\sum_{i=1}^{n} y_i^0 = 1$.

7-6. Games Against Nature [2, 14]

A game problem becomes a statistical-decision problem when one of the human opponents is replaced by circumstances or some impersonal player known as "nature." The following rules have been proposed for use in assigning a value to every pure and to every mixed action. The decision is to select that action which yields the maximum value. Suppose that the mixed action is given by $x = (x_1, \ldots, x_n)$ and let $[a_{ij}]$ be the payoff matrix; then a rule assuming indifference and due to Laplace suggests for the value

$$\sum_j \frac{1}{n} \sum_i x_i a_{ij}$$

A pessimistic rule regarding nature as opposed to the interest of the player, due to Wald, suggests for the value

$$\min_j \left[\sum_j y_j \left(\sum_i x_i a_{ij} \right) \right]$$

A "minimum regret" rule, due to Savage, gives $\min_j \left(\sum_i x_i a_{ij} - \max_k a_{kj} \right)$.

Hurwicz has suggested $a \max_j \left(\sum_i x_i a_{ij} \right) + (1 - a) \min_j \left(\sum_i x_i a_{ij} \right)$ with $0 \leq a \leq 1$ as a rule for assigning values to a game against nature. By using available information to estimate nature's strategy $y = (y_1, \ldots, y_n)$, the following rule has been suggested as a modification of Wald's rule [14]:

$$\min_y \left[\sum_j y_j \left(\sum_i x_i a_{ij} \right) \right]$$

In many practical applications, it is necessary to decide upon an objective function to maximize or to minimize subject to constraints. The function selected is often analogous in justification to one or the other of the rules given above. It is important to note the philosophical differences among these rules.

REFERENCES

1. Arrow, K. J., M. J. Beckman, and S. Karlin: Game Theory Methods Applied to the Optimal Expansion of the Capacity of a Firm, *Stanford University Department of Economics Technical Report* 27, November, 1955.
2. Blackwell, D., and M. A. Girschick: "Theory of Games and Statistical Decisions," John Wiley & Sons, Inc., New York, 1954.
3. Blackwell, D.: Game Theory, in J. F. McCloskey and F. N. Trefethen (eds.), "Operations Research for Management," Johns Hopkins Press, Baltimore, 1954.
4. Bohnenblust, H. F.: The Theory of Games, chap. 9, in E. F. Beckenbach (ed.), "Modern Mathematics for the Engineer," McGraw-Hill Book Company, Inc., New York, 1956.
5. Brown, G. W.: Iterative Solutions of Games by Fictitious Play, in T. C. Koopmans (ed.), "Activity Analysis of Production and Allocation," Cowles Commission for Research in Economics Monograph 13, John Wiley & Sons, Inc., New York, 1951.
6. Brown, R. H.: The Solution of a Certain Two-person Zero-sum Game, *Operations Research*, vol. 5, no. 1, p. 63, February, 1957.
7. Danskin, J. M., and L. Gillman: A Game over Function Space, *Rivista di Matematica*, University of Parma, vol. 4, pp. 83–94, 1953.
8. Dantzig, G. B.: A Proof of the Equivalence of the Programming Problem and the Game Problem, in T. C. Koopmans (ed.), "Activity Analysis of Production and Allocation," Cowles Commission for Research in Economics Monograph 13, John Wiley & Sons, Inc., New York, 1951.
9. Dresher, M.: Games of Strategy, *Mathematics Magazine*, vol. 25, pp. 93–99, 1951.
10. Gillman, L.: Operations Analysis and the Theory of Games: An Advertising Example, *Journal of the American Statistical Association*, vol. 45, December 1950.
11. Gross O., and R. Wagner: A Continuous Colonel Blotto Game, *RAND Corporation Research Memorandum RM-408*, June, 1950.
12. Hoffman, A. J., M. Mannos, D. Sokolowsky, and N. A. Wiegmann: Computational Experience in Solving Linear Programs, *Journal of the Society for Industrial and Applied Mathematics*, vol. 1, no. 1, pp. 17–33, 1953.
13. Index of Publications, The RAND Corporation, Santa Monica, Calif., February, 1954.
14. Isbell, J., and F. Wagner: Military Evaluation and Statistical Decision, *Ballistic Research Laboratory Report*, Aberdeen Proving Ground, Md.
15. Koopmans, T. C., (ed.): "Activity Analysis of Production and Allocation," Cowles Commission for Research in Economics Monograph 13, John Wiley & Sons, Inc., New York, 1951.
16. Kuhn, H. W.: "Lectures on the Theory of Games," Princeton University Press, Princeton, N.J., 1952.
17. Kuhn, H. W., and A. W. Tucker (eds.): "Contributions to the Theory of Games," vols. 1 and 2, Princeton University Press, Princeton, N.J., 1950 and 1953.
18. McDonald, J.: "Strategy in Poker, Business and War," W. W. Norton & Company, Inc., New York, 1950.

19. McKinsey, J. C. C.: "Introduction to the Theory of Games," McGraw-Hill Book Company, Inc., New York, 1952.
20. Robinson, J.: An Iterative Method of Solving a Game, *Annals of Mathematics*, vol. 54, pp. 296–301, 1951.
21. Von Neumann, J., and O. Morgenstern: "Theory of Games and Economic Behavior," 2d ed., Princeton University Press, Princeton, N.J., 1947.
22. Williams, J. D.: "The Compleat Strategyst," McGraw-Hill Book Company, Inc., New York, 1954.

SOME OPTIMIZATION PROBLEMS

1. Solve by elimination and describe the region defined by:
$$x + y \geq 1$$
$$x \geq y$$
$$x \leq 1$$

2. Maximize and minimize $3x + 2y$ subject to the constraints given in Prob. 1.
3. Minimize the function $x + y$ subject to the constraints $xy \geq 5$, $x \geq 0$, $y \geq 0$. Then minimize $2x + y$ subject to the same constraints.
4. Maximize the function $-x^2 + 6x - 8$ on the intervals $[0, 5/2]$ and $[0, 7/2]$. Prove that the results are the desired maxima.
5. Differentiate
$$\sum_{i=1}^{n} a_i(1 - b_i e^{x_i/y_i})y_i$$
with respect to x_i and y_i, respectively.
6. Find and classify the critical points for the following functions:
   a. $f(x) = x^3 - 3x^2 - 9x + 11$
   b. $f(x) = 12 - 24x - 15x^2 - 2x^3$
   c. $z = (6 - x)(6 - y)(x + y - 6)$
7. Determine the function $x^3 + \lambda x^2 + \mu x$ in order that it should have a maximum at $x = a$ and a minimum at $x = b$.
8. Show that $\bar{x} = \left(\sum_{i=1}^{n} x_i\right)/n$ minimizes $\sum_{i=1}^{n} r_i^2$, where $r_i^2 = (x_i - \bar{x})^2$. The square root of the expression $\left(\sum_{i=1}^{n} r_i^2\right)/n$ is called the *root-mean-square deviation*. The standard deviation is obtained by replacing n by $n - 1$ in the denominator of this expression having taken the square root. When the standard deviation is divided by $\bar{x}$, it yields the fractional standard deviation. The average deviation is given by $\left(\sum_{i=1}^{n} |r_i|\right)/n$ where $|r_i|$ is the absolute value of r_i. When this expression is divided by $\bar{x}$, it yields the fractional average deviation.
9. A farmer estimates that if he digs his potatoes now he will have 120 bushels worth $1.00 per bushel. But, if he waits, the crop will grow 20 bushels per week while the price will drop 10 cents per bushel per week. When should he dig them to obtain the largest cash return?
10. Find the dimensions for a rectangular box, without a top, with maximum capacity and a surface area of 108 square inches.
11. In each of the following cases maximize $f(x_1, x_2)$ as defined by
    a. $a_1 x_1 + a_2 x_2$ $a_i > 0$, $i = 1, 2$

b. $a_1 x_1^{k_1} + a_2 x_2^{k_2}$ $a_i > 0$, $k_i > 1$; then for $0 < k_i < 1$, $i = 1, 2$
c. $a_1(1 - e^{-b_1 x_1}) + a_2(1 - e^{-b_2 x_2})$ $a_i, b_i > 0$, $i = 1, 2$
d. $a_1(1 - e^{-b_1 x_1^{k_1}}) + a_2(1 - e^{-b_2 x_2^{k_2}})$ $a_i > 0$, $0 < k_i < 1$

subject to the constraints $x_i \geq 0$, $i = 1, 2$, and $x_1 + x_2 = c$.

Discuss those cases where the first derivative would yield a maximum in the interior of the region.

12. What are the maximum and minimum values of the function $f(x,y) = 2x^2 - 2xy + 3y^2$ in the region for which $x^2 + y^2 \leq 1$, $x + y \geq 0$?

13. Let $z = (ax + by)e^{-\alpha(cx+dy)}$, and study completely the problem of finding the maximum of z subject to the constraints $0 \leq x \leq A$, $0 \leq y \leq B$.

14. Given $f(x,y) = p(1 - e^{-ax}) + (1 - p)(1 - e^{-by})$ subject to

$$x + y \leq c$$
$$x \geq 0$$
$$y \geq 0$$

Study the problem of maximizing $f(x,y)$ subject to the given constraints by examining values of p, a, and b.

15. Calculate approximately the maximum value of $e^{2x} + e^{3y}$ on that portion of the circle of radius 1, center at the origin, lying in the first quadrant.

16. An interesting problem which occurs in statistics is to minimize $x_2 - x_1$ subject to the constraint

$$\int_{x_1}^{x_2} f(t)\, dt = c$$

Study the solution of this problem, both when t ranges over the real numbers and when it is limited to nonnegative values.

17. Let x_i ($i = 1, \ldots, n$) be the errors which arise in the measurement of a quantity whose true value is x. Thus, if the ith measurement is x_i^*, then $x_i = x_i^* - x$ ($i = 1, \ldots, n$). Find "the error function" $f(x)$ which maximizes

$$\prod_{i=1}^{n} f(x_i)$$

with the condition that

$$\sum_{i=1}^{n} x_i = 0 \qquad \int_{-\infty}^{\infty} f(x)\, dx = 1$$

This important problem, studied by Gauss, Bertrand, and Poincaré, yields the Gaussian distribution for $f(x)$. (HINT: Maximize $\sum_{i=1}^{n} \log f(x_i)$ with respect to x. Assume the series expansion for

$$g(x) \equiv \frac{f'(x)}{f(x)} = a_0 + a_1 x + a_2 x^2 + \cdots$$

Then by substituting for each x_i and taking the sum, conclude that a_1 must be arbitrary since $\sum_{i=1}^{n} x_i = 0$, and all the other coefficients are zero. Finally, solve for $f(x)$ to satisfy the conditions of the problem.)

18. Find the value of c which minimizes

$$\int_0^\infty |x - c| e^{-ax}\, dx$$

19. By applying gradient conditions to the problem of maximizing

$$f(x) = \sum_{i=1}^n c_j x_j$$

subject to the constraints

$$g_i(x) = \sum_{j=1}^n c_{ij} x_j - b_j \leq 0 \qquad i = 1, \ldots, m$$

$$x_j \geq 0 \qquad j = 1, \ldots, n$$

obtain the duality theorem of linear programming.

20. The following are the nutrient values per ounce of three well-known breakfast cereals. The units are percentage of minimum daily requirements.

Nutrient	A	B	C
Vitamin B_1	5%	25%	13%
Iron	10%	10%	10%
Phosphorus	10.8%	11.7%	15%
Niacin	12.5%	12%	15%
Price per ounce	3.12¢	5¢	5¢

Taking the three possible pairs of cereals, determine the quantity of each to be eaten each day in order to give 30 per cent of the minimum daily requirements at minimum cost. State the dual of the problem when A and B only are eaten.

21. The simplex method has been applied to a distribution problem of material produced at four cities in Italy and shipped to 10 other cities whose demands are known. Thus:

Centers of production	Quantity produced	Centers of consumption	Quantity requested
Torino	80	Aosta	35
Milano	120	Alessandria	60
Genova	79	Parma	43
Bologna	130	Verona	58
	409	Imperia	42
		La Spezia	60
		Vercelli	36
		Lucca	20
		Sondrio	18
		Ferrara	37
			409

	Aosta	Alessandria	Parma	Verona	Imperia	La Spezia	Vercelli	Lucca	Sondrio	Ferrara	Total
Torino	15 / 35	11 / 22	30	36	27 / 23	33	9	44	33	46	80
Milano	22	11 / 38	15	20 / 28	33	32	8 / 36	41	16 / 18	32	120
Genova	31	9	25	34	14 / 19	13 / 60	16	23	35	41	79
Bologna	50	27	11 / 43	17 / 30	50	23	34	17 / 20	42	6 / 37	130
Total	35	60	43	58	42	60	36	20	18	37	409

It is desired to calculate the amount shipped from each source to each destination to satisfy the supply and demand condition and to minimize the cost of transportation.

The costs of transportation are given in the corner of each square in the table on page 231, which also gives the solution in the middle. The solution gives the amounts to be shipped from each source to each destination. Note that the total cost is then 5495 which is minimum. Solve this problem.

22. Solve the following games, determining the value of the game and a pair of optimum strategies.

$$a. \begin{bmatrix} 3 & 0 & 2 \\ -4 & -1 & 3 \\ 2 & -2 & -1 \end{bmatrix} \quad b. \begin{bmatrix} 2 & 3 & 6 \\ 1 & 6 & 3 \\ 5 & 1 & 1 \end{bmatrix} \quad c. \begin{bmatrix} 3 & 0 & 2 \\ 4 & 5 & 1 \\ 2 & 3 & -1 \end{bmatrix}$$

23. What are the basic assumptions of game theory? Explain.

24. Consulting references, sketch a proof of the minmax theorem for continuous games.

25. Give examples of three possible practical applications of game theory: two for finite and one for continuous games.

26. Check your knowledge and understanding of the basic concept of analysis: By examining references, define a function, continuity, differentiability, monotonicity (monotone), critical points, roots, asymptotic properties, convexity and concavity, integrability on different intervals, and expandability in a series. For what ranges of values of the variables does the series converge to the function? What is meant by a convergent series, and what are some uses of series expansions? Illustrate by examples. Interpret these properties for a real operation from which data have been gathered and fitted with a function. Discuss these properties for the entire range of values of the variables.

27. Using Euler's equation find the shortest path between two points in the plane.

28. Suggest methods of determining the optimum of a continuous nondifferentiable function subject to constraints.

PART 3

PROBABILITY, APPLICATIONS, STATISTICS, AND QUEUEING THEORY

CHAPTER 8

PROBABILITY

8-1. Introduction

Here we pass from the deterministic models of optimization to the techniques of the theory of probability used in quantitative study of inductive inference. By assigning a number, a probability, between zero and unity to the occurrence of an event, one has a measure of the likelihood of its occurrence. For example, unity indicates that the event will occur with certainty.

Probability theory provides a variety of useful models suitable for the description and interpretation of complex problems. As in other applications of mathematics, there is usually sufficient closeness between the ideal and the real to make the results useful. Repetition of a process with varying outcomes enables the detection of order, the formulation of a model, and subsequently the development of a theory with selected measures of effectiveness. This representation is usually checked by means of actual data. When success, within prescribed limits, has been attained, the theory may then be used for prediction purposes.

In any case an application of the theory of probability generally derives its concepts from the underlying physical situation and proceeds to develop, as any nonabstract correct theory does, a logically sound system from which reliable answers may be deduced.

Roughly, the pursuits of probability and statistics may be characterized as follows: Let S be the set of experimental results (outcomes) of single trials of an experiment; let P be the set of all probability functions on S; let P^* be a subset of P; let S^* be the set of all infinite sequences of elements of S, that is, $S^* = \prod_{n=1}^{\infty} S_n$. Let f be a variate on S^*. The problem of probability is to go from P to the distribution function of f. The problem of statistics is, given an element of S^*, to draw conclusions about the distribution function in P^*.

It is clear that probability is deductive since the variate f is completely known, whereas statistics requires additional assumptions and hence is inductive.

The study of the theory of probability was initiated in the seventeenth century by Pascal and Fermat who were asked by their gambler friends to determine the odds in various games of chance. The first major treatise on probability, *Ars Conjectandi*, or "The Art of Conjecture," written by the Swiss professor Jacob Bernoulli, was published posthumously in 1713. It was followed in 1718 by DeMoivre's "The Doctrine of Chance." Both works extended the study of games of chance. The classical period in the study of probability culminated in the great 1812 work "Théorie Analytique des Probabilités," by Laplace. He thought of the theory as a guide to judgments for protection from illusions. He was primarily concerned with methods and decision rules for weighing the acceptability of assumptions [18, 23].

After the middle of the nineteenth century the word "probability" began to acquire a new meaning, and scientists turned more to statistical concepts, essentially a shift from speculative rationalization to empirical reasoning based upon evidence. More recently, investigations have followed the axiomatic approach to probability, contributing mathematical rigor to the development of the theory. The view here is that the probability of an event is a numerical quantity associated with the event and satisfying certain axioms. By the 1920s, R. A. Fisher in England, Richard von Mises and H. Reichenbach in Germany, and others began to develop new probability theories based on statistical interpretation. They were able to use many of the mathematical theorems of classical probability, which hold equally well in statistical probability. But they had to reject some. One of the rejected principles, called the *principle of indifference*, sharply points up the distinction between inductive and statistical probability. Unlike statistical probability, which refers to the actual frequency of an event, inductive probability enables the evaluation of evidence in relation to a hypothesis. With no information regarding the bias of a die, one can only assume that when the die is thrown any one of its six faces is as likely to turn up as any other, in other words, that each face has the same probability 1/6 of appearing if the die is, in fact, thrown. Gathered evidence will very likely deny a priori assumptions such as this.

It may be interesting at the outset simply to present terms of frequent occurrence in the theory of probability, keeping in mind concepts which one encounters while studying the theory. Such terms are probability of events, a priori probability, random variable, randomness, frequency distribution function, cumulative distribution function, discrete and continuous distributions, operations on distributions, conditional and marginal distributions, moments [e.g., mean (expected value) and variance], mode, median, standard deviation, moment generating functions, and characteristic functions.

8-2. Permutations and Combinations

The following concepts, derived in introductory algebra texts and presented briefly below, will undoubtedly be familiar.

PROBABILITY 237

Property 1: The number of permutations of n different things taken r at a time is given by

$$P_r^n \equiv \frac{n!}{(n-r)!} \qquad (8\text{-}1)$$

Property 2: The number of distinct permutations of n things taken n at a time, n_1 of type $t_1, \ldots, n_k$ of type t_k, is

$$\frac{n!}{n_1! \cdots n_k!} \quad \text{where} \quad \sum_{i=1}^{k} n_i = n \qquad (8\text{-}2)$$

Property 3: The number of combinations of n different things taken r at a time is given by

$$C_r^n \equiv \binom{n}{r} = \frac{n!}{(n-r)!r!} \qquad (8\text{-}3)$$

It follows that

$$C_r^n = C_{n-r}^n \qquad (8\text{-}4)$$
$$C_r^n = C_{r-1}^{n-1} + C_r^{n-1} \qquad (8\text{-}5)$$
$$C_r^n = C_{r-1}^{n-1} + C_{r-1}^{n-2} + \cdots + C_{r-1}^r + C_{r-1}^{r-1} \qquad (8\text{-}6)$$

The binomial formula gives

$$(x + a)^n = x^n + C_1^n a x^{n-1} + C_2^n a^2 x^{n-2} + \cdots \\ + C_r^n a^r x^{n-r} + \cdots + C_n^n a^n \qquad (8\text{-}7)$$

Note that the following relationship holds between two consecutive terms:

$$C_r^n = C_{r-1}^n \frac{n-r+1}{r} \qquad (8\text{-}8)$$

Example: Given 26 cells and 100 identical objects to be distributed among the cells, it is desired to find the probability that exactly three objects will be found in a specified cell. It is assumed that each of the arrangements is equally probable.

Solution: There are 26^{100} ways of placing the objects in the cells, each of which is equally probable, so that the probability of any one occurring is 26^{-100}. The three objects can be chosen in $\frac{100!}{97!3!}$ different ways, and the remaining 97 objects may be placed in the other 25 cells in 25^{97} different ways. Therefore, there are

$$\frac{100!}{97!3!} \times 25^{97}$$

different arrangements in which the specified cell contains three objects, each with probability 26^{-100}. Thus

$$\frac{100!25^{97}}{97!3!26^{100}} \sim .20$$

represents the probability that the specified cell contains exactly three objects, having used Sterling's approximation formula

$$n! \sim n^n e^{-n} \sqrt{2\pi n} \tag{8-9}$$

8-3. Discussion

The classical theory of probability utilized the a priori assumption that one outcome of an event is as likely as any other and hence the probability of such an occurrence is the reciprocal of total occurrences. Some of the defects of this theory lie in its reliance on assumptions which are usually not realizable in practice. For example, a biased die will not produce all falls with equal probability. In addition, this assumption also has theoretical difficulties revolving about the well-known controversial idea of "equal likelihood."

An operational definition of the concepts of probability is provided by the relative-frequency theory which defines the relative frequency of occurrence of an event in a series of trials as the ratio of the number of times in which the event occurs to the total number of trials. Since the frequency of occurrence depends on the number of trials, the probability of occurrence of the event is postulated as the limiting value of this ratio as the number of trials is indefinitely increased. From considerations of the relative frequency of occurrence of two events, the properties needed for calculating with probabilities are then obtained.

In the axiomatic approach [15] (axioms for events with a countable number of outcomes A_i, $i = 1, 2, \ldots$), denote the total outcomes of an event by T. Let A be any subset of the total set of outcomes. Then probability $P(A)$ is a function defined for all sets A and satisfies the following axioms:

1. $0 \leq P(A) \leq 1$.
2. $P(T) = 1$, $P(O) = 0$, where O is a never-occurring event.
3. If A_i are mutually exclusive events, then (complete additivity)

$$P\left(\sum_{i=1}^{\infty} A_i\right) = \sum_{i=1}^{\infty} P(A_i)$$

Two events A_1 and A_2 are said to be independent if the probability of their simultaneous occurrence is equal to the product of the probabilities of their separate occurrences; that is, $P(A_1 A_2) = P(A_1)P(A_2)$. This concept can be readily generalized to n events.

One has immediately the following:

1. The "total-probability" property [22]. The probability for one of the mutually exclusive events $A_1, \ldots, A_n$ to occur is the sum of the probabilities of these events. Thus, using the plus symbols as a direct sum

one has

$$P(A_1 + \cdots + A_n) = P(A_1) + \cdots + P(A_n) \tag{8-10}$$

Hence if an event A has the outcomes $A_1, \ldots, A_n$, then

$$P(A) = \sum_{i=1}^{n} P(A_i)$$

2. The "compound-probability" property. The probability of simultaneous occurrence of two events A_1 and A_2 is given as the product of the unconditional probability of event A_1 and the conditional probability of A_2, assuming that A_1 actually occurred. Thus

$$P(A_1 A_2) = P(A_1) P(A_2 | A_1) \tag{8-11}$$

This is the same as $P(A_2) P(A_1 | A_2)$ since the simultaneous occurrence of A_1 and A_2 is the same as the simultaneous occurrence of A_2 and A_1. For A_1 and A_2 we use $A_1 A_2$.

The above expression gives

$$\begin{aligned} P(A_2 | A_1) &= \frac{P(A_1 A_2)}{P(A_1)} \quad \text{if } P(A_1) > 0 \\ P(A_1 | A_2) &= \frac{P(A_1 A_2)}{P(A_2)} \quad \text{if } P(A_2) > 0 \end{aligned} \tag{8-12}$$

The last two expressions are equal to $P(A_2)$ and $P(A_1)$, respectively, if A_1 and A_2 are independent. This concept may be generalized in the obvious manner; thus, for example,

$$P(A_1 A_2 | A_3) = \frac{P(A_1 A_2 A_3)}{P(A_3)} \tag{8-13}$$

$$P(A_1 | A_2 A_3) = \frac{P(A_1 A_2 A_3)}{P(A_2 A_3)} \tag{8-14}$$

Also

$$\begin{aligned} P(A_1 A_2 A_3) &= P(A_1 A_2 | A_3) P(A_3) = P(A_1 | A_2 A_3) P(A_2 A_3) \\ &= P(A_1 | A_2 A_3) P(A_2 | A_3) P(A_3) \end{aligned} \tag{8-15}$$

3. The general law of addition of probabilities. This gives

$$P(A_1 + \cdots + A_n) = \sum_{i=1}^{n} P(A_i) - \sum_{i}\sum_{j} P(A_i A_j) \\ + \sum_{i}\sum_{j}\sum_{k} P(A_i A_j A_k) - \cdots \pm P(A_1 \cdots A_n) \tag{8-16}$$

whose proof is left as an exercise. It is clear here that, if the A_i ($i = 1$, ..., n) are mutually exclusive, then only the first term on the right is not zero.

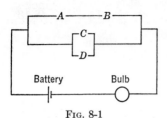

Fig. 8-1

Example 1: In the circuit in Fig. 8-1, what is the probability that the bulb will be lit (i.e., the circuit closed), given that it is equally likely for any of the switches A, B, C, D to be open or closed.

Solution: The bulb will be lit if both switches A and B are closed or if either switch C or switch D is closed. Thus the desired probability is given by

$$P(A \text{ and } B \text{ or } C \text{ or } D) = P(AB) + P(C) + P(D) \\ - P(ABC) - P(ABD) - P(CD) + P(ABCD) = 13/16$$

To see this, note that the probability of any switch being closed is $1/2$. Thus $P(C) = P(D) = 1/2$. On the other hand, $P(AB) = P(CD) = 1/4$, $P(ABC) = P(ABD) = 1/8$, and $P(ABCD) = 1/16$. In each case, because of independence, the probabilities are given by the products of the corresponding single-switch probabilities.

Example 2: In the previous example, given that the bulb is lit, calculate the probability that both switches A and B are closed.

Solution: Here one must calculate a conditional probability. Now

$$P(AB|AB + C + D) = \frac{P[AB(AB + C + D)]}{P(AB + C + D)} \\ = \frac{P(AB + ABC + ABD)}{P(AB + C + D)} = \frac{P(AB)}{P(AB + C + D)} \\ = 4/13$$

since $P(AB) = 1/4$ and $P(AB + C + D) = 13/16$. Note that
$$P(AB + ABC + ABD) = P(AB)$$

since AB comprises ABC and ABD.

Example 3: The ancient game of craps will serve as another illustration. In this game the "passer" rolls out two dice and wins if he makes 7 or 11 on the first roll (these are called "naturals"). He loses if he makes 2, 3, or 12. If he makes 4, 5, 6, 8, 9, 10 on the first roll, then he rolls until he makes the same point or gets 7. If he rolls 7 before getting the same point, he loses; otherwise he wins. What is the probability of winning?

Solution: First, consider the probability of obtaining a particular sum on a single roll of two dice. The probability of turning up a particular point on each die is clearly $1/6$. Then the probability of obtaining a 2 is $1/6 \times 1/6 = 1/36$, by the compound-probability property of independent events. Now for a 3, one can have 1 and 2 or 2 and 1. The probability then is $1/36 + 1/36 = 1/18$, applying the properties of compound and total probability. For 4, one can have 1 and 3, 3 and 1, 2 and 2 for a probability of $1/12$. By considering all possible sums in this fashion, one has the probabilities given in Table 8-1.

TABLE 8-1

No.	Probability	No	Probability
2	1/36	7	1/6
3	1/18	8	5/36
4	1/12	9	1/9
5	1/9	10	1/12
6	5/36	11	1/18
		12	1/36

The sum of all probabilities is one, as it should be, since one is certain to obtain one of these sums. The passer wins if he obtains one of the following:
1. A 7 or an 11 on the first roll, with probability P_A.
2. A repetition of any other number (a 4, 5, 6, 8, 9, or 10) appearing on the first roll, before a 7 is rolled. Let this probability be P_B. Now

$$P_{\text{win}} = P_A + P_B$$

where
$$P_A = \frac{6}{36} + \frac{2}{36} = \frac{8}{36} = \frac{2}{9}$$

To calculate P_B, note that
1. Having obtained a 4 on the first roll, the probability of obtaining another 4 on a subsequent roll before casting a 7 is the sum of the probabilities of obtaining a 4 on the first roll and on the second, obtaining a 4 on the first roll, not obtaining a 4 or a 7 on the second, then obtaining a 4 on the third roll, etc. This gives

$$\frac{3}{36} \times \frac{3}{36} + \frac{3}{36}\left(1 - \frac{3}{36} - \frac{6}{36}\right)\frac{3}{36} + \frac{3}{36}\left(1 - \frac{3}{36} - \frac{6}{36}\right)$$
$$\left(1 - \frac{3}{36} - \frac{6}{36}\right)\frac{3}{36} + \cdots$$

Simplification gives

$$\left(\frac{3}{36}\right)^2 \left(1 + \frac{3}{4} + \left(\frac{3}{4}\right)^2 + \cdots\right) = \frac{1}{36}$$

Interpreted differently, there are nine ways of obtaining 4 or 7, three of which give 4; since the probability of the latter is $3/36$, one has

$$\frac{3}{36} \times \frac{3}{9} = \frac{1}{36}$$

2. The probability of winning on 5 is

$$\frac{4}{36} \times \frac{4}{10} = \frac{8}{5 \times 36}$$

3. The probability of winning on 6 is

$$\frac{5}{36} \times \frac{5}{11} = \frac{25}{11 \times 36}$$

4. The probability of winning on 8 is the same as in 3.
The probability of winning on 9 is the same as in 2.
The probability of winning on 10 is the same as in 1.

$$P_B = 2\left(\frac{1}{36} + \frac{8}{5 \times 36} + \frac{25}{11 \times 36}\right) = \frac{2}{36}\left(1 + \frac{8}{5} + \frac{25}{11}\right) = \frac{1}{18} \times \frac{268}{55}$$

$$P_{\text{win}} = \frac{2}{9} + \frac{1}{18} \times \frac{268}{55} = \frac{220 + 268}{990} = \frac{244}{495}$$

Example 4: Given a chest with three drawers, the first containing two gold coins; the second, a gold and a silver coin; and the third, two silver coins. A drawer is selected at random and a coin is taken out. The coin is a gold one. What is the probability that the second coin in that drawer is a gold one?

Solution: Let S be the set whose elements are the coins. Let G_i and S_i ($i = 1, 2, 3$) denote a gold and a silver coin, respectively. The situation can then be represented as follows:

Drawer	Coin	
I	G_1	Denote this by x_1
I	G_2	Denote this by x_2
II	G_3	Denote this by x_3
II	S_1	Denote this by x_4
III	S_2	Denote this by x_5
III	S_3	Denote this by x_6

Adopting the equally likely assumption on coin selection, there are six possibilities and thus $P(x_i) = 1/6$ ($i = 1, \ldots, 6$). Note that if $A = (x_1, x_2, x_3)$, then $P(A) = P(x_1, x_2, x_3) = P(x_1) + P(x_2) + P(x_3) = 1/2$.

Now S can also be written as

I	$G_1 G_2$	Denote this by y_1
I	$G_2 G_1$	Denote this by y_2
II	$G_3 S_1$	Denote this by y_3
II	$S_1 G_3$	Denote this by y_4
III	$S_2 S_3$	Denote this by y_5
III	$S_3 S_2$	Denote this by y_6

Let $D = (y_1, y_2, y_3)$ and let $C = (y_1, y_2, y_4)$. The problem then asks for $P(C|D)$, the probability of the second coin being gold, given that the first one was gold, i.e., being in the set C having been in the set D. By definition,

$$P(C|D) = \frac{P(CD)}{P(D)} = \frac{P(y_1, y_2)}{1/2} = \frac{1/3}{1/2} = 2/3$$

Example 5: If two integers A and B, $B > A$, are selected at random, what is the probability that they have no common divisor?

Solution: The probability that A has a prime p as a factor is $1/p$. (Note that, in dividing a number by a prime p, the remainders of $0, 1, \ldots, p - 1$ are possible, each occurring with probability $1/p$. In particular, 0 has probability $1/p$ which means that there is a probability $1/p$ that the number be divisible by p.) The probability that both A and B contain p as a factor is $1/p^2$. The probability that both A and B do not contain p is $1 - (1/p^2)$. Now the probability that no prime number is a common factor of A and B is given by $P = p_2 \times p_3 \times p_5 \times \cdots \times p_n \times \cdots$, where p_2 is the probability that 2 is not a common factor of A and B, p_3 the probability that 3 is not a common factor of A and B, etc., the subscripts indicating primes.

Thus the chance that A and B have no common factor is

$$P = \left(1 - \frac{1}{2^2}\right)\left(1 - \frac{1}{3^2}\right)\left(1 - \frac{1}{5^2}\right) \cdots = \frac{6}{\pi^2}$$

Bayes' Theorem: The following important theorem obtains the probability of causes, given the probability of the effects. It often happens that one may wish to isolate the most likely cause to have given rise to an effect when all the possible causes are known and the conditional probability that if a certain cause occurs, then the effect is also known. This probability is determined empirically. For example, it may be desired to decide which of several allergens ingested through food is more likely to give rise to an attack of hives. One has acquired from experience the probability that hives occur on the ingestion of each of these allergens. Such questions are sometimes examined through the use of Bayes' theorem, the derivation of which is straightforward, though its use may not be due to the assumption of a priori probabilities.

Suppose that an event (a cause) B consists of finite or countably infinite independent events B_i. (The B_i may be regarded as the different ways in which an event can occur.) Then $P(B) = \Sigma P(B_i) = 1$. Let A be an event (an effect) and let $P(A)$ be the "a priori" probability of occurrence of A and $P(A|B_i)$ be the conditional probability of its occurrence, given that B_i has occurred; then the conditional probability of the cause B_i, given that A has occurred, is obtained as follows:

$$P(B_iA) = P(B_i)P(A|B_i) = P(A)P(B_i|A)$$

Thus the last two expressions give

$$P(B_i|A) = \frac{P(B_i)P(A|B_i)}{P(A)} \tag{8-17}$$

But $P(A) = P(BA) = P\left(\sum_i B_iA\right) = \sum_i P(B_iA) = \sum_i P(B_i)P(A|B_i)$

Thus the probability of the prior occurrence of cause B_i, given that the effect A has occurred, is

$$P(B_i|A) = \frac{P(B_i)P(A|B_i)}{\Sigma P(B_i)P(A|B_i)} \tag{8-18}$$

which is known as Bayes' theorem.

Example 1: In the example of the chest of drawers with gold and silver coins, Bayes' theorem immediately gives for the answer

$$\frac{1/2 \times 1}{(1/2 \times 1) + (1/2 \times 1/2)} = \frac{2}{3}$$

Example 2: A ball was transferred from an urn containing two white and two black balls to another containing three white and two black balls. A white ball was then drawn from the second urn. What is the probability that the transferred ball was white?

Solution: Let B_1 and B_2 be the two events of transferring a white ball and a black ball, respectively, and let A be the event of drawing a white ball from the second urn.

Now
$$P(B_1|A) = \frac{P(B_1)P(A|B_1)}{\sum_{i=1}^{2} P(B_i)P(A|B_i)}$$

$P(B_1) = 1/2 \qquad P(A|B_1) = 2/3 \qquad P(B_2) = 1/2 \qquad P(A|B_2) = 1/2$

Thus $\qquad P(B_1|A) = \dfrac{1/2 \times 2/3}{(1/2 \times 2/3) + (1/2 \times 1/2)} = \dfrac{4}{7}$

An application of Bayes' theorem to the problem of estimating population parameters from sample information will be given in a later chapter.

8-4. Random Variable and Distribution Function

Suppose that with each outcome of an experiment there is associated a real number. Such a real number, whose value is determined by chance in each try (as are the outcomes of the experiment), is called a *random variable, chance variable, variate,* or *stochastic variable.* Table 8-1, giving the probabilities in the game of craps, assigns to each number a probability with which that number may appear on a throw of two dice. The two dice are assumed unbiased. Each of these dice values constitutes a value of a random variable since they are determined by chance. The probabilities for the entire set of values define a *frequency distribution* or *density function.* Roughly speaking, this distribution theoretically describes the relative frequency with which that value would appear in a large number of throws. In this case, the frequency distribution gradually rises to a maximum at 7 after which it declines in such a way that it is symmetrical. In the case of dice one turns up faces with which numbers are associated and can be used as values of a stochastic variable. Such numbers are usually attached to outcomes of an experiment in studying probabilities. For example, numbers 1, 2, 3 may be attached to selected targets (regions) of a battlefield, and the probability of scoring a hit may be different in each of them. With each of the numbers assigned is associated a probability.

The outcomes of an experiment need not be discrete. In that case one usually specifies the frequency distribution or often fits a set of discrete data with a continuous frequency function if it is known that this is the type of distribution required. For example, the duration of telephone conversations, which may be assumed to take on any value, has a continuous frequency distribution. Conversations with long duration have smaller probabilities associated with them than short ones. With each time value there is associated a probability.

An interesting question at this point is how to determine the probability that a conversation will terminate at any time prior to a prescribed time. This gives rise to the concept of a cumulative distribution function. The discrete example using a pair of dice will again be used to illustrate the ideas.

The cumulative distribution at any of the numbers 2 to 12 gives the probability of obtaining that number or any number less than it. Thus, the cumulative distribution at 3 is $1/36 + 2/36 = 3/36$. At 4, one has the probability of getting a 2, a 3, and a 4, which gives $6/36$. In this manner, one has the probabilities given in Table 8-2.

TABLE 8-2

No.	Cumulative probability	No.	Cumulative probability
2	1/36	7	21/36
3	3/36	8	26/36
4	6/36	9	30/36
5	10/36	10	33/36
6	15/36	11	35/36
		12	36/36

Note that the value beside the number 12 is one which says that it is certain that one of the numbers from 2 to 12 will appear. The number 35/36 gives the probability that one of the numbers 2 to 11 will appear, etc.

With the foregoing explanation, let x be any given number and let X be a random variable. The probability that $X \leq x$ is evidently a function of x, designated by $F(x)$. One writes

$$F(x) = P(X \leq x) \tag{8-19}$$

The function $F(x)$ is called the *cumulative distribution function* of the random variable X. It is known when the law of probability of X has been determined. For example, for the case of the total value appearing on two dice, one has

$$F(x) = 0 \text{ for } x \leq 2 \quad \text{and} \quad F(x) = 1 \text{ for } x > 12$$
$$F(4) = P(x \leq 4) = \frac{1}{36} + \frac{2}{36} + \frac{3}{36} = \frac{1}{6}$$

which is the sum of the probabilities of producing 2, 3, and 4. When the possible values $x_1, x_2, \ldots, x_n \ldots$ of a random variable are finite or countable, it is said to be discretely distributed with $P(X = x_i) = p_i$, $i = 1, \ldots, n, \ldots$. Thus

$$\sum_i p_i = 1 \quad F(x) = \sum_{x_i \leq x} p_i$$

In the continuous case, every random variable X has a cumulative distribution function $F(x)$. However, X has a density function $f(x)$ only if $F(x)$ is differentiable and $f(x) = dF(x)/dx$. In that case,

$$F(x) = \int_{-\infty}^{x} f(t) \, dt \quad P(a \leq X \leq b)$$
$$= \int_a^b f(t) \, dt \quad P(x_0 \leq X \leq x_0 + dx) = f(x_0) \, dx$$

This is called the *probability element* of X. Note that $\int_{-\infty}^{\infty} f(x) \, dx = 1$.

In the discrete case the relation between the density function and the cumulative distribution function is given by

$$p_i \equiv P(X = x_i) = F(x_i) - F(x_{i-1})$$

A random variable is continuously distributed if its density function is everywhere continuous except perhaps at isolated points of discontinuity.

In either case the cumulative distribution function has the following properties:
1. $a \leq b$; then $F(b) - F(a) \geq 0$, that is, is monotone increasing.
2. $F(-\infty) = 0, F(\infty) = 1$.
3. $F(x + 0) = F(x)$.

Note from 1 and 2 that $F(x)$ is nonnegative.

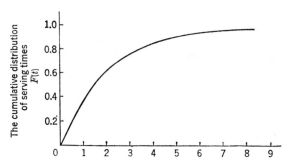

Fig. 8-2. The time axis t. It gives the length of customer-serving time at a single-counter store.

When calculating with probabilities involving a single variable the following properties will be found useful: Let $a \leq b$; then

$$P(X \leq b) = P(X \leq a) + P(a < X \leq b)$$
$$P(a < X \leq b) = F(b) - F(a)$$
$$P(X = x) = F(x) - F(x - 0)$$
$$P(a < X < b) = F(b - 0) - F(a)$$
$$P(a \leq X \leq b) = F(b) - F(a - 0)$$
$$P(a \leq X < b) = F(b - 0) - F(a - 0)$$
$$P(-\infty < X < \infty) = 1$$

The Bivariate Case

The cumulative distribution of the random variables X_1 and X_2 is defined by

$$F(x_1, x_2) \equiv P(X_1 \leq x_1, X_2 \leq x_2) \qquad (8\text{-}20)$$

Their joint density function, if it exists, is defined by

$$f(x_1, x_2) = \frac{\partial^2 F}{\partial x_1 \, \partial x_2} \qquad (8\text{-}21)$$

from which
$$F(x_1, x_2) = \int_{-\infty}^{x_1} \int_{-\infty}^{x_2} f(y_1, y_2) \, dy_1 \, dy_2 \qquad (8\text{-}22)$$

One also has

$$P(a < X_1 \leq b, c < X_2 \leq d) = \int_a^b \int_c^d f(x_1,x_2)\, dx_2\, dx_1 \quad (8\text{-}23)$$

Here, as before, $\int_{-\infty}^{\infty} \int_{-\infty}^{\infty} f(x_1,x_2)\, dx_1\, dx_2 = 1$

As in the single-variable case, $F(x_1,x_2)$ has similar properties with respect to each of the variables. The random variables X_1 and X_2 are said to be independent if $f(x_1,x_2) = f_1(x_1)f_2(x_2)$. In practice it is often difficult to decide whether two variables are independent.

The concepts of marginal and conditional distributions are also employed in probability theory. The marginal distribution of X_1 is given by

$$\int_{-\infty}^{\infty} f(x_1,x_2)\, dx_2$$

and that of X_2 is given by

$$\int_{-\infty}^{\infty} f(x_1,x_2)\, dx_1$$

The conditional frequency distribution $g(x_2|x_1)$ of X_2, given X_1, is obtained from the definition of conditional probability given earlier, which yields

$$f(x_1,x_2) = g(x_2|x_1) \int_{-\infty}^{\infty} f(x_1,x_2)\, dx_2 \quad (8\text{-}24)$$

Thus

$$g(x_2|x_1) = \frac{f(x_1,x_2)}{\int_{-\infty}^{\infty} f(x_1,x_2)\, dx_2} \quad (8\text{-}25)$$

which, when integrated over x_2, yields unity.

The above concepts may be readily extended to several variables.

8-5. Moments

Useful measures in probability theory are obtained through the calculation of moments. If, for example, on the toss of a single die, the chances of the numbers 1 to 6 turning up are all the same, i.e., 1/6, the expected value is the average of all the numbers, each multiplied by its chance of coming up. One has,

$$1 \times \tfrac{1}{6} + 2 \times \tfrac{1}{6} + 3 \times \tfrac{1}{6} + 4 \times \tfrac{1}{6} + 5 \times \tfrac{1}{6} + 6 \times \tfrac{1}{6} = 3\tfrac{1}{2}$$

This tells the die thrower that, over a large number of throws, the numbers he turns up will average in value between 3 and 4.

The expected value or mean μ of a random variable X is also denoted by $E[X]$ and is the abscissa of the center of gravity of the distribution of mass corresponding to X. It is given by the moment about zero

$$E[X] = \Sigma_i x_i p_i \tag{8-26}$$

if X is discrete, and by

$$E[X] = \int_{-\infty}^{\infty} x f(x)\, dx \tag{8-27}$$

if X is continuous with density $f(x)$. Both the series and the integral may diverge in which case the expected value would not exist. The expected value of a function of X, $g(X)$ in the continuous case, for example, is given by

$$E[g(X)] = \int_{-\infty}^{\infty} g(x)f(x)\, dx \tag{8-28}$$

From the definition, note that

$$E[aX + b] = aE[X] + b \tag{8-29}$$

which shows that, by a change of the origin and without changing the unit of measurement or the orientation, it is possible to transform to the expected value

$$E[X] = 0$$

The variance σ^2 of X measures dispersion and is given as the second moment about the mean. It is defined by

$$\sigma^2 \equiv E[(X - \mu)^2] = \Sigma(x_i - \mu)^2 p_i \quad \text{if } X \text{ is discrete} \tag{8-30}$$

$$\sigma^2 \equiv E[(X - \mu)^2] = \int_{-\infty}^{\infty} (x - \mu)^2 f(x)\, dx \quad \text{if } X \text{ is continuous} \tag{8-31}$$

One immediately has

$$\sigma^2 = E[X^2] - \{E[X]\}^2 \tag{8-32}$$

The square root of the variance σ is called the *standard deviation*. Higher moments given by $E[X^n]$ may be similarly defined about selected reference points.

The Cauchy distribution $f(x) = 1/\pi[1 + (x - \mu)^2]$ has a finite mean obtained from an improper integral but no higher moments. The following example due to Khintchine shows how the mean and variance can play an important role in forming a model and deriving results.

Example: This example deals with the length of a single-channel queue in the steady state [16]. Suppose that Poisson arrivals to a waiting line, in statistical equilibrium, before a single counter occur at the rate of λ and are served at a rate of μ with $\lambda/\mu < 1$. Suppose that a customer's departure leaves q in line including the one being served; service time is t. Let r customers arrive during this time t. If the next person leaves q'

customers behind, then

$$q' = \max(q - 1, 0) + r = q - 1 + \delta + r$$

where
$$\delta(q) = \begin{cases} 0 & \text{if } q > 0 \\ 1 & \text{if } q = 0 \end{cases}$$

It is assumed that equilibrium values for $E[q]$ and $E[q^2]$ exist. Now $\delta^2 = \delta$ and $q(1 - \delta) = q$; also $E[q] = E[q']$ and $E[q^2] = E[q'^2]$ since both q and q' are assumed to have the same distribution. Thus

$$E[\delta] = 1 - E[r] = 1 - \lambda/\mu$$

is the probability that a departing customer leaves an empty counter behind him.

Also from the first equation one has

$$q'^2 = q^2 - 2q(1 - r) + (r - 1)^2 + \delta(2r - 1)$$

Taking expectations (r is independent of q and δ) yields for the expected number waiting and being served

$$E[q] = E[r] + \frac{E[r(r-1)]}{2\{1 - E[r]\}} = \frac{\lambda}{\mu} + \frac{\text{variance } (t) + 1/\mu^2}{2/\lambda \, (1/\lambda - 1/\mu)} \quad (8\text{-}33)$$

Other measures useful in studying probability distributions are the median and the mode. In the continuous case the median is that x which yields

$$\int_{-\infty}^{x} f(x) \, dx = 1/2 \quad (8\text{-}34)$$

In the discrete case the median value minimizes

$$\sum_{j=1}^{n} |x - x_j| \quad (8\text{-}35)$$

where x_j ($j = 1, \ldots, n$) are the possible value of x. The median coincides with the mean in the case of a symmetric distribution.

The mode is given by an x which maximizes the density function $f(x)$. In the discrete case it coincides with the largest x_j. The median and the mode may not be unique.

The Moment Generating Function

The moment generating function of a random variable X is defined by

$$m(t) \equiv E[e^{tX}]$$

It facilitates the computation of moments where the kth moment is obtained by differentiating the moment generating function k times and allowing t to tend to zero. Thus, for example, if

$$f(x) = \frac{e^{-\mu}\mu^x}{x!} \qquad x = 0, 1, 2$$

then
$$m(t) = \sum_{x=0}^{\infty} \frac{e^{xt}e^{-\mu}\mu^x}{x!} = e^{-\mu}e^{\mu e^t}$$

The first derivative gives

$$m'(t) = e^{-\mu}\mu e^t e^{\mu e^t}$$

The second derivative gives

$$m''(t) = e^{-\mu}\mu e^t e^{\mu e^t}(1 + \mu e^t)$$

Thus, the first moment is

$$m'(0) = \mu$$

and the second moment about the origin is

$$m''(0) = \mu(1 + \mu)$$

from which the variance is obtained as

$$\sigma^2 = \mu(1 + \mu) - \mu^2 = \mu$$

Moments of continuous variates are analogously obtained.

If two continuous density functions have the same set of moments and if the difference of the density functions has a series expansion about the origin, then the two density functions are identical.

Also, if two random variables have the same moment generating function, then they have the same density function [9, 10].

The Characteristic Function

In the nondiscrete case the formulas for the mean value, the variance, and the moment generating function, for example, do not resolve the problem of calculating these quantities in case the distribution functions are discontinuous. It would be desirable to give a definition by a unique formula which includes all possible cases and without making special assumptions on $F(x)$. This is accomplished by means of the Stieltjes integral.

The simplest definition of a Stieltjes integral $\int_a^b f(x)\, dg(x)$ is that it is the limit of approximating sums:

$$\sum_{i=0}^{n} f(\zeta_i)[g(x_i) - g(x_{i-1})]$$

where the x_i are subdivision points of the interval (a,b) and $x_{i-1} < \xi_i < x_i$. This reduces to the ordinary Riemann integral if $g(x_i) \equiv x_i$ ($i = 0, 1, \ldots, n$). This complicated limit may not exist for arbitrary $f(x)$ and $g(x)$. However, it does exist if, for example, $f(x)$ is continuous in $[a,b]$ and $g(x)$ is a nondecreasing function. The two integrals

$$\int_a^b f(x)\,dg(x) \qquad \text{and} \qquad \int_a^b f(x)g'(x)\,dx$$

are equal if they exist and if $g(x)$ has a derivative $g'(x)$ at every point in $[a,b]$.

Since the cumulative distribution function $F(x)$ is nondecreasing, one may define moments by

$$\int_{-\infty}^{\infty} x^n\,dF(x)$$

This gives a unique representation which, at the same time, includes the different cases which arise.

The moment generating function, although useful, does not exist for every random variable. Note that e^{tx} is unbounded and hence the integral may not exist. A slight modification of the definition of a generating function yields the characteristic function defined by the Fourier Stieltjes transform:

$$\int_{-\infty}^{\infty} e^{itx}\,dF(x) \tag{8-36}$$

which is well defined for all real values of t. Here, of course, $|e^{itx}| = 1$. A characteristic function is continuous, with a value not exceeding unity, and is unity when $t = 0$. It is generally complex but is real when $F(x)$ is absolutely continuous and $f(x)$ is such that $[f(x) = f(-x)]$, that is, symmetric with respect to the origin.

Remark: $F(x)$ is absolutely continuous if the probability density $f(x)$ is defined on the interval $(-\infty, \infty)$:

$$f(x) \geq 0 \qquad \int_{-\infty}^{\infty} f(x)\,dx = 1 \qquad \text{and} \qquad F(x) = \int_{-\infty}^{x} f(t)\,dt$$

The fundamental theorem here is that a random variable is completely determined by its characteristic function. Moments can be obtained from the characteristic function in a way similar to that in which they were obtained from the moment generating function. The first moment, for example, is obtained by differentiating with respect to t, multiplying by $-i$, and setting $t = 0$. There are times when L'Hôspital's well-known rule must be used to take care of a resulting indeterminate form. For example, the characteristic function of the uniform distribution given in a later section requires this step to yield the first moment. If X_1 and X_2 are independent with characteristic functions φ_1 and φ_2, then $\varphi_1\varphi_2$ is the characteristic function of $X_1 + X_2$.

Generalizations

The generalization of the definition of moments to the bivariate case is immediate. In the continuous case the expected value of X_1 is given by (the Stieltjes integral can also be used)

$$\bar{x}_1 = \int_{-\infty}^{\infty} \int_{-\infty}^{\infty} x_1 f(x_1, x_2) \, dx_2 \, dx_1 \tag{8-37}$$

and its variance $\sigma_{x_1}^2$ by

$$\sigma_{x_1}^2 = \int_{-\infty}^{\infty} \int_{-\infty}^{\infty} (x_1 - \bar{x}_1)^2 f(x_1, x_2) \, dx_2 \, dx_1 \tag{8-38}$$

Frequently μ_{x_1} is used instead of $\bar{x}_1$.

Notice that integration with respect to x_2 will yield the marginal distribution of X_1. One similarly defines $\bar{x}_2$ and $\sigma_{x_2}^2$. The covariance of X_1 and X_2 is defined by

$$\sigma_{x_1 x_2} \equiv E[(X_1 - \bar{x}_1)(X_2 - \bar{x}_2)]$$
$$= \int_{-\infty}^{\infty} \int_{-\infty}^{\infty} (x_1 - \bar{x}_1)(x_2 - \bar{x}_2) f(x_1, x_2) \, dx_1 \, dx_2 \tag{8-39}$$

which is easily shown to equal zero if X_1 and X_2 are independent. The converse of this fact is not true. The covariance may be zero without X_1 and X_2 being independent. The correlation coefficient r of the variables X_1 and X_2 is defined by

$$r = \frac{\sigma_{x_1 x_2}}{\sigma_{x_1} \sigma_{x_2}} = \frac{E[X_1 X_2] - \bar{x}_1 \bar{x}_2}{\sigma_{x_1} \sigma_{x_2}} \tag{8-40}$$

Note that

$$E[X_1 + X_2] = E[X_1] + E[X_2] \tag{8-41}$$

If X_1 and X_2 are independent, then

$$E[X_1 X_2] = E[X_1] E[X_2] \qquad E[X_1 - \bar{x}_1)^2 (X_2 - \bar{x}_2)^2]$$
$$= \bar{x}_1^2 \sigma_{x_2}^2 + \bar{x}_2^2 \sigma_{x_1}^2 + \sigma_{x_1}^2 \sigma_{x_2}^2 \tag{8-42}$$

An illustration of the above ideas is provided by the multivariate normal distribution. (The reader may desire to specialize the calculations and obtain the results for the two-variable case.)

$$f(x_1, \ldots, x_n) = \left(\frac{1}{2\pi}\right)^{n/2} \sqrt{|\sigma^{ij}|} \exp\left[-\frac{1}{2} \sum_{i=1}^{n} \sum_{j=1}^{n} \sigma^{ij} (x_i - \bar{x}_i)(x_j - \bar{x}_j)\right] \tag{8-43}$$

where $|\sigma^{ij}|$ is the determinant of the matrix $[\sigma^{ij}]$ which is the inverse of the covariance matrix $[\sigma_{ij}]$ with coefficients defined by

$$\sigma_{ij} = E[(X_i - \bar{x}_i)(X_j - \bar{x}_j)] \tag{8-44}$$

The covariance terms are off the main diagonal, and each diagonal element is a variance. The marginal distribution of a subset of the variables, say the first $m (m < n)$, is given by

$$\left(\frac{1}{2\pi}\right)^{m/2} \sqrt{|\sigma^{kl}|} \exp\left[-\frac{1}{2}\sum_{k=1}^{m}\sum_{l=1}^{m} \sigma^{kl}(x_k - \bar{x}_k)(x_l - \bar{x}_l)\right] \quad (8\text{-}45)$$

In this case the matrix $[\sigma^{kl}]$ is the inverse of the submatrix obtained from $[\sigma_{ij}]$ by striking out the last $n - m$ rows and columns.

The conditional distribution is given by

$$\left(\frac{1}{2\pi}\right)^{(n-m)/2} \frac{\sqrt{|\sigma^{ij}|}}{\sqrt{|\sigma^{kl}|}} \exp\left[-\frac{1}{2}\Biggl(\sum_{i,j=1}^{n} \sigma^{ij}(x_i - \bar{x}_i)(x_j - \bar{x}_j)\right.$$

$$\left.- \sum_{k,l=m+1}^{n} \sigma^{kl}(x_k - \bar{x}_k)(x_l - \bar{x}_l)\Biggr)\right] \quad (8\text{-}46)$$

8-6. Transformations

It is sometimes necessary to obtain the distribution of a variable which is given as a function of another variable whose distribution is known. If, for example, X_1 and X_2 are random variables related by $X_2 = aX_1 + b$, where a and b are constants with $a > 0$, and if $F(x_1)$ and $G(x_2)$ are their cumulative distribution functions, then [24]

$$G(x_2) = P(X_2 \leq x_2) = P(aX_1 + b \leq x_2) = P\left(X_1 \leq \frac{x_2 - b}{a}\right)$$

$$= F\left(\frac{x_2 - b}{a}\right) \quad (8\text{-}47)$$

Let X_1 be a continuous random variable with probability element $f(x_1)\, dx_1$ and let $x_2 = g(x_1)$ be a monotone function with inverse

$$x_1 = g^{-1}(x_2)$$

such that $g'(x_1)$ exists. Then the probability element $h(x_2)\, dx_2$ of X_2 is given by

$$h(x_2) = f(x_1)\frac{dx_1}{dx_2}$$

expressed in terms of x_2. To illustrate, let

$$f(x_1)\, dx_1 = e^{-x_1}\, dx_1 \quad \text{if } x_1 \geq 0$$
$$= 0\, dx_1 \quad \text{if } x_1 < 0$$

Find $P(x_2 < x_1^2 < x_2 + dx_2)$, that is, the probability element $h(x_2)\, dx_2$ of x_2. One has the transformation $x_2 = x_1^2$ (monotone for $x_1 \geq 0$) or

$x_1 = \sqrt{x_2}$ and hence

$$h(x_2)\,dx_2 = e^{-x_1}\left|\frac{dx_1}{dx_2}\right|dx_2 = e^{\sqrt{x_2}}\frac{1}{2\sqrt{x_2}}dx_2 \qquad (8\text{-}48)$$

Remark: If the transformation is not monotone, it can often be divided into monotone parts. The theorem is then applied to each part and the sum taken.

In the case of two random variables, one has

$$P(y_1 < Y_1 \leq y_1 + dy_1;\, y_2 < Y_2 \leq y_2 + dy_2)$$
$$= f(x_1,x_2)\left|\frac{\partial(x_1,x_2)}{\partial(y_1,y_2)}\right|dy_1\,dy_2 \qquad (8\text{-}49)$$

which involves the Jacobian of x_1 and x_2 with respect to y_1 and y_2.

8-7. Distributions of Sums and Products

Frequently one is concerned with the distribution of sums or products of variables whose joint distributions are known. The theorems and illustrations given here provide a basis for such calculations.

Distribution of the Sum of Two Independently Distributed Variables

If X and Y are independently distributed with densities $f(x)$ and $g(y)$, respectively, then the density function of $Z = X + Y$ is given by

$$h(z) = \int_{-\infty}^{\infty} g(z - x)f(x)\,dx$$

If the cumulative distributions are $F(x)$ and $G(y)$, then the cumulative distribution of their sum is

$$H(z) = \int_{-\infty}^{\infty} G(z - x)\,dF(x) \qquad (8\text{-}50)$$

Example: Let

$$f(x) = \frac{1}{\sigma_1\sqrt{2\pi}}e^{-(x^2/2\sigma_1^2)}$$

$$g(y) = \frac{1}{\sigma_2\sqrt{2\pi}}e^{-(y^2/2\sigma_2^2)}$$

Then $\quad h(z) = \int_{-\infty}^{\infty} \frac{1}{\sigma_2\sqrt{2\pi}}\exp\left[\frac{-(z-x)^2}{2\sigma_2^2}\right]\frac{1}{\sigma_1\sqrt{2\pi}}\exp\left(-\frac{x^2}{2\sigma_1^2}\right)dx$

Let $\quad t = \dfrac{\sigma_1^2 z}{\sigma_1^2 + \sigma_2^2}$

Then

$$h(z) = \frac{1}{2\pi\sigma_1\sigma_2}\exp\left[-\frac{z^2}{2(\sigma_1^2+\sigma_2^2)}\right]\int_{-\infty}^{\infty}\exp\left(\frac{-\sigma_1^2-\sigma_2^2}{2\sigma_1^2\sigma_2^2}\right)(x-t)^2\,dx$$

Let $\quad u = \dfrac{\sqrt{\sigma_1^2 + \sigma_2^2}}{\sigma_1\sigma_2}(x - t)$

Then the integral on the right becomes

$$\frac{\sigma_1\sigma_2}{\sqrt{\sigma_1^2+\sigma_2^2}}\int_{-\infty}^{\infty} e^{-u^2/2}\,du = \frac{\sigma_1\sigma_2\sqrt{2\pi}}{\sqrt{\sigma_1^2+\sigma_2^2}}$$

Thus $\quad h(z) = \dfrac{1}{\sqrt{\sigma_1^2+\sigma_2^2}\sqrt{2\pi}}\exp\left[\dfrac{-z^2}{2(\sigma_1^2+\sigma_2^2)}\right]\quad$ (8-51)

Quotient Theorem 8-1 [7]

If the random variables X, Y have the joint frequency function $f(x,y)$, then the distribution of $Z = X/Y$ is given by

$$H(z) = \int_{-\infty}^{z}\left[\int_{-\infty}^{\infty} |v|f(uv,v)\,dv\right]du \quad (8\text{-}52)$$

where $x = uv$, $y = v$, and $|v|$ is the Jacobian of this transformation.

The frequency distribution of Z exists and is given by the formula

$$h(z) = H'(z) = \int_{-\infty}^{\infty} |v|f(zv,v)\,dv$$

If X and Y are independent with respective distribution functions $F(x)$ and $G(y)$, the distribution function of the quotient X/Y is given by the formula

$$H(z) = G(0) + \int_{0}^{\infty} F(zw)\,dG(w) - \int_{-\infty}^{0} F(zw-0)\,dG(w) \quad (8\text{-}53)$$

for all values of z.

Product Theorem 8-2 [7]

If the random variables Y, Z have the joint frequency function $f(y,z)$, then the distribution of $X = YZ$ is given by

$$F(x) = \int_{-\infty}^{x}\left[\int_{-\infty}^{\infty}\left|\frac{1}{v}\right|f\left(\frac{u}{v},v\right)dv\right]du$$
$$= \int_{-\infty}^{x}\left[\int_{-\infty}^{\infty}\left|\frac{1}{v}\right|f\left(v,\frac{u}{v}\right)dv\right]du \quad (8\text{-}54)$$

The frequency function of the distribution of X exists and is given by the formula

$$f(x) = F'(x) = \int_{-\infty}^{\infty}\left|\frac{1}{v}\right|f\left(\frac{x}{v},v\right)dv = \int_{-\infty}^{\infty}\left|\frac{1}{v}\right|f\left(v,\frac{x}{v}\right)dv \quad (8\text{-}55)$$

If Y and Z are independent chance variables with respective distribution functions $G(y)$ and $H(z)$, then the distribution function of X is given by the formula

$$F(x) = \int_{0+}^{\infty} H\left(\frac{x}{v}\right) dG(v) - \int_{-\infty}^{0-} H\left(\frac{x}{v} - 0\right) dG(v) + \begin{cases} G(-0) & x < 0 \\ G(0) & x \geq 0 \end{cases} \quad (8\text{-}56)$$

for all values of x.

Here is an example of the distribution of quotients of variates [6]. If Y and Z are independent variables, where Y has the normal distribution

$$\frac{1}{\sqrt{2\pi}} e^{-v^2/2}$$

and Z has the chi-square distribution with n degrees of freedom

$$\frac{z^{n/2-1} e^{-z/2}}{2^{n/2} \Gamma(n/2)} \quad \text{where } \Gamma(a) \equiv \int_0^{\infty} x^{a-1} e^{-x} \, dx$$

and $-\infty < y < \infty$, $0 < z < \infty$, then the distribution of the variable

$$t = \sqrt{n} \, \frac{Y}{\sqrt{Z}}$$

which is known as "student's t," may be obtained as follows:

Let $H(t)$ be the cumulative distribution of t and let C be an unspecified constant depending on n; then

$$H(x) = P(t \leq x) = P\left(Y \leq \frac{x}{\sqrt{n}} \sqrt{z}\right)$$
$$= C \iint z^{n/2-1} e^{-(y^2/2)-(z/2)} \, dy \, dz \quad (8\text{-}57)$$

with limits $-\infty < y \leq x\sqrt{Z}/\sqrt{n}$ for the inner integral and $0 < z < \infty$ for the outer integral, where the integrand is the joint frequency function of Y and Z (the constant factors are absorbed in C). Note that obtaining the joint distribution of Y and Z—the product in this case due to their independence—is the first step. The above expression can be shown to reduce to

$$H(x) = C \int_0^{\infty} z^{n/2-1} e^{-z/2} \, dz \int_{-\infty}^{x\sqrt{z}/\sqrt{n}} e^{-v^2/2} \, dy$$
$$= C \int_0^{\infty} z^{(n-1)/2} \, dz \int_{-\infty}^{x} e - \left(\frac{u^2 z}{2n} + \frac{z}{2}\right) du$$
$$= C \int_{-\infty}^{x} du \int_0^{\infty} z^{(n-1)/2} \exp\left[-\frac{1}{2}\left(1 + \frac{u^2}{n}\right) z\right] dz$$
$$= C \int_{-\infty}^{x} \frac{du}{(1 + u^2/n)^{(n+1)/2}}$$

The frequency distribution function $h(x)$ of t is obtained from

$$H'(x) = h(x) = \frac{C}{(1 + x^2/n)^{(n+1)/2}}$$

C to be determined from the fact that $\int_{-\infty}^{\infty} h(x) = 1$, which gives

$$C = \frac{1}{\sqrt{n\pi}} \frac{\Gamma[(n+1)/2]}{\Gamma(n/2)} \qquad (8\text{-}58)$$

8-8. Some Distributions of Frequent Occurrence

In this section a brief account is given of a few distributions which have a wider use in practice.

The Uniform or Rectangular Distribution

An application of this distribution is given in the constant-holding-time case of a single-channel queue which is discussed later. In this case,

$$f(x) = \frac{1}{b-a} \qquad a < x < b \qquad (8\text{-}59)$$

Its mean is $(b-a)/2$ and its standard deviation is $(b-a)/\sqrt{12}$. Its characteristic function is

$$\frac{e^{ibt} - e^{iat}}{i(b-a)t} \qquad i = \sqrt{-1}$$

The cumulative distribution is given by

$$F(x) = \begin{cases} \dfrac{x-a}{b-a} & a \leq x \leq b \\ 0 & x < a \\ 1 & x > b \end{cases}$$

Any density for a continuous variate X may be transformed to the uniform density $f(y) = 1$, $0 < y < 1$ by setting $y = G(x)$, where $G(x)$ is the cumulative distribution of X [21].

The Binomial Distribution

The binomial distribution enables one to deal with the occurrence of distinct events. The following is the Bernoulli case

$$\frac{n!}{(n-x)!x!} p^x q^{n-x} \qquad (8\text{-}60)$$

where the probability of success q is independent of the trial of the experiment and $p = 1 - q$ is the probability of failure. To illustrate the ideas,

PROBABILITY 259

the probability of 0, 1, 2, 3, etc., defectives (failures) in a sample of n items drawn at random, i.e., by a uniform distribution, whose proportion of defective items is p and whose proportion of nondefective items is q is given by the successive terms of the binomial expansion of $(q + p)^n$. As a general coefficient in this expansion one obtains the binomial distribution which gives the probability of x defectives in a sample of n items.

The mean of a binomial distribution is given by

$$\mu = \sum_{x=0}^{n} x \frac{n!}{(n-x)!x!} q^{n-x} p^x$$

where p is the probability of failure. To evaluate this, note that

$$\sum_{x=0}^{n} \frac{n!}{(n-x)!x!} q^{n-x} p^x = (q + p)^n$$

Differentiating both sides with respect to p gives

$$\sum_{x=0}^{n} x \frac{n!}{(n-x)!x!} q^{n-x} p^{x-1} = n(q + p)^{n-1}$$

μ is obtained by multiplying this expression by p, which gives

$$\mu = np(q + p)^{n-1} = np$$

since $q + p = 1$.

Similarly, the standard deviation may be shown to be $\sqrt{npq}$. The mode is the integer x satisfying

$$np - q \leq x \leq np + p$$

The characteristic function of this distribution is given by $(q + pe^{it})^n$. The sum of two binomial distributions with the same parameter p is also binomial with p as parameter.

If a sample of n objects of two categories, e.g., good and defective items, is taken from a population of size N, each time an object is taken its category is noted, and then it is replaced, and if in the population there are N_1 objects of the first category and N_2 of the second, $N_1 + N_2 = N$, the probability that the sample contains k objects of the first kind and $n - k$ of the second is given by the binomial distribution, with

$$x = k \qquad p = \frac{N_1}{N} \qquad \text{and} \qquad q = \frac{N_2}{N}$$

If the drawings are made without replacement, the probability is given by

$$\frac{\binom{N_1}{k}\binom{N_2}{n-k}}{\binom{N}{n}}$$

which is the hypergeometric distribution. Its mean is equal to nN_1/N and its variance is $N_1 N_2 n(N-n)/N^2(N-1)$. The distribution may be generalized to

$$\frac{\binom{N_1}{k_1}\binom{N_2}{k_2} \cdots \binom{N_m}{k_m}}{\binom{N}{n}}$$

when there are m categories and no replacements are made.

The multinomial distribution

$$\frac{n!}{n_1! \cdots n_m!} p_1^{n_1} \cdots p_m^{n_m}$$

is used as a generalization of the binomial distribution when there are several categories and replacements are made. Here

$$\sum_{i=1}^{m} n_i = n$$

and the distribution is obtained as a general coefficient in the expansion of $(p_1 + \cdots + p_m)^n$.

The Poisson Distribution

$$f(x) = \frac{e^{-\mu}\mu^x}{x!} \qquad (8\text{-}61)$$

The Poisson distribution is useful when dealing with the occurrence of isolated events ("rare events") in a continuum of time. It is also useful when it is possible to prescribe the number of times an event occurs but not the number of times it does not occur. The Poisson distribution has the characteristic function $e^{\mu(e^{it}-1)}$. Its mean is μ, and its standard deviation is $\sqrt{\mu}$.

The sum of independently distributed variables each with a Poisson distribution is also Poisson-distributed with a mean equal to the sum of the means of the variables. To estimate μ from a sample, calculate np, the mean of the sample, and set it equal to μ.

Applications of this distribution are found in Chap. 11 on queueing theory. The notion of randomness usually refers to the use of the Poisson distribution in one sense, and in another, to the uniform distribution. Let $p = m/n$ be the probability of m heads in a large number n of tosses of a coin where $m > 0$ is fixed. Let X_i indicate heads on the ith toss and let $S_n = \sum_{i=1}^{n} X_i$ indicate the repetition of heads in n tosses. Note that $p = m/n$ is equivalent to requiring the mean $np = n(m/n)$ of S_n to remain constant while n varies. The probability distribution of S_n is the binomial law given by

$$C_x^n p^x q^{n-x}$$

The behavior of this law will now be studied as $n \to \infty$ with fixed x, $p = m/n$, $q = 1 - m/n$. By writing out the above distribution one has

$$\frac{n(n-1) \cdots (n-x+1)}{x!} \frac{m^x}{n^x} \left(1 - \frac{m}{n}\right)^{n-x}$$

But

$$\lim_{n \to \infty} \frac{n!}{(n-x)!} = n^x$$

by using Sterling's formula $n! \sim n^n e^{-n} \sqrt{2\pi n}$. (In using this approximation formula for $n = 1$ an error of 8 per cent is committed; for $n = 10$, the error is .8 per cent; and for $n = 100$ the error is .08 per cent. As $n \to \infty$ the error tends to zero logarithmically.)

$$\lim_{n \to \infty} \left(1 - \frac{m}{n}\right)^x = 1$$

and

$$\lim_{n \to \infty} \left(1 - \frac{m}{n}\right)^n = e^{-m}$$

Therefore as $n \to \infty$ the above expression becomes $e^{-m}(m^x/x!)$, where $m = np$. Thus the law of probability of repeating the tosses of a coin tends to the Poisson law with parameter m with the hypothesis that $p = m/n$, where p is assumed arbitrarily small. Hence, the name "small probabilities" or "rare events" is sometimes used to designate the Poisson law.

The Normal Distribution

The normal distribution is approximated by the sample mean from an arbitrary distribution in the sense of the central-limit theorem: If a population has a finite variance σ^2 and mean μ, then the distribution of the sample mean approaches the normal distribution with variance σ^2/n and mean μ as the sample size n increases [21]. The normal distribution is given by

$$f(x) = \frac{1}{\sqrt{2\pi}\sigma} \exp\left[\frac{-(x-\mu)^2}{2\sigma^2}\right] \quad (8\text{-}62)$$

It has the characteristic function $\exp(i\mu t - \sigma^2 t^2/2)$. Its mean and mode are μ and its standard deviation is σ. The normal curve has a bell shape. Errors in measurement often are assumed in practice to have a normal-distribution law (see Prob. 17, Part 2).

Another form of the central-limit theorem states that the sum of n independently distributed variables X_i with respective means μ_i and variances σ_i^2 as $n \to \infty$ has a normal distribution with mean and variance

$$\mu = \sum_{i=1}^{n} \mu_i \qquad \sigma^2 = \sum_{i=1}^{n} \sigma_i^2$$

The central-limit theorem allows the use of a normal distribution for over-all measurements on effects of independently distributed causes, regardless of the distribution of the measurements of individual causes [22].

The sum (or linear combination) of independent normally distributed variables is normally distributed, with mean and variance as the sums (or linear combinations with the same coefficients) of the means and variances of the separate variables.

An approximation to the Poisson distribution by the normal distribution can be made using Sterling's approximation formula. If m and x are large so that

$$\frac{x-m}{\sqrt{m}} = y$$

is bounded, then

$$\log \frac{e^{-m}m^x}{x!} \sim \log e^{-m}\frac{m^x}{\sqrt{2\pi x}\, e^{-x}x^x} \sim -m + x\log m - \log\sqrt{2\pi} - \tfrac{1}{2}\log x + x - x\log x$$

On replacing x by $m(1 + y/\sqrt{m})$, $y/\sqrt{m}$ being very small, this expression may be further approximated by

$$-\log\sqrt{2\pi m} - \frac{y^2}{2} \quad \text{or} \quad e^{-m}\frac{m^x}{x!} \sim \frac{1}{\sqrt{2\pi m}} e^{-y^2/2}$$

Note that $\sqrt{m}$ would disappear if $x = m + y\sqrt{m}$ is used as a change of unit in the frequency law of probability of y. Thus a Poisson distribution with a large mean may be approximated by a normal distribution. Note that generally, when no information is available on the distribution of a large sample, it may be assumed, in the sense of the central-limit theorem, to be normally distributed. Laplace showed that for large n the binomial distribution loses a great deal of its skewness and is well approximated by the normal curve. One has

$$\sum_{np+\lambda_1\sqrt{npq}}^{np+\lambda_2\sqrt{npq}} C_x{}^n p^x q^{n-x} \sim \frac{1}{\sqrt{2\pi}} \int_{\lambda_1}^{\lambda_2} e^{-t^2/2}\, dt \qquad (8\text{-}63)$$

Thus if the upper limit of the summation is k, $\lambda_2 = (k - np)/\sqrt{npq}$.

Log Normal Distribution

The log normal distribution applies to situations in which several independent factors influence the outcome of an event not additively but according to the magnitude of the factor and the importance of the event at the time in which the factor is applied [6]. X has a log normal distribution if, for some constant a, log $(X - a)$ has a normal distribution. It is given by

$$f(x) = \frac{1}{(x-a)\sqrt{2\pi}\sigma} \exp\left\{-\frac{1}{2\sigma^2}[\log(x-a) - \mu]^2\right\} \qquad \text{for } x > a$$

(8-64)

and is zero otherwise. Its mean is μ, and its standard deviation is σ.

Pearson Type-III Distribution

$$f(x) = A(x - \mu)^{a-1} e^{-b(x-\mu)} \qquad x > \mu \text{ and } a, b > 0 \qquad (8\text{-}65)$$

Special Cases.

1. The gamma distribution

$$f(x) = \frac{b^a x^{a-1} e^{-bx}}{\Gamma(a)} \qquad 0 < x < \infty \qquad (8\text{-}66)$$

arises in a problem such as the distribution of particles impinging on a Geiger counter at an average rate b and accumulating $a - 1$ counts in time x. Its mean is a/b, and its variance is a/b^2. The sum of independent, gamma-distributed random variables is also gamma-distributed. Its characteristic function is given by

$$\left(1 - \frac{it}{b}\right)^{-a}$$

2. The chi-square distribution is widely used in statistics. This dis-

tribution is the density function of the variable

$$x \equiv \chi^2 = \sum_{i=1}^{n} \left(\frac{x_i - \mu_i}{\sigma_i}\right)^2 \qquad (8\text{-}67)$$

with n degrees of freedom, where the x_i are normally distributed with corresponding means and variances μ_i and σ_i. It is given by

$$f(x) = \frac{1}{2^{n/2}\Gamma(n/2)} x^{(n/2)-1} e^{-(x/2)} \qquad x > 0 \qquad (8\text{-}68)$$

Note that if $\mu_i = 0$ and $\sigma_i = 1$ for all i one has the density function of the sum of squares of independent standardized normally distributed variates.

3. The exponential distribution

$$f(x) = be^{-bx} \qquad x \geq 0 \qquad (8\text{-}69)$$

is obtained from the gamma distribution by setting $a = 1$. As applied to telephone conversations, it is interpreted as follows: The probability that a telephone conversation will terminate during a given time interval is constant, and it is independent of the length of the conversation up to that time.

Pascal Distribution

The Pascal distribution (also known as the negative binomial distribution) for the occurrence of n events in $n + x - 1$ equally likely events is given by

$$\binom{n + x - 1}{x} p^x q^n \qquad (8\text{-}70)$$

It has the characteristic function $[q/(1 - pe^{it})]^n$.

The Distribution Function of Pollaczek-Geiringer [14]

This distribution applies in the occurrence of x rare events singly or in pairs, triples, etc., up to k-tuples, e.g., multiple births, accidents with several deaths, counting of blood cells, or insurance against damage done by hail. It is given by

$$\exp\left(-\sum_{j=1}^{k} h_j\right) \sum_{n_1,\ldots,n_k} \frac{h_1^{n_1} h_2^{n_2} \cdots h_k^{n_k}}{n_1! n_2! \cdots n_k!} \qquad (8\text{-}71)$$

where $n_1 + 2n_2 + \cdots + kn_k = x$ and $h_j > 0$. It has the characteristic function

$$\exp\left[\sum_{j=1}^{k} h_j(e^{jit} - 1)\right] \qquad i = \sqrt{-1}$$

8-9. Some Useful Inequality and Limit Theorems

This section provides statements and a brief discussion of ideas which are helpful in dealing with probabilities.

Tchebycheff's Inequality

This inequality is stated as follows:

$$P[g(X) \geq A] \leq \frac{E[g(X)]}{A} \qquad (8\text{-}72)$$

Here X is a random variable with frequency distribution $f(x)$, $g(X) \geq 0$, and A is a constant such that $g(X) \geq A > 0$ is satisfied on a set S. The proof of this inequality follows from the definition.

Bienayme-Tchebycheff Inequality

This inequality is

$$P(|X - \mu| \geq \epsilon) \leq \frac{\sigma^2}{\epsilon^2} \qquad (8\text{-}73)$$

where $\epsilon > 0$ is arbitrary and μ and σ^2 are the mean and variance of X, respectively. This is obtained by putting $g(X) = (X - \mu)^2$ and $A = \epsilon^2$ above. It indicates the measure with which σ defines the dispersion. Note that from the definition of probability one has

$$P(|X - \mu| \leq \epsilon) \geq 1 - \frac{\sigma^2}{\epsilon^2} \qquad (8\text{-}74)$$

Example: In tossing a coin, let the random variable X assume the value unity with probability p in the case of a head and the value zero with probability q in the contrary case $(p + q = 1)$. Then

$$E[X] = p \qquad \sigma^2 = E\{X - E[X]\}^2 = E[X^2] - E[X]^2 = p - p^2 = pq$$

Note that in n tosses of the coin the probabilities of a head may be assumed different, that is, $p_1, p_2, \ldots, p_n$ and those of a tail $q_1, q_2, \ldots, q_n$.

When $p_i = p$ and $q_i = q$ for all i, one has the Bernoulli case.

If in n tosses $X = m/n$ is the proportion of heads, then X is in the random variable given by

$$X = \frac{\sum_{i=1}^{n} X_i}{n}$$

where X_i is unity if the ith toss is a head and zero otherwise.

It is clear that in the Bernoulli case

$$E[X] = p \quad \text{and} \quad \sigma^2 = E\{X - E[X]\}^2 = \frac{pq}{n}$$

If it is desired to estimate p, the proportion of defectives in a lot, one may use the above inequality, which is essentially the law of large numbers, to estimate the sample size, yielding p such that $P(|X - p| \geq .01) \leq .05$, where $\epsilon = .01$ and $.05$ is an accepted risk. Here $\sigma^2 = pq/n$ and one has $pq/(.01)^2 n = .05$, $q = 1 - p$.

One starts out with a sample size n_1 and an estimate p_1 of p and then, using these values, calculates n. Let this value be n_2. Then a sample of size n_2 is taken and p determined. Let its value be p_2. If $n_2 < n_1$ the experiment is continued, using p_2 instead of p_1 as a first estimate, yielding an estimated sample size n_3 and a corresponding p_3. The experiment stops unless $n_3 < n_2$, etc.

The sample size finally obtained to satisfy the inequality is much larger than is obtained by modern statistical techniques described later. This constitutes an older method for estimating sampling sizes.

The law of large numbers has two modified statements in the weak law of large numbers and the strong law of large numbers.

Definition: A sequence of random variables X_i, $i = 1, 2, \ldots$, is said to converge in probability if, given ϵ, $\delta > 0$, there exists an integer $N(\epsilon, \delta)$ such that $P(|X_m - X_n| < \epsilon) > 1 - \delta$ for $m, n > N$. The sequence converges in probability to X if

$$\lim_{n \to \infty} P(|X_n - X| \leq \epsilon) = 1 \tag{8-75}$$

Limit Theorems

The Weak Law of Large Numbers

$$\lim_{n \to \infty} P(|\bar{X} - \bar{\mu}| < \epsilon) = 1 \tag{8-76}$$

where $\quad X = \dfrac{1}{n} \sum_{i=1}^{n} X_i \quad \bar{\mu} = \dfrac{1}{n} \sum_{i=1}^{n} E[X_i] \quad \text{and} \quad \epsilon > 0 \tag{8-77}$

is arbitrary.

A theorem due to Khintchine requires that $\bar{\mu}$ be finite as a necessary and sufficient condition that the independent, identically distributed variables X_i satisfy the weak law of large numbers. If the X_i are uncorrelated with finite means and variances, then another theorem, due to Tchebycheff, gives

$$\lim_{n \to \infty} \frac{1}{n} \sum_{i=1}^{n} \sigma_i^2 = 0 \tag{8-78}$$

as a sufficient condition for this law to be satisfied.

The Strong Law of Large Numbers

$$\lim_{n \to \infty} P(|\bar{X} - \bar{\mu}| = 0) = 1 \tag{8-79}$$

The finiteness of $\bar{\mu}$ is a sufficient condition for this law to hold for a set X_i of independent variables whose distributions are identical.

Bernoulli's Theorem

This theorem follows from the weak law. In a series of independent trials of an experiment let m/n be the ratio of the number of favorable outcomes to n, the number of trials. Let p be the probability of a favorable outcome in any trial; then as $n \to \infty$ the probability of the sequences of trials for which

$$\left| \frac{m}{n} - p \right| \geq \epsilon$$

is satisfied tends to zero; i.e., the difference between the proportion of favorable outcomes in n trials and the probability of a favorable outcome in a single trial exceeding any given number tends to zero as $n \to \infty$.

8-10. Stochastic Processes [3, 8, 9]

In probability theory, one usually deals with a finite number of random variables. However, there are many situations which give rise to an infinite number of random variables. We give a brief discussion of these ideas here and in Chap. 11.

A set of random variables X_t or $X(t)$ which depend on a parameter t (continuous or discrete and usually indicating time) is said to define a *stochastic process*. To every finite set of values of the parameter t, there must correspond a joint probability distribution of the associated variables. Here, as in classical probability, one may calculate moments for a stochastic process. In addition, one can also define a derivative and an integral, using the concept of probability limit instead of the usual concept of limit.

The theory of stochastic processes has a number of important applications of which queueing theory is of frequent use in operations research. Renewal processes provide another illustration. Many important probability distributions obtained as limiting distributions in classical probability constitute exact solutions of appropriate stochastic-process problems. These ideas are illustrated partly in this section and partly in following chapters.

When the values of the parameter t are discrete, one usually denotes the random variables corresponding to values of t by $X_1, X_2, \ldots$. Now each of the X_i may assume a discrete set of values such as $+1$ and

-1, in which case one can write, for the distribution of X_i,

$$P(X_i = 1) = p \quad P(X_i = -1) = q \quad p + q = 1$$

or they may have a continuous distribution associated with them such as a normal distribution. Thus, although the time parameter takes on discrete values, the associated random variables may be discretely (as in a random walk) or continuously (e.g., normally) distributed. Also, the parameter t may be continuous whereas $X(t)$ is discretely distributed for any value of t, as is illustrated by the Poisson process examined below. Finally, the parameter t may be continuous, and the random variable $X(t)$ be also continuously distributed. Note that the distribution of X need not be the same for all values of t. A stochastic process may also be a function of several parameters. Some of these parameters may be discrete, the others continuous.

As a simple illustration of the discrete case, consider a coin-tossing game between two gamblers A and B whose fortunes are a and b, respectively. If a head appears, then A wins one dollar; otherwise he loses the same amount to B. The game is continued until A is ruined, i.e., loses everything, or until B is ruined. Now either fortune increases or decreases by discrete units. Similarly, the tosses of the coin are discrete, no matter how fast the tossing is. Thus, the times associated with each toss are discrete. With a toss of the coin, each gambler's fortune is changed by a unit. The value of A's fortune, for example, moves either toward zero or toward $a + b$ by a single unit at any one of these discrete times. Thus, if the step taken by A's fortune at time $t = t_0$ is denoted by $X(t_0)$, then $X(t_0)$ can have either of the two values $+1$ and -1 with probabilities p and q, respectively, with $p + q = 1$. This game is an instance of a random walk in which transitions are made only to neighboring states. This random walk has an absorbing barrier at zero and at $a + b$. Thus, if either A's fortune is exhausted or he has won B's fortune, the game is terminated. If A and B desire to continue the game, then when A's fortune reaches one unit, on the next toss it either increases to two units with probability p or becomes zero with probability q. In the second case, B lends A one unit and the game continues. Hence, a reflecting barrier may be established at the point $1/2$. A similar reflecting barrier at $a + b - 1/2$ may be established if A would also lend B one unit in case the latter is ruined. In such problems, one is usually interested in the probability of ruin after a certain number of steps. When no barriers are present in such a problem, the random walk is unrestricted and may take on the values 0, ± 1, ± 2, ... In this case it may be desired to determine the probability of returning to the initial position (state) in a given number of steps. In the coin-tossing example where the fortunes may be unlimited, the probability of a first return immediately after the first step is zero since the fortune must either increase or decrease. However, this probability is not zero after the second step, and for a return on the nth step for some n, the probability can be shown to be equal to unity.

The simplest case treated in probability theory is one involving a sequence of possible outcomes (states) $E_1, E_2, \ldots$ of an experiment which are assumed independent. Markoff dropped this assumption and introduced a weak type of dependence in which he assigned a probability to each pair of events, giving the conditional probability of occurrence of E_j when it is known that E_i has occurred, that is, $p_{ij} = P(E_j|E_i)$.

An important role among sequences of random variables is that played by Markoff chains which have this weak-dependence property. Let E_1, $E_2, \ldots$ be the outcomes of a sequence of trials of an experiment. Let p_{ij} be the "transition" probability of outcome E_j, given that in the previous trial outcome E_i has occurred. Suppose that the "initial" probability of being at outcome i is a_i; then the set of outcomes defines a Markoff chain if the probability of any sequence of outcomes E_{i_1}, $\ldots, E_{i_n}$ can be determined from the initial and from the transition probabilities; i.e.,

$$P(E_{i_1}, \ldots, E_{i_n}) = a_{i_1} p_{i_1 i_2} \ldots p_{i_{n-1} i_n} \tag{8-80}$$

The transition probabilities are described by means of a transition matrix

$$T = \begin{bmatrix} p_{11} & p_{12} & \cdots & p_{1n} & \cdots \\ p_{21} & p_{22} & \cdots & p_{2n} & \cdots \\ \cdots & \cdots & \cdots & \cdots & \cdots \end{bmatrix}$$

Note that, in the case of a random walk with n states (e.g., the gambling game described) and with reflecting barriers at the end points where the probability of a transition to the immediately preceding state is q while p is the probability of a transition to the immediately following state, one has all the p_{1j} in the first row equal to zero except p_{12} which is equal to 1. Similarly, all the p_{nj} in the nth row are zero except $p_{n(n-1)}$ which is equal to 1. In the ith row, $p_{i(i-1)}$ equals q and $p_{i(i+1)}$ equals p and the remaining coefficients are zero.

It is clear that, if the trials are independent, then obviously $p_{ij} = a_j$. The probability of reaching state j from state i in n transitions is given by the higher transition probability $p_{ij}^{(n)}$.

Obviously,

$$p_{ij}^{(1)} = p_{ij} \qquad p_{ij}^{(2)} = \sum_k p_{ik} p_{kj}^{(1)} \quad \cdots \quad p_{ij}^{(n)} = \sum_k p_{ik} p_{kj}^{(n-1)} \tag{8-81}$$

Hence, the transition matrix for $p_{ij}^{(1)}$ is the same as T. For $p_{ij}^{(2)}$ it is (from the definition of matrix multiplication) T^2 and for $p_{ij}^{(n)}$ it is T^n.

In a large number of problems there is an interest in the behavior of T^n as $n \to \infty$. This depends only on the transition probabilities and is independent of the initial probabilities a_j.

Example: Suppose that a set of outcomes consists of two states E_1 and E_2 with the transition matrix

$$T = \begin{bmatrix} p & 1-p \\ q & 1-q \end{bmatrix} \qquad 0 < p, q < 1$$

We compute T^n as illustrated in Chap. 4. One has $\lambda_1 = 1$, $\lambda_2 = p - q$. Therefore

$$T^n = \frac{1^n}{1-p+q} \begin{bmatrix} q & 1-p \\ q & 1-p \end{bmatrix} + \frac{(p-q)^n}{1-p+q} \begin{bmatrix} 1-p & p-1 \\ -q & q \end{bmatrix}$$

$$\lim_{n \to \infty} T^n = \begin{bmatrix} \dfrac{q}{1-p+q} & \dfrac{1-p}{1-p+q} \\ \dfrac{q}{1-p+q} & \dfrac{1-p}{1-p+q} \end{bmatrix}$$

since $|p - q| < 1$ and hence $(p - q)^n \to 0$ as $n \to \infty$.

This simple model may be applied to learning theory where the transition matrix corresponds to two possible responses, a correct response and an incorrect one. Here p is the probability of giving a correct response, having given a similar one, and q is the probability of giving a correct response, having given an incorrect one. The two probabilities of producing a correct response are equal in the limit.

It is possible to classify and characterize the states of a Markoff chain by means of the probability of returning to a state after n transitions. A state E_j is said to be *recurrent* if return to it is certain. It is *transient* if return to it is uncertain. If the mean return is infinite, E_j is a *null* state. It is *periodic* with period t if return is possible only in $t, 2t, \ldots$ steps ($t > 1$ is the greatest integer having this property). The state E_j is *ergodic* if it is neither a null state nor a periodic state.

A Markoff process is obtained from a Markoff chain if the probabilities p_{ij} are made to depend on a continuous parameter such as time. It has the property that the probability of a state at any moment depends only on the immediately preceding state. One writes $p(j,t;i,s)$ for the transition probability to state j at time t, given that the system (set of outcomes) was in state i at time s. Of course there are systems occurring in statistical mechanics in which information from all previous states has an effect on the outcome of a new state.

A one-dimensional Markoff process is described by the Chapman-Kolmogorov equation (a generalization of the equations describing

Markoff chains)

$$p(j,t;i,s) = \sum_{k=0}^{\infty} p(j,t;k,u)p(k,u;i,s) \qquad (8\text{-}82)$$

with $s < u < t$ and $i, j = 0, 1, 2, \ldots$. This equation gives rise to two sets of equations:

1. The forward equations which are obtained by taking the derivative with respect to t and then replacing u by t
2. The backward equations which are obtained by taking the derivative with respect to s and then replacing u by s

Thus, for example, if transitions to state i at time t can occur only from the two neighboring states $i - 1$ and $i + 1$, as is the case in a birth and death process, then the backward equations are given by

$$\begin{aligned}p_s(j,t;i,s) = \; & p(j,t;i,s)p_s(i,s;i,s) \\ & + p(j, t; i - 1, s)p_s(i - 1, s; i, s) \\ & + p(j, t; i + 1, s)p_s(i + 1, s; i, s) \qquad (8\text{-}83)\end{aligned}$$

where p_s is the derivative of p with respect to s.

In formulating the forward equations one considers the ways of reaching a state; for the backward equations one considers the ways of getting out of a state. In general, a solution of one of these two sets, with initial conditions, also satisfies the other set. Sometimes it is easier to solve the backward system than the forward system (see exercises).

Among those processes in which the preceding flow of events in a system substantially determines its future are stationary or time-homogeneous processes. In a stationary process, the distribution laws of any two finite sets of variables, where one set is obtained by a given displacement in time from the other, are identical; i.e., the probability distributions do not change by changing the time origin. The first set is obtained by assigning any set of values $t_1, \ldots, t_n$ to t. The second is obtained by adding an arbitrary time interval to the t_i. A stationary process $X(t)$ may satisfy the ergodic hypothesis where averaging with respect to X may be replaced by averaging with respect to the time t. However, there are stationary processes which are not ergodic [4].

Let $x_k(t)$, $k = 1, 2, \ldots$, be a collection of random functions of time called a *time series*. The collection need not be countable, although only a countable collection is considered in this brief discussion. Then an autocorrelation function for a stationary process is given by

$$\lim_{n \to \infty} \frac{1}{n} \sum_{k=1}^{n} x_k(0)x_k(t) \qquad (8\text{-}84)$$

where t is the difference between two times t_1 and t_2. The cross correlation function for two such processes $\{x_k(t)\}$ and $\{y_k(t)\}$ is obtained by replacing $x_k(t)$ by $y_k(t)$ in this formula.

For an ergodic stationary process the autocorrelation function may also be obtained by using a time average

$$\lim_{T \to \infty} \frac{1}{2T} \int_{-T}^{T} x_k(t) x_k(t + u) \, dt \qquad (8\text{-}85)$$

Such a function is of practical use. In general, when studying the variation of the average income of a community in time, for example, one has only a single member of the ensemble. Calculating the autocorrelation function, assuming that the ergodic hypothesis is satisfied, one attempts to determine "a period" u which maximizes this function. If the resulting value of the function is close to ± 1, then periodicity is suspected, sometimes, but very rarely, enabling prediction. Where an entire ensemble is available, the stationary property is tested for different values of t. If the statistical properties for the ensemble at each t are the same, then it is probable that the ensemble is stationary. If it is nonstationary, then averaging can only be taken with respect to the ensemble.

As an example of another type of process, consider a pure birth process in which transitions are of the form $a \to a$, that is, remain in state a, or of the form $a \to a + 1$, that is, move from state a to state $a + 1$. We shall derive the Poisson distribution from a pure birth process. Now in a pure birth process the probability of a transition $a \to a + 1$ is given by $\lambda_n \, dt$. This probability depends on the number of transitions which occurred before. In a Poisson process, $\lambda_n = \lambda$ is independent of n.

At time $t = 0$, no events occur. Let $\lambda \, dt$ be the probability of a transition from an adjacent state. Then $1 - \lambda \, dt$ is the probability of no transition. Let $p(n,t;0,0) \equiv P_n(t)$. The probability of n events occurring in time $t + dt$ is given by

$$P_n(t + dt) = P_n(t)(1 - \lambda \, dt) + P_{n-1}(t) \lambda \, dt \qquad (8\text{-}86)$$

This says that the probability of n transitions in time $t + dt$ equals the probability of n transitions in time t and no transitions during dt plus the probability of $n - 1$ transitions in time t and one transition during dt.

Simplifying and allowing $dt \to 0$ yield

$$\frac{dP_n(t)}{dt} = -\lambda P_n(t) + \lambda P_{n-1}(t) \qquad n > 0$$

For $n = 0$, $\qquad (8\text{-}87)$

$$\frac{dP_0(t)}{dt} = -\lambda P_0(t)$$

Hence $\qquad P_0(t) = e^{-\lambda t}$

As a solution for $n \geq 0$ one has the Poisson distribution

$$P_n(t) = \frac{(\lambda t)^n e^{-\lambda t}}{n!} \tag{8-88}$$

where λt is the expected number of events in time t. This is found by calculating $\sum_{n=0}^{\infty} nP_n(t)$. Note that $\sum_{n=0}^{\infty} P_n(t) = 1$, as it should be.

A Poisson process satisfies the Chapman-Kolmogorov equation. Therefore, a Poisson process is a Markoff process. To see this, let

$$p(j,t;0,0) = \frac{(\lambda t)^j e^{-\lambda t}}{j!} \tag{8-89}$$

If the initial value is (i,s), one has, starting with i transitions at time s instead of zero transitions at time zero, that

$$p(j,t;i,s) = \frac{[\lambda(t-s)]^{j-i} e^{-\lambda(t-s)}}{(j-i)!} \qquad j \geq i \tag{8-90}$$

which depends only on the length of the interval and not on where it began, and hence is stationary. Chapter 11 may be consulted for the method in which this may be obtained. To show that this satisfies the Chapman-Kolmogorov equation for $s < u < t$, one has

$$p(j,t;i,s) = \sum_{k=i}^{j} \frac{[\lambda(t-u)]^{j-k}}{(j-k)!} e^{-\lambda(t-u)} \frac{[\lambda(u-s)]^{k-i}}{(k-i)!} e^{-\lambda(u-s)} \tag{8-91}$$

which readily simplifies to yield the result. A Poisson process can also be shown to satisfy both the forward and the backward equations.

Finally, a birth and death process with transitions of the form $a \to a$, $a \to a-1$, $a \to a+1$, that is, either remaining in a state or moving to an immediately lower or an immediately higher state, has immediate applications to the theory of queues.

REFERENCES

1. Arley, N.: "On the Theory of Stochastic Processes and Their Application to the Theory of Cosmic Radiation," G. E. C. Gad, Copenhagen, 1943.
2. Arley, N., and K. R. Buch: "Introduction to the Theory of Probability and Statistics," John Wiley & Sons, Inc., New York, 1950.
3. Bartlett, M. S.: "An Introduction to Stochastic Processes," Cambridge University Press, London, 1955.
4. Bendat, J.: A General Theory of Prediction and Filtering, *Journal of the Society for Industrial and Applied Mathematics*, vol. 4, no. 3, p. 131, September, 1956.
5. Borel, E., and R. Deltheil: "Probabilités erreurs," Librairie Armand Colin, Paris, 1950.

6. Cramer, H.: "The Elements of Probability Theory and Some of Its Applications," 1st American ed., John Wiley & Sons, Inc., New York, 1955.
7. Curtiss, J. H.: *Annals of Mathematical Statistics*, vol. 12, p. 409, 1941.
8. Doob, J. L.: "Stochastic Processes," John Wiley & Sons, Inc., New York, 1953.
9. Feller, W.: "Probability Theory and Its Applications," vol. I, John Wiley & Sons, Inc., New York, 1950.
10. Fortet, R.: "Calcul des probabilités," Centre National de la Recherche Scientifique, Paris, 1950.
11. Fortet, R.: "Éléments de calcul des probabilités," Centre de Documentation Universitaire, Paris, 1950.
12. Fry, T. C.: "Probability and Its Engineering Uses," D. Van Nostrand Company, Inc., Princeton, N.J., 1928.
13. Gnedenko, B. V., and A. N. Kolmogorov: "Limit Distributions for Sums of Independent Random Variables," Addison-Wesley Publishing Company, Reading, Mass., 1954.
14. Haller, B.: A Summary of Known Distribution Functions, *U.S. Air Force Project RAND Report T-27*, The RAND Corporation, Santa Monica, Calif., January, 1953. (Translated by R. E. Kalaba.)
15. Hildebrand, F. B.: "Method of Applied Mathematics," Prentice-Hall, Inc., Englewood Cliffs, N.J., 1952.
16. Kendall, D. G.: Some Problems in the Theory of Queues, *Journal of the Royal Statistical Society, Series B*, vol. 13, no. 2, 1951.
17. Kolmogorov, A. N.: "Foundations of the Theory of Probability," Chelsea Publishing Company, New York, 1950.
18. Laplace, P. S.: "A Philosophical Essay on Probabilities," Dover Publications, New York, 1951.
19. Loeve, M.: "Probability Theory," D. Van Nostrand Company, Inc., Princeton, N.J., 1955.
20. MIT Summer Course on Operations Research, June 16–July 3, 1953, Technology Press, M.I.T., Cambridge, Mass., 1953.
21. Mood, A. M.: "Introduction to the Theory of Statistics," McGraw-Hill Book Company, Inc., New York, 1950.
22. Munroe, M. E.: "The Theory of Probability," McGraw-Hill Book Company, Inc., New York, 1951.
23. Todhunter, I.: "A History of the Mathematical Theory of Probability," Chelsea Publishing Company, New York, 1949.
24. Uspensky, J. V.: "Introduction to Mathematical Probability," McGraw-Hill Book Company, Inc., New York, 1937.
25. Wax, N. (ed.): "Noise and Stochastic Processes," Dover Publications, New York, 1954.
26. Wilks, S. S.: "Mathematical Statistics," Princeton University Press, Princeton, N.J., 1944.

CHAPTER 9

SOME APPLICATIONS

9-1. Optimization and Probabilities

In this section a few examples are given of the methods of combining the ideas presented in the two main divisions of this book. Such problems are usually formed "probabilistically," and it frequently happens that optimization techniques are required to obtain a solution. These examples have been chosen to illustrate both methods and important models arising from operations.

Example 1: It is desired to develop rules, i.e., a table, for ordering spare parts in order to minimize shortages (subject to weight limitations), where each shortage of part i $(i = 1, \ldots, m)$ is weighted by a criticality factor c_i. The latter is measured either by assuming that all parts are equally critical or by assigning quantitative values to the parts according to relative criticality. It is also assumed that the demand for a part is independent of the quantities stocked, that types of parts cannot be substituted for each other, and finally that it is permissible to record fractions of parts and round off fractions to nearest integers; this assumption is reasonable particularly when the number of parts is large [28].

If $p_i(x)$ is the probability density function for a demand of x units of the ith part during the period of use of the table, if $s_i \geq 0$ is the quantity of the ith part appearing in the table, and if m is the number of different parts to go into the table, then the problem is to minimize the expected total number of shortages over the m different parts given by

$$\sum_{i=1}^{m} c_i \int_{s_i}^{\infty} (x - s_i) P_i(x)\, dx \qquad \text{where } s_i \geq 0 \qquad (9\text{-}1)$$

and subject to a given total weight condition

$$w = \sum_{i=1}^{m} w_i s_i \qquad (9\text{-}2)$$

where w_i is the unit weight of the ith part.

Solution: This problem is solved by the standard methods presented in Chap. 5 on optimization, as follows: Form the Lagrangian

$$F(s_1, \ldots, s_m) = \sum_{i=1}^{m} c_i \int_{s_i}^{\infty} (x - s_i) p_i(x)\, dx - \lambda \left(w - \sum_{i=1}^{m} w_i s_i \right) \quad (9\text{-}3)$$

with $s_i \geq 0$. Note that this function may be written as

$$F(s_1, \ldots, s_m) = \sum_{i=1}^{m} f_i(s_i)$$

The minimum value is achieved when

$$\frac{\partial f_i(s_i)}{\partial s_i} = 0 \quad \text{if } s_i > 0$$

and

$$\frac{\partial f_i(s_i)}{\partial s_i} \geq 0 \quad \text{if } s_i = 0 \quad (9\text{-}4)$$

This gives

$$\lambda = \frac{c_i}{w_i} \int_{s_i}^{\infty} p_i(x)\, dx \quad \text{if } s_i > 0 \quad (9\text{-}5)$$

and

$$\frac{c_i}{w_i} \int_{0}^{\infty} p_i(x)\, dx = \frac{c_i}{w_i} \leq \lambda \quad \text{if } s_i = 0 \quad (9\text{-}6)$$

Thus the ith part is not included in the table if $c_i/w_i \leq \lambda$.

One numerical technique used to determine λ proceeds by selecting an arbitrary λ_1 and computing $s_i > 0$ $(i = 1, \ldots, m)$ from

$$\frac{\lambda_1 w_i}{c_i} = \int_{s_i}^{\infty} p_i(x)\, dx \quad (9\text{-}7)$$

Note that $s_i = 0$ if $\lambda_1 \geq c_i/w_i$. These s_i must satisfy (9-2). In this manner a new weight is obtained from the weight condition and a smaller or a larger value of λ is chosen, according to whether the new weight is less than or greater than the given weight, respectively.

Example 2: In the following supply, demand, and stockpile problem, let y_i $(i = 1, \ldots, n)$ be the amount of the ith commodity stocked and let $f_i(x_i)$ be the probability that an amount x_i $(i = 1, \ldots, n)$ will be demanded during the period to be considered. The expected demand of the ith commodity is given by [29]

$$E(x_i) \equiv \int_0^{\infty} x_i f_i(x_i)\, dx_i \quad i = 1, \ldots, n \quad (9\text{-}8)$$

The expected amount supplied is given by

$$u_i(y_i) \equiv \int_0^{y_i} x_i f_i(x_i)\, dx_i + y_i \int_{y_i}^{\infty} f_i(x_i)\, dx_i \quad i = 1, \ldots, n \quad (9\text{-}9)$$

The unfulfilled demand is given by

$$E(x_i) - u_i(y_i) \quad i = 1, \ldots, n \quad (9\text{-}10)$$

The problem is to maximize the total utility function

$$w(y_1, \ldots, y_n) \equiv \sum_{i=1}^{n} w_i u_i(y_i) \quad (9\text{-}11)$$

where the w_i are relative weights attached to each commodity and subject to the given space condition

$$\sum_{i=1}^{n} c_i y_i = C \quad (9\text{-}12)$$

where c_i is the amount of space occupied by y_i.

Solution: By forming the Lagrangian and applying the necessary condition for a maximum, one has

$$\int_0^{y_i} f_i(x_i)\, dx_i = 1 - \frac{\lambda c_i}{w_i} \quad (9\text{-}13)$$

Together with $\sum_{i=1}^{n} c_i y_i = C$, this is used to determine y_i and λ, assuming that $f_i(x_i) > 0$ for all i and (almost) all values of x_i. The procedure is as in the previous example, selecting an arbitrary value of λ, etc.

Example 3: The following variational example involves inequality constraints but is solved by elementary methods. By analogy with non-variational problems, using variation where gradient was used before, it is easy to derive necessary conditions for a maximum of an integral subject to inequality constraints. Assume that $p(x,y)\,dx\,dy$ is the probability of the existence of a target in an area element $dx\,dy$ of a region R and that $e^{-\phi(x,y)}$ is the probability of not detecting it in that **region, where** $\phi \geq 0$ and

$$\int_R \phi(x,y) = C \quad (9\text{-}14)$$

is the total effort available for search in the region. Find

$$\min_{\phi} \int_R p(x,y) e^{-\phi(x,y)}\, dx\, dy \equiv F[\phi] \quad (9\text{-}15)$$

subject to the above constraints [40]. Note that this problem has a solution if $p(x,y)$ and $\phi(x,y)$ are integrable.

Solution: To solve the problem, let $\phi_0(x,y)$ be the desired ϕ, form

$$G(\alpha) = \int_R p(x,y) \exp\{-\phi_0(x,y) + \alpha[\phi_0(x,y) - \phi(x,y)]\} \, dx \, dy \quad (9\text{-}16)$$

and note that $\quad G(0) = F[\phi_0] \quad G(1) = F[\phi]$

and $\quad\quad\quad\quad\quad\quad G(\alpha) \geq G(0)$

since $\phi_0(x,y)$ is assumed to yield a minimum. Thus

$$\frac{G(\alpha) - G(0)}{\alpha} \geq 0 \quad \text{for all } \phi > 0$$

By passing to the limit, one has

$$\lim_{\alpha \to 0_+} \frac{G(\alpha) - G(0)}{\alpha} \geq 0$$

which is the right-hand derivative $D_+ G(0) \geq 0$ at zero. Thus taking the derivative with respect to ϕ and letting $\phi \to 0$ give

$$\int_R p e^{-\phi_0(x,y)}(\phi_0 - \phi) \, dx \, dy \geq 0 \quad \text{for all } \phi \quad (9\text{-}17)$$

Note that

$$\int_R (\phi_0 - \phi) \, dx \, dy = 0 \quad (9\text{-}18)$$

since $\int_R \phi(x,y) \, dx \, dy = C$ for all ϕ.

By the Du Bois-Reymond lemma [15], which states that if f is piecewise continuous in the region of integration R and if the equation

$$\int_R fg \, dy \, dx = 0 \quad (9\text{-}19)$$

holds for an arbitrary continuous function g satisfying the condition

$$\int_R g \, dx \, dy = 0 \quad (9\text{-}20)$$

then f is a constant, it follows that, if $\phi_0 \not\equiv 0$, $pe^{-\phi_0(x,y)} = A$, where A is to be determined from $\int \phi_0(x,y) = C$. Otherwise, $\phi_0 \equiv 0$, and in this manner ϕ_0 is completely determined.

The proof of the lemma is simple. The lemma is true if f is constant. Otherwise let b be a constant such that

$$\int_R (f - b) \, dx \, dy = 0 \quad (9\text{-}21)$$

Then $\quad\quad\quad\quad \int_R (f - b) g \, dx \, dy = 0 \quad (9\text{-}22)$

since, by assumption, each of the two factors of the integrand vanishes when integrated over the region. In the last equation, let $g = f - b$. Then

$$\int_R (f - b)^2 \, dx \, dy = 0 \tag{9-23}$$

which is true if $f = b$.

Example 4: Suppose that it is necessary to traverse a distance x, where the legal speed is v_0. There is a probability $p(v) \, dt$ of being stopped in the time interval $(t, t + dt)$ and given a ticket [2]. Getting a ticket consumes a fixed time r. (Actually r should be a function of v.) Assume also that $p(v_s) = 1$ for $v = v_s$, the suicidal velocity. At what speed should one travel to minimize the expected time required to cover the distance x?

Solution: Let $T = x/v$ be the actual driving time. The probability of getting n tickets is given by the Poisson distribution

$$p_n = e^{-pT} \frac{(pT)^n}{n!} \tag{9-24}$$

and the time elapsed is $T + nr$. The expected time is

$$\sum_{n=0}^{\infty} (T + nr) e^{-pT} \frac{(pT)^n}{n!} = T(1 + pr) = \frac{x}{v} [1 + p(v)r] \tag{9-25}$$

The velocity v_m for the minimum is therefore obtained by solving

$$-\frac{1 + p(v_m)r}{v_m^2} + \frac{p'(v_m)r}{v_m} = 0 \tag{9-26}$$

for v_m, having equated to zero the derivative of (9-25) with respect to v.

9-2. Some Examples from Inventory Problems

Inventories, "detailed lists or schedules of goods and chattels," are required to meet customer demand. The level of inventories required is related to the demand. If the demand is known exactly, then the inventory required is also determined exactly. The caterer's problem mentioned previously is an example of a situation in which the demands over a period of several days are known and a corresponding schedule of items to be procured can then be drawn.

The case of unknown demand can best be studied by means of data to derive distributions for the demands and to base inventory predictions on these distributions. Complex inventory problems usually consider situations in which not only demands are to be optimumly met but which also include other factors such as storage and delay.

Example 1: Consider a problem of stocking an item to be ordered over a finite number of equal times to fulfill the demand at these times without any delay. Let $g(t) \, dt$ be the probability of a demand between time t and $t + dt$, $k(z)$ be the cost of ordering z items initially to increase stock level, and let $p(z)$ be the "penalty" cost of ordering z items to meet an excess z of demand over supply. Define $f_n(x)$ as the total cost for an n-stage process starting with initial supply x and using optimum ordering policy. Then dynamic programming gives the following formulation for this problem [7]:

$$f_1(x) = \min_{y \geq x} \left[k(y - x) + \int_y^\infty p(t - y)g(t) \, dt \right]$$

$$f_{n+1}(x) = \min_{y \geq x} \left[k(y - x) + \int_y^\infty p(t - y)g(t) \, dt \right. \quad (9\text{-}27)$$
$$\left. + f_n(0) \int_y^\infty g(t) \, dt + \int_0^y f_n(y - t)g(t) \, dt \right]$$

If the specific demand distribution is not known but the general type of distribution is known, then one chooses from the class of ordering policies yielding a maximum expected loss the one which yields the minimum of these maxima. This is the minmax policy.

Example 2: The dynamic approach to inventory is that in which the problem is solved for several periods and an optimum response is then calculated for the last interval with its own initial inventory. Then the loss is computed for the next to the last interval, together with the probability distribution of initial conditions for the last interval as a function of the initial conditions and strategy for next to the last interval [3, 22]. The expected loss for the last interval as a function of the strategy in the next to the last interval may be calculated and added to the previous expected loss. This provides a criterion for choosing an optimum strategy for the next to last interval. The process is continued until the optimum strategy for all the intervals has been computed. Let x_i be the initial condition for the ith period, $y(x)$ the order quantity as a function of the inventory, $l(y_i, x_i)$ the expected loss in the ith period as a function of the order quantity and initial inventory, and let $L[y(x_i), x_i]$ be the expected loss for all periods subsequent to the ith. Then the expected loss is given as the sum of the first-period loss plus the expected loss for all remaining periods, i.e.,

$$L[y(x_1), x_1] = l(y_1 x_1) + \alpha \int_{-\infty}^{\infty} L[y(x_2), x_2] p(x_2) \, dx_2 \quad (9\text{-}28)$$

where the discounting factor α determines present value of future loss and $p(x_2)$ is the probability density of initial inventories in the second period. The problem is to determine an optimum policy $y(x)$ which minimizes the above expression.

Example 3: The (s,S) policy for inventory problems is defined as that policy which minimizes expected loss by specifying two appropriate numbers s and S, with $0 \leq s \leq S$, to order goods when and only when the stock at hand, x, is smaller than s, and then to order the quantity $S - x$ so as to bring the stock back up to S.

One forms

$$V(x,y) = cy + A[1 - F(y)] + \begin{cases} 0 & \text{if } y \leq x \\ K & \text{if } y > x \end{cases} \quad (9\text{-}29)$$

where $\qquad 0 \leq x \leq y < \infty$

and $\quad x =$ initial stock

$\quad y =$ starting stock (obtained by adding amount ordered to initial stock)

$\quad c =$ carrying cost per unit of goods

$\quad A =$ loss involved if impossible to meet demands for goods

$\quad K =$ ordering cost

$V(x,y) =$ expected loss

$F(y) =$ distribution of demand

The problem is then to determine x and y which minimize the expected loss $V(x,y)$. [The distribution functions F of demand for which the above inventory problem satisfies the (s,S) assumption can be characterized [23].]

It has been found possible to cast the preventive-maintenance problem [51] in the above form, with the various terms taking on the following meanings:

$\quad x =$ remaining service life of component

$\quad y =$ starting service life of component

$\quad c =$ cost of component per unit of service life

$\quad A =$ loss involved if component fails

$\quad K =$ component installation cost

$V(x,y) =$ expected loss

$F(y) =$ mortality distribution

Example 4: Another interesting problem has been studied by Whitin and Youngs [61]. The authors calculated the optimum stock y which minimizes the expected loss for carrying parts on shelves and for idletime loss due to waiting for unavailable parts. Let the demand for parts be Poisson-distributed with mean μ, the cost of carrying a unit stock for one day be c, and the daily cost of being out of a part be d and let

$$P(y,\mu) = \sum_{i=y}^{\infty} \frac{e^{-\mu}\mu^i}{i!}$$

be the cumulative Poisson distribution. Then the expected loss is given by the expression

$$y[c - (d + c)P(y + 1, \mu)] - \mu[c - (d + c)P(y,\mu)] \qquad (9\text{-}30)$$

and the optimum choice of y is the value y_0 such that

$$P(y_0,\mu) \geq \frac{c}{c + d} > P(y_0 + 1, \mu) \qquad (9\text{-}31)$$

If y_0 is selected such that the equality on the left holds, then one optimum choice of y is also provided by $y_0 - 1$.

Example 5: Here is an example of a competitive-bidding strategy [27]. Let $h(s)\,ds$ be the probability that the ratio of true cost to the estimated cost is between s and $s + ds$. Let x be the amount bid for a contract. The profit from a winning bid of amount x will be $x - sc$, where c is the estimated cost of fulfilling the contract. If $P(x)$ is the probability that this bid wins, then the expected profit is given by

$$E(x) = \int_0^\infty P(x)[x - sc]h(s)\,ds = P(x)\left[x - c\int_0^\infty sh(s)\,ds\right] \qquad (9\text{-}32)$$

The value of x which maximizes this expression is suggested as the amount to bid. However, it is first necessary to calculate $P(x)$, the probability of winning the bid. If $f(y)$ is the probability density function of the ratio of other bids to the cost estimate, then the probability of a bid x being lower than that of one bidder is given by

$$\int_{x/c}^\infty f(y)\,dy \qquad (9\text{-}33)$$

and the probability of x being less than k other bids is

$$\left[\int_{x/c}^\infty f(y)\,dy\right]^k \qquad (9\text{-}34)$$

If $g(k)$ is the probability of k bidders submitting bids, then

$$P(x) = \sum_{k=0}^\infty g(k)\left[\int_{x/c}^\infty f(y)\,dy\right]^k \qquad (9\text{-}35)$$

9-3. Geometric and Physical Probabilities

In complex operations such as the search for a lost object within time limitations, the method of search, the shape and size of the area searched, and the physical movements of both the searcher and the object sought require a combination of geometric and physical probability considerations to obtain useful models. The examples given here illustrate elementary applications of geometric probability.

SOME APPLICATIONS

First, the assumption of geometric probability may be illustrated by a simple example. If a region A of a plane contains a random point P, the probability that P is a determined portion S of the region A is equal to the ratio s/a, where the area of A is denoted by a and that of S by s [9]. This may be analytically expressed by the quotient of the two double integrals

$$\frac{\iint_S dx\,dy}{\iint_A dx\,dy} \qquad (9\text{-}36)$$

Example 1: A rod of length a is broken into three parts at random. What is the probability that a triangle can be formed from the parts?

Solution: Denote one part by x, the second part by $y - x$, and the third part by $a - y$. If a triangle can be formed from the three parts, then the sum of the lengths of any two sides must exceed the length of the third. This gives the three inequalities

$$x + (y - x) > a - y \qquad (y - x) + (a - y) > x \qquad (a - y) + x > y - x \qquad (9\text{-}37)$$

which, when simplified, yield

$$y > \frac{a}{2} \qquad x < \frac{a}{2} \qquad y - x < \frac{a}{2}$$

that is, $\qquad 0 \leq x \leq \dfrac{a}{2} \qquad$ and $\qquad \dfrac{a}{2} \leq y \leq \dfrac{a}{2} + x$

Hence the desired probability is given by

$$\frac{\int_0^{a/2} \int_{a/2}^{a/2+x} dy\,dx}{\int_0^a \int_0^x dy\,dx} = \frac{1}{4} \qquad (9\text{-}38)$$

which can also be verified by means of a diagram.

Example 2: What is the expected length of tape (in a tape recording machine) between two positions selected at random on the tape?

Solution: Let x_1 and x_2 denote the distance of the two positions from the starting position of the tape. Without loss of generality, assume that the tape has unit length. Then the average length is given by

$$\int_0^1 \int_0^1 |x_1 - x_2|\,dx_1\,dx_2 = \int_0^1 \int_0^{x_1} (x_1 - x_2)\,dx_2\,dx_1$$
$$+ \int_0^1 \int_{x_1}^1 (x_2 - x_1)\,dx_2\,dx_1 = \frac{1}{3} \qquad (9\text{-}39)$$

Example 3: What is the expected value of the square of the distance between two points P and Q selected at random in a square whose side has length L?

Solution: Let (x_1, y_1) be the coordinate of P and (x_2, y_2) that of Q. The square of the distance between the two points is given by

$$(x_1 - x_2)^2 + (y_1 - y_2)^2$$

The desired result is obtained by evaluating

$$\frac{1}{L^4} \left\{ \int_0^L \int_0^L \int_0^L \int_0^L [(x_1 - x_2)^2 + (y_1 - y_2)^2] \, dx_1 \, dy_1 \, dx_2 \, dy_2 \right\} \quad (9\text{-}40)$$

One has, successively,

$$\int_0^L \int_0^L [(x_1 - x_2)^2 + (y_1 - y_2)] \, dx_1 \, dx_2 = \int_0^L \left[\frac{(L - x_2)^3 + x_2^3}{3} \right.$$
$$\left. + L(y_1 - y_2)^2 \right] dx_2$$
$$= \frac{L^4}{6} + L^2(y_1 - y_2)^2 \quad (9\text{-}41)$$

$$\int_0^L \int_0^L \left[\frac{L^4}{6} + L^2(y_1 - y_2)^2 \right] dy_1 \, dy_2 = \frac{L^6}{6} + \frac{L^6}{6} = \frac{L^6}{3} \quad (9\text{-}42)$$

which, on dividing by L^4, yields the desired answer.

Example 4: Calculate the expected distance between two points with polar coordinates (r_1, θ_1), (r_2, θ_2) selected uniformly at random in a circle of radius a and center at the origin.

Solution: It is easy to see that the distance between the two points is given by

$$\sqrt{r_1^2 + r_2^2 - 2r_1 r_2 \cos(\theta_2 - \theta_1)}$$

The expected value is a function of a and is given by

$$L(a) = \frac{1}{\pi^2 a^4} \int_0^{2\pi} \int_0^{2\pi} \int_0^a \int_0^a \sqrt{r_1^2 + r_2^2 - 2r_1 r_2 \cos(\theta_2 - \theta_1)}$$
$$r_1 \, dr_1 \, d\theta_1 \, r_2 \, dr_2 \, d\theta_2 \quad (9\text{-}43)$$

having averaged over the circle twice, since each point can be selected anywhere in the circle. Differentiation with respect to the parameter a, after multiplying through by $\pi^2 a^4$, yields

$$\frac{d}{da}[L(a)\pi^2 a^4] = 2 \int_0^{2\pi} \int_0^{2\pi} \int_0^a \sqrt{r_1^2 + a^2 - 2r_1 a \cos(\theta_2 - \theta_1)}$$
$$r_1 \, dr_1 \, d\theta_1 \, a \, d\theta_2 \quad (9\text{-}44)$$

Note that the derivative of the two integrals with respect to a introduces the factor outside. The integrand is symmetric in r_1 and r_2. By replacing $\theta_2 - \theta_1$ by α and multiplying by 2π, since the averaging can now be taken over α alone, and then replacing r_1 by r, the right side becomes

$$4\pi a \int_0^{2\pi} \int_0^a \sqrt{r^2 + a^2 - 2ra \cos \alpha}\, r\, dr\, d\alpha \qquad (9\text{-}45)$$

This integral arises in calculating the expected value of the distance between any point in the circle and a point on the circumference. Using the point on the circumference as origin for a polar coordinate system (s,β) the integrals for the new problem can be seen to be

$$\int_{-\pi/2}^{\pi/2} \int_0^{2a \cos \beta} s \times s\, ds\, d\beta \qquad (9\text{-}46)$$

which has the value $(32a^3)/9$.

On multiplying by $4\pi a$, one has

$$\frac{d}{da}[L(a)\pi^2 a^4] = \frac{128\pi a^4}{9}$$

which yields the answer

$$L(a) = \frac{128a}{45\pi} \qquad (9\text{-}47)$$

Example 5: What is the average length of the chords of a circle of radius a taken at random in the circle?

Solution: The distance between two points is given by

$$d = \sqrt{(x_2 - x_1)^2 + (y_2 - y_1)^2}$$

The circle is described by $x^2 + y^2 = a^2$. Let

$$x_1 = a \cos \theta_1$$
$$y_1 = a \sin \theta_1$$
$$x_2 = a \cos \theta_2$$
$$y_2 = a \sin \theta_2$$

Then $\qquad d = \sqrt{2a^2 - 2a^2 \cos(\theta_2 - \theta_1)} \qquad (9\text{-}48)$

The average over the circle is given by

$$\frac{1}{2\pi}\frac{1}{2\pi}\int_0^{2\pi}\int_{\theta_1}^{\theta_1+2\pi}\sqrt{2a^2 - 2a^2 \cos(\theta_2 - \theta_1)}\, d\theta_2\, d\theta_1$$

$$= \frac{2a}{4\pi^2}\int_0^{2\pi}\int_{\theta_1}^{\theta_1+2\pi}\sin\left(\frac{\theta_2 - \theta_1}{2}\right) d\theta_2\, d\theta_1$$

$$= \frac{4a}{4\pi^2}\int_0^{2\pi} -\left[\cos\frac{\theta_2 - \theta_1}{2}\right]_{\theta_1}^{\theta_1+2\pi} d\theta_1$$

$$= \frac{a}{\pi^2}\int_0^{2\pi} 2\, d\theta_1 = \frac{4a}{\pi} \qquad (9\text{-}49)$$

Example 6: What is the average length of parallel chords of a circle of radius a?

Solution: Consider the circle $x^2 + y^2 = a^2$. The length of a chord is $2\sqrt{a^2 - x^2}$, and its average length is given by

$$\frac{1}{2a} \int_{-a}^{a} 2\sqrt{a^2 - x^2}\, dx = \frac{2}{a} \int_0^a \sqrt{a^2 - x^2}\, dx$$

$$= \frac{2}{a} \left[\frac{1}{2} \left(x\sqrt{a^2 - x^2} + a^2 \sin^{-1} \frac{x}{2} \right) \right]_0^a = \frac{\pi a}{2} \quad (9\text{-}50)$$

Note that, if the parallel chords are averaged over the circumference of the circle, the answer is $4a/\pi$.

Example 7: This example deals with the change of a distribution with time. Let $f(r)\, dA = (1/2\pi\sigma^2) e^{-r^2/2\sigma^2}\, dA$ be the probability density function that a target is in an area element dA at distance r from a fixed observer O [39]. Let the speed of the target be u, but its direction is not known. Assume that the direction is random (uniform), in that one direction is as likely as any other independently of the target position. It is desired to find the density function $f(r,t)$ after t hours have elapsed.

Solution: Let P_1 be the target position at time zero and P_2 be its position at time t. Denote angle OP_2P_1 by θ. Let r_1 be the distance OP_1 and r_2 the distance OP_2. Then by the cosine law one has

$$r_1^2 = r_2^2 + u^2 t^2 - 2 r_2 u t \cos \theta$$

and

$$f(r_1)\, dA = f(r_2, \theta)\, dA = \frac{1}{2\pi\sigma^2} \exp\left(-\frac{r_2^2 + u^2 t - 2 r_2 u t \cos \theta}{2\sigma^2} \right) dA \quad (9\text{-}51)$$

The probability of an angle change $d\theta$ is given as $d\theta/2\pi$. The product

$$f(r_2, \theta)\, dA\, \frac{d\theta}{2\pi}$$

which is the probability of being at P_2 and arriving there from any direction must be integrated over θ. Integration yields

$$f(r_2, t) = \frac{1}{2\pi\sigma^2} \exp\left(-\frac{r_2^2 + u^2 t^2}{2\sigma^2} \right) I_0 \frac{r_2 u t}{\sigma^2} \quad (9\text{-}52)$$

where $I_0(r_2 u t/\sigma^2) = J_0(i r_2 u t/\sigma^2)$ is the Bessel function of the first kind and zero order and where $i = \sqrt{-1}$.

Example 8: The following elementary example from statistical mechanics [9, 33] is to serve as a reminder of the many excellent applications of probability theory in that field.

Let a collection of N molecules occupy a space of volume V which is divided into a large number s of small partitions of equal volume v.

SOME APPLICATIONS   287

If the molecules were homogeneously distributed, then in each partition there would be $N/s = k_0$ molecules. It is desired to study the fluctuations in the number of molecules in these small partitions. Fluctuation in other characteristics such as density may then be easily studied.

Solution: One first calculates the probability that exactly k molecules are in a given partition. If p is the desired probability, then since there are s partitions, the probability that a molecule is in a given partition is $1/s$. Using the binomial distribution with parameters $1/s$ and $1 - (1/s)$, one seeks the probability of k successes in N trials of an experiment (that is, k given molecules in a partition). This is the coefficient

$$p = \frac{N!}{(N-k)!k!}\left(\frac{1}{s}\right)^k\left(1 - \frac{1}{s}\right)^{N-k} \quad (9\text{-}53)$$

Replacing s by N/k_0 and assuming that N is large (thus also the number of partitions s), one may use Stirling's formula to approximate to $N!$ and $(N-k)!$. Then simplify and replace

$$\left(\frac{N-k_0}{N-k}\right)^{N-k} = \left(1 + \frac{k-k_0}{N-k}\right)^{N-k} \quad (9\text{-}54)$$

by e^{k-k_0} to obtain finally

$$p \sim \frac{k_0^k}{k!} e^{-k_0} \quad (9\text{-}55)$$

If k is large although small when compared with N, one may again use Stirling's approximation with

$$k! \sim k^k e^{-k} \sqrt{2\pi k}$$

which gives $$p \sim \frac{1}{\sqrt{2\pi k}}\left(\frac{k_0}{k}\right)^k e^{k-k_0} \quad (9\text{-}56)$$

Remark: To maximize a function $p > 0$ with respect to k, one sets $p' \equiv dp/dk = 0$. This is the same as maximizing $\log p$ with respect to k by setting $(d \log p)/dk = p'/p = 0$. This elementary fact simplifies calculations.

The most probable value of p in this example may be calculated for large values of k and is easily shown to be attained at k_0 which satisfies $dp/dk = 0$.

Now one determines the distribution of the fluctuations. Taking logarithms after replacing k by $k_0 + x$, where x is the excess of molecules over k_0, gives

$$\log p \sim -\log \sqrt{2\pi(k_0 + x)} - (k_0 + x)\log\left(1 + \frac{x}{k_0}\right) + x \quad (9\text{-}57)$$

Assuming that $-1 < x/k_0 < 1$ and using the series expansion for

$$\log\left(1 + \frac{x}{k_0}\right) = \left[\frac{x}{k_0} - \frac{1}{2}\left(\frac{x}{k_0}\right)^2 + \cdots\right]$$

one has, with a slight change of notation and neglecting terms of higher order in x,

$$p(x)\,dx \sim \frac{1}{\sqrt{2\pi k_0}}\,e^{-(x^2/2k_0)}\,dx \qquad (9\text{-}58)$$

If k_0 is used as the unit of measurement, then this distribution may be written as

$$p(y)\,dy = \sqrt{\frac{k_0}{2\pi}}\,e^{-(k_0 y^2/2)}\,dy \qquad \text{where } y = x/k_0. \qquad (9\text{-}59)$$

This gives the desired distribution of fluctuations.

9-4. Reliability of Systems

The reliability of a system's performance is an integral part of any planning which uses the system to attain an objective, such as destroying a hydrogen-bomb-carrying aircraft with a single missile. If the missile fails to take off or suffers from unsatisfactory operation after taking off, it will fall short of attaining the objective. If it is essential that the target be destroyed by the missile, the latter's reliability must be assured. Otherwise more missiles must be made available, and the target will be "statistically" destroyed. In any case the increased costs of reliability are staggering. Both these points of increasing reliability and the associated cost increase are briefly examined.

In military application of electronic equipment it is of utmost importance that the equipment operate without breakdown for a specific length of time. Rapid developments in electronic equipment have led to greater complexity and lower reliability. Complexity entails a greater number of components and a greater likelihood of equipment and systems failure. Detailed knowledge of the operating conditions would be required in order to build more reliable systems. The effect of weight, maintenance cost, personnel requirements, and availability are important influencing factors. However, reliable operation is as important as the satisfactory solution of other engineering aspects of the problem, such as accuracy, stability, durability, etc. Nature's solution to biological-cell unreliability in living organisms is a complex biological organism consisting of many more cells than are required, most of which are used in parallel linkages [47]. Economy in cost and weight is frequently a limiting factor in duplicating this procedure with equipment. Yet there are instances in which reliability can be increased only by such duplication. The problem of reliability is one of minimizing failure when complexity is increased.

Some Definitions

The simplest object which cannot be further subdivided is called an *element*, e.g., a filament or a relay contact. It is often erroneously called a component. A collection of elements arranged in a prescribed order, e.g., a tube or a resistor, is called a *component*. A collection of components in a prescribed order but not all of which have, as yet, been so arranged is known as an *assembly*, e.g., a terminal board with component parts attached. *Equipment* is a collection of components arranged to operate without need for other components. An *equipment component* is a collection of assemblies packaged in one housing, e.g., a radio transmitter, the central part of a missile, etc. Finally, a *system* is a collection of equipments arranged to perform a function.

Reliability is defined as the probability p that an element or a system will perform satisfactorily for a specified period of time. *Unreliability* is the probability of failure during a specified time and is given by $q \equiv 1 - p$. *Independence* assumes that the failure of one element does not affect the probability of failure of any other element.

Failures

There are two general types of failure [64]: (1) catastrophic or sudden failures and (2) wear-out failures. In the latter case the strength distribution of a part changes with time as the parts wear in use. Both the mean and variance are functions of time.

The problem is, then, to design a part whose failure rate is sufficiently low to ensure that its wear-out failure rate at a later time will not exceed a prescribed amount.

Systems Reliabilities

In the early stages when a system is produced and assembled, reliability study may be conducted on the components by life testing. The components used have presumably been subjected to quality control tests to determine their acceptability. In this manner a frequency of survival distribution may be obtained for the components. This is useful for determining the theoretical reliability of the system. From theoretical analysis one can initially obtain limits on the reliability of the system to determine the feasibility of production and enable design improvement.

The reliability of a system of n elements in series with reliability p_r for the rth element is given by [47]

$$P_n^s = \prod_{i=1}^{n} p_i = p^n \quad \text{if } p_i = p \text{ for all } i \quad (9\text{-}60)$$

For practical reasons, an exponential distribution is frequently assumed for p.

The result for a system of n parallel components is

$$P_n{}^p = 1 - \prod_{i=1}^{n} (1 - p_i) = 1 - (1 - p)^n \quad \text{if } p_i = p \text{ for all } i \quad (9\text{-}61)$$

Remark: This expression also occurs in search theory as the probability of detecting a target at least once in n glimpses [39, 40], where p_i is interpreted as the probability of detecting the target on the ith glimpse. If $p_i = p$, then it is easy to see that the mean number of glimpses $\bar{n}$ is given by

$$\bar{n} = \sum_{n=1}^{\infty} n(1-p)^{n-1}p = \frac{1}{p} \quad (9\text{-}62)$$

where $(1-p)^{n-1}p$ is the frequency distribution of detection in n glimpses obtained as the difference in consecutive cumulative probabilities

$$P_n{}^p - P_{n-1}{}^p$$

Note that $P_n{}^s \leq \min_i p_i$ and $P_n{}^p \geq \max_j p_j$.

System Standby. This is a system of m components in parallel, with each component a system of n components in series. It is easy to see that the reliability of such a system, with $p_i = p$, is given by

$$P_A(n,m) = 1 - (1 - p^n)^m \quad (9\text{-}63)$$

Note that

$$P_A(\infty,m) \to 0 \qquad P_A(\infty,\infty) \to 0$$

Element Standby. This is a system of n components in series, with each component a system of m components in parallel. Its reliability is given by

$$P_B(n,m) = [1 - (1-p)^m]^n \quad (9\text{-}64)$$

Here one has

$$P_B(\infty,m) \to 0 \qquad P_B(n,\infty) \to 1 \qquad P_B(\infty,\infty) \to 1$$

Unlike system standby, the element-standby case approaches reasonable reliability with relatively few parallel elements. It is clear that if $m = n$ the system standby deteriorates rapidly whereas the element standby improves in reliability.

Thus the more series elements that are used in the system, the lower the system's reliability. If it is desired to increase the reliability of a

SOME APPLICATIONS

system by the method of element standby, the system is divided into as many elements as possible before paralleling.

Effect of Switch Reliability. Assume that each element used for paralleling has with it a "sensing and switching" equipment whose function is to switch the standby element into operation in case of a failure. This equipment has a finite reliability p_s. In this case

$$P_B = [1 - (1 - p)(1 - p_s p)^{m-1}]^n \qquad (9\text{-}65)$$

It is not profitable to carry subdivision for paralleling beyond the point where the element reliability equals the switching reliability.

Cost of a System

The conditions for the minimum cost to obtain maximum reliability are determined as follows: Assume a system of n components in series with reliability $p_1, \ldots, p_n$ and costs $c_1, \ldots, c_n$, respectively [47]. Then the reliability of the system and its cost are, respectively,

$$P_n{}^s = \prod_{i=1}^{n} p_i \qquad (9\text{-}66)$$

$$C_0 = \sum_{i=1}^{n} c_i \qquad (9\text{-}67)$$

It is desired to find the minimum cost to obtain a desired reliability P by the method of element standby, assuming that

$$Q = 1 - P \ll 1$$

Suppose that, to obtain P, it is found that the element with reliability p_i ($i = 1, \ldots, n$) must be paralleled m_i times. Then the reliability of the ith group becomes

$$r_i = 1 - q_i{}^{m_i} \qquad \text{where } q_i = 1 - p_i \qquad (9\text{-}68)$$

The problem is then to determine m_i which minimize $\sum_{i=1}^{n} m_i c_i$ and to satisfy the reliability constraint $P = \prod_{i=1}^{n} r_i$. Since each r_i contributes a certain amount to the total reliability P, one may write

$$r_i = P^{a_i} \qquad \text{with } 0 \leq a_i \leq 1 \qquad \sum_{i=1}^{n} a_i = 1$$

Then
$$m_i = \frac{\log (1 - P^{a_i})}{\log q_i} \qquad (9\text{-}69)$$

It is clear that the problem becomes one of finding a_i which minimize the cost, subject to the reliability constraint. Using the Lagrange multiplier method, it is easily verified that the desired a_i are given by

$$a_i = \frac{c_i/\log q_i}{\sum_j c_j/\log q_j} \quad (9\text{-}70)$$

This is substituted in the previous expression to yield m_i, which is then used to calculate the minimum cost.

A similar argument may be used when weight constraints are introduced. In this case the problem is usually one of maximizing the reliability, subject to prescribed weight conditions [20].

9-5. Gaming or Operational Simulation Method and Monte Carlo Method

Simulation [52, 55]

A major difficulty encountered in operations research is that of dealing with a situation so complex that it is impossible to set up an analytical expression, whether deterministic or probabilistic, which can be manipulated to yield immediate answers. Gaming techniques aim at simulating the operation systematically, where the major factors and their interactions are studied. War gaming is an experiment in which an operation is simulated on a gaming board with pieces used for soldiers and ships, etc., in order to develop tactical ideas regarding the operation. Frequently the operation may be divided into parts, each of which is described by a frequency distribution or even by an algebraic formula. Now an actual instance of the operation implies that each part has taken on one numerical value; the numerical values combine to give a numerical result. If several instances of the operation are considered, then several results are obtained. These may then be suitably combined to yield an answer by such means as statistical methods to determine an optimum. If actual data are not available, then reasonable assumptions are made as to what the distribution could be and the results noted.

The advantage here is that in many instances it is the only practicable method available for solving complicated operations-research decision problems. It helps to systematize an operation and simplifies the task of describing its elements and their interrelations, particularly to individuals without the needed technical background for more elaborate analyses. It also yields ideas as to the order of magnitude of the answer. It provides for testing statistically the sensitivity of the result relative to some of the factors involved. Thus a factor whose presence or absence is not appreciably noticeable may be eliminated.

SOME APPLICATIONS

Computers are often used in gaming methods to supply rapidly a larger sample of results. They have the advantages of speed and economy and can be controlled and relied upon.

Monte Carlo Method [43, 45, 58]

It often happens that an equation arises in the formulation of a model (probabilistic or not) which does not yield a quick solution by standard numerical methods or cannot be properly manipulated to obtain desired answers for the problem.

Now a stochastic process with distributions and parameters which satisfy the equation may exist. Instead of using standard methods, it may then be more efficient to construct the stochastic model of the problem and compute the statistics.

Thus an experiment is set up to duplicate the features of the problem under study. The calculation process is entirely numerical and is carried out by supplying numbers into the system and obtaining numbers from it as an answer. Often the numbers supplied are random numbers obtained from a table, dice, a computer, or any (uniform) random-number-producing device such as a roulette wheel; hence the name Monte Carlo. These numbers are fed into the system either directly or through a cumulative distribution (as probabilities) if the latter constitutes a law of entry into the system (e.g., the input distribution in a queue system).

As an illustration of the method, one may wish to evaluate the integral

$$J = \int_0^1 e^{-x} \, dx$$

The actual value of the integral is .63.

To obtain the Monte Carlo value of this integral the following procedure is used:

1. Plot e^{-x} vs. x on the interval $(0,1)$.
2. Use table of random numbers to obtain a column of values of x between zero and unity.
3. Construct neighboring column of random values, also between zero and unity.
4. Using the graph, perform a test to see if, for each x in the first column, the adjacent value in the second column lies above or below the curve.
5. Record the total number of points which are below the curve. Then divide this by the total to obtain the Monte Carlo value of this integral.
6. In $N = 114$ tries there were 68 successes of this experiment, thus yielding a value of .6 for the integral.
7. The probable error p is defined as that value x for which there is

equal choice for a new value to fall on either side of x. It is obtained from

$$\frac{1}{\sqrt{2\pi}\sigma} \int_{-p}^{p} e^{-x^2/2\sigma^2}\, dx = .5$$

which yields $p = .675\sigma$.

Thus the probable error in this case is given by

$$.675 \sqrt{\frac{J(1-J)}{N}} = .03$$

where J is the actual value of the integral and $N = 114$ is the total number of trials. The method of confidence limits given in the next chapter may also be used to estimate the error.

Some examples where Monte Carlo techniques can be used are as follows:

1. Calculation of π by dropping a needle on a checkerboard and counting the number of times it falls across a line.
2. Calculation of probabilities. Suppose that a coin is tossed N times and n heads were observed; i.e., the experimental probability of producing heads is $p' = n/N$. The law of large numbers [25] gives a method of estimating p, the actual probability of heads. It is

$$\lim_{N \to \infty} \sup \left[\frac{|p' - p|}{\left(2pq \dfrac{\log \log N}{N}\right)^{1/2}} \right] = 1 \qquad (9\text{-}71)$$

for almost all sequences of trials. Or suppose that the absolute value of the expression in brackets is less than unity for almost all N. This gives an estimate of the error obtained from a Monte Carlo run of the experiment. It also shows that the error varies as $1/\sqrt{N}$.

3. Numerical solutions of partial differential equations with boundary conditions.
4. Calculation of the vulnerability of a piece of equipment or a target to some type of offensive weapon, e.g., calculation of errors.
5. Evaluation of multiple integrals [35, 58].
6. Industrial repair problems [10].
7. Prediction of a sales pattern from a known distribution of sales.
8. Calculation of matrix eigenvalues and the inversion of matrices [57].

Observations on the Use of This Method.

1. Each problem must be studied individually to determine the manner in which to apply the method. Attention must be focused on the aspects of the problem in which one is really interested and the other aspects ignored.

2. The novelty of the method lies in its applicability to a nonprobabilistic equation which demands numerical solutions not rapidly obtainable by standard numerical methods. It is a contribution to the computing art.

3. A Monte Carlo examination of a problem enables one to eliminate aspects of the problem without significant consequence. This facilitates subsequent analytical treatment. This perhaps is the fundamental use of the method.

4. The use of electronic computers facilitates the job of obtaining quick answers to problems which would otherwise require detailed and time-consuming analysis.

5. If in a problem there are fluctuations due to time, it is advisable to apply the method to the parts which appear to cluster together in time and then combine the results in a suitable manner.

6. For models which cannot be given rigorous mathematical formulation, perhaps because of complexity, but have some form developed by plausible reasoning, the method can be used in a simple fashion to obtain estimates, e.g., upper and lower bounds on the predictions of the theory. Of course, if actual data are available in sufficient amount, standard procedures are then followed.

7. The method should be applied several times to a problem with different sample sizes and an over-all answer computed.

9-6. The Renewal Equation [5, 36]

Renewal-process problems are concerned with sums of random variables where variables are added on until a certain desired result is obtained.

Example 1: Let x_1 be the lifetime of the first of a set of light bulbs; let x_2 be the lifetime of the second, etc. Each x_i is a random variable. The question arises as to what is the expected age of a light bulb in time t and what is the expected number of replacements in time t if the bulbs are in operation.

Solution: Suppose that life exists in discrete units. Let a_k be the probability that the life of a bulb is k units. Suppose that the total time considered is n units. Then the remaining time is $n - k$ units. Let u_n be the expected number of replacements in time n. Then

$$u_n = \sum_{k=0}^{n} a_k(1 + u_{n-k}) = \sum_{k=0}^{n} a_k u_{n-k} + b_n \qquad (9\text{-}72)$$

where $b_n = \sum_{k=0}^{n} a_k$.

In the continuous case, as illustrated in Chap. 4, the renewal equation is given by

$$u(x) = g(x) + \int_{-\infty}^{\infty} u(x - t)f(t) \, dt \qquad (9\text{-}73)$$

For example, let $u(t)$ be the expected number of counts of a Geiger counter at time t, where something about the input distribution $f(t)$ is known. What is the variance of the number of counts? The desired solution of this equation must have specified behavior at $-\infty$ and some growth rate at $+\infty$. In this manner the solution is selected from among an infinite sequence of solutions. If the mean of $f(t)$ is known, then

$$\lim_{x \to \infty} u(x) \to \frac{c}{n} x$$

where c depends on $g(x)$.

Solutions are known to exist by the physical nature of the system. However, one is generally interested in its properties. For the discrete example above, if 10,000 lamps of 2500-hour life expectancy each and a duration x for each are used, then $x/2500$ is the expected number of replacements. The variance is $t\sigma^2/\mu^3$, where μ is the expected life and σ^2 is the variance of the life of a single bulb. The equation

$$u(t) = g(t) + \int_0^t u(t - s)f(s) \, ds \qquad (9\text{-}74)$$

has a unique bounded solution on $[0,T]$ if $g(t) < A$ for some constant A and if $\int_0^t f(s) \, ds$ is convergent.

One proves this assertion by actually constructing a solution. Let

$$u_0(t) = g(t) \quad \text{and} \quad u_{n+1}(t) = g(t) + \int_0^t u_n(t - s)f(s) \, ds \qquad (9\text{-}75)$$

Then one shows that the functions $\{u_n(t)\}$ are uniformly bounded on a subinterval $[0,T_1]$. The uniform convergence of this sequence is then established, which shows that $u(t)$ is the limit providing the solution on $[0,T_1]$. A similar sequence is constructed for an interval $[T_1, 2T_1]$ which converges to a limit for a solution on this interval. Together with $u(t)$, these two functions provide a solution on $[0, 2T_1]$. In this manner the construction is continued to $[2T_1, 3T_1]$ until the entire interval $[0,T]$ has been covered. The solution is absolutely integrable i.e., integrable in absolute value, if $f(t) < A$ and $\int_0^t g(s) \, ds$ is convergent.

Example 2: The following example gives an illustration of renewals and approximations. Suppose that it is desired to estimate renewals of a population of items which are subjected to n types of failures independently distributed according to $g_j(x) = \mu_j e^{-\mu_j x}$ for the jth type of failure. (See Ref. [25], page 276, reference to D. J. Bishop.)

SOME APPLICATIONS

One can assume that the proportion of items used which are subject to the jth type of failure is a_j and that the total number used is N. Then one has for the over-all distribution of failures

$$g(x) = \sum_{j=1}^{n} a_j g_j(x) \qquad \sum_{j=1}^{n} a_j = 1 \qquad (9\text{-}76)$$

To save calculations, this may be adequately approximated by the gamma distribution:

$$f(x) = \frac{b^a}{\Gamma(a)} x^{a-1} e^{-bx} \qquad a, b > 0 \qquad (9\text{-}77)$$

The constants b and a are chosen so that the first and second moments of $f(x)$ agree with the respective moments of $g(x)$. Now the first moment of $g(x)$, using the fact that

$$\int_0^\infty x^a e^{-bx} \, dx = \frac{a!}{b^{a+1}} \qquad (9\text{-}78)$$

is given by

$$m_1 = \int_0^\infty x g(x) \, dx = \sum_{j=1}^{n} \frac{a_j}{\mu_j} \qquad (9\text{-}79)$$

and its second moment is

$$m_2 = \int_0^\infty x^2 g(x) \, dx = 2 \sum_{j=0}^{n} \frac{a_j}{\mu_j^2} \qquad (9\text{-}80)$$

On the other hand, the first moment of the approximating distribution $f(x)$ is given by

$$\frac{b^a}{\Gamma(a)} \frac{\Gamma(a+1)}{b^{a+1}} = \frac{a}{b}$$

having used the relations for the gamma function $\Gamma(a+1) = a\Gamma(a)$, and

$$\Gamma(a) \equiv \int_0^\infty x^{a-1} e^{-x} \, dx$$

Its second moment is

$$\frac{b^a}{\Gamma(a)} \frac{\Gamma(a+2)}{b^{a+2}} = \frac{a(a+1)}{b^2}$$

Hence, one must have

$$\frac{a}{b} = m_1 \quad \text{or} \quad a = b m_1$$

and

$$\frac{b m_1 (b m_1 + 1)}{b^2} = m_2$$

or

$$b = \frac{1}{(m_2/m_1) - m_1}$$

Now we have from the example illustrating integral equations in Chap. 4 that the renewals $u(t)$ at time t may be obtained from

$$L\{u(t)\} = \frac{NL\{f(t)\}}{1 - L\{f(t)\}}$$

Thus
$$L\{f(t)\} = \frac{b^a}{\Gamma(a)} \int_0^\infty t^{a-1} e^{-bt} e^{-xt}\, dt = \frac{b^a}{(b+x)^a} \tag{9-81}$$

and
$$L\{u(t)\} = \frac{Nb^a/(b+x)^a}{1 - b^a/(b+x)^a}$$

$$= N \sum_{i=1}^\infty \left(\frac{b^a}{(b+x)^a}\right)^i \tag{9-82}$$

having used the series expansion $(1-z)^{-1} = \sum_{i=0}^\infty z^i$ for $z < 1$.

The series converges uniformly in x for all $x \geq 0$. Therefore, one can determine, term by term, the functions for which $[b^a/(b+x)^a]^i$ are Laplace transforms. Hence

$$u(t) = N \sum_{i=1}^\infty \left(\frac{b^{ia}}{\Gamma(ia)}\right) t^{ia-1} e^{-bt} \tag{9-83}$$

Note that
$$\frac{1}{\Gamma(a)} \int_0^\infty (t^{a-1} e^{-bt}) e^{-xt}\, dt = \frac{1}{(b+x)^a}$$

One finally has, for the total renewals by time t,

$$U(t) = \int_0^t u(y)\, dy = N \sum_{i=1}^\infty I\left(\frac{tb}{\sqrt{ia}}, ia-1\right) \tag{9-84}$$

where
$$I(u,v) = \frac{\int_0^{u\sqrt{v+1}} x^v e^{-x}\, dx}{\Gamma(v+1)} \tag{9-85}$$

is the incomplete gamma function which may be calculated from K. Pearson's "Tables of the Incomplete Gamma Function" [50].

REFERENCES

1. Alder, B. J., S. P. Frankel, and V. A. Lewison: Radial Distribution Function Calculated by the Monte-Carlo Method for a Hard Sphere Fluid, *Journal of Chemical Physics*, vol. 23, no. 3, pp. 417–419, March, 1955.
2. *American Mathematical Monthly*, vol. 62, p. 737, 1955.
3. Arrow, K. J., I. Harris, and J. Marschak: Optimal Inventory Policy, *Econometrica*, vol. 19, p. 250, 1951.

4. Ball, L. W.: "Management Use of Laboratory Testing to Achieve Reliability," University of Southern California Engineering and Management Course, Jan. 31–Feb. 11, 1955.
5. Bartlett, M. S.: "An Introduction to Stochastic Processes," Cambridge University Press, London, 1955.
6. Beers, Y.: "Introduction to the Theory of Error," Addison-Wesley Publishing Company, Reading, Mass., 1956.
7. Bellman, R.: Dynamic Programming and Multi-stage Decision Processes of Stochastic Type, *Proceedings of the Second Symposium on Linear Programming*, vol. 1, p. 229, sponsored by Directorate of Management Analysis, DCS Comptroller, Hq., U.S. Air Force, and National Bureau of Standards, Washington, D.C., Jan. 27–29, 1955.
8. Bird, G. T.: Controlled Test Status: Evaluation of Incoming Inspection Methods for Missile Tubes, U.S. Air Force Guided Missile Reliability Symposium, Dayton, Ohio, Nov. 2–4, 1955.
9. Borel, E., and R. Deltheil: "Probabilités erreurs," Librairie Armand Colin, Paris, 1950.
10. Bowman, E. H., and R. B. Fetter: Monte Carlo, 1956 *Supplement to Summer Course Notes on Operations Research*, Technology Press, MIT, Cambridge, Mass., 1956.
11. Brown, G. W.: Monte Carlo Methods, chap. 12, in E. F. Bechenbach (ed.), "Modern Mathematics for the Engineer," McGraw-Hill Book Company, Inc., New York, 1956.
12. Bryan, J. E., G. P. Wadsworth, and T. M. Whitin: A Multistage Inventory Model, *Naval Research Logistics Quarterly*, vol. 2, nos. 1 and 2, pp. 25–28, March–June, 1955.
13. Bush, R. R., and C. F. Mosteller: A Stochastic Model with Applications to Learning, *Annals of Mathematical Statistics*, vol. 24, no. 4, December, 1953.
14. Carhart, R. R.: Growing a Reliable Missile, U.S. Air Force Guided Missile Reliability Symposium, Dayton, Ohio, Nov. 2–4, 1955.
15. Courant, R., and D. Hilbert: "Methods of Mathematical Physics," vol. I, Interscience Publishers, Inc., New York, 1953.
16. Curtiss, John H.: Sampling Methods Applied to Differential and Difference Equations, *Proceedings of Seminar on Scientific Computation*, pp. 87–109, International Business Machines Corporation, New York, November, 1949.
17. Cutkosky, R. E.: A Monte Carlo Method for Solving a Class of Intergral Equations, *Journal of Research of the National Bureau of Standards*, vol. 47, pp. 113–116, 1951.
18. Danskin, J. M.: Mathematical Treatment of a Stockpiling Problem, *Naval Research Logistics Quarterly*, vol. 2, nos. 1 and 2, March–June, 1955.
19. Darling, D. A., and W. M. Kincaid: An Inventory Problem, *Journal of the Operations Research Society of America*, vol. 1, p. 80, 1952–1953. Abstract.
20. DiToro, M. J.: Reliability Criterion for Constrained Systems, *Institute of Radio Engineers Transactions on Reliability and Quality Control*, pp. 1–6, September, 1956.
21. Donsker, M. D., and M. Kac: The Monte Carlo Method and Its Applications, *Proceedings of Seminar on Scientific Computation*, pp. 74–81, International Business Machines Corporation, New York, November, 1949.
22. Dvoretzky, A., J. Kiefer, and J. Wolfowitz: The Inventory Problem, *Econometrica*, vol. 20, no. 2, pp. 187–222, April, 1952, and no. 3, pp. 450–466, July, 1952.
23. Dvoretzky, A., J. Kiefer, and J. Wolfowitz: On the Optimal Character of the

(s,S) Policy in Inventory Theory, *Econometrica*, vol. 21, no. 4, pp. 586–596, October, 1953.
24. Feeney, G. J.: A Basis for Strategic Decisions on Inventory Control Operations, *Management Science*, vol. 2, no. 1, pp. 69–82, October, 1955.
25. Feller, W.: "An Introduction to Probability Theory and Its Applications," vol. 1, John Wiley & Sons, Inc., New York, 1950.
26. Forsythe, G. E., and R. A. Leibler: Matrix Inversion by a Monte Carlo Method, in "Mathematical Tables and Other Aids to Computation," pp. 127–129, National Academy of Sciences, Washington, D.C., 1950.
27. Friedman, L.: A Competitive-bidding Strategy, *Operations Research*, vol. 4, p. 104, 1956.
28. Geisler, M. A., and H. W. Karr: The Design of Military Supply Tables for Spare Parts, *Operations Research*, vol. 4, no. 4, p. 431, August, 1956.
29. Gouray, M. H.: An Optimum Allowance List Model, *Naval Research Logistics Quarterly*, vol. 3, no. 3, p. 177, September, 1956.
30. Hall, J. F., and J. M. Cook: Programming a Monte Carlo Problem, *Proceedings of the Second Annual Computer Applications Symposium*, Armour Research Foundation, Chicago, October, 1955.
31. Hill, D. A., H. D. Voegtten, and J. H. Yueh: Parts vs. Systems, The Reliability Dilemma, Western Electronic Show and Conference, San Francisco, April, 1955.
32. Householder, A. S., G. E. Forsythe, and H. H. Germond (eds.): Monte Carlo Method, *National Bureau of Standards Applied Mathematics Series*, No. 12, June 11, 1951.
33. Jeans, Sir James: "An Introduction to the Kinetic Theory of Gases," Cambridge University Press, London, 1948.
34. Johnson, P. C., and F. C. Uffelman: A Punched Card Application of the Monte Carlo Method, *Proceedings of Seminar on Scientific Computation*, pp. 82–88, International Business Machines Corporation, New York, November, 1949.
35. Kahn, H.: Use of Different Monte Carlo Sampling Techniques, *RAND Memorandum*, p. 766, The RAND Corporation, Santa Monica, Calif., November, 1955.
36. Karlin, S.: On the Renewal Equation, *Pacific Journal of Mathematics*, vol. 5, no. 2, p. 229, June, 1955.
37. King, G. W.: Stochastic Methods in Quantum Mechanics, *Proceedings of Seminar on Scientific Computation*, pp. 92–94, International Business Machines Corporation, New York, November, 1949.
38. King, G. W.: The Monte Carlo Method as a Natural Mode of Expression in Operations Research, *Journal of the Operations Research Society of America*, vol. 1, no. 2, February, 1953.
39. Koopman, B. O.: The Theory of Search. Part I: Kinematic Bases, *Operations Research*, vol. 4, no. 3, p. 324, June, 1956.
40. Koopman, B. O.: The Theory of Search. Part II: Target Detection, *Operations Research*, vol. 4, no. 5, pp. 503–531, October, 1956.
41. Kozmetsky, G., and P. Kircher: "Electronic Computers and Management Control," McGraw-Hill Book Company, Inc., New York, 1956.
42. Laderman, J., S. B. Littaur, and L. Weiss: The Inventory Problem, *Journal of the American Statistical Association*, vol. 48, no. 264, pp. 717–732, December, 1953.
43. Metropolis, N., and S. Ulam: The Monte Carlo Method, *Journal of the American Statistical Association*, vol. 44, pp. 335–341, 1949.
44. Meyer, H. A., E. T. Lytle, Jr., L. S. Gephart, and N. L. Rasmussen: Inversion of Matrices by Monte Carlo Methods, *WADC Technical Report* 54-56, Wright Air Development Center, Dayton, Ohio, January, 1954.

45. Meyer, H. A. (ed.): "Symposium on Monte Carlo Method," John Wiley & Sons, Inc., New York, 1956.
46. Milne, W. E.: "Numerical Solution of Differential Equations," John Wiley & Sons, Inc., N. Y., and Chapman & Hall, Ltd., London, 1953.
47. Moskowitz, F., and J. P. McLean: Some Reliability Aspects of Systems Design, *Institute of Radio Engineers Transactions on Reliability and Quality Control*, pp. 7–35, September, 1956.
48. Opler, A.: Application of Computing Machines to Ion Exchange Column Calculations, *Industrial and Engineering Chemistry*, vol. 45, no. 12, pp. 2621–2629, December, 1953.
49. Ott, E. R., and A. B. Mundell: Narrow-limit Gaging, *Industrial Quality Control*, March, 1954.
50. Pearson, K.: "Tables of the Incomplete Gamma Function," Cambridge University Press, London, 1922.
51. *Proceedings of the First Ordnance Conference on Operations Research*, U.S. Army Ordnance Corps, Duke Station, Durham, N.C.
52. Rich, R. P.: Simulation as an Aid to Model Building, *Journal of the Operations Research Society of America*, vol. 3, no. 1, pp. 15–19, February, 1955.
53. Rosenblatt, M.: An Inventory Problem, *Econometrica*, vol. 22, p. 245, 1954.
54. Simon, H. A., and C. C. Holt: The Control of Inventory and Production Rates— A Survey, *Journal of the Operations Research Society of America*, vol. 2, no. 3, pp. 289–301, August, 1954.
55. Thomas, C. J., and W. L. Deemer, Jr.: The Role of Operational Gaming in Operations Research, *Operations Research*, vol. 5, no. 1, February, 1957.
56. Toben, G. J.: Transition from Problem to Card Program: *Proceedings of Seminar on Scientific Computation*, pp. 128–131, International Business Machines Corporation, New York, November, 1949.
57. Todd, J.: "Matrix Inversion by a Monte Carlo Method," prepared for the Office of the Air Comptroller, USAF, U.S. Department of Commerce, National Bureau of Standards, Feb. 2, 1951.
58. U.S. Department of Commerce, National Bureau of Standards: The Monte Carlo Method, Applied Mathematics Seminar 12, June 11, 1951.
59. Uspensky, J. V.: "Introduction to Mathematical Probability," McGraw-Hill Book Company, Inc., New York, 1937.
60. Vassian, H. J.: Application of Discrete Variable Servo Theory to Inventory Control, *Journal of the Operations Research Society of America*, vol. 3, no. 3, pp. 272–282, August, 1955.
61. Whitin, T. M., and J. W. T. Youngs: A Method for Calculating Optimal Inventory Levels and Delivery Time, *Naval Research Logistics Quarterly*, vol. 2, no. 3, pp. 157–174, September, 1955.
62. Woodbury, W. W.: Monte Carlo Calculations, *Proceedings of Seminar on Scientific Computation*, pp. 17–19, International Business Machines Corporation, New York, November, 1949.
63. Yowell, E. C.: A Monte Carlo Method of Solving Laplace's Equation, *Proceedings of Seminar on Scientific Computation*, pp. 89–91, International Business Machines Corporation, New York, November, 1949.
64. Yueh, J. H.: A Developmental Approach to Reliability in Missile System Equipment, *Institute of Radio Engineers Transactions on Reliability and Quality Control*, pp. 44–54, September, 1956.
65. Zimmerman, R. E.: A Monte Carlo Model for Military Analysis, in J. McCloskey and J. Coppinger (eds.), "Operations Research for Management," vol. II, Johns Hopkins Press, Baltimore, 1956.

CHAPTER 10

FUNDAMENTAL STATISTICS

10-1. Introduction

One is nearly awed by the abundance of literature available in the field of statistics. Yet, because of the finiteness of basic methods, one is not discouraged to attempt a brief presentation which, to some extent, should serve to alert the reader to the existence of subject matter and give him some essential auxiliary tools as an aid in applying the scientific method.

Aside from the techniques from the field of statistics, the main method of approaching a problem and the type of analysis required to obtain a solution after an appropriate interpretation of the data are still the scientific method [13]. A good statistical analysis does not stop with calculations but scientifically interprets these results. Simply stated, the handling of statistical methods is a delicate task requiring a considerable degree of scientific maturity. This skill is not achieved until after one has acquired the tools, and their acquisition is a far easier task than making correct use of them.

To avoid continual cautioning regarding the use of these methods, it will be sufficient to say that these tools should be handled only by a trained individual, the danger being that these methods frequently can be twisted to yield unwarranted conclusions [11]. One cannot overemphasize the need for continuous recognition of the assumptions made in the use of specific statistical techniques.

In the remainder of the introduction, a skeletal account of some statistical methods will be given. Although it may be desirable for the sake of continuity to relate, in a logical sequence revolving about a single problem, all the statistical methods to be discussed here, such an attempt could not be made without the appearance of artificiality. Thus, the account will be completed in several stages.

In the research phase of a problem where it may be desired to study a characteristic which has been identified (such as the height of people) one must decide on the population in which this property is to be studied e.g., the United States or China. Having done so, for economy's sake one gathers samples (without exhausting the population) from which one can then make inferences regarding the entire population. The sample may be used for descriptive (textual or verbal, tabular and graphic) purposes or, as mentioned above, for inferences (estimation and generalization) regarding the property under study. One aim in conducting statistical analysis of a problem is to describe the problem by means of a probability distribution. Once the desired distribution has been obtained, one may proceed with probabilistic analysis of the problem, often by forming a theoretical model in which the estimated distribution plays an important role. Then a theoretical analysis enables the search for remedies or the making of predictions. The probability distribution is chosen so as to describe "acceptably" samples taken from the population under study. Thus one must estimate the "best" distribution to describe the sample of a problem (or an experiment). Appropriate samples are taken from the population and are grouped to yield a histogram. To estimate the cumulative distribution function, one arranges the sample values according to size and draws a step function rising by a value $1/n$ at each point, n being the sample size. From Bernoulli's law of large numbers it can be shown that, as $n \to \infty$, this step function converges to the cumulative distribution function.

There are several methods used in practice to approximate the step function by a known distribution and then test the goodness of the approximation. Frequently, one first obtains the frequency distribution and then proceeds with the analysis. If a hypothetical distribution is known to produce a fit, then there are a number of possibilities. Assume that the distribution is normal; then one estimates its parameters μ and σ from the data, often prescribing confidence limits for the estimated values. One asks the question as to whether the data are consistent with the hypothesis that the parameter has a specific value and then proceeds to test this hypothesis. Thus, whatever the method of estimation may be, the result must usually be tested further. Otherwise one may incur the error of rejecting the hypothesis when it is true or accepting it when it is false. The object is to minimize errors. Having estimated the parameters of the normal distribution, one applies the chi-square test to verify that the hypothesized distribution with estimated parameters adequately describes the data. If the hypothesized distribution is completely known without having to estimate its parameters, one applies the Kolmogorov-Smirnov test to verify its adequacy.

There are many other ways in which testing may be used. For example, if one has two samples, one may wish to decide whether the two distributions describing the samples are the same. If both distributions are normal with unknown variances, one uses the t test in order to determine whether there is a significant difference between the two means. However, when the variances are known, one uses the normal distribution. If the distributions are arbitrary, one uses the Wilcoxon test. The F test is used to test the variances and decide whether the two means come from the same normal population.

A standard tool used in prediction is correlation, which is concerned with accounting for variations among factors which are known to interact. The correlation between lung cancer and smoking may be used to predict the effect of smoking on lung cancer.

For some time, the estimation of parameters from samples and the testing of hypotheses were based upon the assumption that the underlying distribution of the population was known, and frequently, because of the central-limit theorem, normality was assumed and methods of estimation and testing revolved about this assumption. More recently, distribution-free methods, frequently called "nonparametric methods," which do not assume that the population distribution is known in estimating parameters and testing hypotheses, have been successfully investigated. The latter theory is particularly useful when sample sizes are small and assumptions of normality become untenable.

10-2. Sampling [5]

The main purpose of sampling is to obtain information about population characteristics. The use of samples avoids costly and time-consuming observations of the entire population, particularly when the experiment destroys the sample taken, and still enables one to reach valid conclusions regarding the population.

The first type of sampling one may consider is pure random sampling when each element has an equal chance of being selected. As an illustration of a random-sampling experiment, suppose that it is desired to measure n diameters taken at random along a cylinder. The cylinder is first divided into serially numbered segments, and diameters are measured in the segments indicated by random numbers taken from a table. In taking a random sample, one relies on the laws of chance to obtain a representative sample.

Suppose that a population characteristic, which is relevant to the characteristic under study, is known. This can be used to divide or stratify the sample in order to improve the information obtained. For example, if a poll is taken among males and females to obtain opinions on a certain subject, one may obtain a random sample of 50 males and another of 50 females and combine the two for study rather than rely on chance to give this proportion from the population of males and females. As another example, if it is known that there is greater variability among males, the sample can be divided in such a way as to include more males. In this way the variability among males is represented through larger samples.

Another type of sampling is systematic sampling. Here the population items are arranged in a sequence and every rth item is selected after a random start. The population may also be divided into subsets of r items each, and from each subset an item is selected at random. Judgment may be used to identify a convenient group as being typical of the population, and hence the study is conducted on the group without sampling from the entire population. In that case, one speaks of "judgment sampling." Some characteristics on which one may wish to base sample studies are the mean,

$$\bar{x} = \frac{1}{n} \sum_{i=1}^{n} x_i \tag{10-1}$$

and the variance,
$$s^2 = \frac{1}{n} \sum_{i=1}^{n} (x_i - \bar{x})^2 \tag{10-2}$$

They are among the simplest characteristics one calculates for a sample. Here x_i is the measurement taken on the ith item in the sample. For large samples, the sample mean is approximately equal to the population mean μ by Tchebycheff's theorem. Furthermore,

$$E[\bar{X}] \equiv E\left[\frac{X_1 + \cdots + X_n}{n}\right] = \mu \tag{10-3}$$

since random variables X_i are identically and independently distributed. Also

$$E[(\bar{X} - \mu)^2] = \frac{\sigma^2}{n} \tag{10-4}$$

However,

$$E[s^2] = E\left[\frac{1}{n}\sum_{i=1}^{n}(X_i - \mu)^2 - (\bar{X} - \mu)^2\right] = \sigma^2 - \frac{\sigma^2}{n} = \frac{n-1}{n}\sigma^2 \tag{10-5}$$

since $\quad E[X_i - \mu]^2 = \sigma^2 \quad$ and $\quad E[\bar{X} - \mu]^2 = \frac{\sigma^2}{n}$

where σ^2 is the population variance.

For calculations with samples of moderate size it is preferable to consider the corrected sample variance

$$\frac{1}{n-1} \sum_{i=1}^{n} (x_i - \bar{x})^2$$

whose expected value is σ^2. The square root of this quantity is the sample standard deviation.

By the central-limit theorem, $\bar{x}$ is normally distributed with mean μ and variance σ^2/n for large n. This fact has a wider applicability to moments. For example, it applies to the sample standard deviation, to measures of skewness and excess, to the median (which is not a moment), etc.

Suppose that random samples of size n are taken from a normal population. As an exercise we shall prove that the sample mean is also normally distributed with mean μ and variance σ^2/n by showing that its moment generating function coincides with that of a normal distribution. Now,

$$m(e^{t\bar{x}}) \equiv \int_{-\infty}^{\infty} \cdots \int_{-\infty}^{\infty} \exp\left(\frac{t}{n}\sum_{i=1}^{n} x_i\right) f(x_1) \cdots f(x_n)\, dx_1 \cdots dx_n$$

since the samples are independent and are identically distributed. But the moment generating function of the sum of n independent variates is the product of their moment generating functions; hence, one has

$$m(e^{t\bar{x}}) = m[e^{(t/n)x_1}] \cdots m[e^{(t/n)x_n}] = \left\{\exp\left[\mu\frac{t}{n} + \frac{1}{2}\left(\frac{t\sigma}{n}\right)^2\right]\right\}^n$$
$$= \exp\left[\mu t + \frac{(t\sigma)^2}{2n}\right]$$

Using series expansion, it is not difficult to show that this moment generating function is also obtained as $n \to \infty$ for the mean of samples taken from a population with an arbitrary distribution.

10-3. Estimation of Parameters [14, 16]

Estimating the characteristics of a population is a prime concern of statistics. There are various ways of obtaining estimators. In estimation theory one also investigates different properties of estimators. For example, there is the unbiased estimator T for which $E[T] = \theta$, where θ is the true parameter value. An estimator is consistent if $P(1T \to \theta 1 < \epsilon) \to 1$ as the sample size tends to infinity. An estimator is not unique since from an estimator T it is possible to obtain an infinity of biased estimators $(n - a)/(n - b)T$, where a and b are constants. In this section some estimation methods will be briefly examined and illustrated.

There are instances in which the type of distribution of the population is known but the specific values of the parameters in the distribution must be estimated from population samples. The accuracy (given in terms of probability) with which the estimated value is close to the true value of the parameter is required to determine the significance of the estimate. If the accuracy is small, the approximation is then crude. This value may be adequate for the purpose, or further sampling may be required to obtain a better estimate.

Note that, when the parameters of the distribution have been estimated, many of the properties of probability distributions can be investigated. However, it is essential that the distribution assumed should be consistent with the data and that the parameters be correctly estimated. In general, different methods of estimation may be used, and some will be better in certain situations than others.

When more than one sample is available from unknown populations, it may be desired to test the hypothesis that the samples are taken from the same population by comparing their corresponding parameters. Significance tests to determine whether the parameters could have come from samples taken from the same population are then applied, and the hypothesis for the common parenthood of the test sample is accepted or rejected, depending on how closely certain predetermined significance levels are met. Procedures for testing such hypotheses will be described later.

Estimation Methods

A Priori Estimation Methods Based on Bayes' Formula. Suppose that a sample of size n is drawn from a lot of mechanical objects and that k of these are found defective. How may the proportion of defectives in the lot be estimated? Let p be the proportion of defectives in the lot. The probability of observing k defectives in a sample of size n is given by the binomial formula

$$P(p|k) = \binom{n}{k} p^k (1-p)^{n-k} \tag{10-6}$$

By Bayes' theorem, $P(k|p)$, the probability of finding k defectives in a sample of size n taken from a population whose defective proportion is p, is given by

$$P(k|p) = \frac{P(p)P(p|k)}{\Sigma P(p)P(p|k)} = \frac{P(p)\binom{n}{k}p^k(1-p)^{n-k}}{\Sigma P(p)\binom{n}{k}p^k(1-p)^{n-k}} \tag{10-7}$$

Once, $P(p)$, the probability that the population has a defective proportion p, has been chosen, $P(k|p)$ is determined.

It is natural to estimate p by a value p_0 which maximizes $P(k|p)$. Instead, it may be desired to determine for p an interval of probability $1 - \alpha$ involving $P(k|p)$. The major drawback of this method is the arbitrary choice frequently made in choosing $P(p)$. In many cases it is difficult to justify any choice of $P(p)$. Hence, the use of Bayes' formula is limited in applicability.

Fisher's Maximum-likelihood Estimation [4, 14]. The principle of maximum likelihood leads to a useful theory of parameter estimation, asserting for simple cases that the best estimates of the parameters are those values that maximize the probability of obtaining the sample.

Let $f(x,\theta)\,dx$ be a probability density, where θ is an unknown parameter. Assume that n independent observations $(x_1, x_2, \ldots, x_n)$ are taken from a population with the above probability density. Since the joint density of independent variables is the product of their densities, one has for the joint density

$$f(x_1,\theta)f(x_2,\theta)\cdots f(x_n,\theta)\,dx_1\,dx_2\cdots dx_n \qquad (10\text{-}8)$$

The product $L = f(x_1,\theta)\cdots f(x_n,\theta)$ is called the *likelihood function* when regarded as a function of θ. The maximum-likelihood estimator is a value $\theta = T$ which maximizes L as a function of θ. Since the logarithm of the likelihood has its maximum at the same point as does the likelihood, for calculation purposes it is convenient to take logarithms and then set the partial derivative with respect to θ equal to zero. Thus by differentiating

$$\log L = \sum_{i=1}^{n} \log f(x_i,\theta) \qquad (10\text{-}9)$$

with respect to θ and setting the result equal to zero, one has

$$\sum_{i=1}^{n} \frac{f_\theta(x_i,\theta)}{f(x_i,\theta)} = 0 \qquad (10\text{-}10)$$

This method poses no a priori hypothesis on θ. However, in turn, it introduces auxiliary assumptions and has its difficulties.

If $f(x_1, x_2, \ldots x_n; \theta_1, \theta_2)$ is the density for a random sample of size n from a population with two parameters θ_1 and θ_2, for example, the maximum-likelihood estimators are obtained from

$$\sum_{i=1}^{n} \frac{f_{\theta_1}(x_i;\theta_1,\theta_2)}{f(x_i;\theta_1,\theta_2)} = 0 \qquad (10\text{-}11)$$

and

$$\sum_{i=1}^{n} \frac{f_{\theta_2}(x_i;\theta_1,\theta_2)}{f(x_i;\theta_1,\theta_2)} = 0$$

which are then solved simultaneously (frequently with difficulty, requiring numerical approximations) for θ_1 and θ_2.

Example: The joint density function of a sample of n observations taken from a normal population with the normal density function

$$\frac{1}{\sqrt{2\pi}\,\sigma}\exp\left[-\frac{1}{2\sigma^2}(x-\mu)^2\right]$$

of unknown parameter μ and σ is given by

$$\prod_{i=1}^{n} \frac{1}{\sqrt{2\pi}\,\sigma} \exp\left[-\frac{1}{2\sigma^2}(x_i - \mu)^2\right] = \left(\frac{1}{2\pi\sigma^2}\right)^{n/2} \exp\left[-\frac{1}{2\sigma^2} \sum_{i=1}^{n}(x_i - \mu)^2\right] \quad (10\text{-}12)$$

It is desired to obtain estimates of μ and σ^2.

Solution: On taking logarithms and setting the partials with respect to μ and σ^2 equal to zero, one has, respectively,

$$\frac{1}{\sigma^2} \sum_{i=1}^{n}(x_i - \mu) = 0$$

$$-\frac{n}{2\sigma^2} + \frac{1}{2\sigma^4} \sum_{i=1}^{n}(x_i - \mu)^2 = 0 \quad (10\text{-}13)$$

Simultaneous solution of the above equations for μ and σ^2 yields, for their respective maximum-likelihood estimators T_1 and T_2,

$$T_1 = \frac{1}{n}\sum_{i=1}^{n} x_i = \bar{x}$$

$$T_2 = \frac{1}{n}\sum_{i=1}^{n}(x_i - \bar{x})^2 \quad (10\text{-}14)$$

These are also the sample moments corresponding to μ and σ^2. Note that the estimation of σ^2 depends on that of μ but not conversely. Since an estimator T of θ is unbiased if $E[T] = \theta$ (a desirable property), note that T_1 is unbiased; however,

$$E[T_2] = \frac{n-1}{n}\sigma^2 \quad (10\text{-}15)$$

which means that T_2 is unbiased only when $n \to \infty$. Maximum-likelihood estimators are either unbiased or have this property as $n \to \infty$.

An interesting theorem is the following: The maximum-likelihood estimators $T_1, T_2, \ldots, T_k$ from large samples of size n, of the parameters in the density

$$f(x;\theta_1, \ldots, \theta_k) \quad (10\text{-}16)$$

are approximately distributed by the multivariate normal distribution

$$\left(\frac{1}{2\pi}\right)^{k/2} \sqrt{|\sigma^{ij}|} \exp\left[-\frac{1}{2}\sum_{i=1}^{k}\sum_{j=1}^{k} \sigma^{ij}(x_i - \theta_i)(x_j - \theta_j)\right] \quad (10\text{-}17)$$

with means $\theta_1, \theta_2, \ldots, \theta_k$ and with matrix $[n\sigma^{ij}]$ for the coefficients of the quadratic form, where

$$\sigma^{ij} = -E\left[\frac{\partial^2}{\partial \theta_i\, \partial \theta_j} \log f(x;\theta_1, \cdots, \theta_k)\right] \qquad (10\text{-}18)$$

The variances and covariances of the estimators are

$$\left[\frac{1}{n}\sigma_{ij}\right] \qquad \text{where } [\sigma_{ij}] = [\sigma^{ij}]^{-1} \qquad (10\text{-}19)$$

The Method of Confidence Intervals. When a parameter θ is estimated by T, it is desirable to obtain information about the accuracy of the estimate. One method is to indicate an interval and a statement of confidence that the interval will include the value θ, such as

$$\text{Prob } \{\theta - h \le T \le \theta + h\} \ge 1 - \alpha \qquad (10\text{-}20)$$

From
$$\theta - h \le T \le \theta + h$$
one has
$$T - h \le \theta \le T + h$$

In other words, the interval $(T - h, T + h)$ covers θ in a proportion of $1 - \alpha$ of the cases. It is called the *confidence interval* and $1 - \alpha$ is called the *confidence coefficient*.

Example 1: Suppose that one wishes to estimate the mean μ of a normal population from $\bar{x}$, the mean of a sample with standard deviation s.

Solution: One assumes a significance level α and obtains a value of the statistic

$$t_\alpha = \frac{\bar{x} - \mu}{s}\sqrt{n - 1}$$

from the t table with $n - 1$ degrees of freedom, where n is the size of the sample. Since the mean of a random sample may be greater or less than the population mean, one has

$$P\left(-t_\alpha \le \frac{\bar{x} - \mu}{s}\sqrt{n - 1} \le t_\alpha\right) = 1 - \alpha \qquad (10\text{-}21)$$

which, when simplified, gives

$$P\left(\bar{x} - t_\alpha \frac{s}{\sqrt{n-1}} \le \mu \le \bar{x} + t_\alpha \frac{s}{\sqrt{n-1}}\right) = 1 - \alpha \qquad (10\text{-}22)$$

The two quantities on either side of μ provide confidence limits for μ; together they define a confidence interval. Note that t_α is not unique, and one may choose as limits any two values t_1 and t_2 such that

$$\int_{t_1}^{t_2} h(t)\, dt = 1 - \alpha$$

where $h(t)$ is the t frequency distribution. By selecting the two values symmetrically, that is, t_α and $-t_\alpha$, one obtains the smallest confidence interval when the distribution is symmetric, as is the case with $h(t)$.

If $(X - \mu)/\sigma$ is known to be normally distributed, one may calculate instead $-n_\alpha$ and n_α from the normal-distribution table in a similar way.

The probability statement given above may be interpreted so that, in successive calculations of confidence intervals at the indicated significance level for samples of equal size taken from the population, the intervals will be found to cover the population mean $100(1 - \alpha)$ per cent of the time. Before the sample was drawn, $1 - \alpha$, a measure of confidence, was the probability that the interval to be constructed would contain the population mean.

For a normal distribution, 95 per cent of the population has a value of the variate $X - \mu$ lying between 1.96σ and -1.96σ, and 99 per cent of the population has a value of the variate $X - \mu$ lying between 2.58σ and -2.58σ.

Example 2: To obtain a confidence interval for the variance of a normal distribution, note that for samples of n observations the expression

$$\chi^2 = \frac{ns^2}{\sigma^2}$$

follows the chi-square law with $n - 1$ degrees of freedom.

Solution: As before, one uses

$$P\left[\frac{\Sigma(x_i - \bar{x})^2}{\chi_\alpha'^2} \leq \sigma^2 \leq \frac{\Sigma(x_i - \bar{x})^2}{\chi_\alpha''^2}\right] = 1 - \alpha \qquad (10\text{-}23)$$

where $\chi_\alpha'^2$ and $\chi_\alpha''^2$ are two values of chi square which yield the value $1 - \alpha$ for $n - 1$ degrees of freedom as obtained from a table. This gives the desired confidence interval for σ^2.

10-4. The Testing of Hypotheses

The techniques used in hypothesis testing are often determined by a concern as to how, for example, on the basis of a sample, an examined lot should be accepted or rejected. For instance, one may wish to introduce an acceptance level of defectives as a percentage and, on the basis of the sample, accept it if the percentage of defectives is less than the acceptable percentage or otherwise reject it. In this process one can make two types of error: The first type consists of rejecting a lot when it is good, and the second type consists of accepting a lot when it is bad.

The Neyman Pearson theory [14] provides methods for testing hypotheses when the risk of making the first type of error is specified and minimizing the risk of making a second type of error. To provide a rule for determining whether the proportion of defectives in a lot is equal to 2 or to

4 per cent under the hypothesis that one or the other of these two percentages of defectives is in the lot, one chooses a probability α of erroneously refusing the lot if it presents 2 per cent of defectives and minimizes β, the probability of accepting the lot if it contains 4 per cent of defectives. The rule consists of rejecting the lot if the result of the sample falls in a region W or accepting it in the contrary case [16]. The probability of a sample of n observations $x_1, x_2, \ldots, x_n$ may be given as

$$dF(x_1)\, dF(x_2) \cdots dF(x_n) \tag{10-24}$$

where $dF(x)$ is the frequency distribution of the examined population. Now the distribution is estimated on the basis of w, the percentage of defectives in the lot. If $w = 2$ per cent, let

$$dF_1(x_1)\, dF_1(x_2) \cdots dF_1(x_n) = d\psi_1(x)$$

and for $w = 4$ per cent let

$$dF_2(x_1)\, dF_2(x_2) \cdots dF_2(x_n) = d\psi_2(x)$$

Let α be the probability of finding the result of the sample in W when the lot presents 2 per cent defectives; then

$$\alpha = \int_W d\psi_1(x) \tag{10-25}$$

Let β be the probability of finding the result of the sample in the region complementary to W (or $1 - \beta$ be the probability of finding the result in W) when the lot presents 4 per cent of defectives. Thus

$$\beta = 1 - \int_W d\psi_2(x) \tag{10-26}$$

The problem is to determine W with the aid of the two conditions

$$\begin{aligned} \alpha &= \int_W d\psi_1(x) \\ 1 - \beta &= \int_W d\psi_2(x) = \max \end{aligned} \tag{10-27}$$

This is a well-defined problem in the calculus of variations. The region W may be determined from the condition that

$$\lambda\, d\psi_1(x) \leq d\psi_2(x) \tag{10-28}$$

where λ is obtained from the first relation. The general procedure may be illustrated by an example.

Example: Let $w_1 = 2$ per cent, $w_2 = 4$ per cent, $\alpha = 5$ per cent, and $n = 400$. It is desired to determine a rule for accepting w_1 or w_2.

Solution: Since the proportion of defectives follows a binomial law which for large n may be approximated by the normal distribution with mean w and $\sigma = \sqrt{w(1-w)/n}$, one has for $d\psi_1$ the approximation

$$\frac{1}{\sigma_1 \sqrt{2\pi}} \exp\left[-\frac{(x-w_1)^2}{2\sigma_1^2}\right] dx \quad \text{with } w_1 = .02 \quad \sigma_1 = \sqrt{\frac{.02(1-.02)}{400}}$$

and for $d\psi_2$ the approximation

$$\frac{1}{\sigma_2 \sqrt{2\pi}} \exp\left[-\frac{(x-w_2)^2}{2\sigma_2^2}\right] dx \quad \text{with } w_2 = .04 \quad \sigma_2 = \sqrt{\frac{.04(1-.04)}{400}}$$

and the region of rejection is defined by $w \geq w_0$.

To determine w_0 one writes

$$\alpha = .05 = \frac{1}{\sqrt{2\pi}} \frac{\int_{w_0-w_1}^{\infty}}{\sqrt{w_1(1-w_1)/400}} e^{-(u^2/2)} du$$

using the normal approximation to the binomial. From a table for the cumulative normal distribution, the lower limit of the integral must approximately equal 1.64 in order to yield the value .05. This gives

$$\frac{w_0 - .02}{\sqrt{.02 \times .98/400}} = 1.64$$

or $w_0 = .0315$. Thus the rule requires that one rejects the lot if the proportion of defectives in the sample exceeds 3.15 per cent and accepts the lot otherwise. To calculate β, the probability of making an error of the second type, i.e., accepting the lot with 4 per cent defectives, one considers the expression

$$1 - \beta = \frac{1}{\sqrt{2\pi}} \frac{\int_{w_0-w_2}^{\infty}}{\sqrt{w_2(1-w_2)/400}} e^{-(u^2/2)} du$$

By using the previously calculated value of w_0, one has

$$\frac{.0315 - .0400}{\sqrt{.04 \times .96/400}} = -.87$$

Again from the table, this lower limit yields the value .81 to the integral. Hence $\beta = 1 - .81 = .19$. For practical reasons, it may be assumed more desirable to reject a good lot than to accept a bad one.

Note that the rule is given in terms of n, w_1, w_2, and α, and then a minimum β is calculated; or, if preferred, one may prescribe w_1, w_2, α, and β and then find n. Operating in this manner, there is a probability $p_1 \leq \alpha$ of making an error when $w \leq w_1$ and a probability $p_2 \leq \beta$ of making an error when $w \geq w_2$. It is not possible to make both α and β arbitrarily small for a fixed sample size n.

Some Ideas from Sequential Analysis

The above scheme requires the use of a sample of fixed size; then a decision is made as a function of the results obtained. Wald suggested a progressive- or sequential-sampling scheme in which, instead of determining beforehand the sample size, one specifies w_1, w_2, α, and β which are used to determine three regions: W_1 for accepting the lot, W_3 for rejecting it, and W_2 for continuing the sampling process.

To illustrate, suppose that among n items initially examined one has n_1 defectives. The probability of such a result is given by

$$p_n^{(1)} = \binom{n}{n_1} w_1^{n_1}(1 - w_1)^{n-n_1} \qquad (10\text{-}29)$$

under the hypothesis of a percentage of w_1 defectives, and

$$p_n^{(2)} = \binom{n}{n_1} w_2^{n_1}(1 - w_2)^{n-n_1} \qquad (10\text{-}30)$$

under the hypothesis of a percentage of w_2 defectives. The regions W_i ($i = 1, 2, 3$) are determined as follows: W_1 is defined by

$$\frac{p_n^{(2)}}{p_n^{(1)}} = \left(\frac{w_2}{w_1}\right)^{n_1} \left(\frac{1 - w_2}{1 - w_1}\right)^{n-n_1} \leq \frac{\beta}{1 - \alpha} \qquad (10\text{-}31)$$

and W_3 is defined by

$$\frac{p_n^{(2)}}{p_n^{(1)}} = \left(\frac{w_2}{w_1}\right)^{n_1} \left(\frac{1 - w_2}{1 - w_1}\right)^{n-n_1} \geq \frac{1 - \beta}{\alpha} \qquad (10\text{-}32)$$

whereas W_2 is the region between W_1 and W_3. More explicitly, by taking logarithms, one has W_1 defined by

$$n_1 \leq -h_1 + un \qquad (10\text{-}33)$$

and W_3 defined by

$$n_1 \geq h_2 + un \qquad (10\text{-}34)$$

(plotted with n_1 as ordinate and n as abscissa) with

$$h_1 = \frac{\log [(1 - \alpha)/\beta]}{\log (w_2/w_1) + \log [(1 - w_1)/(1 - w_2)]}$$

$$h_2 = \frac{\log [(1 - \beta)/\alpha]}{\log (w_2/w_1) + \log [(1 - w_1)/(1 - w_2)]} \qquad (10\text{-}35)$$

and

$$u = \frac{\log [(1 - w_1)/(1 - w_2)]}{\log (w_2/w_1) + \log [(1 - w_1)/(1 - w_2)]}$$

The rule consists of counting the n_1 obtained, rejecting the lot if the representative point (n, n_1) is above the upper line defined by the second inequality above, accepting it if it lies below the lower line defined by the first inequality, and continuing the sampling if it lies between the two regions, i.e., in W_2 which is between the two lines. By plotting the sample size as a function of the number of defectives, one obtains information as to an economic size of the sample one should use in nonsequential methods.

10-5. Student's, the Chi-square, and the F Tests

Student's t

The ratio of the deviation of the mean $\bar{x}$ of a particular sample from the population mean to its standard deviation $s/\sqrt{n-1}$ is known as Student's t, where

$$t = \frac{\bar{x} - \mu}{s/\sqrt{n-1}} \qquad (10\text{-}36)$$

By computing t from

$$\frac{\bar{x} - \mu}{s/\sqrt{n-1}}$$

and then also entering the t table (see Table 10-6) at a prescribed level with $n-1$ degrees of freedom, one compares the value of t from the table with the calculated value. The hypothesis that the sample and population means are the same at the prescribed significance level is rejected if the computed result exceeds the value from the table; otherwise, it is accepted. In a similar manner the t test serves to determine whether two sample means $\bar{x}$ and $\bar{\bar{x}}$ come from the same population by calculating

$$t = \frac{|\bar{x} - \bar{\bar{x}}|}{s} \sqrt{\frac{n_1 n_2}{n_1 + n_2}} \qquad (10\text{-}37)$$

where n_1 and n_2 are the respective sample sizes and s is the standard deviation of the two samples combined. For large samples, e.g., exceeding 30 items, the normal distribution may be used as an approximation to the t distribution.

The Chi-square Test

Chi square is defined as the sum of the squares of a number of independent variates from a normal population with zero mean and unit standard deviation. The number of independent variables determines the number of degrees of freedom $n-1$ associated with chi square. The frequency distribution was given in Chap. 8.

It is used to decide whether the estimate of variance from a sample of normally distributed observations is consistent with their having been drawn from a population of specified variance. It is also used to test whether the numbers of observations found in different classes agree (within the limits of random error) with the numbers expected in those classes.

As a goodness-of-fit test it is usually used as follows: Suppose that it is desired to test the hypothesis that the arrivals to a waiting line are Poisson-distributed with a mean arrival rate per unit time equal to .4. The intervals for counting the number of arrivals are fixed at 2 minutes. In each such interval, one counts the total number of arrivals. Then one counts the frequency of intervals in which a given number of arrivals occurs. This gives the first two columns of Table 10-1. When a number in

TABLE 10-1

Number of arrivals n	Number of 2-minute intervals observed	Value from Poisson distribution	Number of 2-minute intervals expected
0	21	.449	26.9
1	23	.359	21.5
2	10	.144	8.6
3	4 ⎫		
4	1 ⎪		
5	0 ⎬ 6	.048	2.9
6	0 ⎪		
7	1 ⎭		
	60		

the frequency column is small, e.g., less than five, it is grouped with others to increase the sample size in a group; e.g., the frequencies for three to seven arrivals are grouped to yield a total frequency of 6. Using the Poisson distribution

$$\frac{(.4t)^n e^{-.4t}}{n!}$$

with $t = 2$, one computes the third column for values of n from the first column. This gives the probabilities for the indicated arrivals. By multiplying each probability by the total frequency 60, one has the number of 2-minute intervals expected for the number of arrivals indicated. Finally one forms the expression

$$\chi^2 = \Sigma \frac{(\text{number observed} - \text{number expected})^2}{\text{number expected}}$$

where the differences are taken between the numbers in the second and last columns, squared, and divided by the last-column number. The sum is then taken for all corresponding pairs. This gives

$$\frac{(26.9 - 21)^2}{26.9} + \frac{(21.5 - 23)^2}{21.5} + \frac{(8.6 - 10)^2}{8.6} + \frac{(2.9 - 6)^2}{2.9} = 4.94$$

Entering the chi-square table with $n - 1 = 3$ degrees of freedom (there are four groups considered, and the degrees of freedom is this number reduced by unity), one finds that this number is less than that indicated at the highest significance levels. Hence, one may conclude that, if the hypothesis that the mean of the Poisson distribution is .4 is accepted, the observed data do not differ significantly from values obtained from the "fitted" Poisson distribution. If the computed value exceeds the value from the table, the fitness of the curve is rejected at that significance level.

When there is only one degree of freedom, a correction due to Yates is applied. It consists of adding one-half a unit to the observed frequencies which are less than the corresponding expected value and subtracting it from those which are greater than the corresponding expected value. There is a risk in applying the χ^2 test to small samples, for then a change of a few units gives a relatively large change in chi square and makes the value unstable.

The F Test (Variance-ratio Test)

The ratio of two estimates of variance obtained from two independent samples drawn from a normal population has the F distribution [14]. This distribution may be used to decide whether two independent estimates of variance can be accepted as being the variance of a single normally distributed population.

Thus one calculates for the two samples of sizes n_1 and n_2 and variances s_1^2 and s_2^2, respectively, the expression

$$F = \frac{n_1 s_1^2/(n_1 - 1)}{n_2 s_2^2/(n_2 - 1)} \tag{10-38}$$

This value is compared with that obtained from the F table with corresponding degrees of freedom ($n_1 - 1$ and $n_2 - 1$) for the two samples and at a prescribed significance level to determine whether the variances are significantly different or whether they may be taken as estimates of the variance of a normally distributed population.

10-6. Analysis of Variance [8, 9]

There is a simple application of the analysis of variance. Assume that one has a problem requiring the estimation of the differences in capabilities of three individuals at a given task. Assume that the task is to type standard correspondence. Assume that the individuals are given four different tests and their performance is recorded. This yields Table 10-2. The objective is to decide from the results of the tests whether there

TABLE 10-2

Tests \ Man	A	B	C
I	10	11	15
II	8	6	4
III	7	3	12
IV	15	10	8

is a significant difference among the individuals in performing a task. Assuming that there are no differences among them, one would expect to find that the averages of the columns would be different since the columns would correspond to three samples of four observations from the same population and hence there would be random variation among them. The problem then is to decide whether the differences are due to random variations or whether the differences are actually significant.

Again assuming no significant differences among the column means, one can utilize the variation within the columns as a measure of random variation in the sense that the between-column variation may be assumed to arise from the same random causes as the within-column variation. Thus if the between-column variation is significantly different from the within-column variation, one would reject the hypothesis that the means are from the same population. The F distribution is used to test the hypothesis of the homogeneity of the means. It is assumed that all other influencing factors, such as differences of typewriters, time of day, etc., are randomized between individuals and thus contribute to the within-column variance.

Note that the two-dimensional test can be extended to higher dimensions, for example, by including the typewriters, hours of the day, etc., as varying factors. Latin squares are utilized for this purpose.

The analysis of the variance of this randomized design can be summarized as in Table 10-3 by the arrangement of the analysis of n rows, m columns, and a total of N elements, where x_{ij} is the observation in the ith row and jth column, $\bar{x}_j$ is the mean of the jth column, and $\bar{x}$ is the mean of all x_{ij}.

The results of the example can then be summarized as in Table 10-4.

One forms the hypothesis that the m-column means are from the same normal population. If this were true, then σ_1^2, σ_2^2, and σ_3^2 should vary one from the other only by an amount that could be caused by sampling error.

If these mean squares are significantly different, then one must reject the hypothesis.

Adopting the hypothesis of normality of the distributions and that the variances of the various columns are taken from populations with the

TABLE 10-3

Source of variance	Sum of squares	Degrees of freedom	Mean square
Between columns....	$n \sum_{j=1}^{m} (\bar{x}_j - \bar{x})^2$	$m - 1$	$\sigma_1^2 = n \sum_{j=1}^{m} \frac{(\bar{x}_j - \bar{x})^2}{m - 1}$
Within columns....	$\sum_{j=1}^{m} \sum_{i=1}^{n} (x_{ij} - \bar{x}_j)^2$	$N - m$	$\sigma_2^2 = \sum_{j=1}^{m} \sum_{i=1}^{n} \frac{(x_{ij} - \bar{x}_j)^2}{N - m}$
Total.............	$\sum_{j=1}^{m} \sum_{i=1}^{n} (x_{ij} - \bar{x})^2$	$N - 1$	$\sigma_3^2 = \sum_{j=1}^{m} \sum_{i=1}^{n} \frac{(x_{ij} - \bar{x})^2}{N - 1}$

TABLE 10-4

Source of variation	Sum of squares	Degrees of freedom	Mean square
Between columns................	15.16	$3 - 1 = 2$	7.58
Within columns..................	147.75	$12 - 3 = 9$	16.42
Total...........................	162.92	11	

same variances, one can use the F test to determine whether the ratio of the two component variances is significantly different from unity. In this case,

$$F = \frac{7.58}{16.42} = .46$$

and from the F table one finds that this is not significant. Thus one accepts the hypothesis that the three typists' abilities do not differ significantly.

10-7. Regression and Correlation [15]

Correlation analysis provides information for prediction and control. For example, it may be desired to study the bookkeeping performance of a student in relation to other fields of activity, such as arithmetic or reading, so that bookkeeping ability may be predicted from performance in arithmetic or reading. It is usually desirable to find measurable and preferably independent factors which are suspected to interact with the factor under study. In the above example, one calculates the correlation coefficient r from x_i and y_i, the respective bookkeeping and arithmetic grades of the students. One has

$$r = \frac{(1/n) \sum_{i=1}^{n} x_i y_i - \bar{x}\bar{y}}{\sigma_x \sigma_y} \tag{10-39}$$

where n is the number of students, $\bar{x}$ is the average bookkeeping grade for the class, $\bar{y}$ is the average grade for arithmetic, and σ_x and σ_y are the corresponding standard deviations. One has $-1 \leq r \leq 1$, and $r = 0$, generally indicating no correlation. The regression

$$x - \bar{x} = \frac{\sigma_x}{\sigma_y}(y - \bar{y}) \tag{10-40}$$

may be used for prediction.

Whether the correlation could have arisen by chance or is significant can be determined by using the t test, where

$$t = \frac{r\sqrt{n-2}}{\sqrt{1-r^2}}$$

is calculated and then compared with the value obtained from the t table at $n - 2$ degrees of freedom and at the desired significance level. Another test for large n consists of comparing r with its standard error

$$\frac{1 - r^2}{\sqrt{n}} \tag{10-41}$$

If it is several times its standard error, then it is significant. Significant simple correlation often requires delicate interpretation. It has been found, for example, that there is a significant correlation between crime rate and church going. A hasty conclusion might be that church going implies increased crime!

If several correlation coefficients r_i $(i = 1, \ldots, k)$ are available from k samples with size n_i for the ith sample, then one calculates

$$z_i = 1.15 \log_{10} \frac{1 + r_i}{1 - r_i} \tag{10-42}$$

called *Fisher's* z transformation, and obtains a pooled correlation coefficient by solving for r in

$$z = 1.15 \log_{10} \frac{1 + r}{1 - r} \tag{10-43}$$

where

$$z = \frac{\sum_{i=1}^{k} z_i(n_i - 3)}{\sum_{i=1}^{k} (n_i - 3)} \tag{10-44}$$

Then z is compared with its standard error, which in the case of two samples is given by

$$\sqrt{\frac{1}{n_1 - 3} + \frac{1}{n_2 - 3}} \tag{10-45}$$

When several factors are found to be significantly correlated with the factor under study, then it is often necessary to use multiple correlation. Simple-correlation analysis may indicate no correlation, whereas a correlation of higher order involving interaction with more than a single factor at a time is actually significant. To simplify computations, one may select the two factors, say x_2 and x_3, with the largest simple correlation coefficients and then proceed with multiple-correlation analysis, using these two factors with x_1, the factor under study. The partial-regression equation assumed to fit the data is then given by

$$x_1 - \bar{x}_1 = a(x_2 - \bar{x}_2) + b(x_3 - \bar{x}_3) \tag{10-46}$$

where a and b are partial-regression coefficients obtained from two equations; these equations are obtained by first multiplying Eq. (10-46) by $(x_2 - \bar{x}_2)$ and summing, and then multiplying it by $(x_3 - \bar{x}_3)$ and summing. When a and b have been determined, the equation can be used only for predicting x_1, given x_2 and x_3. If desired, similar calculations can be carried out when the roles of the variables are interchanged, in order to predict, for instance, x_2 from x_1 and x_3. A multiple-correlation coefficient is calculated from the actual values of x_1 and from the predicted values obtained from the partial-regression equation, using values of x_2 and x_3. The t test can then be used to determine whether this multiple correlation is significant. If, after having determined the significance of the multiple-correlation coefficient and of the partial-correlation coefficients given by

$$r_1 = a \sqrt{\frac{\Sigma x_2^2}{\Sigma x_3^2}} \quad \text{and} \quad r_2 = b \sqrt{\frac{\Sigma x_3^2}{\Sigma x_1^2}} \tag{10-47}$$

322 PROBABILITY, APPLICATIONS, STATISTICS, QUEUEING THEORY

it is desired to test further the remaining factors, a regression equation with a greater number of variables may be adopted to include new variables or simply to predict only from new variables, having found the two examined factors to yield not very significant multiple correlation [9].

10-8. Nonparametric Methods

When a sample size n is small, for example, $n = 6$, or when the underlying distribution is not known, one uses nonparametric (not concerned with population parameters) methods for hypothesis testing. Particularly in the behavioral sciences [19] one is often interested in ranking attributes in order of preference and assigning appropriate numbers to each. In this case, studies using the mean and variance are not as meaningful as those using the median, which is the mid-point of the ordered sample. The population median is estimated by the sample median. In addition, there are nonparametric tests, used to study samples taken from different populations, which do not require making the assumptions about the underlying distributions which are made in parametric methods. Of course, if all the assumptions could be shown to be satisfied, parametric methods are preferred. Nonparametric methods are not as yet extensive enough to include, for example, tests of interaction in the analysis of variance. (*Text continues on page* 328.)

TABLE 10-5. THE NORMAL DISTRIBUTION*

P†	.00	.01	.02	.03	.04	.05	.06	.07	.08	.09
.0	∞	2.575829	2.326348	2.170090	2.053749	1.959964	1.880794	1.811911	1.750686	1.695398
.1	1.644854	1.598193	1.554774	1.514102	1.475791	1.439521	1.405072	1.372204	1.340755	1.310579
.2	1.281552	1.253565	1.226528	1.200359	1.174987	1.150349	1.126391	1.103063	1.080319	1.058122
.3	1.036433	1.015222	.994458	.974114	.954165	.934589	.915365	.896473	.877896	.859617
.4	.841621	.823894	.806421	.789192	.772193	.755415	.738847	.722479	.706303	.690309
.5	.674490	.658838	.643345	.628006	.612813	.597760	.582841	.568051	.553385	.538836
.6	.524401	.510073	.495850	.481727	.467699	.453762	.439913	.426148	.412463	.398855
.7	.385320	.371856	.358459	.345125	.331853	.318639	.305481	.292375	.279319	.266311
.8	.253347	.240426	.227545	.214702	.201893	.189118	.176374	.163658	.150969	.138304
.9	.125661	.113039	.100434	.087845	.075270	.062707	.050154	.037608	.025069	.012533

P........	.001	.000,1	.000,01	.000,001	.000,000,1	.000,000,01	.000,000,001
x........	3.29053	3.89059	4.41717	4.89164	5.32672	5.73073	6.10941

* Reprinted from Table I of Fisher and Yates, "Statistical Tables for Biological, Agricultural, and Medical Research," published by Oliver & Boyd, Ltd., Edinburgh and London, by permission of the authors and publishers.

† The value of P for each entry is found by adding the column heading to the value in the left-hand margin. The corresponding value of x is the deviation such that the probability of an observation falling outside the range from $-x$ to $+x$ is P. For example, $P = .03$ for $x = 2.170090$, so that 3 per cent of normally distributed values will have positive or negative deviations exceeding the standard deviation in the ratio 2.170090 at least.

TABLE 10-6.* TABLE OF t
(The integral is taken outside $[-t,t]$)

Degrees of freedom	Significance level				
	0.10	0.05	0.02	0.01	0.001
1	6.31	12.71	31.82	63.66	636.62
2	2.92	4.30	6.97	9.93	31.60
3	2.35	3.18	4.54	5.84	12.94
4	2.13	2.78	3.75	4.60	8.61
5	2.02	2.57	3.37	4.03	6.86
6	1.94	2.45	3.14	3.71	5.96
7	1.90	2.37	3.00	3.50	5.41
8	1.86	2.31	2.90	3.36	5.04
9	1.83	2.26	2.82	3.25	4.78
10	1.81	2.23	2.76	3.17	4.59
11	1.80	2.20	2.72	3.11	4.44
12	1.78	2.18	2.68	3.06	4.32
13	1.77	2.16	2.65	3.01	4.22
14	1.76	2.15	2.62	2.98	4.14
15	1.75	2.13	2.60	2.95	4.07
16	1.75	2.12	2.58	2.92	4.02
17	1.74	2.11	2.57	2.90	3.97
18	1.73	2.10	2.55	2.88	3.92
19	1.73	2.09	2.54	2.86	3.88
20	1.73	2.09	2.53	2.85	3.85
21	1.72	2.08	2.52	2.83	3.82
22	1.72	2.07	2.51	2.82	3.79
23	1.71	2.07	2.50	2.81	3.77
24	1.71	2.06	2.49	2.80	3.75
25	1.71	2.06	2.48	2.79	3.73
26	1.71	2.06	2.48	2.78	3.71
27	1.70	2.05	2.47	2.77	3.69
28	1.70	2.05	2.47	2.76	3.67
29	1.70	2.04	2.46	2.76	3.66
30	1.70	2.04	2.46	2.75	3.65
40	1.68	2.02	2.42	2.70	3.55
60	1.67	2.00	2.39	2.66	3.46
120	1.66	1.98	2.36	2.62	3.37
∞	1.65	1.96	2.33	2.58	3.29

* Abridged from Table III of Fisher and Yates, "Statistical Tables for Biological, Agricultural, and Medical Research," published by Oliver & Boyd, Ltd., Edinburgh and London, by permission of the authors and publishers.

TABLE 10-7.* TABLE OF χ^2
(The integral is taken on $[\chi^2, \infty]$)

Degrees of freedom	Significance level									
	0.99	0.98	0.95	0.90	0.50	0.10	0.05	0.02	0.01	0.001
1	0.000	0.001	0.004	0.016	0.46	2.71	3.84	5.41	6.64	10.83
2	0.020	0.040	0.003	0.211	1.39	4.60	5.99	7.82	9.21	13.82
3	0.115	0.185	0.352	0.584	2.37	6.25	7.82	9.84	11.34	16.27
4	0.297	0.429	0.711	1.06	3.36	7.78	9.49	11.67	13.28	18.46
5	0.554	0.752	1.145	1.61	4.35	9.24	11.07	13.39	15.09	20.52
6	0.87	1.13	1.64	2.20	5.35	10.64	12.59	15.03	16.81	22.46
7	1.24	1.56	2.17	2.83	6.35	12.02	14.07	16.62	18.48	24.32
8	1.65	2.03	2.73	3.49	7.34	13.36	15.51	18.17	20.09	26.12
9	2.09	2.53	3.32	4.17	8.34	14.68	16.92	19.68	21.67	27.88
10	2.56	3.06	3.94	4.86	9.34	15.99	18.31	21.16	23.21	29.59
11	3.05	3.61	4.58	5.58	10.34	17.28	19.68	22.62	24.72	31.26
12	3.57	4.18	5.23	6.30	11.34	18.55	21.03	24.05	26.22	32.91
13	4.11	4.76	5.89	7.04	12.34	19.81	22.36	25.47	27.69	34.53
14	4.66	5.37	6.57	7.79	13.34	21.06	23.68	26.87	29.14	36.12
15	5.23	5.98	7.26	8.55	14.34	22.31	25.00	28.26	30.58	37.70
16	5.81	6.61	7.96	9.31	15.34	23.54	26.30	29.63	32.00	39.25
17	6.41	7.26	8.67	10.08	16.34	24.77	27.59	31.00	33.41	40.79
18	7.02	7.91	9.39	10.86	17.34	25.99	28.87	32.35	34.80	42.31
19	7.63	8.57	10.12	11.65	18.34	27.20	30.14	33.69	36.19	43.82
20	8.26	9.24	10.85	12.44	19.34	28.41	31.41	35.02	37.57	45.32
21	8.90	9.92	11.59	13.24	20.34	29.62	32.67	36.34	38.93	46.80
22	9.54	10.60	12.34	14.04	21.34	30.81	33.92	37.66	40.29	48.27
23	10.20	11.29	13.09	14.85	22.34	32.01	35.17	38.97	41.64	49.73
24	10.86	11.99	13.85	15.66	23.34	33.20	36.42	40.27	42.98	51.18
25	11.52	12.70	14.61	16.47	24.34	34.38	37.65	41.57	44.31	52.62
26	12.20	13.41	15.38	17.29	25.34	35.56	38.88	42.86	45.64	54.05
27	12.88	14.12	16.15	18.11	26.34	36.74	40.11	44.14	46.96	55.48
28	13.56	14.85	16.93	18.94	27.34	37.92	41.34	45.42	48.28	56.89
29	14.26	15.57	17.71	19.77	28.34	39.09	42.56	46.69	49.59	58.30
30	14.95	16.31	18.49	20.60	29.34	40.26	43.77	47.96	50.89	59.70

* Abridged from Table IV of Fisher and Yates, "Statistical Tables for Biological, Agricultural, and Medical Research," published by Oliver & Boyd, Ltd., Edinburgh and London, by permission of the authors and publishers.

FUNDAMENTAL STATISTICS 325

TABLE 10-8.* TABLE OF VARIANCE RATIO F
(The integral is taken on $[F, \infty]$)

n_2 \ n_1†	0.05 Significance level								
	1	2	3	4	5	6	12	24	∞
1	161.4	199.5	215.7	224.6	230.2	234.0	243.9	249.0	254.3
2	18.5	19.0	19.2	19.2	19.3	19.3	19.4	19.4	19.5
3	10.1	9.6	9.3	9.1	9.0	8.9	8.7	8.6	8.5
4	7.7	6.9	6.6	6.4	6.3	6.2	5.9	5.8	5.6
5	6.6	5.8	5.4	5.2	5.1	5.0	4.7	4.5	4.4
6	6.0	5.1	4.8	4.5	4.4	4.3	4.0	3.8	3.7
7	5.6	4.7	4.4	4.1	4.0	3.9	3.6	3.4	3.2
8	5.3	4.5	4.1	3.8	3.7	3.6	3.3	3.1	2.9
9	5.1	4.3	3.9	3.6	3.5	3.4	3.1	2.9	2.7
10	5.0	4.1	3.7	3.5	3.3	3.2	2.9	2.7	2.5
11	4.8	4.0	3.6	3.4	3.2	3.1	2.8	2.6	2.4
12	4.8	3.9	3.5	3.3	3.1	3.0	2.7	2.5	2.3
13	4.7	3.8	3.4	3.2	3.0	2.9	2.6	2.4	2.2
14	4.6	3.7	3.3	3.1	3.0	2.8	2.5	2.4	2.1
15	4.5	3.7	3.3	3.1	2.9	2.8	2.5	2.3	2.1
16	4.5	3.6	3.2	3.0	2.8	2.7	2.4	2.2	2.0
17	4.5	3.6	3.2	3.0	2.8	2.7	2.4	2.2	2.0
18	4.4	3.6	3.2	2.9	2.8	2.7	2.3	2.2	1.9
19	4.4	3.5	3.1	2.9	2.7	2.6	2.3	2.1	1.9
20	4.4	3.5	3.1	2.9	2.7	2.6	2.3	2.1	1.8
22	4.3	3.4	3.1	2.8	2.7	2.6	2.2	2.0	1.8
24	4.3	3.4	3.0	2.8	2.6	2.5	2.2	2.0	1.7
26	4.2	3.4	3.0	2.7	2.6	2.5	2.2	2.0	1.7
28	4.2	3.3	3.0	2.7	2.6	2.4	2.1	1.9	1.6
30	4.2	3.3	2.9	2.7	2.5	2.4	2.1	1.9	1.6
40	4.1	3.2	2.8	2.6	2.4	2.3	2.0	1.8	1.5
60	4.0	3.2	2.8	2.5	2.4	2.3	1.9	1.7	1.4
120	3.9	3.1	2.7	2.4	2.3	2.2	1.8	1.6	1.2
∞	3.8	3.0	2.6	2.4	2.2	2.1	1.8	1.5	1.0

* Abridged from Table V of Fisher and Yates, "Statistical Tables for Biological, Agricultural, and Medical Research," published by Oliver & Boyd, Ltd., Edinburgh and London, by permission of the authors and publishers.
† n_1 must always correspond with the greater variance.

TABLE 10-9.* TABLE OF VARIANCE RATIO F
(The integral is taken on $[F, \infty])$

0.01 Significance level

n_2 \ n_1	1	2	3	4	5	6	8	12	24	∞
1	4052	4999	5403	5625	5764	5859	5981	6106	6234	6366
2	98.5	99.0	99.2	99.2	99.3	99.3	99.4	99.4	99.5	99.6
3	34.1	30.8	29.5	28.7	28.2	27.9	27.5	27.1	26.6	26.1
4	21.2	18.0	16.7	16.0	15.5	15.2	14.8	14.4	13.9	13.5
5	16.3	13.3	12.1	11.4	11.0	10.7	10.3	9.9	9.5	9.0
6	13.7	10.9	9.8	9.2	8.8	8.5	8.1	7.7	7.3	6.9
7	12.2	9.6	8.4	7.9	7.5	7.2	6.8	6.5	6.1	5.6
8	11.3	8.6	7.6	7.0	6.6	6.4	6.0	5.7	5.3	4.9
9	10.6	8.0	7.0	6.4	6.1	5.8	5.5	5.1	4.7	4.3
10	10.0	7.6	6.6	6.0	5.6	5.4	5.1	4.7	4.3	3.9
11	9.6	7.2	6.2	5.7	5.3	5.1	4.7	4.4	4.0	3.6
12	9.3	6.9	6.0	5.4	5.1	4.8	4.5	4.2	3.8	3.4
13	9.1	6.7	5.7	5.2	4.9	4.6	4.3	4.0	3.6	3.2
14	8.9	6.5	5.6	5.0	4.7	4.5	4.1	3.8	3.4	3.0
15	8.7	6.4	5.4	4.9	4.6	4.3	4.0	3.7	3.3	2.9
16	8.5	6.2	5.3	4.8	4.4	4.2	3.9	3.6	3.2	2.8
17	8.4	6.1	5.2	4.7	4.3	4.1	3.8	3.4	3.1	2.6
18	8.3	6.0	5.1	4.6	4.3	4.0	3.7	3.4	3.0	2.6
19	8.2	5.9	5.0	4.5	4.2	3.9	3.6	3.3	2.9	2.5
20	8.1	5.8	4.9	4.4	4.1	3.9	3.6	3.2	2.9	2.4
22	7.9	5.7	4.8	4.3	4.0	3.8	3.4	3.1	2.8	2.3
24	7.8	5.6	4.7	4.2	3.9	3.7	3.4	3.0	2.7	2.2
26	7.7	5.5	4.6	4.1	3.8	3.6	3.3	3.0	2.6	2.1
28	7.6	5.4	4.6	4.1	3.8	3.5	3.2	2.9	2.5	2.1
30	7.6	5.4	4.5	4.0	3.7	3.5	3.2	2.8	2.5	2.0
40	7.3	5.2	4.3	3.8	3.5	3.3	3.0	2.7	2.3	1.8
60	7.1	5.0	4.1	3.6	3.3	3.1	2.8	2.5	2.1	1.6
100	6.8	4.8	4.0	3.5	3.2	3.0	2.7	2.3	2.0	1.4
∞	6.6	4.6	3.8	3.3	3.0	2.8	2.5	2.2	1.8	1.0

* Abridged from Table V of Fisher and Yates, "Statistical Tables for Biological, Agricultural, and Medical Research," published by Oliver & Boyd, Ltd., Edinburgh and London, by permission of the authors and publishers.

TABLE 10-10.* RANDOM NUMBERS

03 47 43 73 86	36 96 47 36 61	46 98 63 71 62	33 26 16 80 45	60 11 14 10 95
97 74 24 67 62	42 81 14 57 20	42 53 32 37 32	27 07 36 07 51	24 51 79 89 73
16 76 62 27 66	56 50 26 71 07	32 90 79 78 53	13 55 38 58 59	88 97 54 14 10
12 56 85 99 26	96 96 68 27 31	05 03 72 93 15	57 12 10 14 21	88 26 49 81 76
55 59 56 35 64	38 54 82 46 22	31 62 43 09 90	06 18 44 32 53	23 83 01 30 30
16 22 77 94 39	49 54 43 54 82	17 37 93 23 78	87 35 20 96 43	84 26 34 91 64
84 42 17 53 31	57 24 55 06 88	77 04 74 47 67	21 76 33 50 25	83 92 12 96 76
63 01 63 78 59	16 95 55 67 19	98 10 50 71 75	12 86 73 58 07	44 39 52 38 79
33 21 12 34 29	78 64 56 07 82	52 42 07 44 38	15 51 00 13 42	99 66 02 79 54
57 60 86 32 44	09 47 27 96 54	49 17 46 09 62	90 52 84 77 27	08 02 73 43 28
18 18 07 92 46	44 17 16 58 09	79 83 86 19 62	06 76 50 03 10	55 23 64 05 05
26 62 38 97 75	84 16 07 44 99	83 11 46 32 24	20 14 85 88 45	10 93 72 88 71
23 42 40 64 74	82 97 77 77 81	07 45 32 14 08	32 98 94 07 72	93 85 79 10 75
52 36 28 19 95	50 92 26 11 97	00 56 76 31 38	80 22 02 53 53	86 60 42 04 53
37 85 94 35 12	83 39 50 08 30	42 34 07 96 88	54 42 06 87 98	35 85 29 48 39
70 29 17 12 13	40 33 20 38 26	13 89 51 03 74	17 76 37 13 04	07 74 21 19 30
56 62 18 37 35	96 83 50 87 75	97 12 25 93 47	70 33 24 03 54	97 77 46 44 80
99 49 57 22 77	88 42 95 45 72	16 64 36 16 00	04 43 18 66 79	94 77 24 21 90
16 08 15 04 72	33 27 14 34 09	45 59 34 68 49	12 72 07 34 45	99 27 72 95 14
31 16 93 32 43	50 27 89 87 19	20 15 37 00 49	52 85 66 60 44	38 68 88 11 80
68 34 30 13 70	55 74 30 77 40	44 22 78 84 26	04 33 46 09 52	68 07 97 06 57
74 57 25 65 76	59 29 97 68 60	71 91 38 67 54	13 58 18 24 76	15 54 55 95 52
27 42 37 86 53	48 55 90 65 72	96 57 69 36 10	96 46 92 42 45	97 60 49 04 91
00 39 68 29 61	66 37 32 20 30	77 84 57 03 29	10 45 65 04 26	11 04 96 67 24
29 94 98 94 24	68 49 69 10 82	53 75 91 93 30	34 25 20 57 27	40 48 73 51 92
16 90 82 66 59	83 62 64 11 12	67 19 00 71 74	60 47 21 29 68	02 02 37 03 31
11 27 94 75 06	06 09 19 74 66	02 94 37 34 02	76 70 90 30 86	38 45 94 30 38
35 24 10 16 20	33 32 51 26 38	79 78 45 04 91	16 92 53 56 16	02 75 50 95 98
38 23 16 86 38	42 38 97 01 50	87 75 66 81 41	40 01 74 91 62	48 51 84 08 32
31 96 25 91 47	96 44 33 49 13	34 86 82 53 91	00 52 43 48 85	27 55 26 89 62
66 67 40 67 14	64 05 71 95 86	11 05 65 09 68	76 83 20 37 90	57 16 00 11 66
14 90 84 45 11	75 73 88 05 90	52 27 41 14 86	22 98 12 22 08	07 52 74 95 80
68 05 51 18 00	33 96 02 75 19	07 60 62 93 55	59 33 82 43 90	49 37 38 44 59
20 46 78 73 90	97 51 40 14 02	04 02 33 31 08	39 54 16 49 36	47 95 93 13 30
64 19 58 97 79	15 06 15 93 20	01 90 10 75 06	40 78 78 89 62	02 67 74 17 33
05 26 93 70 60	22 35 85 15 13	92 03 51 59 77	59 56 78 06 83	52 91 05 70 74
07 97 10 88 23	09 98 42 99 64	61 71 62 99 15	06 51 29 16 93	58 05 77 09 51
68 71 86 85 85	54 87 66 47 54	73 32 08 11 12	44 95 92 63 16	29 56 24 29 48
26 99 61 65 53	58 37 78 80 70	42 10 50 67 42	32 17 55 85 74	94 44 67 16 94
14 65 52 68 75	87 59 36 22 41	26 78 63 06 55	13 08 27 01 50	15 29 39 39 43

* Abridged from Table XXXIII of Fisher and Yates, "Statistical Tables for Biological, Agricultural, and Medical Research," published by Oliver & Boyd, Ltd., Edinburgh and London, by permission of the authors and publishers.

Nonparametric methods are used by the industrial statistician in controlling the quality of products. Three nonparametric tests will now be described. Frequently (for small samples, for example), in order to determine the goodness of fit of a hypothetical distribution, one uses the Kolmogorov-Smirnov test. (As previously pointed out, the chi-square test is inadequate for small samples.) This test consists of obtaining the maximum vertical difference between the sample cumulative distribution function (a step function) and the hypothetical distribution. A table is then used to determine the goodness of fit [19].

The Wilcoxon test serves the purpose of deciding whether the effect of a certain factor on a population is significant. The qualities to be affected are assigned relative values before and after the factor is introduced; the absolute differences between each pair are then calculated and assigned ranks. Each rank is multiplied by 1 or -1, depending on whether the differences are positive or negative. The ranks with the less frequent sign are then summed, and the test is continued with the aid of a table [19].

Here is an example of a simple nonparametric test. Assume that a set of n objects receives two rankings $1, 2, \ldots, n$, where the differences between two consecutive ranks in each ranking is unity but the two rankings do not necessarily assign the same integer to the same object. It is desired to derive a test to determine whether the two rankings are significantly different. Let x_i be the integer value received by the ith object in one ranking and y_i its rank in the second ranking. Let the differences be

$$d_i = x_i - y_i$$

Squaring both sides and summing, one has

$$\sum_{i=1}^{n} d_i^2 = \sum_{i=1}^{n} x_i^2 - 2\sum_{i=1}^{n} x_i y_i + \sum_{i=1}^{n} y_i^2 = 2\frac{n(n+1)(2n+1)}{6} - 2\sum_{i=1}^{n} x_i y_i$$
(10-48)

This follows from the fact that

$$1^2 + 2^2 + \cdots + n^2 = \frac{n(n+1)(2n+1)}{6}$$

Now the last term on the right appears in the calculation of a simple-correlation coefficient. Thus

$$\sum_{i=1}^{n} x_i y_i = \frac{n(n+1)(2n+1)}{6} - \frac{1}{2}\sum_{i=1}^{n} d_i^2$$

But the correlation coefficient is given as

$$r = \frac{(1/n) \sum_{i=1}^{n} x_i y_i - \bar{x}\bar{y}}{\sigma_x \sigma_y} \qquad (10\text{-}49)$$

and $\quad \bar{x} = \bar{y} = \dfrac{1}{n}(1 + 2 + \cdots + n) = \dfrac{n+1}{2}$

and $\quad \sigma_x{}^2 = \dfrac{1}{n} \sum_{i=1}^{n} x_i{}^2 - \bar{x}^2 = \dfrac{(n+1)(2n+1)}{6} - \left(\dfrac{n+1}{2}\right)^2 = \sigma_y{}^2 \qquad (10\text{-}50)$

Substituting in r, one has

$$r = 1 - \frac{6 \sum_{i=1}^{n} d_i{}^2}{n^3 - n}$$

To test whether or not r could have arisen by chance, one uses the distribution of r derived from random samples drawn from permutations of ranks of n objects. By prescribing a significance level one can test whether the computed value of r is significant. Often for $n \geq 10$, for example, the t test is used and the test becomes a parametric one.

REFERENCES

1. Brownlee, K. A.: "Industrial Experimentation," 2d ed., Chemical Publishing Company, Inc., New York, 1952.
2. Cochran, W. G., and G. M. Cox: "Experimental Designs," John Wiley & Sons, Inc., New York, 1950.
3. Cramer, H.: "Mathematical Methods of Statistics," Princeton University Press, Princeton, N.J., 1951.
4. Darmois, G.: "Statistique mathématique," Cours de L'Institut de Statistique de L'Université de Paris.
5. Deming, W. E.: "Some Theory of Sampling," John Wiley & Sons, Inc., New York, 1950.
6. Dixon, W. J., and F. J. Massey, Jr.: "Introduction to Statistical Analysis," McGraw-Hill Book Company, Inc., New York, 1951.
7. Fisher, R.: "The Design of Experiments," Hafner Publishing Company, New York, 1947.
8. Freeman, H. A.: "Industrial Statistics," John Wiley & Sons, Inc., New York, 1942.
9. Hald, A.: "Statistical Theory with Engineering Applications," John Wiley & Sons, Inc., New York, 1952.
10. Hoel, P.: "Introduction to Mathematical Statistics," John Wiley & Sons, Inc., New York, 1947.
11. Huff, D.: "How to Lie with Statistics," W. W. Norton & Company, Inc., New York, 1954.

12. Kendall, M. G.: "The Advanced Theory of Statistics," vols. 1 and 2, Charles Griffin & Co., Ltd., London, 1948.
13. Mayne, J. W.: Role of Statistics in Scientific Research, *The Scientific Monthly*, vol. 84, no. 1, p. 26, January, 1957.
14. Mood, A. M.: "Introduction to the Theory of Statistics," McGraw-Hill Book Company, Inc., New York, 1950.
15. Moroney, M. J.: "Facts from Figures," Penguin Books, Inc., Baltimore, 1951.
16. Mothes, J.: "Les Apports de la statistique au contrôle industriel (II)," Institut Henri Poincaré, Paris, 1951.
17. Pookes, G. L.: Statistical Quality Control, *Journal of the Society of Plastic Engineers*, vol. 71, no. 5, May, 1955.
18. Savage, L. J.: "The Foundations of Statistics," John Wiley & Sons, Inc., New York, 1954.
19. Siegel, S.: "Nonparametric Statistics," McGraw-Hill Book Company, Inc., New York, 1956.
20. Tucker, R. A.: "An Introduction to Statistical Decision Functions," Thesis, U.S. Naval Postgraduate School, Monterey, Calif., 1955.
21. Wald, A.: "Statistical Decision Functions," John Wiley & Sons, Inc., New York, 1950.
22. Wallis, W. A., and H. V. Roberts: "Statistics," Free Press, Glencoe, Ill., 1956.
23. Wilcoxon, F.: "Some Rapid Approximate Statistica Procedures," American Cyanamid Co., New York.
24. Wilks, S. S.: "Mathematical Statistics," Princeton University Press, Princeton, N.J., 1944.

CHAPTER 11

RÉSUMÉ OF QUEUEING THEORY

11-1. Introduction

This chapter summarizes some of the results and uses of queueing theory which apply to operations research. It does not attempt to discuss in detail the techniques used in deriving results; it merely points out the fields of activity to which the theory may be applied.

The subject of waiting lines will be introduced by an example. Consider the operation of the "minute car wash," where cars wait in a line to be washed by passing through a system of sprays, brushes, blowers, and driers. Each car is served in the order in which it arrives—first come, first served. Assume that the timing of the washing machine is automatic. Thus the servicing or *holding time* of all cars is of the same duration, $1/\mu$. The number of cars entering the waiting line per unit time is a variable. By means of arrival-rate samples, it is possible to estimate the distribution of the number of arrivals per unit time, or the distribution of the time between arrivals. Thus, the number of cars waiting in the line fluctuates. After having waited in line, the cars are served at a constant rate μ (cars per unit time). They also leave the system at a constant rate as long as there is a line. The owner of the enterprise may be interested in the average number of cars in the waiting line in order to provide waiting-line space and to determine whether the system is adequate to handle the cars. A customer may become impatient and decide to leave because there are too many customers ahead of him; he is able to multiply the number of cars ahead of him by the holding time $1/\mu$ and obtain an interval of time challenging to his patience. The owner, having observed customers leaving the waiting line, decides to open another service channel so that both channels may cooperate in handling the line of waiting cars. Thus, the average length of the line may be considerably shortened by using two parallel independent channels. The probability of a customer's becoming impatient and leaving may thereby be reduced.

If the owner desired to include a lubrication service such that each car leaving the wash service may wait in a line for the additional service, the wash channel and the lubrication channel operate in series. Each may have a waiting line before it. Priorities may be assigned to customers, if they are willing to pay more in order to be served first.

The owner may decide to select his customers for service by the roll of multisided dice, making the selection of each individual as likely as that of any other (random selection for service). In the car-wash enterprise, this gamble might wreck the business. However, it is the only means which used to be available to the long-distance telephone operator—the latter's indifference formed a substitute for the use of dice.

Thus, as indicated above, a knowledge of the input- and servicing-rate distributions enables both the owner and the customer to make useful decisions.

To specify a queue completely, the distribution function for arrivals, i.e., the input distribution, must be given. The type of queue discipline, whether first come, first served, priority, random selection for service, etc., must also be described. The number of channels and the distribution of service times for each channel are further information needed to specify a queueing situation.

11-2. Historical Remarks

Queueing theory was developed in order to provide a model to predict the behavior of a system that attempted to provide services for randomly arising demands. The earliest problem studied was that of telephone-traffic congestion. The pioneer investigator was A. K. Erlang who in 1909 published "The Theory of Probabilities and Telephone Conversations." In later works he observed that the telephone system was characterized by (1) Poisson-input, exponential-holding-time, multiple-service channels or (2) a Poisson-input, constant-holding-time, single channel. Work on applying the theory to telephone problems continued. Thus Molina in 1927 published his "Application of the Theory of Probability to Telephone Trunking Problems," and Fry in 1928 expanded Erlang's work in a book, "Probability and Its Engineering Uses." In 1930 Pollaczek studied the Poisson-input, arbitrary-holding-time, single-channel case. In 1934 he studied the Poisson-input, arbitrary-holding-time, multiple-channel problem. Further work in this field was done by Kolmogorov in 1931, Sur le problème d'attente; Khintchine in 1932 studied the Poisson-input, arbitrary-holding-time, single-channel problem; and Crommelin in 1932 published "Delay Probability Formulae When the Holding Times Are Constant." A greater variety of problems has been encountered and solved by recent investigators working principally in operations research. A brief survey of the dates in the references will enable the reader to deduce the history of more recent developments.

11-3. Notation

Some of the notations used in queueing theory are as follows:

λ = mean arrival rate (number of arrivals per unit time)
μ = mean service rate per channel

c = number of service channels
c_f = mean number of free service channels
n = number of units (customers) in system
k = number of phases in Erlang service case
ρ = utilization factor for service facility; $\rho = \lambda/c\mu$
$P_n(t)$ = probability that there be, at time t, exactly n units in system, both waiting and in service
p_n = steady-state (time-independent) probability that there be n units in system, both waiting and in service:

$$\sum_{n=0}^{n=\infty} P_n(t) = \sum_{n=0}^{n=\infty} p_n = 1$$

$c\rho$ = traffic intensity in erlangs:

$$c\rho = c - c_f = \sum_{n=0}^{n=c-1} np_n + \sum_{n=c}^{n=\infty} cp_n$$

$P(=0)$ = probability of no waiting
$P(>0)$ = probability of any waiting
$P(>\tau)$ = probability of waiting greater than time τ
L = average number of units in system, both waiting and in service:

$$L = \sum_{n=0}^{n=\infty} np_n$$

L_q = average number of units waiting in queue:

$$L_q = \sum_{n=c}^{\infty} (n-c)p_n = L - c + c_f$$

W = average waiting time in system:

$$W = -\int_0^\infty \tau \, dP(>\tau)$$

$A(t)$ = cumulative distribution of times between arrivals with density function $a(t)$
$B(t)$ = cumulative distribution of service or holding times with density function $b(t)$
$b_k(t)$ = probability density for kth Erlang distribution

11-4. Input and Service or Holding-time Distributions

Arrivals or inputs into a queueing system may occur at intervals of regular length, which means that the cumulative distribution of time intervals between arrivals is given by the constant-input distribution

$$A(t) = 0 \text{ for } t < \tau \qquad A(t) = 1 \text{ for } t \geq \tau$$

If the input distribution is of the Poisson type (see below), the time intervals between arrivals are exponentially distributed. The cumulative distribution is then given by

$$A(t) = 1 - e^{-\lambda t}$$

where λ is the arrival rate per unit time.

An intermediate type of input which bridges the gap between these two arrival-time distributions may be described by the Erlangian frequency distribution of times between arrivals:

$$b_k(t) = \frac{(\lambda k)^k}{\Gamma(k)} e^{-\lambda k t} t^{k-1} dt$$

This yields the constant-input distribution as $k \to \infty$ and the exponential distribution when $k = 1$. If a Poisson input is filtered in such a way that only every kth item is admitted into the system, the resulting frequency distribution of arrival times is $b_k(t)$.

A normal distribution has been observed to produce a good fit to arrival data in some practical problems. Since a Poisson distribution with a large mean may be approximated by a normal distribution, the frequent appearance of the latter distribution as a fit to input data is not surprising.

Generally, one is concerned with the input distribution of arrival *times*. Note that a regular input gives rise to a constant-input distribution of arrival times; a Poisson input, to an exponential distribution of arrival times; and a filtered Poisson input, to an Erlangian-type distribution of arrival times. There have been some attempts to study queues with an arbitrary distribution of times between arrivals [28]. The results are not sufficiently explicit to present here without elaborate development.

An entirely similar set of distributions of service or holding times is analogously defined. If there is always a line before the service mechanism, service-time distributions and distributions describing output from the facility are of interest. The output distribution of a queue with Poisson input and exponential holding time in the steady state is also Poisson.

An Erlangian-type distribution may also be assumed for the service times. If theoretical solutions of queueing problems involving the Erlangian are obtained, then, by assigning a proper value to k, a satisfactory approximation to the solution of a practical problem is obtained.

Because of the frequent occurrence of Poisson inputs and exponential service times in practice, a few mathematical remarks leading to these distributions and describing them are given below.

A Stochastic Process

Let X be an arbitrary set. The elements of X are called *states*, where the kth state is denoted by E_k. Any function defined on $[0, \infty]$ with values in X is called a *history*. A change of state is called a *transition*. Let F be the set of all histories. A probability measure P on F defines a stochastic process.

Suppose that X is the set of nonnegative integers $(0,1,2, \ldots)$. Let F' be the set of all monotone increasing histories: $f \in F'$ if and only if $f(t_2) \geq f(t_1)$ for all t_1 and t_2 such that $t_2 > t_1$.

If $P(F') = 1$, so that $P(A) = 0$ for any set A contained in the complement of F', the stochastic process is called a *pure birth process*.

Let $A(t,n)$ be the set of all histories having the value n at time t. Then $P[A(t,n)]$ is referred to as the *probability* that the system is in state n at time t.

Let B be the smallest Boolean algebra of sets containing all $A(t,n)$. B contains the empty set, the complement of every $A(t,n)$, and the union and intersection of any denumerable collection $A(t_1 n_1)$, $A(t_2, n_2)$, ...

P is defined for each $S \in B$, $P(S) \geq 0$, $P(F') = 1$, and P is completely additive. If $P(S) \neq 0$, and V is any set in B, $P_S(V)$ or $P(V;S)$, called the *conditional probability* of V, given S, is defined by $P(V) = P(S)P_S(V)$.

P is said to be *stationary* (i.e., for any time interval the state of the system does not depend on the initial point of the interval but only on its length) if $P[A(t,n)] = P[A(t + s,n)]$ for every s and t.

This yields the fact that the average number of arrivals in an interval of time is proportional to the length of the interval and is independent of the position of the interval on the time axis. For a nonstationary process, the average number of arrivals becomes a function of the time t.

P is said to be *Markovian* (i.e., the probability that the system is in state n at time t depends only on information from the previous state and is independent of the preceding states) if

$$P[A(t,n); \prod_{i=1}^{k} A(s_i, m_i)] = P[A(t,n); A(s_k, m_k)]$$

with $s_1 < s_2 < \cdots < s_k < t$ and $m_1 \leq m_2 \cdots \leq m_k$. In a non-Markovian process the dependence of a state may extend to several of the preceding ones. A Poisson process is a pure birth, a stationary, and a Markovian process.

The Assumptions Leading to a Poisson Input [18]

One has a Poisson input when both the following assumptions are satisfied:

1. The total number of arrivals during any given time interval is independent of the number of arrivals that have already occurred prior to the beginning of the interval.

2. For any interval $(t, t + dt)$, the probability that exactly one arrival will occur is $\lambda\, dt + O(dt^2)$, where λ is a constant, while the probability that more than one arrival will occur is of the order of dt^2 and may be neglected. λ represents the mean arrival rate (number per unit time) of incoming traffic. It follows that $1/\lambda$ is the mean length of the time interval between two consecutive arrivals as shown below.

The Assumptions Leading to an Exponential-holding-time Distribution

If a channel is occupied at time t, the probability that it will become free during the following time element dt is $\mu\, dt$, where μ is a constant. It follows that the frequency function of the service times is $\mu e^{-\mu t}$ while the mean duration of service is $1/\mu$, since the expected value of t is

$$E(t) = \mu \int_0^\infty t e^{-\mu t}\, dt = \frac{1}{\mu}$$

Properties of a Poisson Process

Let $p(n,t;m)$ be the probability that the system be in *state n* at time t, given that it was in state m at time zero.

Theorem 11-1: $p(n,t;n)$ must be of the form $e^{-\lambda t}$.

Theorem 11-2: $p(n + 1, t; n) = \lambda t e^{-\lambda t}$, and by induction

$$p(n + k, t; n) = (\lambda t)^k e^{-\lambda t}/k!$$

(the Poisson probability function). This theorem gives the probability of k transitions in any interval of length t.

Some properties of the Poisson process are as follows:

1. The expected number of transitions is given by

$$\sum_{k=0}^{k=\infty} k \left[\frac{(\lambda t)^k}{k!} \right] e^{-\lambda t} = \lambda t$$

Therefore, λ is the expected number of transitions in an interval of unit length. Note that the expected number of transitions in an interval of length t is proportional to t.

2. The density function for the time of transition τ is $\lambda e^{-\lambda \tau}$.

3. $E(\tau) = 1/\lambda$. Thus, one arrival is expected in an interval of length $1/\lambda$. This can also be seen from the fact that the expected number of arrivals in an interval of length t is λt.

4. The probability of more than one arrival is given by

$$1 - [e^{-\lambda t} + \lambda t\, e^{-\lambda t}] = O(t^2)$$

where the first term in the brackets is the probability of zero arrivals, the second term the probability of one arrival. This expression is obtained using series expansion. This property simplifies the differential-difference-equation approach to the Poisson input and exponential-holding-time problem.

11-5. Measures of Effectiveness

In searching for measurable quantities that would serve as criteria for studying a queue, studying averages immediately comes to mind as a possible measure of effectiveness. However, there are many errors committed in direct study from averages. Refined approaches to the theory have led to measures of effectiveness that are not only theoretically sound but have proved effective in providing remedies for practical problems.

The following are examples of measures of effectiveness used in queueing theory. Thus, for example, knowledge of the length of the line enabled the owner in the introductory example to decide upon enough space for the waiting cars.

Obviously, a useful measure of effectiveness is that of the probability of having n people waiting at time t, given the initial state of the system. It may be obtained as the solution of the following difference equation:

$$P_n(t + s) = \sum_{j=0}^{j=\infty} P_j(t) R_{n-j}(s|j,t)$$

where $R_{n-j}(s|j,t)$ is the probability that the number of arrivals minus the number of departures during an interval of length s is equal to $n - j$, given that the system was in state j at time t.

The probability $P_n(t)$ of having a given number of items in the system (waiting and in service) at time t is needed for the calculation of the average number of cars in the system in the transient case. The probability of waiting less than a given time τ for a car picked at random is needed to calculate the average waiting time. If the probability of waiting more than τ is large, customers may be discouraged from attempting to join the queue; thus this probability can itself be used as a measure of effectiveness.

The expression for the expected waiting time

$$W = \int_0^\infty \tau \, dP(<\tau)$$

where $P(<\tau) = 1 - P(>\tau)$, may be used, for example, to determine whether it is cheaper to increase the number of channels by k or improve the service facility (decreasing the mean holding time) by $\Delta\mu$. Thus, starting with c channels, one may equate $W(c + k, \mu, \lambda) = W(c, \mu + \Delta\mu, \lambda)$ and obtain the relative improvement in facilities $\Delta\mu/\mu$ corresponding to an increase of k channels. Then a decision may be made, using cost considerations, as to which improvement should be adopted.

11-6. Some Theoretical Queueing Models

Thus far there have been essentially two different theoretical approaches to the subject of queues: One through differential difference equations is due to Erlang; the other, through integral equations, was successfully studied by Lindley. In addition to a simple example, this section aims at illustrating these two methods.

Suppose that arrivals at a single channel occur at a constant rate λ. Also suppose that the service takes place at a constant rate μ. For example, in an automobile assembly line, arrivals and assembling take place at constant rates. Obviously there will be no waiting if $\lambda \leq \mu$ except if an item arrives while another is still in service, which can happen depending on when the operation starts. However, if $\lambda > \mu$, the number of items waiting will grow indefinitely.

Differential Difference Equations

Assuming that the operation starts with no items waiting in line, then the following equation provides a representation for a Poisson-input, exponential-holding-time, first-come, first-served, single-channel queue.

$$P_n(t + dt) = P_n(t)[1 - (\lambda + \mu) dt] + P_{n-1}(t)\lambda dt + P_{n+1}(t)\mu dt \qquad n \geq 1$$
$$P_0(t + dt) = P_0(t)(1 - \lambda dt) + P_1(t)\mu dt \qquad n = 0$$

These equations state that the probability of n items in the system at time $t + dt$ equals the probability of n items in the system at time t multiplied by the probability of no arrivals and no departures plus the probability of $n - 1$ items in the system at time t multiplied by the probability of one arrival and no departures plus the probability of $n + 1$ items in the system at time t multiplied by the probability of a single departure and no arrivals.

Note that the probability of no arrivals and no departures is given by $(1 - \lambda dt)(1 - \mu dt)$. The term involving $(dt)^2$ drops out in forming the differential equation. Thus one may use $1 - (\lambda + \mu) dt$. For the remaining two terms above, also note that $\lambda dt(1 - \mu dt) \sim \lambda dt$ and $\mu dt(1 - \lambda dt) \sim \mu dt$.

Transposing and passing to the limit with respect to dt, these equations become

$$\frac{dP_n(t)}{dt} = -(\lambda + \mu)P_n(t) + \lambda P_{n-1}(t) + \mu P_{n+1}(t) \qquad n \geq 1$$
$$\frac{dP_0(t)}{dt} = -\lambda P_0(t) + \mu P_1(t) \qquad n = 0$$

The time-independent steady-state solution is obtained either by solving the time-dependent transient equations given above and letting $t \to \infty$

in the solution, or by setting the derivatives with respect to time equal to zero and solving the resulting steady-state equations. Both these ideas will be illustrated below. Transient solutions are particularly useful when the utilization factor $\rho = \lambda/\mu \geq 1$, since in this case no steady state occurs. It has been shown [31] that the distribution of waiting times does not exist in the steady state for $\rho \geq 1$, except when special conditions are satisfied in the case of equality. However, the waiting time may be obtained for the transient case. Here, an expression for the expected number waiting and the expected waiting time will be derived in the steady state with $\lambda/\mu < 1$. If $\lambda/\mu \geq 1$, the number waiting would be infinite.

Setting the time derivatives equal to zero and eliminating time from the above equations, there result after transposing

$$(\lambda + \mu)p_n = \lambda p_{n-1} + \mu p_{n+1} \quad n \geq 1$$
$$\lambda p_0 = \mu p_1 \quad n = 0$$

Let $\rho = \lambda/\mu$; then these equations become

$$(1 + \rho)p_n = p_{n+1} + \rho p_{n-1} \quad n \geq 1$$
$$p_1 = \rho p_0 \quad n = 0$$

Let $n = 1$ in the first equation. Then $(1 + \rho)p_1 = p_2 + \rho p_0$. Substituting for p_1 from the second equation, this becomes $p_2 = \rho^2 p_0$. Repetition of this process yields $p_n = \rho^n p_0$. Now $\sum_{n=0}^{n=\infty} p_n = 1$, since the sum gives the total probability that there are no items, one item, two items, ... , in the system. This total probability must yield certainty since it accounts for all the possible states of the system. Thus

$$\sum_{n=0}^{n=\infty} \rho^n p_0 = 1$$

or

$$\sum_{n=0}^{n=\infty} p_0 \rho^n = p_0 \sum_{n=0}^{\infty} \rho^n = \frac{p_0}{1 - \rho} = 1$$

or
$$p_0 = 1 - \rho$$
Hence
$$p_n = \rho^n(1 - \rho)$$

Now the expected number in the system is given by

$$L = \sum_{n=0}^{n=\infty} n p_n = (1 - \rho) \sum_{n=0}^{n=\infty} n \rho^n = \frac{\rho}{1 - \rho}$$

as can be readily verified. Note that L is an expected value, and fluctuations in the number waiting can, in fact, occur. This can best be seen by calculating the variance:

$$\sum_{n=0}^{\infty}(n-L)^2 p_n = \sum_{n=0}^{\infty} n^2 p_n - \left(\frac{\rho}{1-\rho}\right)^2$$

But

$$\sum_{n=0}^{\infty} n^2 p_n = (1-\rho)\sum_{n=0}^{\infty} n^2 \rho^n = (1-\rho)\rho \frac{d}{d\rho}\rho \frac{d}{d\rho}\sum_{n=0}^{\infty}\rho^n = \frac{\rho}{1-\rho} + \frac{2\rho^2}{(1-\rho)^2}$$

Therefore the variance is given by

$$\frac{\rho}{1-\rho} + \frac{\rho^2}{(1-\rho)^2} = L + \left(\frac{\rho}{1-\rho}\right)^2$$

The expected number in the line is given by

$$L_q = L - \rho = \frac{\rho}{1-\rho} - \rho = \frac{\rho^2}{1-\rho}$$

since ρ is a measure of the expected number in the service channel at any time. Note that there are times when the channel is not occupied.

A useful device used as a quick check for the possible incorrectness of a formula is dimensional analysis. Thus, for example, in the last expression above, both the expressions on the left and on the right are dimensionless.

As a point of interest, it will now be shown that the transient solution given also by Eq. (11-26) yields the steady-state solution for $\lambda < \mu$ when the operation is studied with no items in the system; i.e., the initial state is E_0. The solution of the transient equations without derivation is given by

$$P_n(t) = e^{-(\lambda+\mu)t}\left[\left(\sqrt{\frac{\mu}{\lambda}}\right)^{-n} I_{-n}(2\sqrt{\lambda\mu}\,t)\right.$$
$$\left. + \left(\sqrt{\frac{\mu}{\lambda}}\right)^{1-n} I_{1+n}(2\sqrt{\lambda\mu}\,t) + (1-\rho)\rho^n \sum_{k=n+2}^{\infty}\left(\sqrt{\frac{\mu}{\lambda}}\right)^k I_k(2\sqrt{\lambda\mu}\,t)\right]$$

where $I_n(x) \equiv i^{-n}J_n(ix)$ is the modified Bessel function of the first kind. For fixed λ and μ as $t \to \infty$ the asymptotic behavior of this function is given by

$$I_n(2\sqrt{\lambda\mu}\,t) \sim \frac{\exp(2\sqrt{\lambda\mu}\,t)}{\sqrt{2\pi(2\sqrt{\lambda\mu})t}}$$

which is independent of n. When this expression is substituted for I_n in the first two terms in brackets and the latter multiplied by the outside factor, each of the resulting two expressions is readily seen to approach 0 as $t \to \infty$, having used the fact that $\lambda + \mu > 2\sqrt{\lambda\mu}$. The latter fact follows from the assumption $\lambda/\mu < 1$. As for the limiting behavior of the last factor, note that

$$\sum_{k=n+2}^{\infty} \left(\sqrt{\frac{\mu}{\lambda}}\right)^k I_k(2\sqrt{\lambda\mu}\,t) = \sum_{k=0}^{\infty} \left(\sqrt{\frac{\mu}{\lambda}}\right)^k I_k(2\sqrt{\lambda\mu}\,t) - \sum_{k=0}^{n+1} \left(\sqrt{\frac{\mu}{\lambda}}\right)^k I_k(2\sqrt{\lambda\mu}\,t)$$

The finite sum on the right when multiplied by $e^{-(\lambda+\mu)t}$ tends to zero as $t \to \infty$, since each term tends to zero separately. The series will now be examined. From standard textbooks,

$$\exp\left[\tfrac{1}{2}x\left(y + \frac{1}{y}\right)\right] = \sum_{n=-\infty}^{n=\infty} y^n I_n(x) = \sum_{n=0}^{n=\infty} y^n I_n(x) + \sum_{n=1}^{n=\infty} y^{-n} I_n(x)$$

since $I_{-n}(x) = I_n(x)$ for integral n.

The last sum on the right with $y = \sqrt{\mu/\lambda}$ and $x = 2\sqrt{\lambda\mu}\,t$, when multiplied by $e^{-(\lambda+\mu)t}$, tends to zero as $t \to \infty$ when the asymptotic expression of $I_n(x)$ is used together with the fact that $\lambda/\mu < 1$. On the other hand, the first sum on the right is the desired series under study. Thus finally

$$e^{-(\lambda+\mu)t} \sum_{n=0}^{n=\infty} \left(\sqrt{\frac{\mu}{\lambda}}\right)^k I_k(2\sqrt{\lambda\mu}\,t)$$

$$= e^{-(\lambda+\mu)t} \exp\left[\tfrac{1}{2}t(2\sqrt{\lambda\mu})\left(\sqrt{\frac{\mu}{\lambda}} + \sqrt{\frac{\lambda}{\mu}}\right)\right] = 1$$

Therefore $\qquad P_n(t) \to \rho^n(1 - \rho) \quad \text{as} \quad t \to \infty$

which is the desired result.

To obtain the expected waiting time, let $w_n(\tau)$ be the probability that an arrival will be delayed less than or equal to time τ when the system is in state E_n. Now the delay for this arrival consists of n consecutive intervals exponentially distributed corresponding to the service times of n items ahead. Thus if the new arrival is the nth in line, then the probability that a delay is less than τ is

$$w_n(\tau) = 1 - \sum_{i=0}^{n-1} \left[\frac{(\mu\tau)^i}{i!}\right] e^{-\mu\tau}$$

(there are $n - 1$ items waiting ahead in the line and one in service), but the probability that an arrival will be nth in the line or $(n + 1)$st in the system is p_n and thus the probability that an arbitrary arrival waits a time less than τ is given by

$$P(<\tau) = \sum_{n=0}^{n=\infty} p_n w_n(\tau) = 1 - \rho e^{-\mu\tau(1-\rho)}$$

Finally the expected waiting time in the steady state is given by

$$W = \int_0^\infty \tau \, dP(<\tau) = \frac{\rho}{\mu(1-\rho)}$$

Integral Equations

Integral equations will be illustrated by a single-channel application, with an input of independently distributed arrival times and an arbitrary-holding-time distribution with mean $1/\mu$. Let w_{n-1} be the waiting time of the $(n - 1)$st item, and the nth item arrives at a time t later and waits w_n. If the service time of the $(n - 1)$st item is s, then

$$\begin{aligned} w_n &= w_{n-1} + u & \text{if} & & w_{n-1} > 0 \\ w_n &= u & \text{if} & & w_{n-1} \leq 0 \end{aligned} \quad (11\text{-}1)$$

where $u = s - t$. The delay here is the negative of the time the channel has been free. Note that the same arrival distribution and the same service distribution are assumed for every item.

The formula relating the joint probability density functions for the waiting time of n items gives

Prob $(w_n = \tau)$ = Prob $(w_n = \tau, w_{n-1} > 0)$
$\qquad\qquad\qquad\qquad$ + Prob $(w_n = \tau, w_{n-1} \leq 0)$ $\quad (11\text{-}2)$

which, using Eq. (11-1), becomes

Prob $(w_n = \tau)$ = Prob $(w_{n-1} + u = \tau, w_{n-1} > 0)$
$\qquad\qquad\qquad\qquad$ + Prob $(u = \tau, w_{n-1} \leq 0)$ $\quad (11\text{-}3)$

Let $w_n(\tau) \equiv \text{Prob}(w_n = \tau)$; then (11-3) becomes, on using the formula for the distribution of the sum of two independently distributed variates,

$$w_n(\tau) = \int_0^\infty w_{n-1}(x) u(\tau - x) \, dx + u(\tau) \int_{-\infty}^0 w_{n-1}(x) \, dx \quad (11\text{-}4)$$

From the definition of $u(\tau)$ as the difference of two variates one has for its distribution

$$u(\tau) = \int_0^\infty b(y) a(y - \tau) \, dy = \int_0^\infty b(y + \tau) a(y) \, dy \quad (11\text{-}5)$$

where $b(t)$ and $a(s)$ are the holding- and arrival-times frequency distributions, respectively.

It can be shown (Lindley [38]) that a steady-state solution of this problem exists if $\lambda < \mu$; that is, when this relationship holds, $w_{n-1}(\tau)$ and $w_n(\tau)$ tend to a limit $w(\tau)$ as $n \to \infty$. Thus Eq. (11-4) becomes the integral equation of the Wiener-Hopf type:

$$w(\tau) = \int_0^\infty w(x)u(\tau - x)\,dx + u(\tau)\int_{-\infty}^0 w(x)\,dx \qquad (11\text{-}6)$$

Now introducing cumulative distributions by integrating (11-6) and (11-5) from $-\infty$ to τ, interchanging integrals and integrating by parts with respect to x, the probability of waiting less than τ is given by the integral equation

$$W(\tau) = \int_0^\infty W(x)u(\tau - x)\,dx \qquad (11\text{-}7)$$

where $W(\tau)$ is to be determined by solving this equation, and

$$U(\tau) = \int_0^\infty B(y+\tau)a(y)\,dy = \int_0^\infty B(y)a(y-\tau)\,dy$$
$$= 1 - \int_0^\infty b(y)A(y-\tau)\,dy = 1 - \int_0^\infty b(y+\tau)A(y)\,dy \qquad (11\text{-}8)$$

where $B(y)$ and $A(y)$ are the cumulative service time and input distributions. It is sometimes preferable to use (11-8) instead of (11-5) and obtain (11-5) by differentiation.

To justify (11-7) note that

Therefore
$$W(z) = \int_{-\infty}^z w(x)\,dx$$
$$W(0) = \int_{-\infty}^0 w(x)\,dx$$

Integrating Eq. (11-6) from $-\infty$ to τ yields

$$W(\tau) = \int_0^\infty \left[\int_{-\infty}^\tau u(t-x)\,dt\right] w(x)\,dx + U(\tau)W(0)$$

where
$$U(\tau) = \int_{-\infty}^\tau u(t)\,dt$$

Integrating the first expression on the right with respect to x by parts gives

$$W(\tau) = \left\{\left[\int_{-\infty}^\tau u(t-x)\,dt\right]\left[\int w(x)\,dx\right]\right\}_0^\infty$$
$$- \int_0^\infty \left[\frac{d}{dx}\int_{-\infty}^\tau u(t-x)\,dt\right]\left[\int w(x)\,dx\right] dx + U(\tau)W(0)$$

The first integral above is $U(\tau - x)$; the second is $W(x)$. But
$$U(\tau - \infty) = U(-\infty) = 0$$
since $U(x)$ is a cumulative distribution. Hence
$$W(\tau) = -U(\tau)W(0) - \int_0^\infty W(x)(-1)u(\tau - x)\,dx + U(\tau)W(0)$$
$$= \int_0^\infty W(x)u(\tau - x)\,dx$$

Application. Consider the case of arrivals at constant intervals of length T. The cumulative distribution of input times is then given by [24]
$$A(\tau) = \begin{cases} 0 & \text{for } \tau < T \\ 1 & \text{for } \tau \geq T \end{cases} \tag{11-9}$$

Suppose that the service times are exponentially distributed with
$$b(\tau) = \mu e^{-\mu\tau} \qquad \tau \geq 0 \tag{11-10}$$
and zero otherwise. Then, from Eq. (11-8),
$$U(\tau) = 1 - \int_T^\infty b(y + \tau)\,dy = B(T + \tau) \tag{11-11}$$
Thus
$$u(\tau) = b(T + \tau) \tag{11-12}$$
and substituting in (11-7) yields
$$W(\tau) = \int_0^{T+\tau} W(x)b(T + \tau - x)\,dx \tag{11-13}$$

[Note that $b(T + \tau - x)$ vanishes for negative argument.] This may be easily transformed into
$$W(\tau - T) = \int_0^\tau W(\tau - y)b(y)\,dy \tag{11-14}$$
by replacing τ by $\tau' - T$ and then putting $y = \tau' - x$, and finally using τ for τ'.

Substituting from (11-10) into (11-14) and trying as solution
$$W(\tau) = 1 + c_1 e^{c_2 \tau} \tag{11-15}$$
the following conditions, which must be satisfied, are obtained:
$$\frac{1}{\mu} + \frac{c_1}{\mu + c_2} = 0 \tag{11-16}$$
$$\frac{\mu}{\mu + c_2} = e^{-c_2 T} \tag{11-17}$$
which give
$$c_1 = -[1 - W(0)] = -[1 - p_0] \tag{11-18}$$
$$c_2 = -\mu p_0$$

Here p_0 is the nonzero root of $1 - p_0 = e^{-\mu p_0 T}$. Thus the probability $P(\leq \tau) \equiv W(\tau)$ of waiting less than time τ is given by

$$P(\leq \tau) \equiv 1 - (1 - p_0)e^{-\mu p_0 \tau} \tag{11-19}$$

Finally, the waiting time is given by

$$W = \int_0^\infty \tau \, dP(\leq \tau) = \frac{1 - p_0}{\mu p_0} \tag{11-20}$$

QUEUEING RESULTS

In the following sections, some queueing results available in the literature are given. Queue discipline is first come, first served unless indicated otherwise. Often when the queue discipline is changed, the distribution of waiting times is also changed but not the expected waiting time. Similarly, all inputs are Poisson unless indicated otherwise. A brief mention of the holding-time distribution is made. A few transient-state results are given. Steady-state results are given for the unsaturated case $\rho < 1$ when available.

The results given here do not apply to cases where items enter or leave in parties of two or more. Some of the results may be adjusted to cover special cases. For a brief discussion see Ref. [10].

11-7. Single Channel ($c = 1$)

Arbitrary Holding Time

The expected number waiting in queue and in service is given by [28, 30]

$$L = \frac{\lambda}{\mu} + \frac{\text{variance}(t) + (1/\mu)^2}{(2/\lambda)(1/\lambda - 1/\mu)} \tag{11-21}$$

where t has the holding-time distribution. Note that the numerator in the second expression on the right is the second moment of the holding-time distribution. This may also be expressed by

$$L = \left(W + \frac{1}{\mu}\right)\lambda \tag{11-22}$$

Thus the expected waiting time in the line is

$$W = \frac{\rho}{2\mu(1 - \rho)}[1 + (s\mu)^2] \tag{11-23}$$

where s is the standard deviation of the service time. The expected waiting time of those delayed, i.e., for whom $P(>0)$ holds, is given by

$$\frac{W}{P(>0)} \tag{11-24}$$

Note that these formulas hold using any holding-time distribution, as shown in the following:

Exponential-holding-time Distribution. The transient equations that give the probability of n items in the system at time t given that there were i items in the system at time zero (i.e., the system was in state E_i), are

$$\begin{aligned} P_0'(t) &= -\lambda P_0(t) + \mu P_1(t) \\ P_n'(t) &= -(\lambda + \mu) P_n(t) + \lambda P_{n-1}(t) + \mu P_{n+1}(t) \qquad n \geq 1 \end{aligned} \tag{11-25}$$

The solution of the above, if the initial state is E_i (first given by A. B. Clarke), is

$$P_n(t) = e^{-(\lambda+\mu)t} \left[\left(\sqrt{\frac{\mu}{\lambda}}\right)^{i-n} I_{i-n}(2\sqrt{\lambda\mu}\, t) + \left(\sqrt{\frac{\mu}{\lambda}}\right)^{i-n+1} I_{i+n+1}(2\sqrt{\lambda\mu}\, t) + \left(1 - \frac{\lambda}{\mu}\right)\left(\frac{\lambda}{\mu}\right)^n \sum_{k=i+n+2}^{k=\infty} \left(\sqrt{\frac{\mu}{\lambda}}\right)^k I_k(2\sqrt{\lambda\mu}\, t) \right] \tag{11-26}$$

where i and n take on the values $0, 1, 2, \ldots$, and I_k is the modified Bessel function of the first kind defined by $I_k(x) \equiv i^{-k} J_k(ix)$, with $i = \sqrt{-1}$ in this last expression. Another form of this result, if the initial state is E_i, is given by [37]

$$P_n(t) = \frac{\mu}{\pi} \left(\sqrt{\frac{\lambda}{\mu}}\right)^{n-i} \int_0^{2\pi} \left[\sin i\theta - \sqrt{\frac{\lambda}{\mu}} \sin(i+1)\theta\right] \\ \left[\sin n\theta - \sqrt{\frac{\lambda}{\mu}} \sin(n+1)\theta\right] \frac{e^{-wt}}{w} d\theta \tag{11-27}$$

where $\qquad w = \mu + \lambda - 2\sqrt{\lambda\mu}\cos\theta$

When there is dependence on E_i, $P_n(t)$ is frequently written as $P_{i,n}(t)$. The steady-state equations, which are independent of the initial state E_i, are given by

$$\lambda p_0 = \mu p_1 \qquad (\lambda + \mu)p_n = \lambda p_{n-1} + \mu p_{n+1} \tag{11-28}$$

Their solution is given by [14, 18]

$$p_n = \left(\frac{\lambda}{\mu}\right)^n \left(1 - \frac{\lambda}{\mu}\right) \tag{11-29}$$

The expected number waiting in the system (steady-state) is given by [14]

$$L = \frac{\rho}{1 - \rho} \quad (11\text{-}30)$$

and in the line by

$$L_q = \frac{\rho^2}{1 - \rho} \quad (11\text{-}31)$$

Also,

$$P(>\tau) = \rho e^{\mu(\rho-1)\tau} \quad (11\text{-}32)$$

and

$$W = \frac{\lambda}{\mu(\mu - \lambda)} = \frac{\rho}{\mu(1 - \rho)} = \frac{L_q}{\lambda} = \frac{L}{\mu} \quad (11\text{-}33)$$

The expected number waiting, of those delayed, is

$$\frac{1}{1 - \rho} \quad (11\text{-}34)$$

The expected waiting time of those delayed is

$$\frac{W}{P(>0)} = \frac{1}{\mu(1 - \rho)} \quad (11\text{-}35)$$

Constant-holding-time Distribution. The steady-state equations are [24]

$$\begin{aligned}
p_0 &= 1 - \rho \\
p_1 &= (1 - \rho)(e^\rho - 1) \\
p_n &= (1 - \rho) \sum_{k=1}^{k=n} (-1)^{n-k} e^{k\rho} \left[\frac{(k\rho)^{n-k}}{(n-k)!} + \frac{(k\rho)^{n-k-1}}{(n-k-1)!} \right] \quad n \geq 2
\end{aligned} \quad (11\text{-}36)$$

ignoring the second factor for $k = n$. Here

$$P(>\tau) = \rho \sum_{i=0}^{k} e^{\rho(\mu\tau - i)} \frac{[-\rho(\mu\tau - i)]^i}{i!} \quad (11\text{-}37)$$

where k is the largest integer less than or equal to $\mu\tau$. Also,

$$W = \frac{\lambda}{2\mu^2(1 - \lambda/\mu)} = \frac{\rho}{2\mu(1 - \rho)} \quad (11\text{-}38)$$

The expected waiting time of those delayed is

$$\frac{1}{2\mu(1 - \rho)} \quad (11\text{-}39)$$

Poisson-input, Erlangian Holding-time Distribution.

$$b_k(t) = \frac{(\mu k)^k}{\Gamma(k)} e^{-\mu k t} t^{k-1} \quad (11\text{-}40)$$

The steady-state equations are

$$\lambda p_0 = \mu p_1$$
$$(\lambda + \mu)p_n = \mu p_{n+1} + \lambda p_{n-k} \quad n \geq 1 \tag{11-41}$$

Here
$$L = \frac{\rho(\rho + 2k - \rho k)}{2k(1 - \rho)} \tag{11-42}$$

$$W = \frac{\rho(k + 1)}{2k\mu(1 - \rho)} \tag{11-43}$$

Results for W with $k = 2, 3$ are available in Ref. [24].

Priority-discipline, Arbitrary-holding-time, Nonpreemptive Service [11]

Finite Number of Priorities N. This is a case of a Poisson input for the kth priority with arrival rate λ_k, arbitrary holding time with service rate μ_k, and priority queue discipline. Items of different types enter the system with assigned priorities for service. Whenever the system is free to acquire an item, it selects items of highest priority on a first-come, first-served basis. However, if an item of higher priority enters the system while one of lower priority is in service, this service is not preempted, i.e., sent back to the waiting line. In this case,

$$W_k = \frac{W_0}{(1 - \sigma_{k-1})(1 - \sigma_k)} \tag{11-44}$$

where
$$\rho_i = \frac{\lambda_i}{\mu_i} \quad \text{and} \quad \lambda = \sum_{i=1}^{i=N} \lambda_i$$

$$\sigma_k = \sum_{i=1}^{i=k} \rho_i < 1 \tag{11-45}$$

$$W_0 = \tfrac{1}{2}\lambda \int_0^\infty t^2 \, dF(t) \tag{11-46}$$

$$F(t) = \frac{1}{\lambda} \sum_{i=1}^{N} \lambda_i F_i(t) \tag{11-47}$$

$F_k(t)$ is the cumulative-holding-time distribution function for the kth priority. The expected length of the line is

$$L = \sum_{i=1}^{i=N} \lambda_i W_i \tag{11-48}$$

The steady-state distribution of waiting times for the kth priority is given in Ref. [62]. Let

$$G_k(t) = \begin{cases} 0 & \text{for } t < 0 \\ 1 - e^{-\lambda_k t} & \text{for } t \geq 0 \text{ with } \lambda_k > 0 \end{cases}$$

be the cumulative distribution of arrival times for the kth priority in a set of N priorities, and let $F_k(t)$ be the cumulative-holding-time distribution for that priority in a single-channel queueing operation. Assume independence among all arrival intervals and among arrivals in each priority and among service times. Let

$$\lambda = \sum_{i=1}^{i=N} \lambda_i \qquad \mu_k^{(l)} = \int_{0-}^{\infty} t^l \, dF_k(t)$$

$$\varphi_k(\alpha) = \int_{0-}^{\infty} e^{-\alpha t} \, dF_k(t) \qquad \psi_k(\alpha) = \int_{0-}^{\infty} e^{-\alpha t} \, dH_k(t)$$

where $K_k(t)$ is the cumulative-waiting-time distribution function for the kth priority. Note that if $\psi_k(\alpha)$ is determined, then by taking inverse transforms $H_k(t)$ will also be determined. In the nonsaturated case

$$\sum_{i=1}^{i=N} \lambda_i \mu_i^{(1)} < 1$$

$$\psi_k(\alpha) = \frac{\left(1 - \sum_1^N \lambda_i \mu_i^{(1)}\right)\left(-\sum_1^{k-1} \lambda_i + z_k^* - \alpha\right) - \sum_{k+1}^N \lambda_i \left[1 - \varphi_i\left(\sum_1^{k-1} \lambda_j - z_k^* + \alpha\right)\right]}{\lambda_k - \alpha - \lambda_k \varphi_k \left(\sum_1^{k-1} \lambda_j - z_k^* + \alpha\right)}$$

where $z_1^* = 0$, and $z_k^* = z_k^*(\alpha)$ satisfies

$$z_k^* - \sum_1^{k-1} \lambda_i \varphi_i \left(\sum_1^{k-1} \lambda_j - z_k^* + \alpha\right) = 0 \qquad k \geq 2$$

Thus, for example,

$$\psi_1(\alpha) = -\frac{\left(1 - \sum_1^N \lambda_i \mu_i^{(1)}\right)\alpha + \sum_2^N \lambda_1[1 - \varphi_i(\alpha)]}{\lambda_1 - \alpha - \lambda_1 \varphi_1(\alpha)}$$

The expected waiting time is given by

$$W_k \equiv E(w_k) = \frac{\sum_1^N \lambda_i \mu_i^{(2)}}{2\left[1 - \sum_1^{k-1} \lambda_i \mu_i^{(1)}\right]\left[1 - \sum_1^k \lambda_i \mu_i^{(1)}\right]} \qquad (11\text{-}48a)$$

and the second moment is given by

$$E(w_k{}^2) = \frac{\sum_{1}^{N} \lambda_i \mu_i^{(3)}}{3\left[1 - \sum_{1}^{k-1} \lambda_i \mu_i^{(1)}\right]^2 \left[1 - \sum_{1}^{k} \lambda_i \mu_i^{(1)}\right]}$$

$$+ \frac{\sum_{1}^{N} \lambda_i \mu_i^{(2)} \sum_{1}^{k-1} \lambda_j \mu_j^{(2)}}{2\left[1 - \sum_{1}^{k-1} \lambda_i \mu_i^{(1)}\right]^2 \left[1 - \sum_{1}^{k} \lambda_i \mu_i^{(1)}\right]^2}$$

$$+ \frac{\sum_{1}^{N} \lambda_i \mu_i^{(2)} \sum_{1}^{k-1} \lambda_j \mu_j^{(2)}}{2\left[1 - \sum_{1}^{k-1} \lambda_i \mu_i^{(1)}\right]^3 \left[1 - \sum_{1}^{k} \lambda_i \mu_i^{(1)}\right]} \qquad k = 1, \ldots, N$$

In the saturation case $\sum_{1}^{N} \lambda_i \mu_i^{(1)} \geq 1$, and one can find an integer s ($0 \leq s < N$) such that $\sum_{i=1}^{s} \lambda_i \mu_i^{(1)} < 1$, $\sum_{1}^{s+1} \lambda_i \mu_i^{(1)} \geq 1$,

$$W_k \equiv E(w_k) = \frac{\sum_{i=1}^{s} \lambda_i \mu_i^{(2)} + [\mu_{s+1}^{(2)}/\mu_{s+1}^{(1)}]\left[1 - \sum_{1}^{s} \lambda_i \mu_i^{(1)}\right]}{2\left[1 - \sum_{1}^{k-1} \lambda_i \mu_i^{(1)}\right]\left[1 - \sum_{1}^{k} \lambda_i \mu_i^{(1)}\right]}$$

$$k = 1, 2, \ldots, s$$

$$W_k \equiv E(w_k) = \infty \qquad k = s+1, \ldots, N$$

Two Priorities, Preemptive Service, Exponential Holding Time [6]. Priority 1 and 2 calls arrive at a single channel with arrival rates λ_1 and λ_2, respectively. Both priorities have Poisson arrival distribution. Priority 1 calls in the queue enter the channel before all priority 2 calls in queue and replace any priority 2 calls in the channel on their arrival. The priority 2 call in the channel then reenters the queue. It is unnecessary to make any assumptions about whether servicing on priority 2 calls is wanted or not, for the exponential distribution is self-renewing.

Priority 1 and 2 calls have exponential-service-time distribution with service rates μ_1 and μ_2, respectively. Service is preempted in this case.

Let $\rho_1 = \lambda_1/\mu_1$, $\rho_2 = \lambda_2/\mu_2$, where

$$\frac{\lambda_1}{\mu_1} + \frac{\lambda_2}{\mu_2} < 1 \tag{11-49}$$

Let $p_n{}^m$ be the probability that n priority 1 calls and m priority 2 calls are in the queue. The steady-state equations are

$$\begin{aligned}
0 &= \mu_1 p_{n+1}^m - (\mu_1 + \lambda_1 + \lambda_2) p_n{}^m \\
&\quad + \lambda_1 p_{n-1}^m + \lambda_2 p_n^{m-1} \quad && m, n > 0 \\
0 &= \mu_1 p_1{}^m + \mu_2 p_0^{m+1} - (\mu_2 + \lambda_1 + \lambda_2) p_0{}^m \\
&\quad + \lambda_2 p_0^{m-1} \quad && n = 0, m > 0 \\
0 &= \mu_1 p_{n+1}^0 - (\mu_1 + \lambda_1 + \lambda_2) p_n{}^0 + \lambda_1 p_{n-1}^0 \quad && m = 0, n > 0 \\
0 &= \mu_1 p_1{}^0 + \mu_2 p_0{}^1 - (\lambda_1 + \lambda_2) p_0{}^0 \quad && m = n = 0
\end{aligned} \tag{11-50}$$

Put
$$F(z, u) = \sum_{n=0}^{n=\infty} \sum_{m=0}^{m=\infty} p_n{}^m z^m u^n \tag{11-51}$$

Then
$$F(z, u) = \frac{1 - \rho_1 - \rho_2}{1 - \alpha - \rho_2 z} \frac{1 - \alpha}{1 - \alpha u} \tag{11-52}$$

where
$$\alpha = \frac{\mu_1 + \lambda_1 + \lambda_2(1 - z) - \sqrt{[\mu_1 + \lambda_1 + \lambda_2(1 - z)]^2 - 4\lambda_1 \mu_1}}{2\mu_1} \tag{11-53}$$

The expected number waiting, first priority, is

$$\bar{n} = \left.\frac{\partial F}{\partial u}\right]_{z=u=1} = \frac{\rho_1}{1 - \rho_1} \tag{11-54}$$

The expected number waiting, second priority, is

$$\bar{m} = \left.\frac{\partial F}{\partial z}\right]_{z=u=1} = \rho_2 \left[\frac{1 + (\mu_1/\mu_2)\bar{n}}{1 - \rho_1 - \rho_2}\right] \tag{11-55}$$

$$\left.\frac{\partial^k F}{\partial u^k}\right]_{z=1, u=0} = \rho_1{}^n (1 - \rho_1) \tag{11-56}$$

the probability that n priority 1 calls are in the queue; hence

$$p_n{}^m = \frac{1}{n! m!} \left.\frac{\partial^{m+n} F}{\partial z^m \, \partial u^n}\right]_{z=u=0} \tag{11-57}$$

Continuous Number of Priorities [53]. Results for a single-channel priority queueing system with application to machine breakdowns are given. The number of machines available is assumed to be infinite. Priorities are assigned according to the length of time needed to service a machine, higher priorities to shorter jobs. Since the length of service time may correspond to any real number, so do the priorities, i.e., a continuous number of priorities.

Suppose that new repair jobs are generated by a Poisson law with an average of λ jobs per unit time. Also, suppose that the arrival of jobs of priority t (the amount of time required to service) is according to a Poisson law with mean λ_t. Let $F(t)$ be the average cumulative job repair time distribution for units of all priorities. Then

$$F(t) = \frac{1}{\lambda} \int_0^t \lambda_t F_t(s)\, ds \qquad (11\text{-}58)$$

Here $F_t(s)$ is the cumulative job repair time for units of priority t, and s is the integration dummy variable.

The expected waiting time W_t of a job of duration t is

$$W_t = \frac{W_0}{\left[1 - \lambda \int_0^t s\, dF(s)\right]^2} \qquad t \leq \tau$$
$$W_t = \infty \qquad t > \tau \qquad (11\text{-}59)$$

where
$$W_0 = \tfrac{1}{2}\lambda \int_0^\tau t^2\, dF(t) \qquad (11\text{-}60)$$

and τ, the initial job duration, is obtained from the equation

$$\lambda \int_0^\tau t\, dF(t) = 1 \qquad (11\text{-}61)$$

The expected number of jobs in the waiting line L is given by

$$L = \frac{W_0 \int_0^\tau dF(t)}{\left[1 - \lambda \int_0^t s\, dF(s)\right]^2} \qquad (11\text{-}62)$$

If, for example, $\qquad F(t) = 1 - e^{-\mu t} \qquad (11\text{-}63)$

corresponding to exponential repair times, then

$$W_0 = \frac{\lambda}{\mu^2} \qquad \lambda < \mu \text{ (unsaturated case)} \qquad (11\text{-}64)$$

$$W_0 = \frac{1}{\mu}(1 - \lambda\tau e^{-\mu\tau}) \qquad \lambda \geq \mu \text{ (saturated case)} \qquad (11\text{-}65)$$

The expected waiting time is

$$W_t = \frac{W_0}{\{1 - (\lambda/\mu)[1 - e^{-\mu t}(1 + \mu t)]\}^2} \qquad (11\text{-}66)$$

and the expected number of jobs waiting is

$$L = \frac{\lambda^2}{\mu} \int_0^\infty \frac{e^{-\mu t}\, dt}{\{1 - (\lambda/\mu)[1 - e^{-\mu t}(1 + \mu t)]\}^2} \qquad (11\text{-}67)$$

L is infinite in the saturated case.

Random Selection for Service, Impatient Customers, Exponential Holding Time [5]

This is a case of Poisson input, exponential holding time, random selection for service, where impatient customers leave after a wait of time T_0. If items are available for service, the probability that one will be successfully serviced and discharged from the system during any time interval $(t, t + h)$ is μh + terms of higher order in h. These items are selected for service at random from among those in the system. Thus the distribution of successful servicing is exponential.

The steady-state equations are given by

$$-\lambda p_0 + (\mu + C_1)p_1 = 0$$
$$\lambda p_{n-1} - (\lambda + \mu + C_n)p_n + (\mu + C_{n+1})p_{n+1} = 0 \quad (11\text{-}68)$$

where we obtain p_0 from $\sum_{0}^{\infty} p_n = 1$. C_n is the average rate at which customers become "lost" when there are n customers in the system.

The solution of the above is given by

$$p_n = \lambda^n p_0 \prod_{k=1}^{k=n} (\mu + C_k)^{-1} \quad n = 1, 2, \ldots \quad (11\text{-}69)$$

$$C_n = \frac{\mu e^{-\mu T_0/n}}{1 - e^{-\mu T_0/n}} \quad (11\text{-}70)$$

Limited Source, Exponential Holding Time [18]

This is a case of input from a source having only a finite number m of customers, exponential service time, and a single channel (servicing of m machines).

The probability that a machine calls for service before $t + dt$ is $\lambda\, dt + O(dt^2)$, given that it is in a working state at time t. The transient equations are

$$P_0'(t) = -m\lambda P_0(t) + \mu P_1(t)$$
$$P_n'(t) = -[(m - n)\lambda + \mu]P_n(t) + (m - n + 1)\lambda P_{n-1}(t)$$
$$\qquad\qquad + \mu P_{n+1}(t) \quad 1 \le n \le m - 1 \quad (11\text{-}71)$$
$$P_m'(t) = -\mu P_m(t) + \lambda P_{m-1}(t)$$

The steady-state equations are

$$m\lambda p_0 = \mu p_1$$
$$[(m - n)\lambda + \mu]p_n = (m - n + 1)\lambda p_{n-1} + \mu p_{n+1} \quad (11\text{-}72)$$
$$\mu p_m = \lambda p_{m-1}$$

The solution of the steady-state equations is given by

$$p_n = m(m-1) \cdots (m-n+1) \left(\frac{\lambda}{\mu}\right)^n p_0 \qquad (11\text{-}73)$$

$$p_0 = 1 - \sum_{n=1}^{n=m} p_n$$

$$L_q = m - \frac{\lambda + \mu}{\lambda}(1 - p_0) \qquad (11\text{-}74)$$

$$L = m - \frac{\mu}{\lambda}(1 - p_0)$$

Constant Input, Exponential Holding Time [24]

For constant input at intervals of length δ,

$$p_n = p_0(1 - p_0)^n \qquad (11\text{-}75)$$

where p_0 is obtained from

$$1 - p_0 = e^{-\mu p_0 \delta} \qquad (11\text{-}76)$$
$$P(>\tau) = (1 - p_0)e^{-\mu p_0 \delta} \qquad (11\text{-}77)$$
$$W = -\int_0^\infty \tau \, dP(>\tau) = \frac{1 - p_0}{\mu p_0} \qquad (11\text{-}78)$$

Queue-length-dependent Parameters

In this case the transient equations are given by the birth- and death-process equations:

$$\frac{d}{dt} P_{i,n}(t) = a_{n-1} P_{i,n-1}(t) - (a_n + b_n) P_{i,n}(t) + b_{n+1} P_{i,n+1}(t) \qquad (11\text{-}79)$$

where $P_{i,n}(t)$ is the probability that there are n items in the system at time t, given that there were i items in it a time zero, and $a_n \, dt$ and $b_n \, dt$ are the probabilities of an arrival and of a departure, respectively, during an interval of length dt when there are n items in the system. The solution of the steady state equation is given by

$$p_n = p_0 \prod_{i=1}^{n} \frac{a_{i-1}}{b_i}$$

In Ref. [54] there is a brief discussion of the solution of this problem. A recent paper [55] gave attention to a part of this problem as shown below.

Assuming that a newly arrived customer joins a queue with a probability a_i when he has found i customers waiting or being served and that $a_0 = 1$, $0 \leq a_{i+1} \leq a_i \leq 1$ ($i = 0, 1, 2, \ldots$), then the transient equations for a Poisson-input, exponential-holding-time, single-channel queue

$$P'_n(t) = \lambda\{a_{n-1}P_{n-1}(t) - a_nP_n(t)\} - \mu\{P_n(t) - P_{n-1}(t)\} \quad n \geq 1$$
$$P'_0(t) = \mu P_1(t) - \lambda P_0(t) \quad (11\text{-}80)$$

Let $w(t)$ be the waiting time of a customer joining the queue at time t. Then

$$F(x,t) \equiv P[w(t) \leq x] = \int_0^x \sum_{n=0}^{\infty} P_n(t) e^{-\mu y} \frac{(\mu y)^{n-1}}{(n-1)!} \mu \, dy \quad (11\text{-}81)$$

The Laplace transform of $dF(x,t)$ is given by

$$\phi(s,t) = \int_0^{\infty} e^{-sx} \, dF(x,t) = \sum_{n=0}^{\infty} P_n(t) \left(\frac{\mu}{\mu + s}\right)^n \quad (11\text{-}82)$$

On taking the derivative with respect to time and substituting for $P'_n(t)$ from the transient equations, then putting

$$\frac{\partial \phi(s,t)}{\partial t} = 0$$

to obtain the equilibrium of waiting time, there results

$$\phi(s,t) = \frac{\lambda}{\mu + s} \sum_{n=0}^{\infty} a_n P_n(t) \left(\frac{\mu}{\mu + s}\right)^n + P_0(t) \quad (11\text{-}83)$$

By putting $s = 0$ this gives

$$P_0(t) = 1 - \rho \sum_{n=0}^{\infty} a_n P_n(t) \quad (11\text{-}84)$$

The mean $E(w)$ and variance $\sigma^2(w)$ of the waiting time w are obtained by using the coefficients of s and s^2 of $\phi(s,t)$. That is,

$$E(w) = \frac{1 - P_0(t)}{\mu} + \frac{\rho}{\mu} \sum_{n=1}^{\infty} n a_n P_n(t) \quad (11\text{-}85)$$

$$\sigma^2(w) = \frac{\rho}{\mu^2} \sum_{n=0}^{\infty} (n+1)(n+2) a_n P_n(t) - [E(w)]^2 \quad (11\text{-}86)$$

Note that by putting $a_n = 1 \ (n \geq 0)$ the result for the mean waiting time in the steady state is obtained, i.e.,

$$E(w) = \frac{\rho}{\mu(1 - \rho)} \quad (11\text{-}87)$$

since
$$\lim_{t \to \infty} P_n(t) \equiv p_n = (1 - \rho)\rho^n \qquad n \geq 0 \qquad (11\text{-}88)$$

and
$$\sum_{n=1}^{\infty} n p_n = \frac{\rho}{(1 - \rho)} \qquad (11\text{-}89)$$

Time-dependent Parameters [57, 58]

The case where $\lambda(t)$ and $\mu(t)$ are arbitrary functions of the time t is of interest. Let $\tau = \int_0^t \mu(s)\,ds$. Then the Poisson-input, single-channel, exponential-holding-time equations, if the initial state is E_i, are

$$\begin{aligned}\frac{dP_{i,n}(t)}{dt} &= -[\lambda(t) + \mu(t)]P_{i,n}(t) + \lambda(t)P_{i,n-1}(t) + \mu(t)P_{i,n+1}(t) \\ \frac{dP_{i,0}(t)}{dt} &= -\lambda(t)P_{i,0} + \mu(t)P_{i,1}\end{aligned} \qquad (11\text{-}90)$$

and have the solution

$$P_{i,n}(t) = \exp\{-\tau[1 + R(\tau)]\}[A_{i-n}(0,\tau,\tau) + \int_0^\tau A_{-n}(\sigma,\tau,\tau)f_i(\sigma)\,d\sigma] \qquad (11\text{-}91)$$

where
$$R(\tau) = \frac{\int_0^t \lambda(s)\,ds}{\int_0^t \mu(s)\,ds}$$

$$A_n(\sigma,\tau,Z) = Z^{n/2}[R(\tau)\tau - R(\sigma)\sigma]^{-n/2} I_n(2\{[R(\tau)\tau - R(\sigma)\sigma]Z\}^{1/2})$$
$$n = 0, \pm 1, \pm 2, \ldots$$

$$f_i(\tau) = B_i(0,\tau) + \int_0^\tau B_0(\sigma,\tau)f_i(\sigma)\,d\sigma$$
$$B_n(\sigma,\tau) = A_n(\sigma,\tau,\tau) - \rho(\tau)A_{n+1}(\sigma,\tau,\tau)$$
$$I_n(y) = i^{-n}J_n(iy) \qquad i = \sqrt{-1}$$

is the modified Bessel function of the first kind, and

$$\rho(\tau) = \frac{\lambda(\tau)}{\mu(\tau)}$$

The expected length of the line is given by

$$L_i(\tau) = [R(\tau) - 1]\tau + \int_0^\tau P_{i,0}(\sigma)\,d\sigma + i \qquad (11\text{-}92)$$

$$\sigma_i^2(\tau) = 2\int_0^\tau \{\rho(\sigma) + [\rho(\sigma) - 1]L_i(\sigma)\}\,d\sigma - M_i(\tau)$$
$$- [M_i(\tau)]^2 + i + i^2 \qquad (11\text{-}93)$$

When $R(\tau) \equiv \rho$ is a constant, then the solution is given by

$$P_{i,n}(\tau) = e^{-\tau(1+\rho)}[A_{i-n}(0,\tau,\tau) + \rho^n A_{n+i+1}(0,\tau,\tau)$$
$$+ (1 - \rho)\rho^n \sum_{k=n+i+2}^{\infty} A_k(0,\tau,\tau)] \qquad (11\text{-}94)$$

In this case, if $\rho > 1$, then

$$L_i(\tau) = (\rho - 1)\tau + \frac{1}{\rho^i(\rho - 1)} + i + o(1) \qquad (11\text{-}95)$$

where $o(1)$ is a function which tends to zero for large τ, and

$$\sigma_i(\tau) = \sqrt{\tau(\rho + 1)} + o(\sqrt{\tau}) \qquad (11\text{-}96)$$

where $o(\sqrt{\tau})$ is a function which when divided by $\sqrt{\tau}$ tends to zero for large τ. If $\rho = 1$,

$$L_i(\tau) = 2\sqrt{\frac{\tau}{\pi}} + O(1) \qquad (11\text{-}97)$$

where $O(1)$ is a bounded function, and

$$\sigma_i(\tau) = \sqrt{2\tau\left(1 - \frac{2}{\pi}\right)} + O(1) \qquad (11\text{-}98)$$

Glen D. Camp [10] describes an interesting difficulty that arises with regard to the meaning of the statement "the probability of having n items in the system at time t." Camp points out that if the history can be proved to be a stationary time series, then everything is simple; one can generate other "statistically equivalent" histories and then the "probability of n at time t" is the fraction of such equivalent histories for which, at time t, n has any particular value. However, if the series is nonstationary, it is difficult to distinguish between "long-term" trends and a fluctuation, a fact necessary for generating equivalent sequences. The above statement is interpreted by him as "the fractional time (which one is interested in, in practice) of a moving interval T, centered at t, during which n has a particular value." T must be selected "long" in order to give the system a chance to occupy most possible states for a good approximation to the appropriate expected times. It should be "short" in order to avoid smoothing out time trends. It is of the order of the longest relaxation time.

The conclusion is that in many practical problems the difficulty of the mathematically time-dependent coefficients is soluble since a succession of steady states adequately describes the situation.

11-8. Two Channels in Series [26]

Exponential Holding Times Different for Each

This is a case of Poisson input (mean λ) and two channels in series with exponential holding times μ_1 and μ_2, respectively. After finishing service at the first gate the customer moves on to the second gate.

Unlimited Input. Assuming that there is an unlimited number of customers available, the average distribution of customers throughout the system is given in Table 11-1.

TABLE 11-1. AVERAGE DISTRIBUTION OF CUSTOMERS THROUGHOUT THE SYSTEM

	Channel 1	Channel 2	Total system
Average number of customers waiting for service	$\dfrac{x_1}{1-x_1}$	$\dfrac{x_2}{1-x_2}$	$\dfrac{x_1^2}{1-x_1} + \dfrac{x_2^2}{1-x_2}$
Average number of customers being served	x_1	x_2	$x_1 + x_2$
Average total number of customers	$\dfrac{x_1}{1-x_1}$	$\dfrac{x_2}{1-x_2}$	$\dfrac{x_1}{1-x_1} + \dfrac{x_2}{1-x_2}$

The steady-state solution [26] giving the probability that there are n_1 customers waiting at the first gate and n_2 at the second is given by

$$p(n_1,n_2) = (x_1)^{n_1}(x_2)^{n_2}(1-x_1)(1-x_2) \qquad (11\text{-}99)$$

where $\qquad x_1 = \dfrac{\lambda}{\mu_1} < 1 \quad \text{and} \quad x_2 = \dfrac{\lambda}{\mu_2} < 1 \qquad (11\text{-}100)$

When the queue is in equilibrium, λ is the same for both channels.

The probability of having n customers waiting at the first channel and at the second channel is, respectively,

$$\begin{aligned} p_1(n) &= x_1^n(1-x_1) \\ p_2(n) &= x_2^n(1-x_2) \end{aligned} \qquad (11\text{-}101)$$

Limited Input. If the number of customers is limited to N, and it is assumed that arrivals are random until the total number of potential customers is N, then no arrivals occur until this number drops to $N-1$ or less.

$$p(n_1,n_2) = x_1^{n_1} x_2^{n_2} \left[\frac{(x_1-x_2)(1-x_1)(1-x_2)}{(x_1-x_2) - (x_1^{N+2} - x_2^{N+2}) + x_1 x_2 (x_1^{N+1} - x_2^{N+1})} \right] \qquad (11\text{-}102)$$

Denote the quantity in brackets by $p(0,0)$. The average number of customers in the system is given by

$$\frac{1}{x_1 - x_2} \left\{ \frac{x_1^2[1 - (N+1)x_1^N + N x_1^{N+1}]}{(1-x_1)^2} - \frac{x_2^2[1 - (N+1)x_2^N + N x_2^{N+1}]}{(1-x_2)^2} \right\} p(0,0) \qquad (11\text{-}103)$$

The average number of customers being served is given by
$$\frac{(x_1 + x_2)[(x_1 - x_2) - (x_1^{N+1} - x_2^{N+1}) + x_1x_2(x_1^N - x_2^N)]}{(1 - x_1)(1 - x_2)(x_1 - x_2)} p(0,0) \quad (11\text{-}104)$$

The average number of customers waiting is given by
$$\left\{ \frac{x_2^2(1 + x_1)[1 - Nx_1^{N-1} + (N - 1)x_1^N]}{(1 - x_1)^2} \right.$$
$$+ \frac{x_1^2(1 + x_2)[1 - Nx_2^{N-1} + (N - 1)(x_2^N)]}{(1 - x_2)^2}$$
$$+ \left\{ \frac{x_1 x_2}{x_1 - x_2} \left[\frac{1 - (N - 1)x_1^{N-2} + (N - 2)x_1^{N-1}}{(1 - x_1)^2} \right. \right.$$
$$\left. \left. \left. - \frac{1 - (N - 1)x_2^{N-2} + (N - 2)x_2^{N-1}}{(1 - x_2)^2} \right] \right\} \right\} p(0,0) \quad (11\text{-}105)$$

Similar results for the three-channels-in-series case are available in Ref. [26].

11-9. Multiple Channels in Parallel [18], Poisson Input

Finite Number of Channels $c > 1$, $\rho = \lambda/\mu c$

Identical Exponential Holding Times [18]. The transient equations are

$$P'_0(t) = -\lambda P_0(t) + \mu P_1(t) \qquad n = 0$$
$$P'_n(t) = -(\lambda + n\mu)P_n(t) + \lambda P_{n-1}(t)$$
$$\qquad\qquad + (n + 1)\mu P_{n+1}(t) \qquad 1 \leq n < c \quad (11\text{-}106)$$
$$P'_n(t) = -(\lambda + c\mu)P_n(t) + \lambda P_{n-1}(t) + c\mu P_{n+1}(t) \qquad c \leq n$$

The steady-state equations are [14]
$$\lambda p_0 = \mu p_1 \qquad n = 0$$
$$(\lambda + n\mu)p_n = \lambda p_{n-1} + (n + 1)\mu p_{n+1} \qquad 1 \leq n < c \quad (11\text{-}107)$$
$$(\lambda + c\mu)p_n = \lambda p_{n-1} + c\mu p_{n+1} \qquad c \leq n$$

The solution of the above is given by
$$p_n = \frac{p_0(\lambda/\mu)^n}{n!} \qquad 0 \leq n \leq c$$
$$p_n = \frac{p_0(\lambda/\mu)^n}{c! c^{n-c}} \qquad n \geq c \quad (11\text{-}108)$$

where
$$p_0 = \frac{1}{\sum_{n=0}^{c-1} \frac{(c\rho)^n}{n!} + \frac{(c\rho)^c}{c!(1 - \rho)}}$$

$\left(\text{Note that } p_0 \text{ is obtained from } \sum_{n=0}^{n=\infty} p_n = 1.\right)$

$$L = \frac{(\rho c)^{c+1}}{(c-1)! \sum_{n=0}^{c} \frac{(\rho c)^n}{n!} [(c-n)^2 - n]} \tag{11-109}$$

or simply
$$L_q = \frac{\rho(c\rho)^c}{c!(1-\rho)^2} p_0 \tag{11-110}$$

$$P(>0) = \frac{(c\rho)^c}{c!(1-\rho)} p_0 \tag{11-111}$$

$$P(>\tau) = \exp[-c\mu\tau(1-\rho)]P(>0) \tag{11-112}$$

The expected waiting time (not including the waiting time in service) is

$$W = -\int_0^\infty \tau\, dP(>\tau) = \frac{L_q}{\lambda} \tag{11-113}$$

It may also be obtained by the following plausibility argument:

Since the system is in equilibrium, on an average, the line must move up one for each new arrival and hence the total time waited by a customer starting at the end of the line must be the average time $1/\lambda$ between arrivals multiplied by L_q.

The expected number waiting of those delayed is given by

$$\frac{1}{1-\rho} \tag{11-114}$$

The expected waiting time of those delayed is

$$\frac{W}{P(>0)} \tag{11-115}$$

The probability $P_n(t)$ that exactly n channels are busy, given that all c channels are free at $t = 0$, is given by [14]

$$P_n'(t) = -(\lambda + n\mu)P_n(t) + \lambda P_{n-1}(t) + (n+1)\mu P_{n+1}(t) \qquad 0 < n < c \tag{11-116}$$

For $n = 0$ the second term on the right drops out. For $n = c$ the last term on the right drops out, and the coefficient of $P_c(t)$ is $-c\mu$.

The solution of the steady-state equation resulting from the above is due to Erlang and is given by

$$p_n = \frac{\rho^n/n!}{1 + \rho/1! + \rho^2/2! + \cdots + \rho^c/c!} \tag{11-117}$$

with
$$\sum_{n=0}^{c} p_n = 1 \quad \text{and} \quad \rho = \frac{\lambda}{\mu}$$

Identical Constant Holding Times [13]. This case has Poisson input, constant holding time $1/\mu$, and $c > 1$ channels. Crommelin gives

$$P(>0) = 1 - \exp\left[-\sum_{i=1}^{\infty} \frac{e^{-i\rho c}}{i} \sum_{j=ic}^{\infty} \frac{(i\rho c)^j}{j!}\right] \quad (11\text{-}118)$$

$$W = \sum_{i=1}^{\infty} e^{-i\rho c}\left[\sum_{j=ic}^{\infty} \frac{(i\rho c)^j}{j!} - \frac{c}{\rho c}\sum_{j=ic+1}^{\infty} \frac{(i\rho c)^j}{j!}\right] \quad (11\text{-}119)$$

As an approximation to the above, Pollaczek gives for large c

$$P(>0) = 1 - P(=0) = \frac{1}{1-\rho}\frac{\rho^c e^{(1-\rho)c}}{\sqrt{2\pi c}} \quad (11\text{-}120)$$

An approximation due to Molina [36] is

$$W = P(>0)\frac{1}{\mu c(1-\rho)}\frac{c}{c+1}\frac{1-\rho^{c-1}}{1-\rho^c} \quad (11\text{-}121)$$

where $\displaystyle P(>0) = \frac{(\lambda^c e^{-\lambda}/c!)c/(c-\lambda)}{1 - e^{-\lambda}\sum_{n=c}^{\infty} \lambda^n/n! + (\lambda^c e^{-\lambda}/c!)c/(c-\lambda)} \quad (11\text{-}122)$

It is assumed here that $\mu = 1$. To get a result involving μ, replace λ by λ/μ.

Priority Discipline, Different Poisson Inputs, Finite Number of Priorities with the Same Exponential Holding Time (Nonpreemptive) [11]. This case has Poisson input for the kth priority with arrival rate λ_k, exponential holding time all having the same rate μ, and $c > 1$ channels:

$$W_k = \frac{\pi/c\mu}{\left(1 - 1/c\mu \sum_{i=1}^{k-1}\lambda_i\right)\left(1 - 1/c\mu \sum_{i=1}^{k}\lambda_i\right)} \quad (11\text{-}123)$$

where

$$\pi = \frac{(\lambda/\mu)^c}{c!\left(1 - \dfrac{\lambda}{c\mu}\right)\left[\displaystyle\sum_{j=0}^{c-1}\frac{(\lambda/\mu)^j}{j!} + \sum_{j=c}^{\infty}\frac{(\lambda/\mu)^j}{c!c^{j-c}}\right]} \quad \lambda = \sum_{i=1}^{n}\lambda_i \quad (11\text{-}124)$$

Limited Source, Exponential Holding Time [18]. This is a case of servicing m machines, with the same assumptions as in the exponential-holding-time single-channel case. The steady-state equations are

$$m\lambda p_0 = \mu p_1$$
$$[(m-n)\lambda + n\mu]p_n = (m-n+1)\lambda p_{n-1} + (n+1)\mu p_{n+1} \qquad 1 \le n < c \quad (11\text{-}125)$$
$$[(m-n)\lambda + c\mu]p_n = (m-n+1)\lambda p_{n-1} + c\mu p_{n+1} \qquad c \le n \le m$$

with the solution given by[1]

$$p_n = p_0(c\rho)^n \binom{m}{n} \qquad 0 \le n < c$$
$$p_n = p_0(c\rho)^n \binom{m}{n} \frac{n!}{c!\,c^{n-c}} \qquad c \le n \le m \quad (11\text{-}126)$$
$$p_0 = 1 - \sum_{n=1}^{m} p_n = \left[\sum_{n=0}^{m} \frac{p_n}{p_0}\right]^{-1}$$

$$P(>0) = \sum_{n=c}^{m} p_n \qquad (11\text{-}127)$$

$$W = \frac{1}{\lambda} \sum_{c+1}^{m} (n-c) p_n \qquad (11\text{-}128)$$

The expected waiting time of those delayed is given by

$$\frac{W}{P(>0)} \qquad (11\text{-}129)$$

Infinite Number of Channels $c = \infty$ [18]

Exponential Holding Time. For the case of Poisson input, exponential holding time, and $c = \infty$ number of channels, the transient equations are

$$\begin{aligned} P_0'(t) &= -\lambda P_0(t) + \mu P_1(t) \\ P_n'(t) &= -(\lambda + n\mu)P_n(t) + \lambda P_{n-1}(t) + (n+1)\mu P_{n+1}(t) \qquad n \ge 1 \end{aligned} \quad (11\text{-}130)$$

---

[1] The λ used in (11-126) in this and in the following section (limited-source case) and also in the definition of ρ is different from the λ of the notation section. Here it is a characteristic of the bids of a particular source, giving the constant probability density of a bid arriving at the service facility during the time the source is not being served or waiting for service. On the other hand, λ itself is defined as the equilibrium arrival rate from all sources which actually obtains when the system of sources and the servers interact. With this reinterpretation of the ρ in (11-126), λ itself can be determined from the notation section, using the formula

$$\lambda = (c\rho)\mu = \mu \left(\sum_{n=0}^{c-1} n p_n + c \sum_{n=c}^{m} p_n \right)$$

The solution of the above if the initial state is E_i is given by

$$P_n(t) = \frac{\exp[-\lambda/\mu(1 - e^{-\mu t})]}{n!} \sum_{k=0}^{n} \binom{n}{k} \frac{i!}{(i-k)!} \left(\frac{\lambda}{\mu}\right)^{n-k} (1 - e^{-\mu t})^{n+i-2k} e^{-\mu t k}$$

for $n < i$. This is obtained by differentiating

$$P = \exp\left[\left(-\frac{\lambda}{\mu}\right)(1-s)(1-e^{-\mu t})\right][1-(1-s)e^{-\mu t}]^i \quad (11\text{-}131)$$

with respect to s, n times, dividing by $n!$ and setting $s = 0$. $P_n(t)$ for $n > i$ may be similarly obtained.

The steady-state equations are

$$\begin{aligned} \lambda p_0 &= \mu p_1 \\ (\lambda + n\mu)p_n &= \lambda p_{n-1} + (n+1)\mu p_{n+1} \end{aligned} \quad n \geq 1 \quad (11\text{-}132)$$

The solution of this equation is given by

$$p_n = e^{-\lambda/\mu} \frac{(\lambda/\mu)^n}{n!} \quad n \geq 0 \quad (11\text{-}133)$$

The expected number waiting in the transient state if the initial state is E_i is

$$L_{\infty,t} = \frac{\lambda}{\mu}(1 - e^{-\mu t}) + ie^{-\mu t} \quad (11\text{-}134)$$

The expected number waiting in the steady state is

$$L_\infty = \frac{\lambda}{\mu} \quad (11\text{-}135)$$

Limited Source, Exponential Holding Time [18]. In this case there is an input with a source having a finite number m of customers, exponential service time with the average service time proportional to the number waiting, and a single channel (servicing of m welders from one power circuit). If at time t a welder uses current, the probability that he ceases to use it at time $t + dt$ is $\mu \, dt + o(dt)$; if at time t he requires no current, the probability that he calls for current before $t + dt$ is $\lambda \, dt + o(dt)$. The welders work independently. Here, $P_n(t)$ is the probability that n welders are using current at time t.

The transient equations are

$$\begin{aligned} P_0'(t) &= -m\lambda P_0(t) + \mu P_1(t) \\ P_n'(t) &= -[n\mu + (m-n)\lambda]P_n(t) + (n+1)\mu P_{n+1}(t) \\ &\quad + (m-n+1)\lambda P_{n-1}(t) \quad 1 \leq n \leq m-1 \\ P_m'(t) &= -m\mu P_m(t) + \lambda P_{m-1}(t) \end{aligned} \quad (11\text{-}136)$$

The generating function of this system is given by

$$P(s,t) = \sum_{n=0}^{m} P_n(t)s^n \quad (11\text{-}137)$$

and satisfies the partial differential equation

$$(\mu + \lambda s)\frac{\partial P}{\partial s} = m\lambda P \quad (11\text{-}138)$$

with solution

$$P = \left(\frac{\mu + \lambda s}{\lambda + \mu}\right)^m \quad (11\text{-}139)$$

To obtain $P_n(t)$, differentiate P, n times with respect to s, divide the result by $n!$, and then set $s = 0$.

The solution of the steady-state equations obtained from this is given by

$$p_n = \binom{m}{n}\left(\frac{\lambda}{\lambda + \mu}\right)^n \left(\frac{\mu}{\lambda + \mu}\right)^{m-n} \quad (11\text{-}140)$$

11-10. Activities to Which Queueing Theory Has Been Applied

In commerce today, where queueing theory has been applied, many congestion problems have been reduced to a manageable scale. Among the recent applications of queueing theory of particular interest, in addition to the work of telephony, is the work done by Edie [64] of the New York Port Authority. He analyzed traffic delays at toll booths at Port Authority tunnels and bridges. The result of his study was the recommendation of an optimum number and a schedule for the toll collectors and the number of toll booths required at any time of day. Use of his method permitted savings in toll-collection expenses and better services. Brigham [56], of Boeing Airplane Company, obtained the optimum number of clerks to be assigned to tool crib counters in use in the Boeing factory area. The cribs stored a variety of tools required by mechanics in shops and assembly lines. The problem resulted from complaints of foremen who felt their mechanics were waiting too long in line. This would have led the company to consider assigning more clerks. However, management was under pressure to reduce overhead and assign fewer clerks. Queueing analysis helped solve this problem.

Several important industrial applications have been made to problems of machine breakdown and repair. A company has to assign mechanics to repair machines so that production losses caused by the unavailability of machines are kept to a minimum. The machines essentially form a queue waiting for repairs from the mechanics who service them. There is a point at which the salaries of the mechanics assigned to be ready in case of machine failure become greater than the potential production loss. Here again, one seeks an optimum number of mechanics by use of queueing theory.

In the many processes which are worthy of study by operations analysts, a surprising number are very well described by a queueing model. The theory would be valuable in obtaining information similar to that discussed in the car-wash service and in the measures of effectiveness section.

Telephone Conversations [13, 14, 20, 36, 45, 46]

In the case of a telephone trunking system, calls are initiated by individuals. The frequency of initiation or attempted initiation of calls to a given trunk line may be characterized by a frequency distribution (input distribution). Calls are not identical in length, and so a distribution is again needed to characterize the length of a call or its "holding time." There is a maximum number of calls that may be handled at one time, and there is a fixed number of channels. Because of statistical fluctuation in input and holding times, at times all channels will be occupied and a queue will form. The service system used has been that of random selection for service, though a first-come, first-served system might also be applicable in some situations.

The Landing of Aircraft [9, 41]

Planes approaching an airport can land on prescribed runways if there is more than a single runway. If these runways are being used by other aircraft, the planes are stacked over the airport at prescribed altitudes until runways become free. A stacked plane is in a queue. Although most flights have definite arrival times, the variation in flight time caused by weather, etc., the arrival of unscheduled flights, flights transferred to one field from another closed by weather, and planes in trouble leads to their arrival early or late by independent random time errors.

The Loading and Unloading of Ships [16]

Ships arrive at a port by some input distribution, which has been assumed Poisson in some studies. It is not clear whether data have justified this assumption. The time required to load or unload a ship depends on its size, cargo, labor disputes, equipment available, etc. Surprisingly, this has been assumed exponential. Servicing facilities may consist of a single or several docks (single channel or multiple channel).

The Scheduling of Patients in Clinics [2, 3, 35]

Patient arrival may be random in the time during which the clinic is open. The holding time required to treat each patient varies from one patient to the next, with most being treated in a short time and a few for a very long time, and hence an exponential servicing time is typical. There may be only one doctor, or a number of doctors, any of whom can see the patient. Patient arrivals may be scheduled by appointment, to keep the queue down to a minimum. In this case the input is regular.

Customers and Taxis at a Stand

Kendall [28] gives this example to illustrate that expected values alone can be misleading when used as measures of effectiveness in queueing situations in which supply and demand balance. Suppose, on the one hand, that single customer arrivals at a taxi stand can be described by a Poisson distribution, and that of taxi arrivals, on the other hand, by another Poisson distribution. Let α and β be the respective expected number of arrivals per unit time. Starting from zero, the size of the queue at time t will be $q = u - v$, where u and v are Poisson variables with means αt and βt, respectively. If $q < 0$, then there is a line of taxis, and if $q > 0$, there is a line of customers. If $\alpha = \beta$, then the expected value of q is zero, whereas its variance is $2\alpha t$, which becomes infinite with time, so that the expected value alone does not provide adequate information about the system.

Impatient Customers at a Store (Random Selection for Service) [5]

Customers arrive at a store at random. Some become impatient and leave the line. The problem is to design the service facility to keep the waiting time sufficiently low to forestall leaving by a high percentage of the impatient customers. The holding time is exponential, for most customers are taken care of rapidly, and a few with long orders take a long time. The operation is single- or multiple-channel, depending on the number of clerks available.

Radio Communications

Messages to be transmitted arrive at random. Holding time, which is the time required to read the message, is exponential. There may be only a single frequency for transmission or several. Priorities are often assigned to messages to speed up the processing of urgent ones.

Railroad Classification Yards [12]

In a hump yard, uncoupled cars are pushed over the top of the hump and released to roll down to a track corresponding to the proper destination of the car. Cars accumulate on classification tracks until a train consisting of the cars on one or more tracks is to be made up.

A small number of switch engines is used to push cars over the hump and to do other jobs around the classification yard, which interrupt the operation of pushing cars over the hump. Thus the cars are in a queue to be serviced by a finite number of channels—the switch engines. The Monte Carlo method has been used in computing average delay times of cars.

Flow in Production

Items in a production line are in a queue. They may arrive at some stage of the line at a constant rate. The holding time is also constant. An item may have a number of parallel channels to go through or only a single channel. Operations may be carried on in parallel or in several channels, or a single channel may consist of a series of operations.

Other Applications

In passing, we mention other activities to which the theory is applicable or has been applied. Some of these are the passage of people through customs; machine breakdown and operation [1, 7, 18, 19, 23, 26, 27, 34, 51]; the timing of traffic lights; the movement of aircraft along taxiways; car washing; restaurant service; queueing at automobile parks, supermarkets, and bakeries; and theater tickets.

11-11. How to Generate the Needed Information; The Use of the Monte Carlo Method

In collecting data for the estimation of the distribution of the number of arrivals or of holding times, the number of arrivals and the intervals of time in service are recorded as a function of time. These data may then be grouped in time intervals, after which statistical tests for independence of the means in successive time intervals may be made, e.g., by the variance-ratio test and other appropriate methods found in standard texts on statistics. Should there prove to be lack of independence, the data must be subdivided into time intervals within which a steady state is a reasonable approximation. A nonsteady-state situation arises, for example, in a restaurant operation before and after the meal hours.

With the data, or portion of the data which appears to be homogeneous, a further grouping is made and a frequency histogram constructed taking 10 to 20 subgroups. The appearance of this histogram will suggest certain standard distributions such as the normal, Poisson exponential, or Erlangian.

The parameters of the hypothesized distribution are then estimated from the data. The observed and model distributions are then compared using the χ^2 test. When slight but statistically significant deviations are found, a decision as to whether or not the model should be altered can be made only with a knowledge of the objectives of the particular problem.

Another consideration must be borne in mind when the data are not very numerous. Parameters estimated from such data may vary quite markedly from the time population values, even though a χ^2 test shows no significant deviation from the model.

The considerations of the preceding two paragraphs are less important if the model can be tested in the actual operations. Ingenuity may sometimes be required to verify the model experimentally, such as closing one lane of a multilane highway in order to verify the applicability of the model and subsequently to predict the effect of opening an additional lane.

In exceptional circumstances where no experimental verification is possible, greater care should be taken in fitting the available data with the model. Unfortunately, no work has appeared discussing the effect upon predicted quantities of departures of the model from reality. The problem may perhaps be approached analytically by varying the parameters of the distribution of arrivals and holding times over a range corresponding to the expected uncertainty, and observing the effects of the changes upon the predictions.

It is worthy of note that the Poisson distribution has the desired properties for many practical problems. There have been some applications in which, for example, a normal distribution gives a better fit for the data. The Erlangian may be used on specializing values for k in fitting the data. Work is in progress on solving queueing problems with this type of distribution.

The Monte Carlo method has been frequently applied to queueing problems to obtain numerical solutions. Application of the method may be illustrated by an example for obtaining a single-channel first-come, first-served waiting-time result.

A plot of the cumulative input and service-time distributions is made either from data or from formulas. Then a column of two-digit random numbers is formed. As decimals they are used as ordinates in the figure for the input distribution, and the corresponding abscissas are then obtained. The latter provide a column of the length of time intervals between arrivals. Similarly, two adjacent columns are included to provide service times. In a fifth column the waiting time for every item is calculated as the sum of the holding time and waiting of the previous item minus the delay in the arrival of the new item. If this difference is negative, then the waiting time is zero.

The average total wait is calculated in a straightforward manner. Note that for the steady-state results $\lambda/\mu < 1$, which must be taken into consideration when using the method. The accuracy of the method improves roughly as $\sqrt{N}$, where N is the number of trials. This will now be illustrated by a numerical example.

In a production-line operation it was calculated that speeding up operation would bring items to the inspection line at the constant rate of one every 5 minutes. The present inspection service times in minutes when grouped and fitted with a curve were observed to be exponentially distributed according to

$$b(t) = .5e^{-.5t}$$

The Monte Carlo method was applied to this problem to calculate the average waiting times for possible increase in service capacity caused by the increase in production rate.

TABLE 11-2

Intervals between consecutive arrivals	Random numbers taken as decimals	The corresponding service time obtained from the figure using the decimals	The waiting time
5	0.10	0.20	0
5	0.22	0.50	0
5	0.24	0.55	0
5	0.42	1.05	0
5	0.37	0.90	0
5	0.77	2.90	0
5	0.99	9.00	0
5	0.96	6.30	4.00
5	0.89	4.50	5.30
5	0.85	3.80	4.80
5	0.28	0.65	3.60
5	0.63	2.00	0
5	0.09	0.20	0
...	...	...	...

Obviously the input is fixed at constant intervals and any experiment must use this information as it is. On the other hand, the service times for each item must be selected (at random) without any discrimination; i.e., one service time is as likely as any other. The cumulative service time distribution was plotted and the columns in the table were generated as described above. This was carried out for 100 arrivals and the average waiting time was calculated to be .25 minute, which did not warrant additional inspection service to maintain smooth operation.

11-12. A Method of Solution

As an illustration, we give here a method of solving the infinite-channel unlimited-input problem. Using the generating function and having multiplied by s^n and summed over n, one has the following linear first-order partial differential equation in the generating function:

$$\frac{\partial P}{\partial t} = (1-s)\left(-\lambda P + \mu \frac{\partial P}{\partial s}\right)$$

or, on putting it in standard form, one has

$$\frac{\partial P}{\partial t} - (1-s)\mu \frac{\partial P}{\partial s} = -\lambda(1-s)P$$

The solution is obtained in the usual manner by forming

$$\frac{dt}{1} = \frac{ds}{-(1-s)\mu} = \frac{dP}{-\lambda(1-s)P}$$

where the coefficient under dt is that of $\frac{\partial P}{\partial t}$, under ds is that of $\frac{\partial P}{\partial s}$, and under dP is the right side of the equation.

We use the first equation to obtain a one-parameter family of surfaces $u(s,t,P) = c_1$ and the second equation for another, $v(s,t,P) = c_2$. The general solution is an arbitrary function f of u and v such that $f(u,v) = 0$. One can then solve for v and obtain it as a function of u, that is, $v = g(u)$. The form of g can then be determined from the given initial conditions.

For our example

$$\frac{dt}{1} = \frac{ds}{-(1-s)\mu}$$

yields

$$u(s,t,P) \equiv (1-s)e^{-\mu t} = c_1$$

and

$$\frac{ds}{-(1-s)\mu} = \frac{dP}{-\lambda(1-s)P}$$

yields

$$v \equiv Pe^{-(\lambda/\mu)s} = c_2$$

and hence the general solution is given by

$$P = e^{(\lambda/\mu)s}g[(1-s)e^{-\mu t}]$$

Now when $t = 0$, $P = s^i$, since all the $P_n(0)$ are zero, except that $P_i(0) = 1$ due to the fact that the initial state is E_i.

This gives

$$s^i = e^{(\lambda/\mu)s}g(1-s)$$

Let $y = 1 - s$; then

$$g(y) = e^{-(\lambda/\mu)(1-y)}(1-y)^i$$

But for arbitrary t, the argument of g is $(1-s)e^{-\mu t}$, and in that case we must replace y in the right side of the above equation by this argument to obtain $g[(1 - s)e^{-\mu t}]$.

When this is substituted in the general solution, we obtain the desired expression for P given in the results section.

REFERENCES

(For additional references, see J. F. McCloskey and J. M. Coppinger (eds.), "Operations Research for Management," vol. II, Johns Hopkins University Press, Baltimore, 1956.)

1. Ashcroft, H.: The Productivity of Several Machines under the Care of One Operator, *Journal of the Royal Statistical Society, Series B*, vol. 12, pp. 145–155, 1950.
2. Bailey, N. T. J.: A Study of Queues and Appointment Systems in Hospital Outpatient Departments, With Special Reference to Waiting-Times, *Journal of the Royal Statistical Society, Series B*, vol. 14, p. 185, 1952.
3. Bailey, N. T. J.: Queueing for Medical Care, *Applied Statistics*, vol. 3, p. 137, 1954.
4. Bailey, N. T. J.: On Queueing Processes with Bulk Service, *Journal of the Royal Statistical Society, Series B*, vol. 16, p. 80, 1954; *ibid.*, p. 288.
5. Barrer, D. Y.: A Waiting Line Problem Characterized by Impatient Customers and Indifferent Clerks, *Journal of the Operations Research Society of America*, vol. 3, p. 360, 1955.
6. Barry, J. Y.: A Priority Queueing Problem, *Operations Research*, vol. 4, p. 385, June, 1956.
7. Benson, F., and D. R. Cox: The Productivity of Machines Requiring Attention at Random Intervals, *Journal of the Royal Statistical Society, Series B*, vol. 13, p. 65, 1951.
8. Borel, E.: "Sur l'emploi du théorème de Bernoulli pour faciliter le calcul d'une infinité de coefficients. Application au problème de l'attente a un guichet," *Comptes rendus de l'academie des sciences de Paris*, vol. 214, p. 452, 1942.
9. Bowen, E. G., and T. Pearcy: Delays in Air Traffic Control, *Journal of the Royal Aeronautical Society*, pp. 251–258, 1948.
10. Camp, G. D.: Bounding the Solutions of Practical Queueing Problems by Analytic Methods, Johns Hopkins Informal Seminar in Operations Research, Paper 22, 1955; also J. F. McCloskey and J. M. Coppinger (eds.), "Operations Research for Management," vol. II, Johns Hopkins Press, Baltimore, 1956.
11. Cobham, A.: Priority Assignment in Waiting Line Problems, *Journal of the Operations Research Society of America*, vol. 2, pp. 70–76, 1954; *ibid.*, vol. 3, p. 547, 1955.
12. Crane, R. R., F. B. Brown, and R. O. Blanchard: An Analysis of Railroad Classification Yard, *Journal of the Operations Research Society of America*, vol. 2, p. 262, 1955.
13a. Crommelin, C. D.: Delay Probability Formulae When the Holding Times Are Constant, *Post Office Electrical Engineers Journal*, vol. 25, p. 51, 1932.
13b. Crommelin, C. D.: Delay Probability Formulae, *Post Office Electrical Engineers Journal*, vol. 26, p. 266, 1934.
14. Brockmeyer, E., H. L. Halstrom, and Arne Jensen: The Life and Works of A. K. Erlang, *Transactions of the Danish Academy of Technical Sciences*, vol. 2, 1948.
15. Everett, J. L.: State Probabilities in Congestion Problems Characterized by Con-

stant Holding Times, *Journal of the Operations Research Society of America*, vol. 1, p. 279, 1953.
16. Everett, J. L.: Seaport Operations as a Stochastic Process, *Journal of the Operations Research Society of America*, vol. 1, p. 76, 1953. Abstract.
17. Fagen, R. E., and J. Riordan: Queueing Systems for Single and Multiple Operation, *Journal of the Society for Industrial and Applied Mathematics*, vol. 3, p. 73, 1955.
18. Feller, W.: "An Introduction to Probability Theory and Its Applications," John Wiley & Sons, Inc., New York, 1950.
19. Field, J. W.: Machine Utilization and Economical Assignment, *Factory Management and Maintenance*, vol. 104, p. 288, 1946.
20. Fry, T. C.: "Probability and Its Engineering Uses," D. Van Nostrand Company, Inc., Princeton, N.J., 1928.
21. Garber, H. N.: A Class of Queueing Problems, Ph.D. Thesis, Massachusetts Institute of Technology, Cambridge, Mass., 1956.
22. Gaver, D. P.: The Influence of Servicing Times in Queuing Processes, *Journal of the Operations Research Society of America*, vol. 2, p. 139, 1954.
23. Gross, M.: Leistungsvorrechnung bei Mehrmaschinenbedienung, *Textil-Praxis*, vol. 4, p. 113, 1949.
24. Raymond, Haller, and Brown, Inc.: "Queueing Theory Applied to Military Communication Systems," State College, Pa., 1956.
25. Holley, J. L.: Waiting Line Subject to Priorities, *Journal of the Operations Research Society of America*, vol. 2, p. 341, 1954.
26. Jackson, R. R. P.: Queueing Systems with Phase Type Service, *Operational Research Quarterly*, vol. 5, p. 109, 1954.
27. Jones, D.: A Simple Way to Figure Machine Down Time, *Factory Management and Maintenance*, vol. 104, p. 118, 1948.
28. Kendall, D. G.: Some Problems in the Theory of Queues, *Journal of the Royal Statistical Society, Series B*, vol. 13, no. 2, 1951.
29. Kendall, D. G.: Stochastic Processes Occurring in the Theory of Queues, *Annals of Mathematical Statistics*, vol. 24, pp. 338–354, 1953.
30. Khintchine, A.: Studied Poisson-input arbitrary-holding-time, single-channel problems, *Matematicheskiĭ Sbornik*, vol. 39, p. 73, 1932.
31. Kiefer, J., and J. Wolfowitz: On the Theory of Queues with Many Servers, *Transactions of the American Mathematical Society*, January, 1955.
32. Kiefer, J., and J. Wolfowitz: The General Queueing Process, *Annals of Mathematical Statistics*, vol. 27, no. 1, 1956.
33. Kolmogorov, A.: Sur le problème d'attente, *Matematicheskiĭ Sbornik*, vol. 38, p. 101, 1931.
34. Kronig, R.: On Time Losses in Machinery Undergoing Interruptions, *Physica*, vol. 10, p. 215, 1943.
35. Lindley, D. V.: The Theory of Queues with a Single Server, *Proceedings of the Cambridge Philosophical Society*, vol. 48, part 2, 1952.
36. Molina, E. C.: Application of the Theory of Probability to Telephone-trunking Problems, *Bell System Technical Journal*, vol. 6, p. 461, 1927.
37. Morse, P. M.: Stochastic Properties of Waiting Lines, *Journal of the Operations Research Society of America*, vol. 3, p. 255, 1955.
38. Morse, P. M., H. N. Garber, and M. L. Ernst: *Journal of the Operations Research Society of America*, vol. 2, p. 444, 1954.
39. O'Brien, G. G.: The Solution of Some Queueing Problems, *Journal of the Society for Industrial and Applied Mathematics*, vol. 2, no. 3, 1954.

40. Palm, C.: The Distribution of Labor for the Service of Automatic Machines (in Swedish), *Industritidningen Norden*, vol. 75, pp. 75–123, 1947.
41. Pearcey, T.: Delays in Landing of Air Traffic, *Journal of the Royal Aeronautical Society*, pp. 799–812, 1948.
42. Pollaczek, F.: Studied the Poisson-input, arbitrary-holding-time, single- and multiple-channel problem, 1930, 1934; for reference, see Kendall's 1951 paper.
43. Pollaczek, F.: La loi d'attente des appels téléphoniques, *Comptes rendus de l'academie des sciences de Paris*, vol. 222, p. 353, 1946.
44. Reich, E.: "Birth-Death Process and Tandem Queues," p. 863, The RAND Corporation, Santa Monica, Calif., 1956.
45. Riordan, J.: Delay Curves for Calls Served at Random, *Bell System Technical Journal*, vol. 32, no. 1, 1953.
46. Riordan, J.: Telephone Traffic Time-Averages, *Bell System Technical Journal*, vol. 30, pp. 1129–1144, 1951.
47. Smith, W. L.: On the Distribution of Queueing Times, *Proceedings of the Cambridge Philosophical Society*, vol. 49, 1953; studied the general independent- (the intervals between arrivals are statistically independent random variables with the same arbitrary distribution) input, arbitrary-holding-time, and single-channel case.
48. Takacs, L.: Investigation of Waiting-time Problems by Reduction to Markov Processes, *Acta Math. Hung.*, vol. 6, 1956.
49. Taylor, J., and R. R. P. Jackson: An Application of the Birth and Death Process to the Provision of Spare Machines, *Operational Research Quarterly*, vol. 5, p. 95, 1954.
50. "A Collection of Delay Probability Formulas and Curves," Port of New York Authority, Operations Standards Division, March, 1956.
51. Weir, W. F.: Figuring Most Economical Machine Assignment, *Factory Management and Maintenance*, vol. 102, p. 100, 1944.
52. Phipps, T. E., Jr.: Machine Repair as a Priority Waiting Line Problem, *Operations Research*, vol. 4, pp. 76–86, 1956.
53. Koenigsberg, E.: Queueing with Special Service, *Operations Research*, vol. 4, pp. 213–220, 1956.
54. Koopman, B. O.: New Mathematical Methods in Operations Research, *Journal of the Operations Research Society of America*, vol. 1, pp. 3–9, 1952.
55. Homma, T.: On a Certain Queuing Process, *Reports of Statistical Application Research, Union of Japanese Scientists and Engineers*, vol. 4, no. 1, 1955.
56. Brigham, G.: On a Congestion Problem in an Aircraft Factory, *Journal of the Operations Research Society of America*, vol. 3, pp. 412–428, 1955.
57. Clarke, A. B.: The Time-dependent Waiting Line Problem, *University of Michigan Engineering Research Institute Report* M720-1 R 39, 1953.
58. Clarke, A. B.: A Waiting Line Process of Markov Type, *Annals of Mathematical Statistics*, vol. 27, no. 2, June, 1956.
59. Ledermann, W., and G. E. Reuter, Spectral Theory for the Differential Equations of Simple Birth and Death Process, *Philosophical Transactions of the Royal Society of London, Series A*, vol. 246, 1954.
60. Luchak, George: The Solution of the Single-channel Queuing Equations Characterized by a Time-dependent Poisson-distributed Arrival Rate and a General Class of Holding Times, *Operations Research*, vol. 4, pp. 711–732, 1956.
61. Burke, P. J.: The Output of a Queuing System, *Operations Research*, vol. 4, pp. 699–704, 1956.
62. Kasten, H., and J. Th. Runnenburg: Priority in Waiting-line Problems, *Mathematisch Centrum*, Amsterdam, Holland, December, 1956.

63. Cox, R. E.: Traffic Flow in an Exponential Delay System with Priority Categories, *Proceedings of the Institute of Electrical Engineers*, November, 1955.
64. Edie, L. C.: Traffic Delays at Toll Booths, *Journal of the Operations Research Society of America*, vol. 2, pp. 107–138, 1954.
65. Downton, F.: On Limiting Distributions Arising in Bulk Service Queues, *Journal of the Royal Statistical Society, Series B*, vol. 18, pp. 265–274, 1956.
66. Jackson, R. R. P., and D. G. Nickols: Some Equilibrium Results for the Queueing Process $E_k/M/1$, *Journal of the Royal Statistical Society, Series B*, vol. 18, pp. 275–279, 1956.
67. Naor, P.: On Machine Interference, *Journal of the Royal Statistical Society, Series B*, vol. 18, pp. 280–287, 1956.
68. Homma, T.: On the Many Server Queuing Process with a Particular Type of Queue Discipline, *Reports of Statistical Application Research, Union of Japanese Scientists and Engineers*, vol. 4, pp. 20–31, 1956.
69. Saaty, T. L.: Résumé of Useful Formulas in Queuing Theory, *Operations Research*, vol. 5, pp. 162–187, April, 1957.
70. Morse, P. M.: "Queues, Inventories and Maintenance," John Wiley & Sons, New York, 1958.

SOME PROBABILITY AND STATISTICS PROBLEMS

1. A group of riflemen who are defending a position fire simultaneously on N riflesquad attackers. The defenders act independently and select their attackers at random. An attacker is assumed dead when hit. The attackers do not return the fire; thus the number of defenders remains constant. The process is repeated in a sequence of volleys. Show that $P_{n_2 n_1}{}^k$, the probability that there are n_1 attackers surviving the $(k-1)$st volley and n_2 surviving the kth volley fired, is given by

$$p_{n_2 n_1}{}^k = \left[\frac{1}{\binom{n_1}{n_2}}\right] \sum_{i=n_2}^{n_1} (-1)^{i-n_2} \binom{n_1 - n_2}{i - n_2} \left(\frac{1 - ip_k}{n_1}\right)^S$$

where p_k is the hit probability of each shot fired on the kth volley and S is the total number of shots fired on each volley (assumed constant on all volleys).

2. Find the probability that a tank will be knocked out by either the first or second shot from an antitank gun, assumed to possess a constant kill probability.

3. Given n men who are ranked $1, \ldots, n$ according to ability, and given n tasks ordered according to difficulty from 1 to n in descending order of difficulty. Each man is best suited to perform the task corresponding to his rank. However, each man assumes that one task is equal in difficulty to any other. They successively select the tasks by drawing from a bag numbers corresponding to the tasks. What is the probability that at least one man selects a task whose order is equal to the man's rank?

4. What is the probability of extracting two white balls simultaneously from a box with a white and b black balls?

5. Each of n urns contains a white and b black balls. One ball is transferred from the first urn into the second, then one ball from the latter into the third, and so on. Finally, one ball is taken from the last urn. What is the probability of its being white? What is this probability if it is known that the first ball transferred was white?

6. Two players, each possessing $2.00, agree to play a series of games. The probability of winning a single game is 1/2 for both, and the loser pays $1.00 to his adversary after each game. Find the probability for each of them to be ruined at or before the nth game? Also solve the problem if each one starts with $3.00.

7. Assume that a hypothetical guided missile has an electronic guidance system made up of two distinct parts, A and B, *both* of which must function successfully in order that the missile itself perform satisfactorily. In the laboratory, tests simulating actual operating conditions are performed on large numbers of these parts separately, leading to the conclusion that part A by itself performs successfully 90 per cent of the time and part B by itself 80 per cent of the time. Thus the tests indicate that the probabilities of success for these two parts are, respectively,

$$P(A) = .9 \quad P(B) = .8$$

a. What are the respective probabilities of failure of the two parts?

b. Assume that there exist no causes for failure of the complete missile other than the failure of either A or B or both and that the probabilities of failure of A and B are independent of each other and unaffected by the process of assembly inside the missile (i.e., same values as in the laboratory tests). Under these assumptions, what is the probability of failure of the complete missile?

c. In actual field tests of the complete missile, it is observed that the missile fails 25 per cent of the time. Assuming that the values of $P(A)$ and $P(B)$ are .9 and .8, as before, what are the implications of this field-test result on the assumption of the "independence" of failure of A and B? Compute $P(A/B)$ and $P(B/A)$.

8. Suppose that X and Y are independently distributed according to

$$f(x) = 12x^2(1 - x) \quad 0 \leq x \leq 1$$
$$g(y) = 2y \quad 0 \leq y \leq 1$$

Find the expected value of $y/x^2 + x/y$ and the distribution of y/x.

9. Assuming known that an enemy is in one of two regions, A or B, their areas, the rates (possibly different) at which they can be searched by one search unit, the available number of searching units, and the probability that the enemy is in, say, region A, discuss the optimum distribution of searching effort.

10. Consider two forces, I and II. Assume that whenever there are m and n units on the two sides, respectively, the probability that the next casualty is on the second side is $m/(m + n)$. Let $P(m,n)$ denote the probability that the first side wins, if initially there are m and n units in I and II, respectively.

a. Calculate $P(4,3)$ directly from the equation

$$P(m,n) = \frac{m}{m+n} P(m, n - 1) + \frac{n}{m+n} P(m - 1, n)$$

b. Calculate $P(4,3)$ approximately, using the approximation involving the normal integral.

c. Calculate $P(400, 300)$ approximately.

d. Discuss the difference in the answers to parts *b* and *c*.

11. A shot is fired from each of three guns at a single target, the probability of a hit being .1, .2, and .3, respectively. Find the probability of the possible numbers of hits.

12. Three machines in a plant produce light bulbs as follows:

Machine	% of total production	% production of each machine as defectives
A	20	3
B	55	5
C	25	4

What is the probability that a bulb taken at random and found to be defective is produced by machine B?

13. Rainfall in inches in corresponding months of two seasons I and II was as follows:

I	27	29	35	32	37	33
II	30	23	20	25	29	35

By testing hypotheses, determine whether the two means are equal. Do the same for the variances. Find a 95 per cent confidence interval for the difference in the mean falls and a corresponding one for the ratio of the variance of the first season's fall to that of the second.

14. Samples are taken from a normally distributed population whose variance is 2. Determine the sample size required to test the hypothesis that the population mean is 7 with a 10 per cent risk if the probability is .15 of accepting the hypothesis that the mean is 6.

15. A certain large retailer decides to test his methods of advertising men's suits as follows: He normally advertises by direct mail to his charge customers and in an evening newspaper. To advertise to all charge customers costs $400. He decides to estimate the increase in sales from $400 worth of newspaper advertising and to estimate the effect of the same value of direct mail, using a simple factorial experiment laid out as shown below. The advertisements are run in random order, advertising whatever stock he has, with pricing according to his normal policy.

SALES OF MEN'S SUITS IN THOUSANDS OF DOLLARS

	No newspaper advertising	$400 in newspaper advertising
No direct-mail advertising	19.6	20.3
	19.0	21.1
$400 in direct-mail advertising	22.6	23.2
	22.1	22.7

Point out the good and bad features of the design of this experiment.

Are the sales significantly greater when newspaper advertising is used than when it is not? Are sales significantly greater when direct mail is used than when it is not? HINT: Using the t test, compare the two row means of four individual and then the column means of four individual observations. The error estimates are obtained from the pooled estimate of variance obtained from each of the four pairs of duplicates.

The diagonal means may be compared also in this manner. In this case they are not significantly different. This difference between the diagonals is called an *interaction*. Interpret the meaning of an "interaction" in an experiment of this kind.

16. Consider a queue with Poisson input and two service channels with identical exponential holding times in the steady state. Compare the length of the line with that obtained by doubling the service rate at a one-channel queue with the same input and exponential service time. Interpret the result.

17. Using the generating function method, solve the difference equations of the single channel, Poisson input exponential holding time steady state queue.

18. Solve the partial differential equation arising through the use of the generating function in the infinite number of channels limited input queue. Obtain the mean and the variance by differenting the generating function.

19. The backward equations corresponding to the queue length dependent parameter case are given by

$$P_{i,n}(t) = -(a_i + b_i)P_{i,n}(t) + a_iP_{i+1,n}(t) + b_iP_{i-1,n}(t)$$

obtained by considering the possible transitions from state E_i. Write down and solve the backward equations corresponding to the infinite-channel unlimited-input queue with nonzero initial input. Compare the solution you obtain with that of the forward equations. Do the same for a Poisson process.

20. To set up an absorbing barrier at E_0 one is allowed to go from E_1 to E_0, but not conversely. This alters the equation in $P'_0(t)$. Solve one of the several sets of equations given in queueing theory if an absorbing barrier is established at E_0. Note that the equations are given with a reflecting barrier at E_0.

21. Write down the forward equations of the birth and death process (queue-length-dependent parameters) in vector notation. Do the same for the backward equations. Note the interesting relationship between the two coefficient matrices. Using the vector method of solving differential equations given in Chapter 4, it is possible to give a solution to the problem at least in theory by considering finite sections of the matrix which tend to the given matrix in the limit. By considering absorbing barriers at both ends one has a finite problem. Set up and solve a numerical problem of this kind where the largest value of n is 4.

PART 4

AN ESSAY

CHAPTER 12

SOME THOUGHTS ON CREATIVITY

Any individual can acquire the tools of a trade. As he learns to use these tools he becomes technically competent. It is, however, only when he contributes of himself that the individual becomes more than a technician. This giving of oneself is essentially what we mean by creativity. The process of developing and applying one's originality is seldom adequately emphasized in any technical course.

A few factors play a direct role; others are perhaps less obvious and require some elaboration. Among the essentials for creativity which are considered in this essay are motivation and the need for it and for freedom of individual expression; training in technical methods and for independent, imaginative thought; and the development of creativity through individual analysis of problems.

In the past there were those who held that the educated man was one familiar with Latin and with Greek literature. Those versed in other branches of knowledge were popularly regarded, in the words of Thomas Huxley, as "more or less educated specialists." There are still those who attempt to judge which branch of knowledge is superior. Some say that philosophy is above science; others, that the spirit of the problem solver is inferior to that of the poet. This is a futile argument. There is little doubt, however, that the creative thinkers in any field represent the culmination of achievement in that field, for without them little progress can be made anywhere.

The creative power which gives life its values is basic and necessary. Whether it is a religious expression, an artistic creation, or a scientific discovery, it is, for modern society and modern man, the very element upon which depends the judgment of history. To be civilized or "educated" is not adequate; the creative spirit must also be present. Today it cannot even be regarded as a luxury. It is an integral part of our ability to survive. In the broad sense, all man's creative activities may be regarded as attempts at solving problems. Creativity is that element of the human personality which provides direction. It supplies ideas which are "autocatalytic" by creating the atmosphere for the generation and expansion of other ideas. Creative ideas are "pregnant" ideas.

The ability to create exists in every individual in varying degree and according to inclination. With some effort, these inclinations and innate talents can be discovered and utilized in attacking the problems for which they are best adapted.

The domains where worthwhile effort may be exerted are sufficiently varied to offer each individual an opportunity to exhibit originality. The growth of mental potentialities through experimentation with the creative process broadens the individual's capability of original contribution. These potentials are imperfectly demonstrated in history in the form of a civilization's challenge, growth, and decay. The last stage results from inability to develop the necessary creative expression to meet new challenges. If it were possible to define progress, then creativity might be evaluated in terms of its contribution to the progress of society.

History records man's creativity in the way he approaches a problem. In earlier times, man's forebears, having observed a problem, admonished "thou shalt not" in order to prevent others from encountering it. Some of the great religions began by first setting down laws concerning what people should not do. In modern society, however, considerable thought is devoted to probing for cause before judgment is passed. This more tolerant approach has been gained through the compilation and analysis of cumulative experience. Although there are still "thou shalt nots," there are an increasing number of "thou mayests." This attitude makes possible daring and enterprise.

The techniques previously studied in this book are auxiliary tools. To employ these tools successfully requires not only creative ability but also an environment conducive to the generation and development of new ideas. It is generally understood that severe regimentation stifles creative talent. This, of course, is not to say that there should be no goals, no objectives toward the attainment of which human effort may be applied. Goals and objectives are sometimes necessary for the expression of creativity. When they are not apparent, they are often intrinsically present. However, new fields of knowledge are frequently opened and explored as a result of the revolt of creative individuals against regimentation. One need not labor this point in order to arrive at the conclusion that even in fields of practical endeavor a compromise is required to provide an atmosphere sympathetic to both applied and basic research. Today both government and industry have become acutely aware of such a need and here and there have set out to create an atmosphere conducive to the generation of ideas. That our ability to compete successfully in the world of ideas thus will be enhanced is only a simple corollary of the basic requirements of creativity.

There remains the problem of keeping certain tentative objectives in mind and evolving new ones as required by the times. That this is necessary follows from the fact that our creative potential is not infinite. Although some of it is required for free explorations in the "spirit of the creative mind," a considerable portion of our potential is required to attain objectives related to successful competition. The analogy with the survival of individuals is obvious. One usually turns to leisurely creativity when one has ensured, by means of creativity or otherwise, that the basic needs are looked after. On the other hand, a life devoted only to the satisfaction of immediate needs is a dull one. Such devotion to needs alone would kill the urge toward creativity, and in the end the practical goals also would suffer.

An essential element in education is the stimulation of creativity, a delicate task which requires special attention on the part of a teacher to the scholarly presentation of ideas in the form of lectures on methods. Successful stimulation usually requires adequate motivation. An example of motivation is the encouragement of curiosity for its own sake in modern European culture. This is reinforced by the prestige which accompanies successful scholarship and research. Without motivation there can be little inclination to explore, to discover, to develop. Stimulation, however, requires correct responses. The method of stimulation therefore must be developed independently for each subject.

For example, early mathematical training may be carried out by guiding the child toward abstractions of many concrete instances with which he is familiar. But the final degree of abstraction, together with the intermediate stages, should be carried out by the youth himself. The results need not coincide with the conceptions of the teacher. The latter provides guidance for continued explorations and abstractions. On the other hand, when complete explanations are given, in any avenue of culture, the thrill of discovery is lost. Having acquired, through exploration, habits of independent thinking, the student still requires continuous encouragement and is ready for exposition on facts and on methods.

To develop creativity in an individual it is essential to encourage him from youth in habits of patient and independent thinking. This can best be accomplished in youth. Experience indicates that it is difficult to cultivate in the later stages of education the independence required for creativity; it must already be present. Without this independence the acquisition of facts and details becomes the goal of the mind, and their use is limited to established avenues of expression.

The varied manifestations of creativity, be they ingenious and revolutionary ideas or artistic masterpieces, are not generated spontaneously and without an appropriate background. Suggestions for stimulating creativity may be found by studying the training, methods, and accomplishments of great men. The popular conception that Newton, on being struck by a falling apple, spontaneously discovered the law of gravitation is incorrect. Newton had accumulated a vast store of knowledge in one of the finest institutions of learning of his time, and he had received profound and detailed scientific training. His discoveries were a consequence of much thought and experimentation, and his knowledge of Kepler's results provided a basis for the discovery of the law of gravitation. Simply stated, a creative individual is also most likely to be a trained individual. Training provides exposure to creative ideas as well as the facts and discipline which permit concrete results from creative activity.

Each surviving individual may be assumed to have enough creative potential to meet environmental emergencies. The reliable performance of the complex and fragile human body is evidence that man is equipped with a flexible physical structure. That man's sociological and psychological potentials also have a certain degree of flexibility in responding to changes is evidenced by the survival and success of a large class of individuals who have radically changed their social and intellectual environment, thus exhibiting man's ability to make appreciable adaptations. The ability to adapt is desirable in prescribing solutions for operational problems. There is some doubt that a highly specialized individual can adapt readily.

There is no unique method of formulating and attacking an operational problem. It would, indeed, be naïve to assume that one could give an exhaustive account of problems and methods of solution. Many problems are encountered for which well-defined methods cannot be prescribed. Sometimes it is difficult even to form a coherent framework for the problem. The individual must, frequently by violating the syllogism, make his assumptions, formulate the problem, and find his own solution.

It is not necessary to limit oneself to the field of one's academic specialty in order to tackle a problem creatively. The inclination toward independent thought can easily be stunted by overspecialization. One's talents are frequently so thoroughly absorbed in techniques that little opportunity remains for the broader creative abilities which one might possess. Bernard Koopman [21] aptly points out that "a specialist who works all his life in one field may produce a great succession of papers. But too often it happens that a sameness of thought sets in and every result looks like every other: the morning light is lost. A little of the spirit of the amateur may give much of the spirit of adventure and is . . . closer to the freshness in the air of discovery."

Several studies show the opportunities for productivity at different ages. There is wide agreement that the productivity of an individual's more advanced years is usually based on the expansion and development of ideas conceived in earlier years of research and discovery. Ensuing success frequently results in an individual's undertaking responsibilities of a different sort. To illustrate, R. C. Coile [8], by considering their publications, has examined the ages at which electronic engineers were most creative. A tabulation of the ages of 474 contributors to the *Proceedings of the Institute of Radio Engineers* in 1931, 1941, and 1951 was statistically adjusted to make allowance for the smaller number of engineers at older age levels. The most creative ages were found to be between 30 and 40, as is illustrated in Fig. 12-2. The results were in agreement with a broader study by H. C. Lehman, who found that the two newest fields in electronics, transistors and digital computers, were being developed by people in this 30-year-old age bracket, Lehman [24] also finds, for example, that heightened creativity in most scientific fields and in mathematics, resulting in significant contributions, occur roughly between the

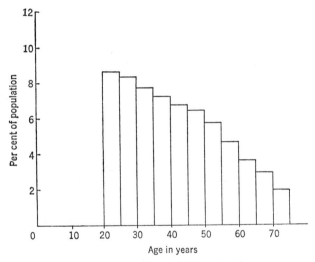

FIG. 12-1. Percentage of population at various ages.

ages of 30 to 36, except perhaps in astronomy, where a long period is required to collect data before publication; in this field the age is 40 to 44. These findings indicate that there are periods of maximum productivity in various fields. Generally, rising pay rewards a scientist for increased output until the top brackets carry him to administrative jobs. Often an individual is delegated with responsibilities which deprive him of the opportunity to be prolific in his specialty. He may, nevertheless, actually be highly creative in a field in which creativity is much more difficult to measure.

An excellent example of creativity in a difficult field is the rules for success derived by Machiavelli, who had little success in putting them into effect himself. However, having derived the rules, he was successful in imparting them to his prince. Through the ages some men, more than others, have been able to bend with circumstances and to utilize changes of plan to good advantage. Machiavelli's "The Prince" is concerned with discovering from history and from contemporary events how principalities are won, how they are held, and how they are lost. The book has an empirical foundation, since it drew upon a multitude of examples from fifteenth-century Italy [25]. The object lesson to be learned from Machiavelli is his statesmanship, which in part is clever gamesmanship. It is futile to pursue a purpose, political or otherwise, by methods that are bound to fail. If the end is held good, one must find adequate moral means for its achievement. Machiavelli indicates that the science of success, which he defined as attaining one's purpose, can be studied as well from the wicked,

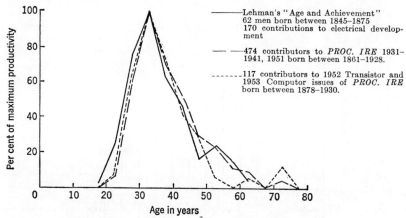

FIG. 12-2. Age and creativeness. (*Courtesy of R. C. Coil.*)

of whom there are more, as it can from the saints, the former having no qualms in utilizing methods not approved by the latter.

Machiavelli's theme has been propounded and extended in recent times. For example, versatility in handling situations is not only written about but also taught by England's Stephen Potter [38]. He points out in his books "Gamesmanship," or the art of winning games without actually cheating, and "Lifemanship," or the art of "getting away with it" without being an absolute "plonk," that it is not always necessary to be an expert to obtain a solution. Thus, it may be possible to win a chess game by frustrating the opponent by taking much time in making a move, humming, filling the room with smoke, reading a newspaper, and assuming an air of nonchalance. When the opponent has relaxed his efforts as a consequence of this gamesmanship, one may be able to move in for the kill.

Some will assume that the intent of these methods is evil. They will point to the unpopularity of Machiavelli but forget that there are ruthless competitors who do not hesitate to use his relentless methods. However, few will deny the importance of knowing how others plan their strategies. If one tolerates liberal expressions of the mind, one can perhaps also tolerate varied expressions of personality. It would be naïve to assume that the entire world consists of honest intellectuals bent on utilizing their talents. One must admit the usefulness of a flexible personality which can be adapted to different circumstances.

Examples of thinking which leads to successful action may be encountered in everyday activity where individuals are driven to invent a solution from limited resources. Following are a few elementary examples. A persistent Neapolitan, whose new watch had been stolen, stopped people at random for three years, asked the time, and carefully examined the watches they consulted. He finally located his watch. His method was one of exhaustion. Living in an age when timing devices were not readily available, Galileo used his pulse beat to measure the period of a swinging lamp of a cathedral in Pisa. Alexander the Great, passing by in his conquering sweep across Asia Minor, tried unsuccessfully to untie the Gordian knot. Having little time to waste, impatiently he drew his sword and slashed it in two.

Here one discerns, perhaps more than anything else, the element of enterprise and the desire to accept a challenge. Enterprise need not always consist of herculean tasks, such as cleaning out the Augean stables. The thoughts and plans which the dexterous craftsman applies to the making of a single piece of beautiful jewelry are not unlike those of an industrialist planning an undertaking from its birth to its completion.

The achievement of most goals requires concrete conviction; otherwise one is at the mercy of chance and circumstance. The Neapolitan could hardly have found his watch, which he apparently wanted very badly, if he had not proceeded with a plan and a conviction. All too often distraught hesitation is fatal to enterprise; courage is of the essence of it. Any action is sometimes preferable to hesitation or deferment of action until the perfect solution, the perfect moment, is found. Still, there is greater risk in activity, and perhaps only the inert desire escape.

While it is helpful to understand the various methods that may be employed in creatively attacking a problem, it is additionally important to be aware of individual motivations and how they may affect expressions of creativity. Any one of several factors may be responsible for the expression of individual creativity. Expression may arise out of a traumatic experience, as was the case of Beethoven. Realizing that he was growing deaf, he reacted, not by quitting, but by dynamically and persistently communicating his feelings through even greater compositions, such as the Ninth Symphony, the Great Fugue, or the Hammerclavier Sonata —all exhibiting increased depth of expression [46].

The need for a mortally urgent solution may result in creativity. The medical corpsman who cut a hole in the trachea of a wounded soldier, broke off the top of his fountain pen, and inserted it within the man's throat in order that he might breathe, graphically demonstrated great presence of mind during a moment of pressure.

A state of excitement may often be the start of a creative surge. Vincent Van Gogh, who, more often than not, worked in a state of frenzy, admitted that his emotions were sometimes so strong that he worked without being aware of working but that, even so, the strokes of his brush came with a sequence and coherence, like words in a book [43].

A burning desire to communicate may drive the painter to expression by the use of shapes and figures; the writer, by symbolism, experience, and analysis.

Whatever the incentive—challenge, excitement, competition, the desire for improvement, or a simple impulse to create for self-expression—no person is motivated solely by a single stimulus but rather by a combination. For many individuals there is no one great moment of sudden discovery, but only patient grinding, analysis, and finally teleological conclusions derived from the findings.

"Genius is 1 per cent inspiration and 99 per cent perspiration" is an oft-quoted maxim. Marie Curie, in her search for bodies possessing the power of radiation, undertook to examine all known chemical bodies, not only the simple compounds, salts, and oxides, but also samples of minerals from the collection at the School of Physics. Under her husband's direction, almost anything she could lay her hands on was subjected to analysis. Continuous questioning, testing, exhausting, and relating finally brought her the results she sought, and she was able to announce the presence of a new element.

Another method is that of trial and error. "We must try other experiments," said Louis Pasteur when, in his search for a cause and cure of rabies, experiment after experiment failed.

Out of many possible contingencies may come one ingenious solution. This does not imply that all possible contingencies must be tried out, but rather that the aspects or experiments which seem to have direct bearing on a hypothesis may be selected and verified. In the case of failure, other related experiments may be considered. Frequently, a reshaping of the hypothesis may be necessary in order to have the approximate answer fit the statement of the hypothesis. In other words, verification of a given hypothesis may be obtained by successive adaptation and approximation until the final results are achieved.

Leonardo Da Vinci [10] suggests the following:[1]

[1] The passage cited is from "The Notebooks of Leonardo da Vinci," p. 65, edited by E. MacCurdy, published by George Braziller, Inc., New York, 1956.

When you wish to produce a result by means of an instrument do not allow yourself to complicate it by introducing many subsidiary parts but follow the briefest way possible, and do not act as those do, who, when they do not know how to express a thing in its own proper vocabulary, proceed by a method of circumlocution and with great prolixity and confusion.

From extensive production of a large number of works come one or more that are superior and imaginative. It sometimes happens that a style which manifests itself in a number of productions is often better appreciated. As in the case of Johann Sebastian Bach, some productions will show more power than others. Bach, in his frenzy to produce the required Sunday quota of cantatas for the church for which he was *Kappelmeister*, composed more than 200 church cantatas, oratorios, masses, preludes, and hymns. Of these, some survive and are still performed. Bach was daring as well as productive. The church authorities of his day often reprimanded him for his daring improvisations, for he had little compunction about moving a section from a secular cantata into a sacred one if it suited his purposes. Untiring experiments, trying for new effects, and indulgence in his own ideas of form—all are indicated in his compositions.

Eureka, eureka—within all creative persons a period of incubation precedes the point of illumination or discovery. Henri Poincaré, in describing his own discoveries, explains how, after a long period of previous subconscious work and only after such a period of effort, flashes of sudden illumination made their appearance with striking and fruitful effect [13, 33, 34]. Gauss described how he arrived at a proof of a theorem which had baffled him for two years, not by painful effort but like a sudden streak of lightning. He could not say what was the connecting thread between his previous knowledge and success.

Creative results are often obtained from the re-creation and representation of already familiar things and ideas. Presentation of a familiar object from several points of view at once—cutting the whole into pieces and putting them back together again in an unconventional way—is the basis of cubistic art. Immanuel Kant, working from and consequently rejecting the previous Cartesian and Lockean theories, developed a new philosophy—that logic alone cannot reach the ultimate truth—in his "Critique of Pure Reason." His pupil, Johann Gottlieb Fichte, in turn, went one step further and constructed a philosophy of pure idealism. Here is a sequence of philosophical ideas, each one based upon the previous one, and pushing beyond. Another illustration of reconstruction, as found in music, is the fugal form, which states and restates its principal theme by variations in form and harmony.

Any work, whether created in terms of words, paint, music, metal, or clay, is more forceful when presented with a mastery of expression. Vincent Van Gogh, employing individual brush strokes used by no artist before him, communicated his own excitement. Form and color were to him only means of expressing his emotions, and to him green and red were expressive of men's passions [43].

Another type of creativity is a talent for organization. Some of the world's great religious leaders have been organizers. To organize into a coherent community thousands of fleeing, often disgruntled slaves lost in a wilderness and to maintain that community, as did Moses, is one example of creative organization. Charles Darwin used initiative and presented in a masterly and orderly way the vast array of scientific facts upon which he based his theory of evolution. John D. Rockefeller's capacity for organization and leadership enabled him to build one of the greatest oil trusts in the world and to expand his influence and direct control to many other industries.

In philosophical thought, as well as in daily conversation, one frequently has ideas that are remote from the original subject. One idea leads to the next, and an entirely new chain of ideas is formed. This is referred to as "serendipity" and is a useful tool in creative thinking. The Merriam-Webster Dictionary[1] defines serendipity as "The gift of finding valuable or agreeable things not sought for;—a word coined by Walpole, in allusion to a tale, 'The Three Princes of Serendip,' who in their travels were always discovering, by chance or by sagacity, things they did not seek." During the consideration of a subject under study, one's mind frequently wanders to ideas which are rather removed from the problem but which turn out to have bearing on other problems. Thus, while considering one problem, a solution to another problem may arise.

Creativity and originality are enhanced by other mental tools and operations, such as comprehension, memory, etc., but do not coincide with them.

1. Comprehension is a necessary tool in the development of creativity. However, developing comprehension alone is not sufficient; it may result in habits of rigidity and the ability to think only in strictly defined channels, thus becoming a deterrent. On the other hand, the inability to apply learned techniques to problems which bear a resemblance to the example studied is not sufficient proof of the lack of creativity.

Although comprehension enables an individual to pass certain scholastic tests and meet social standards, it may produce a complacency resulting from a delusion that meeting these tests and standards is sufficient and that there is no need to exert himself toward higher standards which may be more subtle or unclearly defined. It often happens that maturity, together with the added responsibility of adulthood, brings one

[1] By permission. From Webster's New International Dictionary, 2d ed., copyright 1934, 1939, 1945, 1950, 1953, 1954, 1957 by G. & C. Merriam Co., Springfield, Mass.

into an active role, in comparison with the passive role connected with learning to comprehend.

2. Memory is the term used for the conscious recall of past events and for the recognition and retention of knowledge or skill. It provides a storehouse of information from which patterns of thinking are formed. It is the mind's power to revive past experiences. Imagination brings together pieces of memories to build up a combination of images and ideas. Experiments have shown that one's impressions are, in fact, much more accurate and intense than experience usually reveals. It has been estimated that an individual retains about 10 billion "snapshots" during a lifetime. Each of these individual memories is broken into bits and stored in the impressionable molecules of hundreds of cells. Dissection of the parts continues to the finest units. Objects and scenes are observed to have a multitude of properties such as shape, size, color, texture, smell, etc.; ideas have subtle properties. It is not known how storage or reconstruction takes place. Usually one encourages memory which involves logical understanding rather than the less analytical verbal, or rote, memory.

Among the errors of memory are the exaggeration of certain impressions and the fading of others—remembrance of the desire instead of the fact. Logical events are sometimes increased in their rigor. For the improvement of memory and its contribution to creative thinking, mnemonics experts recommend the following techniques:

a. The close involvement of oneself, one's senses, emotions, and aesthetics in the situation one wishes to retain.

b. Learning things as a whole, rather than in parts, unless they are too long and complex. In this manner, better use is made of the relationship between the parts, and wasteful repetition eliminated.

3. Judgment is interpretation aided by previously acquired ideas, or concepts. Complex thinking is a synthesis of many judgments, and reasoning involves judgment which makes use of previous judgments. Discrimination, the process of arranging things into groups or classes, is one of the simpler forms of judgment. Good judgment is indispensable in evaluating alternative hypotheses and selecting optimum conclusions and should therefore be cultivated.

4. Imagery also plays an important part in the thinking of all people. Although evidence indicates that one can think without imagery (imageless thoughts), the use of images as models for thinking is a valuable mental tool. Individual differences in imagery are so great that any statement about them would be controversial. The most common types of imagery are visual and auditory. All the senses are associated with imagery of some type. Many people, for instance, hear sounds in their "mind's ear." Since imagery is unstable and fleeting, it is an unsure guide

for rigorous thinking. However, in the hypothesis-forming stage, recalled images can suggest patterns which may prove useful.

Perspectives acquired through independent habits of thought and expression provide the individual with an extended awareness, allowing him to perceive a wide range of contingencies. The value of this is not that he can always select a successful solution from the outer perspectives which he can provide, but rather that, when working over a wide range of phenomena, he can occasionally present an "ingenious" solution or idea which has been missing in the consideration of the problem by others. This is closely related to the ability to make many assumptions in solving a problem. The capacity to select the important assumption from the many is also necessary, although there is the danger of becoming attached to a less-expected idea and assigning too much importance to it. Judgment is required to discern the probability of occurrence and usefulness of a remote contingency from a workable one and to be able to accord the proper credit.

The inability to select the proper mode of action is demonstrated by an imaginative individual who mailed a check to his bank for deposit and did not receive a receipt. He concluded that either the check was not received by the bank or that the bank had been robbed. After worrying about the problem for a few days, he felt that action was necessary for the relief of his worry and proceeded to the police to inform them that the bank had been robbed.

Talent is made evident by expression, and satisfactory expression requires enterprise and understanding. To gain access to human minds conditioned by past experiences, a new message usually requires continuous repetition in various forms. Thus, a flexible method of reciprocal communication is necessary to obtain desired responses. Realizing the communication difficulty, one will encourage the communicants to present their ideas in a variety of ways in order to have better access to those ideas and be able to interpret them. A negative approach to problems is easily developed as a consequence of inadequate communication and, therefore, inadequate understanding.

Inasmuch as the world of experience consists of people as well as of ideas and objects, it is essential for the individual to expand not only his communicative capacity but also his set of values, the first to convey his ideas and induce action and the second to know when to resist mediocrity and when to place personal relations above ideas. Success in action depends partially on one's ability to control a situation without sacrificing people.

Conventional methods for originating ideas and solutions of problems have been found by some companies to be unnecessarily restrictive both in scope and in usefulness. Rigidity of established patterns of thinking may too often lead to failure to utilize (or even consider) as many of the possible solutions of a problem as are conceivable. The main objective of a committee working on the solution of a problem may finally become that of determining reasons why suggested ideas will not work rather than how they should be modified to aid in the solution. Traditional conference discussions tend to break down into debates over the pros and cons of a single course of action. Realizing that such is often the case, an increasing number of concerns have begun to adopt methods of group creative thinking to facilitate the origination of ideas and concepts oriented toward the solution of problems. Usually questions of logic and rigor do not enter into the initial work; several possible answers toward solving a problem are amassed from the group. In general, more facets of a problem can be seen by such a group than by a single individual. As M. Morse states [26]: "The creative scientist lives in the wilderness of logic, where reason is the handmaiden and not the master. I shun all movements that are coldly legible. I prefer the world where the images turn their faces in every direction, like the masques of Picasso." The first act of imagination sometimes violates reason and is subsequently subjected to rational analysis in order to make it useful and to integrate it into the logical framework of the mind. In addition to the knowledge that is available to him from others, each individual is at liberty to satisfy his own curiosity, to seek his own answers. The situation is better described by Immanuel Kant, who shrewdly observed that, rather than drawing its laws from nature, man's intellect imposes its laws on nature, where "laws" presumably means "models." Thurber's rooster put this idea in homelier language when he boasted that if it were not for his crowing the new day would never dawn [47].

Some limitations of logical thinking are that (1) it generally assumes that there is a single right answer (in the new approach, instead of looking for one "right" answer, a large number of possible answers is amassed) and (2) it stifles the use of valuable intuitive insights. The rigor associated with logic is postponed to the analytical phase of the creative process.

The following three methods of group creative thinking have been used [27]: First is brainstorming. To encourage young men with ideas, this procedure emphasizes in the initial group thinking the quantity of answers rather than their quality. Evaluation comes afterwards. The main advantage of this "free-wheeling" method is that it points out many avenues to the solution of a problem. Criticism of the method centers about the superficial answers it produces, which require considerable time for evaluation and selection. It does not present a good opportunity for studying the creative process. Participants realize that few very original ideas are presented in a session. There is no way of telling to what extent an idea may have been taken from previous suggestions. A single person attempting to use the method may start out by jotting down and accumulating ideas in the hope of obtaining his own suggestions for possible methods of solution. It has been observed that the simple questions, "How would I do it?" and "How many alternative solutions are there?" when remembered and applied, contribute to the origination of constructive ideas. Another procedure is the definition of a problem in many different ways, searching for possible solutions, evaluating the ideas for a best approach toward an answer, and, finally, obtaining the solution by producing the structure yielding an answer to the problem. This procedure is effective and leads to a better understanding of the problem, together with the technical principles involved. It also avoids superficial answers. However, it is long and costly and does not allow a full use of creative imagination in finding inspired answers. The third technique examines the underlying concepts of the problem rather than the problem itself. It postpones early solutions and enables the use of old techniques. It is time-consuming and requires broad knowledge and a great deal of rapport among the participants.

The following are suggestions for leading a group session (preferably no more than 15): One starts by selecting a significant problem and taking time to define it. Avoiding the single-answer deadlock, one phrases questions in terms of "how many" possible ways there are for producing a solution. It is usually advisable to avoid technical terms. When a number of answers have been amassed, a pattern is formed by properly classifying the suggested answers. During the process one may encourage the adaptation and borrowing of ideas and guide the group according to the objectives of the session. It is preferable if no promise to apply any of the ideas is made. The result of the foregoing is to encourage individuals and stimulate their enthusiasm for submitting useful ideas. When applied to industrial problems, in addition to giving the employee the advantage of closer identification with the organization and its goals, such methods improve interdepartmental communications and enhance the receptiveness of supervisors to new ideas.

Interactions among individuals in a group meeting have been studied [4]. On an average, about 56 per cent of the acts during a group session are problem-solving attempts. The remaining 44 per cent are distributed among positive and negative reactions and, in a ratio of 2:1, asking questions. A speaker's first remark is likely to be a reaction; the second, a problem-solving attempt. After a negative reaction the members of the group seem to feel that they must make another problem-solving attempt which meets with a positive reaction. The participation of each individual declines gradually and not suddenly. The program is successful when unopposed acceptance of a problem-solving attempt is finally reached; in other words, a hypothetical steady state of a feedback process which alternates between the problem-solving attempts of one person and the social-emotional reaction of another. The individual who succeeds in the problem-solving process does more of it and acquires a rank order by task ability. There is some confusion initially in a group without consensus on the best idea man, and it frequently requires time to build such a consensus, a task which is easier in a group with similar backgrounds. In some groups the members reach a high degree of consensus on their ranking of "who had the best ideas." In one set of experiments the top idea man had about an even chance of also being best liked at the end of the first meeting, but by the end of the fourth meeting his chances were about one in ten. The best-liked man is usually second or third in the participation hierarchy. The task leader "locks onto" and addresses the most responsive person. The best-liked man talks to and agrees more with the top-ranking idea specialist than with any other member. They form a supporting pair. The original rank order of task ability is finally restored after some oscillation.

Creative thinking is most important in the early phase of discovery where daring hypotheses and their testing are required. The mind's tendency to think along the prescribed channels in which it has been trained is resisted and a recourse made to free imagination.

A broad outlook and an extensive experience add to one's versatility in problem handling. It is not a waste of time for an individual to present his own ideas on the solution of a problem before plunging into someone else's. Because of time limitations, one cannot always afford to exhaust the literature or, frequently, even to start a fruitful search for the solution. The trained mind should have some useful suggestions of its own.

The ability to discriminate among results can be cultivated by applying oneself to a problem, regardless of its complexity. General principles for handling a wide variety of problems can then be developed. Thus, one may conclude that the best way to learn how to solve problems is by exposure and solution—sophisticated or not. However, while this is a recommended and valuable procedure, it should be remembered that individual talents should be properly channeled and energies harnessed. The child is taught how to cope with problems of every-day life as he encounters them. His natural curiosity also leads to questions of all sorts and to both possible and impossible problems for the parent. It is hoped that with maturity and a certain amount of training, the individual will not lose the awareness of the existence of problems everywhere and an interest in their creation and solution. The study of any science consists in the acquisition of useful reflexes and independent habits of thought. One needs constantly to distinguish between what is assumed and what is proved, endeavoring to assume as little as possible at every stage. The acquisition of useful reflexes should not, however, be separated from the perception of their usefulness. For example, the usefulness of problem solving is twofold: to train the individual in the application of some important method and to develop his originality by guiding him along some new path [49].

A certain degree of self-confidence is helpful in announcing one's ideas without a feeling of embarrassment or incertitude. Originality is one test of creative thinking, and convention is a great discourager of originality.

Among the social constraints which are fed into every human mind are the moral ones. Occasionally they may even be a hindrance in the use of certain notions of optimality. Everyone is familiar with occasions when truth, which is demanded by our moral structure, would cause suffering, which is unacceptable in the same moral structure. The better one understands these dilemmas, the more direct and careful will be one's effort in solving problems. A useful method for determining and understanding constraints is to examine extremes and then gradually introduce limitations and note their effect. If the constraints are flexible, greater freedom in the use of imagination becomes available. One frequently may be interested in rational justification of a mode of action rather than in exact numbers. In that case, both technical and nontechnical methods may logically be used to derive the desired orders of magnitude. Free use of the imagination is helpful. However, imaginative creations, in order to materialize, should satisfy certain physical or material restrictions, i.e., should be traceable to real life or be an abstraction ultimately derived from it. Thus there are reasons why, for instance, a man conceived as being 20 feet tall may be doomed to crawl. A cross section of his leg would show that it does not increase in the required proportion to his height and weight; his greater size would cause breakage. Stories of giant mutant ants developing as a result of atomic explosions constitute splendid fiction. Again, if considered geometrically, the surface area of the ant must grow as a square and the volume as a cube. Such ants, therefore, would not have sufficient apparatus for the diffusion of oxygen to their tissues and possibly could not survive. The speed of light taken as a limiting speed is one of the corollaries of the theory of relativity. Thus far, there is no evidence of a motion faster than the speed of light. Consequently, useful speculations cannot violate this assumption without pitfalls.

Creativity, as such, is no more teachable than Plato found virtue (or the square root of 2) to be in his dialogue "Meno." If nothing else, Plato proved that what we cannot define we cannot learn or teach formally.

There are courses in which the creative talent in individuals is brought out by presenting bizarre situations and asking for solutions of problems for which the student can see solutions in his own environment but not in the special environment for which he is required to find a solution. Such a course sets up a model. It demands the abandonment of conventional thinking in order to solve problems within new, unfamiliar limits established in the model.

Yet even here, no definition of creativity is given. The approach used is to require creative engineering without ever saying what creativity is.

Thus, in the preceding pages, examples of creativity and conditions from which creative results flowed have been presented. The reader must react to those examples which have the greatest meaning. Then he must try to understand why others had no meaning. When he can comprehend the latter and see where they contribute to creativity, he will have performed a creative act which will contribute to the development of creativity in himself.

Creativity is essentially the act of an individual. It can come solely from within or it can be generated by a team which so works together that it takes on a team consciousness to which the individual members contribute. Creativity then becomes the result of team individuality.

Materials and models have been presented herein which may help the reader to understand creativity. If, in making his analysis to reach this understanding, he finds that he must contribute of himself, he will then practice creativity, which is more important than understanding it.

REFERENCES

1. Ashby, W. R.: "Design for a Brain," Chapman & Hall, Ltd., London, 1952.
2. Ball, W. W. R.: "Mathematical Recreations and Essays," The Macmillan Company, New York, 1947.
3. Barzun, J.: "Teacher in America," Doubleday & Company, Inc., New York, 1954.
4. Bates, R. F.: How People Interact in Conference, *Scientific American*, vol. 192, no. 3, March, 1955.
5. Bliven, B.: Your Brain's Unrealized Powers, *Reader's Digest*, October, 1956.
6. Bronowski, J.: How Scientific Ideas Are Born, *Science Digest*, March, 1957.
7. Cellini, B.: "The Autobiography of Benvenuto Cellini," Penguin Books, Inc., Baltimore, 1956.
8. Coile, R.: Ages of Creativeness of Electronic Engineers, Letter to the Editor, *Proceedings of the Institute of Radio Engineers*, vol. 42, no. 8, pp. 1322–1323, August, 1954.
9. Courant, R., and H. Robbins: "What Is Mathematics?" Oxford University Press, New York, 1941.
10. MacCurdy, E. (ed.): "The Notebooks of Leonardo da Vinci," p. 65, George Braziller, Inc., New York, 1956.
11. Denis, W.: Articles on productivity, *American Psychologist*, May, 1954, and *The Scientific Monthly*, September, 1954.
12. Gordon, W. J. J.: Operational Approach to Creativity, *Harvard Business Review*, vol. 34, no. 6, November-December, 1956.
13. Hadamard, J.: "The Psychology of Invention in the Mathematical Field," Princeton University Press, Princeton, N.J., 1945.

14. Harrison, G. R.: How the Brain Works, *The Atlantic Monthly*, September, 1956.
15. Heath, R. V.: "Mathemagic," Dover Publications, New York, 1953.
16. Huff, D., "How to Lie with Statistics," W. W. Norton & Company, Inc., New York, 1953.
17. Hunt, M. M.: How to Overcome Mental Blocks, *Reader's Digest*, January, 1957.
18. Hunt, M. M.: Where Students Lose Earthly Shackles, *Life Magazine*, Spring, 1955.
19. Jones, E.: Nature of Genius, *The Scientific Monthly*, February, 1957.
20. Kasner, E., and J. Newman: "Mathematics and the Imagination," Simon and Schuster, Inc., New York, 1952.
21. Koopman, B. O.: New Mathematical Methods in Operations Research, *Journal of the Operations Research Society of America*, vol. 1, no. 1, November, 1952.
22. Kraitchik, M.: "Mathematical Recreations," Dover Publications, New York, 1953.
23. Kramer, E. E.: "The Main Stream of Mathematics," Oxford University Press, New York, 1952.
24. Lehman, H. C.: "Age and Achievement," Princeton University Press, Princeton, N.J., 1953.
25. Machiavelli, N.: "The Prince," New American Library of World Literature, Inc., New York, 1952.
26. Morse, M.: Mathematics and the Arts, *Yale Review*, no. 4, p. 604, June, 1951.
27. Nicholson, S.: Group Creative Thinking, *Management Record*, vol. 18, no. 7, pp. 234–237, July, 1956.
28. Osborn, A. F.: "Applied Imagination," Charles Scribner's Sons, New York, 1953.
29. Osborn, A. F.: Brainstorming, *Time Magazine*, Feb. 18, 1957.
30. Osborn, A. F.: "Your Creative Power," Charles Scribner's Sons, New York, 1948.
31. Overstreet, H. A.: "The Mature Mind," W. W. Norton & Company, Inc., New York, 1949.
32. Pheiffer, J.: "The Human Brain," Harper & Brothers, New York, 1955.
33. Poincaré, H.: "Science and Hypothesis," Dover Publications, New York, 1955.
34. Poincaré, H.: "Science and Method," Dover Publications, New York, 1955.
35. Polya, G.: "How to Solve It," Princeton University Press, Princeton, N.J., 1948.
36. Polya, G.: "Mathematics and Plausible Reasoning," vols. I and II, Princeton University Press, Princeton, N.J., 1954.
37. Porterfield, A.: "Creative Factors in Scientific Research," Duke University Sociological Series, Duke University Press, Durham, N.C., 1941.
38. Potter, S.: "Gamesmanship"; "One-Upmanship," 1951; "Lifemanship," 1950; Henry Holt and Company, Inc., New York.
39. Saaty, T. L., R. Smith, and R. S. Titchen: Notes for a course in Operations Research, given at the U.S. Department of Agriculture Graduate School, Washington, D.C., 1955–1957.
40. Sawyer, W. W.: "Prelude to Mathematics," Penguin Books, Inc., Baltimore, 1955.
41. Shapley, H., S. Rapport, and H. Wright (eds.): "A Treasury of Science," Harper & Brothers, New York, 1954.
42. Shockley, W.: Article on Creativity, *Newsweek Magazine*, p. 72, Dec. 6, 1954.
43. The Nineteenth Century, in "The Great Centuries of Painting," Skira, Inc., Publishers, New York, 1951.
44. Steinhaus, H.: "Mathematical Snapshots," Oxford University Press, New York, 1950.
45. Stone, M. H.: Mathematics and the Future of Science, *Bulletin of the American Mathematical Society*, vol. 63, no. 2, March, 1957.

46. Sullivan, J. W. N.: "Beethoven's Spiritual Development," New American Library of World Literature, Inc., New York, 1949.
47. Thurber, J.: "Further Fables for Our Time," Simon and Schuster, Inc., New York, 1956.
48. Walter, W. G.: "The Living Brain," W. W. Norton & Company, Inc., New York, 1953.
49. Weil, A.: *American Mathematical Monthly*, vol. 61, p. 34, 1954.
50. Wiener, N.: "Cybernetics," John Wiley & Sons, Inc., New York, 1948.

ELEMENTARY ANALYTICAL PROBLEMS AND PROJECTS

Why do we think that a variety of problems presented as puzzles are helpful in imaginative thinking? Creativity is a relative thing. Constructive actions of individuals in survival problems are creative expressions. If someone had at his disposal a collection of puzzles which contained all the niceties of thinking, one would consider education in those puzzles as enriching and valuable in solving other problems. A real-life problem may be considered as a chain of puzzles for which the solution of one puzzle is an assumption of the next. By studying the different subtleties developed through cumulative experience, one essentially acquires a greater capacity for inventiveness. Whether a person lacking in imagination can be enriched by solving such problems can only be answered individually. There must be incentive and interest.

Some problems posed here are aimed at encouraging fluency in imaginative expression of ideas. Logical thinking alone would not be sufficient for solving problems whose complete statement cannot be given until some investigation has been made. Imagination is the only tool which provides the link with broader perspectives. It, in turn, can be trained along the constructive lines required by the problems of a civilization. Its value is frequently judged by both its artistic and nonartistic contributions which expand experience.

It is difficult to teach people through formal devices to use imaginative thinking. The "problem" approach seems to be a possible alternative in the training of an individual. These arguments, though inconclusive, are further supported by classroom experiments with which several people have had success.

The problems selected are meant only to serve as a challenge and a guide to the reader in stimulating his interests. It is hoped that he will develop his own techniques in expanding and analyzing each problem separately, pointing out differences in the method of thinking and in the perspective gained from each. In addition to these problems, projects are suggested. Despite a division of the problems into classes, some overlapping cannot be avoided. The classification is by no means rigid.

OPERATIONAL PROBLEMS

1. On some distant planet, in another of the many possible solar systems about which astronomers tell us, live creatures whose state of knowledge is highly developed. We are able to communicate with them by radio. The signals are too diffuse to determine their direction. We wish to carry out an experiment with them in which our clockwise, counterclockwise sense plays an important role. We cannot be sure that our clockwise sense is not their counterclockwise sense. We wish to settle this point before proceeding with the experiment. How should we do it?

2. Suggest an experiment to measure the volume of blood of a living human being (without killing him).

3. With your acquaintance across a river and with a yardstick in your hand, you wish to estimate the width of the river. How?

4. The sense of smell adds to the pleasure of eating and of appreciating perfumes and flowers. Suggest three problems of the operational type in which smell plays an important role. For instance, used-car dealers spray the insides of used cars to give them a "new-car smell"; mercaptans are used in natural gas to aid leak detection. Differentiation between odors may depend on differences between the infrared spectrum of the substances involved. How might this theory be used in some operation where odor plays a significant role?

5. It was learned that much machinery sent by the United States to be used for agriculture in India was abandoned because of lack of parts and of know-how. Suggest a manner in which this type of waste might have been avoided and the Indian farmers aided in utilizing the machines.

6. How might one study electronic equipment used in aircraft communication in order to discover facts about the reliability of its performance?

7. An industrial warehouse is needed for storing stock to control inventory. The total cost of stocking may be minimized by studying what factors regarding inventory?

8. Given four men and four tasks with a score indicating the relative performance of each man at each task, how might one study the problem of assigning the men to the tasks?

9. Thirty per cent of the dollar volume of the gross sales of a mail-order company represented returned goods. There were two types of returns: unclaimed returns and actual customer returns. The mail-order business has great advantages despite this setback; however, it is desired to minimize these types of loss. How?

10. How would one analyze the profitability of keeping department stores open at night?

11. Advertising in newspapers is common for a department store. How may one test its profitability?

12. How may one organize the operation of traffic toll booths to minimize waste and at the same time provide efficient service?

13. Describe some methods by means of which one may simulate a naval battle. What useful lessons can be learned about the real situation?

14. Point out some weaknesses of choosing the maximum output per man-hour as a measure of effective production of an industry.

15. A certain wine importer noticed that his sales of wine were not what they should be in comparison with other types of liquor. He hired you as a consultant to look into this problem with the intention of improving the wine business. What would you do?

16. Your college classmates have for you that particular type of contempt that is bred by familiarity, and you are their supervisor for a week in the testing department of a manufacturing plant. Your esprit de corps influences production, etc. They will, in turn, each supervise for one week. What plan of action would you follow to gain their respect?

17. What can you do to gain a sales interview with a purchasing agent who buys heavily from your competitors and who has consistently refused to see you?

18. If your car were badly stuck in heavy mud in a wooded area, far from assistance, what are some methods you would use to free yourself? In addition to a jack and spare tire, you have a sizable length of stout rope and a few tools.

19. A large tractor-trailer, while traveling an unfamiliar route, became wedged beneath a low bridge over the road. The driver and several passing motorists were baffled in their attempts to free the trailer. A creative bystander then suggested a simple solution that enabled the driver to free the trailer within a few minutes. What was the suggestion?

20. A lively three-year-old boy was suffering from an infected arm that had to be treated at home by soaking the injured arm in a medicinal solution for a two-hour period each day. His mother, who had four other young children to care for without help, could not give the boy continuous or even close attention during the two-hour periods. Faced with this problem, the mother devised a gentle means of accomplishing the prescribed treatment. How?

21. You are in a row boat; you have dropped anchor at a fairly deep spot in the lake, and the anchor is stuck in the mud. No amount of hauling has helped to bring the anchor into the boat. What can be done to save the anchor?

22. The captain of a sailing vessel wishes to leave a port. He has only a row boat. There is no wind. It is desired to move the sailboat out of port, using the row boat and without pulling. How?

MECHANICAL INVENTION

23. Suggest superior equipment or methods to permit a blind individual to carry on normal human activity such as walking without a cane, reading a normal printed page, seeing a movie. Do not worry about engineering technicalities. Use any method that makes some sense to you.

24. Everyone who has eaten a soft-boiled egg has, on occasion, been plagued by bits of shell which, because of human error, fell into the dish. What effective method or instrument can you suggest for shelling an egg or for minimizing this unappetizing hazard?

25. Describe a robot whose abilities are basically the same as those of a human being and are achieved by known mechanical devices. List some of the uses to which such a robot could be put. Extend your thinking to a community of these robots.

UNINHIBITED IMAGINATION

26. C. K.'s wife had left the car windows completely open, and it began to rain. C. K. was in his pajamas. He put on his raincoat, walked to the car, and found all the doors and the trunk closed. He succeeded, without opening any door of his new sedan, in closing the windows and locking the doors. He did not remove any part of the car to do this. How was the task accomplished?

27. Connect nine points arranged in a square array by a continuous path consisting of four straight-line segments.

28. Jules Verne wrote to his father, "Everything one man is capable of imagining,

ELEMENTARY ANALYTICAL PROBLEMS AND PROJECTS 403

other men will be capable of realizing." Indicate the extent to which this statement is true. Give some counterexamples.

29. Find extended future uses of television.

30. It is desirable and useful, in the course of the examination of an object, to consider the object as an entity, unrelated to any surroundings, and to determine its various mathematical, physical, chemical, biological, psychological, aesthetic, etc., properties (wherever applicable); then to relate it to its surroundings; and to imagine its utility in every conceivable manner. In this manner, characterize an ordinary melon.

31. See if you can mention 20 different ways in which light can be produced.

32. What is the object whose silhouettes from three perpendicular directions are a circle, a square, and a triangle, respectively?

33. With the assumption that it is possible to find a rational framework which can be used to relate any two ideas (or objects) in a feasible manner (not always scientific), select any two ideas, or objects, which have no apparent connection and attempt to provide such a framework. Also, select one such idea and ask someone else to provide an idea which he thinks is unrelated to it. Show him how they can be related. Distinguish between scientific and pseudoscientific explanations. This is a first step to providing theories. Carry out this exercise with people of different backgrounds and note the variety of approaches.

MATHEMATICAL PROBLEMS

34. Three planes, each with a tank capacity of 1000 gallons and capable of making 1 air mile per gallon, take off simultaneously from a certain base. It is desired to send the first of these planes to a distant base where it will land. The other two planes are equipped to refuel this plane (and also each other) from their own supply but must return safely to their original base on the remainder. Assume that no fuel is used for warm-up, take-off, or landing reserve, that the figure of 1 mile per gallon applies at all stages of the flight, and that refueling from one plane to another takes place with no appreciable loss of time.
What is the maximum distance to which the first plane can be sent?
At what distances should refueling take place?

35. It is desired to divide a cube of side three units into 27 cubic pieces of side one unit. Give an a priori lower bound on the number of cuts required; a cut consists of a single plane division of the piece, or pieces piled on top of each other, into two parts. Is this lower bound sufficient for solving the problem? Give proof.

36. As is well known, Lower Slobbovia is too poor a country to afford its own mint. There are n coiners engaged in making Rasbukniks, the official currency, to government specifications. However, it is suspected that some of them may be counterfeiting by introducing some base metal into the alloy. Any form of counterfeit will weigh the same, although slightly different from the weight of a good coin. Each coiner produces all good coins or all counterfeit. With one guaranteed good coin, a set of infinitely refinable weights, a beam balance, and as many coins from each coiner as may be needed, determine in three weighings whether any of the coiners is dishonest, and which ones. (*Taken from the Am. Math. Monthly.*)

37. Identify a counterfeit coin in a set of 12 coins in three weighings, using a simple balance. No weights are required in the operation. Also tell how the counterfeit coin may be distinguished as heavier or lighter.

38. Given n holes on a line beside each other and n identical marbles. It is desired to place the marbles in the holes in such a way that at no time does one place a marble in a hole without having previously placed a marble next to it (to its left or right). **No gaps** are left in placing the marbles. In how many ways can this be done?

39. A wise and aging father wished to divide his estate as equally as possible between his two greedy and suspicious sons. He devised a simple means of achieving his wish with the maximum degree of fairness and minimum possibility of suspicion between the sons by asking one of them to divide the property and the other to make the first choice. Solve the problem for three sons.

40. A salesman wishes to travel by shortest total distance from his home to each of $n - 1$ specified cities and then return home. State the problem, using the matrix of the distances, and discuss the solution of the cases $n = 3$ and $n = 4$.

41. Given a 2 by 2 matrix with randomly selected integer coefficients, what is the probability of the determinantal value being even?

42. Using a chess board with the squares in two opposite corners removed, and enough dominoes, each sufficiently large to cover completely two squares at a time, it is desired to cover all the remaining squares of the board with dominoes. Can this be done? Explain.

43. A man has a map which gives the following instructions: "At the island go to the gallows and from there walk to the pine tree, measuring the distance. At the pine tree turn right and walk the same distance. Again, from the gallows, walk to the oak tree, measuring the distance; turn left and walk the same distance. Join these two end points and you will find the treasure at the mid-point." When you go to the island, you find that the gallows is gone, but the trees are still there. How can you find the treasure?

44. A census taker knocked on the door of a house and obtained some information from a man living there. He asked if anyone else lived there. The man said, "Yes, there are three other people of different ages living here. The sum of their ages is the number of this house (which the census taker knew) and the product of their ages is 1296." The census taker, after some calculations, asked a question and left. He had determined the ages. What is the number of the house?

45. A census taker and a public-opinion pollster both approach the house at 900 Main Street at the same time, each wanting to know the ages of its occupants. The owner gives his own age and says three other people live there also and that the product of their ages (three different whole numbers) is the same as the house number.

The owner says he will tell the census taker the age of the middle person. He whispers this age to the census taker, who then says he is unable to determine the other two ages. Then the owner says he will tell the pollster the sum of the ages of the oldest and of one of the other two. He whispers this sum to the pollster, who says he, too, is unable to figure out the ages.

The owner asks each in turn, and the census taker says he cannot determine the ages. The pollster says he cannot either. The census taker says he still cannot. The pollster says he cannot yet. The census taker says he still does not know. But the pollster says, "Now I know all the ages."

What are the three ages? All the information needed to solve the problem is here.

46. A passenger train moves nine times as fast as a freight train. The schedule at a certain station calls for one passenger train per hour and one freight train per hour, not at the same time, but at a fixed time for all hours. A man goes at *random* times to the station. He waits for the first train to arrive, and noting its arrival time and the type of train it is, he immediately leaves, observing that nine times out of ten the first train to enter is a passenger train. Explain.

47. If V denotes the number of vertices of a polyhedron, E the number of its edges, and F the number of its faces, by using samples of polyhedra and multiple-correlation methods, obtain the Euler formula which relates V, E, and F linearly.

48. Suggest a method of analysis with data collection which provides a measure for the accuracy of a person shooting darts at a dart board. Use hypothetical numbers and carry out the analysis. Obtain the mean position of the fall of darts and the standard deviations. Determine the distribution of fall of darts (by actually carrying out an experiment) and the probability of scoring a hit. Give the expected number of throws needed to hit the bull's eye, which is a small circle of radius r.

49. There are thousands of persons engaged in economic forecasting. If all of them simply guessed, hundreds would be correct in their forecasts time and time again. The record does not separate the sophisticated forecaster, who has used a scientific system, from the guesser and charlatan. Among 1000 coin-tossing forecasters, about how many forecasters, on an average, would be correct in nine consecutive forecasts. Discuss and compare with the above statement.

50. How would you estimate the number of board-feet (1 inch thick, 1 foot wide, any length) in a stand, i.e., a forest, of timber of 10,000 acres?

51. Three runners, A, B, and C, participate in a 100-yard dash. B was 10 yards behind A when the latter finished, and C was 10 yards behind B when B finished. How far was C behind A when A finished? Assume that they run at constant rates.

52. A sack of potatoes was observed to weigh 100 pounds when 99 per cent of the total weight was water. Later, dehydration reduced the moisture content to 98 per cent. What did the sack weigh?

53. Hanging over a pulley there is a rope with a weight at one end. At the other end hangs a monkey of equal weight. The rope weighs 4 ounces per foot. The combined ages of the monkey and its mother are four years, and the weight of the monkey is as many pounds as its mother is years old. The mother is twice as old as the monkey was when the mother was half as old as the monkey will be when the monkey is three times as old as its mother was when she was three times as old as the monkey was. The weight of the rope and the weight is half as much again as the difference between the weight of the weight and the weight of the weight plus the weight of the monkey. What is the length of the rope?

54. A man rows at constant speed whether going upstream or downstream. He tosses his hat into the stream at some spot and rows up the stream for 15 minutes, after which he decides to retrieve his hat. He turns around with no loss of time and rows downstream and recovers his hat which had moved one mile down the stream. What is the speed of the stream?

55. A and B swim at the same speed V in a river of constant speed v. A swims a hundred yards up the stream and then swims back to the starting point. B swims a hundred yards across the stream and back to the starting point. If they start out at the same time from the same point, which one arrives first at the starting point? Give proof.

56. How might one measure the radius of the earth, using only instruments which measure angular dip and distance? How would one measure the distance to the moon?

57. When a man and his wife approached a very long escalator in a London underground, she inquired from him how they might count the total steps on the escalator which it would be necessary to walk up if the escalator were not moving. He provided a quick scheme whereby they could deduce the number of steps after having walked up the moving escalator once at different constant rates. How?

58. As an army 5 miles long began to march, a courier left the rear for the front. Upon reaching the front, he immediately started for the rear and reached there after the army had traveled 5 miles. How far did the courier travel?

59. The bank made a mistake in cashing a check and gave dollars for cents and cents for dollars. After spending $3.50 a man finds he has remaining twice the true amount of the check. What is the true amount?

60. Determine A from the two division exercises, where each x is an integer from zero to 9:

$$\begin{array}{r} xxxxx \leftarrow A \\ xxx \overline{\smash{\big)}\,xxxxxxxx} \\ \underline{xxx} \\ xxxx \\ \underline{xxx} \\ xxx \\ \underline{xxx} \\ xxxx \\ \underline{xxxx} \\ 0000 \end{array}$$

$$\begin{array}{r} xxxxx \\ xx \overline{\smash{\big)}\,xxxxxx} \leftarrow A \\ \underline{xx} \\ xxx \\ \underline{xx} \\ xxx \\ \underline{xxx} \\ xxx \\ \underline{xxx} \\ 000 \end{array}$$

61. Give an upper and a lower bound to the area of a unit circle by means of the circumscribed and inscribed squares. Generalize to three dimensions and then to n dimensions. Can you find better approximations? What are the errors incurred in each of these approximations?

62. In attempting to assign an order of preference to several items, one may associate a different symbol with each item and obtain between every pair comparisons of the form $a > b$, $a < b$, or $a = b$, depending on whether a is more preferable than b or conversely, or whether they are equally preferred. These relations are used to obtain an ordering of preference among all the items. Obtain a set of pairwise preferences on five items for which your preference ordering is not immediately obvious. The symbols may now be assigned numerical values to indicate the relative magnitude of preference. If five seems too small, try six items.

LOGIC PROBLEMS

63. Given a cup of tea and a cup of milk, each having five spoonfuls. A spoonful is taken from the milk and added to the tea. After stirring, a spoonful of the homogeneous mixture is returned to the milk. Is there now more milk in the tea or more tea in the milk? Test your intuition on this problem.

64. Is there another place on the earth, besides the North Pole, where one can travel 100 miles south, 100 east, and 100 north, and be back at the starting point?

65. The two hands of a clock cross at 12:00. At what time does the next crossing occur?

66. In his book "Patterns of Plausible Inference," Polya asks the question: To which language is English more closely related—Hungarian or Polish? He gives an interesting analysis of the problem. Find possible invariants and measures of effectiveness which would enable a comparison.

67. A test on "Are you a good parent?" appeared in a Sunday newspaper. A high score was to indicate good parenthood, and conversely. Does one who theoretically knows the answers necessarily succeed in practicing good parenthood? How about the good parent; is he supposed to answer correctly? Derive some conclusions on correspondence between thought and action.

68. Give an explanation of how one may go about learning to walk on hot coals, such as is done by some experts in India.

69. If A, B, C, and D each speak the truth once in three times (independently) and A affirms that B denies that C declares that D is a liar, what is the probability that D was telling the truth? Prove that the correct probability is 13/41, and that this is also the probability that A, B, and C each told the truth.

70. A cable running from London to New York is made up of n independent wires. A single individual desires to identify the ends of these n wires by means of three operations. Any number of short connecting cords is available, along with one continuity meter to measure flow of the current. An operation consists of patching any number of wires together, checking continuity of the others, and labeling as is necessary at one side. How?

71. Two trains meet head on near a siding. Given the facts that train A is 30 cars long and train B is 20 cars long, that the siding can hold only 10 cars, and that trains may uncouple at any car, describe how the trains may pass without disturbing the order and direction of the cars.

72. Tom, Dick, and Harry take a sequence of tests. On each test only one of them received A, another B, and the third C. After all tests were completed, three different nonzero integer values were assigned to A, B, and C. It was found that Tom's total score was 22; Dick's, 9; and Harry's 9. The question is: If Dick was first in spelling, who was second in arithmetic?

73. In a distant land dwelt the members of two political parties. The Mendacians were inveterate liars, while the Veracians were unfailingly truthful. Once a stranger visited the land, and on meeting a group of three inhabitants inquired as to which political party they belonged. The first murmured something that the stranger failed to hear. The second remarked, "He said he was a Mendacian." The third said to the second, "You're a liar!" To which political party did the third belong?

74. Three candidates for membership in a society were given the following test of logic. They were told that each would be blindfolded and a hat would be put on his head. The hat might be either black or white. Then the blindfolds would be removed, so that each might see the colors of the hats worn by the other two. Each man who saw a black hat was to raise a hand. The first to infer correctly the color of his own hat would be admitted to membership.

The test was duly carried out. Black hats were put on all three men. The blindfolds were removed, and of course all three raised a hand. Presently one man said, "My hat must be black." He was admitted to membership when he proved his assertion to the satisfaction of the judges. Explain his reasoning.

75. A man comes to a fork in the road. Not knowing which to take, he sees two men, one a liar and the other honest. (They know each other but he does not know which one is which, except that one is a liar and the other one is honest.) He asks only one question from one man. The answer he receives is a "yes" or a "no." The other man is present but does not speak. With this, he takes the right road. What does he ask, and how does he solve the problem? What question does he ask if there is only one man, whose veracity cannot be ascertained?

76. Three cannibals, only one of whom can row, and three missionaries, only one of whom can row, are faced with crossing a river in a boat which can take no more than two people. How is this done so that at no place would there be left more cannibals than missionaries (the former, out of habit, would naturally eat the latter)?

77. Given the following premises:
Heavy armored planes with a short range are fighters.

Either long-range planes or fighter planes have light armor.
Either conventional-engine planes or short-range planes have heavy armor.
Using truth tables, test the validity of the conclusion that a jet (i.e., a nonconventional-engine) plane has long range and light armor.

78. Discuss the "reality" in the operational sense of the following:
The *sound* when a tree falls in the forest with no ears within hearing distance
Euclidean plane geometry
Genes (carriers of hereditary characteristics)

79. Distinguish as either *valid* or *true* and explain:
Light travels *in vacuo* at a speed of approximately 186,000 miles per second.
The sum of the length of any two sides of a triangle exceeds the length of the third side.
Nothing can travel faster than light.

80. In the defense of shipping from submarine attack, the value of various defensive tactics might be examined in the light of three different measures of effectiveness:
1. The number of enemy subs sunk per month at fixed cargo tonnage
2. The total cargo tonnage safely delivered per month
3. The ratio of cargo tonnage safely delivered to the total shipped

Comment on each of these with respect to their merits for long- and short-term war. Which would most likely be applicable to a war expected to last only three months?

81. Discuss the nature of the *causal* relation in the following statements:
The cause of ignorance is lack of school facilities.
Because you remained too long in the hot sun, you have an uncomfortable sunburn.
The spontaneous expansion of a gas is the result of the second law of thermodynamics.

82. Suggest how one might quantify the following:
The emotional stability of mice
The ability to write clear and readable prose
The musical talent of a violinist
The ability to distinguish odors

83. Point out the fallacies in the following statements:
This man has written inflammatory pamphlets against the United States government. He has demonstrably disturbed the tranquility of the nation . . . He should, therefore, be silenced.

If I am the devil, then the devil and I are one. I am not the devil; therefore we are two. Hence $2 = 1$.

To allow every man an unbounded freedom of speech must always be, on the whole, advantageous to the State, for it is highly conducive to the interests of the community that each individual should enjoy a liberty, perfectly unlimited, of expressing his sentiments.

84. State explicitly and resolve the following paradoxes:
He shaves all those men of Seville and only those men of Seville who do not shave themselves. Does he shave himself?

Epimenides, the Cretan, said, "All Cretans always lie." How about Epimenides, himself?

OPERATIONS-RESEARCH PROJECTS

There are many activities in which one may apply operations-research techniques, as well as some areas where it is not appropriate. The following assignment is intended to test your knowledge of the techniques as well as to give you a full opportunity to use your imagination and common sense.

Below are listed several fields of activity with suggested special problems which might reasonably be encountered. You should select one field of activity which interests you and explain how operations research might profitably be employed. A suggested method of presentation is to write as if you were a small consultant who had been approached to undertake an investigation. This would be your suggested program submitted for approval to the prospective client. The following suggestions should help you to make a good report.

The client has specifically mentioned the special problems shown below. If you believe some of them are not suitable for investigation, explain why. In any event, you should suggest a certain order of attacking the problems based on your knowledge of the relative gain to be expected from the problem's solution and the cost of solving it. This will require the introduction of a few basic concepts of operations research, such as measure of effectiveness, the appropriate fields of application of these methods, quantification, etc.

Having selected and carefully formulated one or two specific problems, you should explain in some detail the kind of data required; the way in which they are to be obtained; the way in which they are to be analyzed; and the possible outcomes, that is, what type of decision can be made on the basis of the data which you will collect. Make sure that this agrees with the original statement of the problem. You may be able to answer some other questions as well with very little extra effort. The client is intelligent and wants enough information to know whether you have a reasonable chance of solving his problems by the methods you outline.

85. Municipal Affairs
   School building program
   Water-supply facilities
   Fire-department location and size
   Municipal zoning laws
86. Administration of a Large Group of Production Workers
   Wage policy, including fringe benefits, leave, hospitalization, etc.
   Communication between management and labor
   Selection of supervisory personnel
   Hiring of new employees
   Work standards
   Inventory policy
   Production scheduling
87. Medical Services
   Scheduling of a private doctor's activities
   Planning of new hospital facilities
   Selection of one among several hospitalization plans for a large group of employees

88. Activities of an Individual
　　Efficient purchase of important items (car, home, university education, etc.)
　　Personal investment policy: distribution of funds among stocks, bank deposits, and life insurance, or other investments

89. Management of a Large Chain of Restaurants
　　Optimization of food purchasing
　　Prevention of pilfering or other losses of food, cash, and small items by employees
　　Decision as to whether to open a new restaurant; if so where, when, and how to begin

Only the problems are mentioned here, and many "facts" will have to be supplied as a starting point for the problems which follow. Spend time planning an operations-research program around the "facts."

90. Listed below are situations in which there are several choices of a course of action. The choices will presumably be made on the basis of some profit (in the most general sense). State the most appropriate measure of effectiveness, with reasons supporting your opinion.

You inherit $5000.

You are responsible for the operation of a police force in a small community of about 5000 inhabitants.

You are the proprietor of a small retail store.

You wish to replace your present automobile. Would this problem be different if you were the editor of "Consumers Research Bulletin" considering the problem for the "average buyer"?

91. Order the following recommended facilities to be introduced into supermarkets according to priority. Give pertinent reasons, considering costs and occupied space as compared with possible increase in business. Discuss the choice of other measures of effectiveness. At each step in this ordering give plausibility arguments as to why a facility occupies the relative position with regard to its neighboring ones. Suggest further facilities if possible.

Parking Facilities
　　Approximately 93 per cent of supermarkets have parking lots.
　　The average size of the parking lot is 25,510 square feet.
　　The ratio of the parking area to the size of the supermarket is 2.13.
　　Ninety-four and four-tenths per cent of the colossal supermarkets have parking lots.

Air Conditioning
　　$18,000,000 was spent in 1954 on grocery store air conditioning.
　　The cost of air-conditioning equipment in 1953 accounted for more than 10 per cent of the total expenditure made by supermarkets for construction and modernization.
　　Approximately 61 per cent of supermarkets are equipped with air-conditioning units.
　　More than three-fourths of the large and colossal markets are air-conditioned while only 40 to 50 per cent of the smaller markets possess air-conditioning units.

Nursery and Children Play Facilities
　　Baby sitting while the customer is shopping.
　　Provision of children's books and toys.
　　Visiting Santas at Christmas.
　　Entertainments at openings and holiday occasions.

ELEMENTARY ANALYTICAL PROBLEMS AND PROJECTS

Circuses, pony rides, children's play equipment, and free gifts available for children accompanying their parents.

Over 60 per cent of large and colossal supermarkets provide kiddie corners.

92. Study a system of two categories of components (perhaps by using an array), one category being propulsion devices and the other the medium of motion. By subdividing each into components such as jet propulsion, wheels, claws, wings, etc., for the first category and water surface, under water, land surface, air, space, etc., show how combinations of the two categories suggest mechanical systems for transportation or for any other useful purpose which one can suggest.

93. Do the same for a type of target (e.g., people, ships, cities, etc.) vs. means-of-destruction system (e.g., gas, bacteriological warfare, etc.). Can you suggest a bacteriological means of crippling a submarine target? Give an example of a three-category system. In this case, the three-dimensional array may be replaced by several two-dimensional arrays.

94. A certain small town (population 50,000) lies at the intersection of two major U.S. highway routes, which currently pass through the heart of the downtown shopping center, with only a "Stop" sign at the intersection. Because of increased traffic volume in recent years, the town is plagued with a host of traffic problems, including vehicle accidents, pedestrian deaths, shortage of parking space, impairment of shopping facilities, and complaints from the AAA for obstruction of traffic flow. You are called in by the Mayor, who knows only that he has a "problem," and asked for advice. Answer the following:

a. What kinds of data would you seek for such a study?

b. List some operational factors in the existing situation in which you might expect to find room for improvement without undertaking a major financial investment.

c. List some possible "drastic" changes.

d. List some possible measures of effectiveness to test proposed solutions of the various aspects of the problem. Is it likely you could find a single measure for the entire problem?

e. List two mathematical models likely to be used in such a study.

95. Select from your own work experience one (or preferably two) problems which are amenable to operations-research solutions. Discuss them in the same way as you have the previous problems.

PART 5

DECISION MAKING

CHAPTER 13

MULTICRITERIA DECISION MAKING: THE ANALYTIC HIERARCHY PROCESS

13-1. Perspective

The purpose of using mathematics in complex situations is to help us understand these situations. If it takes a great deal of effort to learn this mathematics we become brainwashed by the tool itself and may fall victim to the many assumptions made to facilitate the approach. The central purpose of all explanation is understanding, not playing with numbers. The question then is, what is the connection of our understanding, judgment, and feeling with numbers, and how do these numbers reflect the strength of our feelings? Is it possible to optimize without losing touch with the problem as we see it? What if there are many things to choose from? What about probabilities, can we estimate them in a meaningful way without guessing at numbers? How do we represent our perception of complexity and project ahead? What about risk and uncertainty?

Below we offer the reader a new tool for this purpose, that of using mathematics to understand our own feelings and help us in the process of decision making. Space limitation prevents us from elaborating on all the areas to which this tool can be applied. Still we are confident that the reader will try to discover for himself the power of the method, which is a reflection of how an intelligent person would deal with difficult problems in the absence of extended indoctrination in techniques. This approach itself is easy to understand and use and does not require too much time to learn.

13-2. The Analytic Hierarchy Process

The Analytic Hierarchy Process (AHP) is a general theory of measurement. It is used to derive ratio scales from both discrete and continuous paired comparisons. These comparisons may be taken from actual measurements or from a fundamental scale that reflects the relative strength of preferences and feelings. The AHP has a special concern with departure from consistency and the measurement of this departure, and with dependence within and between the groups of elements of its structure. It has found its widest applications in multicriteria decision making, in planning and resource allocation, and in conflict resolution [3, 4, 5, 6, 7, 9]. In its general form the AHP is a nonlinear framework for carrying out both deductive and inductive thinking without use of the syllogism by taking several factors into consideration simultaneously and allowing for dependence and for feedback, and making numerical tradeoffs to arrive at a synthesis or conclusion.

We cover the subject here as follows. First we introduce the notions of absolute and relative measurement, followed by two hierarchically structured examples (in contrast with more elaborate feedback-network structures). Next we discuss some useful ideas with regard to scaling and to benefits and costs. This is followed by a section on the axioms and some of the central theoretical underpinnings of the theory. Then an elaborate section is included to show the importance of the eigenvector method for deriving scales of measurement; we also give measures of consistency for an entire hierarchy and for a network system. The final three sections deal with rank preservation and reversal, with group judgments, and with topics for investigation.

For a long time people have been concerned with the measurement of both physical and psychological events. By physical we mean the realm of what is fashionably known as the tangibles insofar as they constitute some kind of objective reality outside the individual conducting the measurement. By contrast, the psychological is the realm of the intangibles, comprising the subjective ideas, feelings, and beliefs of the individual and of society as a whole. The question is whether there is a coherent theory that can deal with both these worlds of reality without compromising either. The AHP is a method that can be used to establish measures in both the physical and social domains.

In using the AHP to model a problem, one needs a hierarchic or a network structure to represent that problem, as well as pairwise comparisons to establish relations within the structure. In the discrete case these comparisons lead to dominance matrices and in the continuous case to kernels of Fredholm Operators [15], from which ratio scales are derived in the form of principal eigenvectors, or eigenfunctions, as the case may be. These matrices, or kernels, are positive and reciprocal, e.g., $a_{ij} = 1/a_{ji}$. In particular, special effort has been made to characterize these matrices [5]. Because of the need for a variety of judgments, there has also been considerable work done to deal with the process of synthesizing diverse judgments [1].

13–3. Absolute and Relative Measurement and Structural Information

Cognitive psychologists have recognized for some time that there are two kinds of comparisons, absolute and relative. In absolute comparisons alternatives are compared with a standard in one's memory that has been developed through experience; in relative comparisons alternatives are compared in pairs according to a common attribute. The AHP has been used with both types of comparisons to derive ratio scales of measurement. We call such scales absolute and relative measurement scales. Relative measurement in the AHP is well developed and has already been used, as will be seen in the school-selection example, below. Here is a brief description of absolute measurement. Incidentally the software package Expert Choice [2] also includes this method of measurement under the name of "ratings."

Absolute measurement (sometimes called scoring) is applied to rank the alternatives in terms of the criteria or else in terms of ratings (or intensities) of the criteria; for example: excellent, very good, good, average, below average, poor, and very poor; or A, B, C, D, E, F, and G. After setting priorities for the criteria (or subcriteria, if there are any), pairwise comparisons are also made between the ratings themselves to set priorities for them under each criterion. Finally, alternatives are scored by checking off their respective ratings under each criterion and summing these ratings for all the criteria. This produces a ratio scale score for the alternative. The scores thus obtained of the alternatives can be normalized.

Using absolute measurement, no matter how many new alternatives are introduced, or old ones deleted, the ranks of the alternatives cannot reverse. This idea has been used to rank cities in the United States according to nine criteria as judged by six different people [11]. Another example of an appropriate use for absolute measurement is that of schools admitting students. Most schools set their criteria for admission independent of the performance of the current crop of students seeking admission. Their priorities are then used to determine whether a given student meets the standard set for qualification. In that case absolute measurement should be used to determine which students qualify for admission.

Absolute measurement needs standards, often set by society for its convenience, and sometimes having little to do with the values and objectives of the judge making the comparisons. In completely new decision problems or in old problems where no standards have been established, we must use relative measurement comparing alternatives in pairs to identify the best. The question now is, what happens to rank when using relative measurement, and alternatives are added or deleted?

When relative measurement is used, for example, to buy a car, even when the priorities of the criteria are set in advance independent of the alternatives, the car that qualifies in the end depends on the number of cars examined. Adding a new car to the collection being examined may cause reversal in the ranks of the original cars. This phenomenon can be accounted for by considering the normalization operation as a structural criterion that has to do with information generated in the measurement process. With relative measurement the priority of such a criterion changes when new alternatives are added or old ones deleted and hence the priorities of the old alternatives, which depend on all the criteria, including this structural criterion, change and a different rank order may occur among the old alternatives.

13-4. EXAMPLES

Relative Measurement: Choosing the Best House

When a family of average income is being advised on buying a house, the family identifies eight factors that they think they have to look for in a house. These factors fall into three categories: economic, geographic, and physical. Although one might begin by examining the relative importance of these clusters, the family feels they want to prioritize the relative importance of all the factors without working with clusters. The problem is to decide which of three candidate houses to choose. In applying the AHP, the first step is *decomposition*, or the structuring of the problem into a hierarchy. (See Figure 13-1.) On the first (or top) level is the overall goal of *Satisfaction with House*. On the second level are the eight factors or criteria that contribute to the goal, and on the third (or bottom) level are the three candidate houses that are to be evaluated in terms of the criteria on the second level. The definitions of the factors and the pictorial representation of the hierarchy follow.

The factors important to the individual family are:

1. *Size of House:* Storage space; size of rooms; number of rooms; total area of house.

2. *Transportation:* Convenience and proximity of bus service.

3. *Neighborhood:* Degree of traffic, security, view, taxes, physical condition of surrounding buildings.

4. *Age of House:* Self-explanatory.

5. *Yard Space:* Includes front, back, and side space, and space shared with neighbors.

6. *Modern Facilities:* Dishwashers, garbage disposals, air conditioning, alarm system, and other such items.

7. *General Condition:* Extent to which repairs are needed; condition of walls, carpet, drapes, wiring; cleanliness.

8. *Financing:* Availability of assumable mortage, seller financing, or bank financing.

The next step is *comparative judgment*. Arrange the elements on the second level into a matrix and elicit from the people buying the house judgments about the relative importance of the elements with respect to the overall goal, *Satisfaction with House*. The scale to use in making the judgments is given in Table 13-1. This scale has been validated for effectiveness, not only in many applications by a number of people, but also through theoretical comparisons with a large number of other scales.

The questions to ask when comparing two criteria are of the following kind: of the two alternatives being compared, which is considered more important by the family buying the house and how much more important is it with respect to family satisfaction with the house, which is the overall goal?

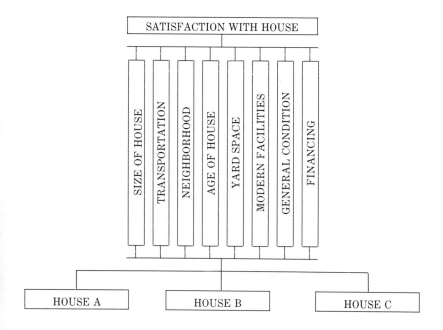

Fig. 13-1. Decomposition of the problem into a hierarchy.

The matrix of pairwise comparisons of the factors given by the homebuyers in this case is shown below, along with the resulting vector of priorities. The vector of priorities is the principal eigenvector of the matrix. This vector gives the relative priority of the factors measured on a ratio scale. That is, these priorities are unique to within a positive similarity transformation. However, if one insures that they add up to unity, they are always unique. In this case financing has the highest priority, with 33 percent of the influence.

In Table 13-2, instead of naming the criteria, we use the number previously associated with each.

TABLE 13–1: SCALE OF RELATIVE IMPORTANCE

Intensity of Relative Importance	Definition	Explanation
1	Equal importance	Two activities contribute equally to the objective.
3	Moderate importance of one over another	Experience and judgment slightly favor one activity over another.
5	Essential or strong	Experience and judgment strongly favor one activity over another.
7	Very strong importance	An activity is strongly favored and its dominance is demonstrated in practice.
9	Extreme importance	The evidence favoring one activity over another is of the highest possible order of affirmation.
2,4,6,8	Intermediate values between the two adjacent judgments	When compromise is needed.
Reciprocals of above nonzero numbers	If activity i has one of the above nonzero numbers assigned to it when compared with activity j, then j has the reciprocal value when compared to i.	

We now move to the pairwise comparisons of the elements on the bottom level. The elements to be compared pairwise are the houses with respect to how much better one is than the other in satisfying each criterion on the second level. Thus there will be eight 3 × 3 matrices of judgments since there are eight elements on level two, and three houses to be pairwise compared for each element. The matrices (Table 13–3) contain the judgments of the family involved. In order to facilitate understanding of the judgments, a brief description of the houses is given below.

House A: This house is the largest of them all. It is located in a good neighborhood with little traffic and low taxes. Its yard space is comparably larger than that of houses B and C. However, its general condition is not very good and it needs cleaning and painting. Also, the financing is unsatisfactory because it would have to be financed through a bank at a high rate of interest.

House B: This house is a little smaller than House A and is not close to a bus

MULTICRITERIA DECISION MAKING

TABLE 13–2: PAIRWISE COMPARISON MATRIX FOR LEVEL 1

	1	2	3	4	5	6	7	8	Priority Vector
1	1	5	3	7	6	6	1/3	1/4	.173
2	1/5	1	1/3	5	3	3	1/5	1/7	.054
3	1/3	3	1	6	3	4	6	1/5	.188
4	1/7	1/5	1/6	1	1/3	1/4	1/7	1/8	.018
5	1/6	1/3	1/3	3	1	1/2	1/5	1/6	.031
6	1/6	1/3	1/4	4	2	1	1/5	1/6	.036
7	3	5	1/6	7	5	5	1	1/2	.167
8	4	7	5	8	6	6	2	1	.333
								λ_{max} =	9.669
								C.I. =	.238
								C.R. =	.169

route. The neighborhood gives one the feeling of insecurity because of traffic conditions. The yard space is fairly small and the house lacks the basic modern facilities. On the other hand, its general condition is very good. Also, an assumable mortgage is obtainable, which means the financing is good with a rather low interest rate.

House C: House C is very small and has few modern facilities. The neighborhood has high taxes, but is in good condition and seems secure. The yard space is bigger than that of House B, but is not comparable to House A's spacious surroundings. The general condition of the house is good, and it has a pretty carpet and drapes. The financing is better than for A but not better than for B.

Following (Table 13–3) are the matrices of the houses and their local priorities with respect to the elements on level two.

The next step is to apply the *principle of composition of priorities* (or *synthesis of priorities*). In order to establish the composite or global priorities of the houses we lay out in a matrix (Table 13–4) the local priorities of the houses with respect to each criterion and multiply each column of vectors by the priority of the corresponding criterion and add across each row, which results in the composite or global priority vector of the houses. House A, which was the least desirable with respect to financing (the highest-priority criterion), contrary to expectation was the winner. It was the house that was bought.

While there is no interdependence among the factors in the house-buying example just given, occasionally the problem of interdependence arises at different levels of the hierarchy. When there is such interdependence among a set of factors on a particular level, for each factor we measure the importance of the contribution of each of the other factors to it. In this manner we generate a matrix of vectors of dependence. We weight each column of this matrix by

TABLE 13-3: COMPARISON MATRICES AND LOCAL PRIORITIES

Size of House	A B C	Priority Vector
A	1 6 8	.754
B	1/6 1 4	.181
C	1/8 1/4 1	.065
	λ_{max} = 3.136	
	C.I. = .068	
	C.R. = .117	

Yard Space	A B C	Priority Vector
A	1 5 4	.674
B	1/5 1 1/3	.101
C	1/4 3 1	.226
	λ_{max} = 3.086	
	C.I. = .043	
	C.R. = .074	

Transportation	A B C	Priority Vector
A	1 7 1/5	.233
B	1/7 1 1/8	.055
C	5 8 1	.713
	λ_{max} = 3.247	
	C.I. = .124	
	C.R. = .213	

Modern Facilities	A B C	Priority Vector
A	1 8 6	.747
B	1/8 1 1/5	.060
C	1/6 5 1	.193
	λ_{max} = 3.197	
	C.I. = .099	
	C.R. = .170	

Neighborhood	A B C	Priority Vector
A	1 8 6	.745
B	1/8 1 1/4	.065
C	1/6 4 1	.181
	λ_{max} = 3.130	
	C.I. = .068	
	C.R. = .117	

General Condition	A B C	Priority Vector
A	1 1/2 1/2	.200
B	2 1 1	.400
C	2 1 1	.400
	λ_{max} = 3.000	
	C.I. = .000	
	C.R. = .000	

Age of House	A B C	Priority Vector
A	1 1 1	.333
B	1 1 1	.333
C	1 1 1	.333
	λ_{max} = 3.000	
	C.I. = .000	
	C.R. = .000	

Financing	A B C	Priority Vector
A	1 1/7 1/5	.072
B	7 1 3	.650
C	5 1/3 1	.278
	λ_{max} = 3.065	
	C.I. = .032	
	C.R. = .056	

the importance of the dependent factors and add across the rows to obtain a measure of interdependence. The result is to shift the weights among these factors from the more dependent ones to the less dependent ones. The weights of the factors outside the set remain unaffected. The resulting vector of interdependence is then used to smooth the vector of independence priorities by multiplying corresponding priorities and renormalizing.

TABLE 13-4: LOCAL PRIORITIES AND THE FINAL VECTOR

	1 (.173)	2 (.054)	3 (.188)	4 (.018)	5 (.031)	6 (.036)	7 (.167)	8 (.333)		
A	.754	.233	.754	.333	.674	.747	.200	.072	=	.396
B	.181	.055	.065	.333	.101	.060	.400	.650		.341
C	.065	.713	.181	.333	.226	.193	.400	.278		.263

Absolute Measurement: Evaluating Employees for Raises

Employees are evaluated for raises. The criteria are Dependability, Education, Experience, and Quality. Each criterion is subdivided into intensities, standards, or subcriteria as shown in Figure 13-2. Priorities are set for the criteria by comparing them in pairs, and these priorities are then given in a matrix. The subcriteria are then pairwise compared according to priority with respect to their parent criterion (as in Table 13-5) and their priorities are then weighted by the priority of the criterion. Finally, each individual is rated in a table (see Table 13-6) by assigning the subcriterion that applies to him or her under each criterion. The scores of these subcriteria are summed to derive a total score for the individual. This approach can be used whenever it is possible to set priorities for intensities of criteria, which is often possible when sufficient experience with a given operation has been accumulated.

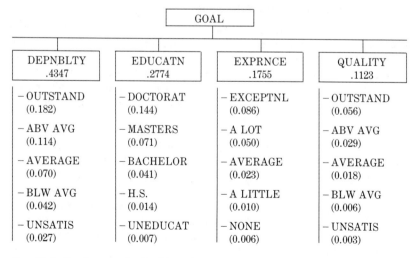

FIG. 13-2. Employee evaluation hierarchy.

TABLE 13–5: Judgments and Priorities with Respect to Dependability < Goal

	OUTSTAND	ABV AVG	AVERAGE	BLW AVG	UNSATIS	PRIORITIES
OUTSTAND	1.0	2.0	3.0	4.0	5.0	0.419
ABV AVG	1/2	1.0	2.0	3.0	4.0	0.263
AVERAGE	1/3	1/2	1.0	2.0	3.0	0.160
BLW AVG	1/4	1/3	1/2	1.0	2.0	0.097
UNSATIS	1/5	1/4	1/3	1/2	1.0	0.062

INCONSISTENCY RATIO = 0.015

TABLE 13–6: Ranking Alternatives

	DEPNBLTY .4347	EDUCATN .2774	EXPERNCE .1755	QUALITY .1123	TOTAL
1 ADAMS, V	OUTSTAND	BACHELOR	A LITTLE	OUTSTAND	0.289
2 BECKER, L	AVERAGE	BACHELOR	A LITTLE	OUTSTAND	0.177
3 HAYAT, F	AVERAGE	MASTERS	A LOT	BLW AVG	0.197
4 KESSELMAN, S	ABV AVG	H.S.	NONE	ABV AVG	0.163
5 O'SHEA, K	AVERAGE	DOCTORAT	A LOT	ABV AVG	0.293
6 PETERS, T	AVERAGE	DOCTORAT	A LOT	AVERAGE	0.282
7 TOBIAS, K	ABV AVG	BACHELOR	AVERAGE	ABV AVG	0.208

13–6. Further Comments on the Scale

There are various ways of carrying out measurement in particular pairwise comparisons. When the elements being compared share a measurable property such as weight, they can be measured directly on an absolute scale, and pairwise comparisons become unnecessary. However, if one uses the AHP and forms ratios of the measurable factors (resulting in a consistent matrix), then solving for the principal eigenvector to obtain the priorities (see below) gives the same result obtained by normalizing the numbers.

It is rare that numbers should be used in this way. It is almost always the case that measurement only indicates some kind of arithmetical accuracy that does not reflect the actual value that an individual would assign to the numbers to reflect the satisfaction of his needs. In some situations habituation and familiarity may cause people to use the numbers as they are. It must be understood that these people should be able to justify how such numbers correspond to their own judgments of the relative intensities of importance. There are situations when people use their judgment to estimate numerical magnitudes. They should do so by means of comparison. Numbers or ranges of numbers can be arranged into intensity-equivalence classes and then compared either directly, by using representative numbers from each range, or indirectly—qualitatively according to intensity. Without these means of comparison, numbers in the millions, billions, or trillions may mean practically the same thing to an individual who is unfamiliar with very large numbers, does not know what they apply to, or does not understand their real significance in a particular instance.

Here we need to remember that the reason for a hierarchic structure is to decompose complexity in stages to enable us to scale its smallest elements gradually upward in terms of its largest elements. If it were possible to assign meaningful numerical values to the smallest elements, there would be no need for the elaborate process of careful decomposition.

Thus the question remains as to how responsive individual judgments can be to make it possible to discriminate between elements sharing a property, or properties sharing a higher property. One of the axioms of the AHP relates to how disparate the elements are allowed to be.

In making paired comparisons, the accuracy of the judgments and hence also the derived scale may be improved by ordering the elements according to rank in descending or ascending order. In addition one may compare the largest element with the smallest one first, or, alternatively, take the middle element as the basis of the comparisons.

Two problems arise in teaching people to use the AHP: first, that of enabling them to specify qualitative intensities of judgment and feeling that facilitate spontaneous response without elaborate prior training; and, second, that of enabling them to associate appropriate linguistic designations with these expressions. Numerical scale values must systematically be associated with those verbal expressions that lead to meaningful outcomes; these values, particularly in known situations, must be capable of being tested for their accuracy. Small changes in the words (or the numbers) should lead to small changes in the answer derived from them.

Finally we use consistency arguments, along with the well-known work of Fechner in psychophysics, to derive and substantiate the scale and its range.

In 1860 Fechner [5] considered a sequence of just-noticeable increasing stimuli. He denoted the first one by s_o, the next just-noticeable stimulus by

$$s_1 = s_o + \Delta s_o = s_o + \frac{\Delta s_o}{s_o} s_o = s_o (1+r)$$

having used Weber's law. Similarly

$$s_2 = s_1 + \Delta s_1 = s_1 (1+r) = s_o (1+r)^2 \equiv s_o \alpha^2$$

In general

$$s_n = s_{n-1} \alpha = s_o \alpha^n \ (n = 0, 1, 2, ...)$$

Thus stimuli of noticeable differences follow sequentially in a geometric progression. Fechner felt that the corresponding sensations should follow each other in an arithmetic sequence occurring at the discrete points at which just-noticeable differences occur. But the latter are obtained when we solve for n. We have

$$n = (\log s_n - \log s_o)/\log \alpha$$

and sensation is a linear function of the logarithm of the stimulus. Thus if M denotes the sensation and s the stimulus, the psychophysical law of Weber-Fechner is given by

$$M = a \log s + b, a \neq 0$$

We assume that the stimuli arise in making pairwise comparisons of relatively comparable activities. We are interested in responses whose numerical values are in the form of ratios. Thus $b = 0$, from which we must have $\log s_o = 0$ or $s_o = 1$, which is possible by calibrating a unit stimulus. The next noticeable response is due to the stimulus

$$s_1 = s_0 \alpha = \alpha$$

This yields a response $\log \alpha / \log \alpha = 1$. The next stimulus is

$$s_2 = s_o \alpha^2$$

which yields a response of 2. In this manner we obtain the sequence 1, 2, 3, ... For the purpose of consistency we place the activities in a cluster whose pairwise-comparison stimuli give rise to responses whose numerical values are of the same order of magnitude. In practice, qualitative differences in response to stimuli are few. Roughly, there are five distinct ones as listed in Table 13-1, with additional ones that are compromises between adjacent responses. (The notion of compromise derives from the thinking, judgmental process as distinct from the sensory process.) This brings the total up to nine, which is compatible with the order-of-magnitude assumption made earlier.

Now we examine the impact of consistency on scaling.

The mind absorbs new ideas by contrasting them through scanning or through concentration and analysis with other ideas and by understanding how they are similar to familiar ideas. They are also related to current or future activities and applied to concrete situations to test their compatibility with what is already believed to be workable.

The ideas may be accepted as a consistent part of the existing understanding or they may be inconsistent with what is already known or accepted. In that case the system of understanding and practice is extended, expanded, and adjusted to include the new ideas.

Growth implies such expansion. If the adjustment of old ideas to accommodate a new one is drastic, then the inconsistency of the new idea is great and may require considerable adjustments in the old ideas and beliefs, whose old relations are no longer intuitively recognizable, and may call for a new theory or interpretation if at all possible. But such major changes cannot be made every hour, every day, or even every week, because it takes time to interpret and assimilate relations. Thus inconsistency arising from exposure to new ideas or situations can be threatening, unsettling, and painful. Our biology has recognized this and developed for us ways to filter information.

When we are exposed to a new idea, we usually make minor adjustments in what we already know. We absorb new ideas gradually, by interpreting them from the vantage point of already established relations. In this process, our emphasis on consistency exceeds our desire for exposure and readjustment. As a result, maintaining consistency is considered to be of higher priority than changing, because of the disturbing impact of recently encountered concepts or experiences. Yet these are also considered genuine—though slightly less important—concerns. One conclusion is that our tendency toward consistency differs by one order of magnitude from our tendency to change—a split of 90 percent and 10 percent.

In addition, in order to maintain the identity of old ideas, their significance must be visibly greater than that of the adjustments we would make in them as a result of exposure to the new. In other words, the 90 percent effort to maintain consistency can at best be divided among a few entities, each of which would receive in the understanding an emphasis or priority of the order of 10 percent so that slight readjustment (i.e., by an order of 1 percent) would not change the old relations significantly. With such gradual adjustment new relations can be systematically made with little stress, and change is achieved at a certain level of experience. In order to accomplish this, the mind in establishing relations according to priority at a certain level needs the ability to distinguish the relative importance of a few entities in any issue. The number of objects whose average priority is 10 percent (to conform with the desire for consistency) is at most nine.

13-7. How to Structure a Hierarchy

Perhaps the most creative part of decision making that has a significant effect on the outcome is the structuring of the decision as a hierarchy. The basic principle to follow in creating this structure is always to see if one can answer the following question: "Can I compare the elements on a lower level in terms of some or all of the elements on the next higher level?"

A useful way to proceed is to come down from the goal as far as one can and then go up from the alternatives until the levels of the two processes are linked in such a way as to make comparison possible. Here are some suggestions.

1. Identify overall goal. What are you trying to accomplish? What is the main question?

2. Identify subgoals of overall goal. If relevant, identify time horizons that affect the decision.
3. Identify criteria that must be satisfied to fulfill subgoals of the overall goal.
4. Identify subcriteria under each criterion. Note that criteria or subcriteria may be specified in terms of ranges of values of parameters or in terms of verbal intensities such as high, medium, low.
5. Identify actors involved.
6. Identify actor goals.
7. Identify actor policies.
8. Identify options or outcomes.
9. For yes-no decisions take the most preferred outcome and compare benefits and costs of making the decision with those of not making it.
10. Do benefit/cost analysis using marginal values. Because we are dealing with dominance hierarchies, ask which alternative yields the greatest benefit; for costs, which alternative costs the most.

13-8. Scale Comparisons

Some people have suggested alternative scales to the 1-9 scale used here. We have experimented with a wide variety of such scales (given in Table 13-7) to compare the answers obtained by using the following pairwise discrimination attributes: E, equal; M, moderate; S, strong; VS, very strong; and EX, extreme. B is used to indicate a value between a pair of these. Two applications are included here, in Tables 13-8 and 13-9. One is made to a weight-comparison example and one to a beverage-consumption example. The actual answers are given at the bottom of each table. They should be compared with the solutions obtained for a variety of scales in Table 13-7, e.g., (1) 1-3, (2) 1-5, (3) 1-7, and so on. The best results for the examples below are obtained for the scales (4), (12), and (26). There is a counterexample to the validity of the exponential scale (26).

The use of the scale 1-9 has been justified and demonstrated by many examples illustrated in "The Analytic Hierarchy Process" [5] (which describes the AHP in detail). However, the following simple optics illustration, carried out with little children, shows that perceptions, judgments, and these numbers lead to results which can be validated by laws of physics. In this example, four identical chairs were placed at distances of 9, 15, 21, and 28 yards from a floodlight. The children stood by the light, looked at the line of chairs, and compared the first with the second, with the third, and then with the fourth. They repeated this procedure until each chair had been compared with every other. Each time, the children expressed as best they could the relative brightness of the two chairs being compared.

Their judgments were entered in a matrix (Table 13-10) to record the relative

TABLE 13-7: SCALE CONVERSION

Scales	Equal	Betwn	Modt	Betwn	Strg	Betwn	Very Strg	Betwn	Extrm
1) 1–3	1	2	2	2	2	3	3	3	3
2) 1–5	1	2	2	3	3	4	4	5	5
3) 1–7	1	2	2	3	4	5	6	6	7
4) 1–9	1	2	3	4	5	6	7	8	9
5) 1–11	1	3	4	5	7	8	9	10	11
6) 1–13	1	3	4	6	7	9	10	12	13
7) 1–15	1	3	5	7	8	9	11	13	15
8) 1–17	1	3	5	7	9	11	13	15	17
9) 1–18	1	4	6	8	10	12	14	16	18
10) 1–26	1	5	8	11	14	17	20	23	26
11) 1–90	1	20	30	40	50	60	70	80	90
12) .9	1	.9 times corresponding values in 1–9 scale above							
13) .7	1	.7 times corresponding values in 1–9 scale above							
14) .5	1	.5 times corresponding values in 1–9 scale above							
15) .3	1	.3 times corresponding values in 1–9 scale above							
16) .1	1	.1 times corresponding values in 1–9 scale above							
17) $1 + 0.x$	1	$1 + 0.x$ where x is the corresponding value in 1–9 above							
18) $2 + 0.x$	1	$2 + 0.x$ where x is the corresponding value in 1–9 above							
19) $3 + 0.x$	1	$3 + 0.x$ where x is the corresponding value in 1–9 above							
20) $4 + 0.x$	1	$4 + 0.x$ where x is the corresponding value in 1–9 above							
21) $\sqrt{x}$	1	$\sqrt{x}$ where x is the corresponding value in 1–9 scale above							
22) x^2	1	x^2 where x is the corresponding value in 1–9 scale above							
23) x^3	1	x^3 where x is the corresponding value in 1–9 scale above							
24) x^4	1	x^4 where x is the corresponding value in 1–9 scale above							
25) x^5	1	x^5 where x is the corresponding value in 1–9 scale above							
26) $2^{n/2}$	$2^0=1$	$2^{.5}=1.414$	$2^1=2$	$2^{1.5}=2.828$	$2^2=4$	$2^{2.5}=5.657$	$2^3=8$	$2^{3.5}=11.31$	$2^4=16$
27) $9^{x/8}$	1	$9^{1/8}$	$9^{2/8}$	$9^{3/8}$	$9^{4/8}$	$9^{5/8}$	$9^{6/8}$	$9^{7/8}$	9
28) $\log n$, $n \geq e$	$\log e$ (1)	$\log 3$ (1.099)	$\log 4$ (1.386)	$\log 5$ (1.609)	$\log 6$ (1.792)	$\log 7$ (1.946)	$\log 8$ (2.079)	$\log 9$ (2.197)	$\log 10$ (2.303)

brightness of the chairs. The reciprocals were used in the transpose position. The results are in the table.

The inverse-square law of optics is now used to test these judgments. Since the distances are 9, 15, 21, and 28 yards, we square these numbers and take their reciprocals. This gives .0123, .0044, .0023, and .0013 respectively. If we normalize these values we get .61, .22, .11, and .06, which are very close to the brightness ratios obtained in our test using the AHP.

TABLE 13-8: WEIGHT COMPARISON EXAMPLE

	Radio	Typewriter	Large Attaché Case	Projector	Small Attaché Case
Radio	E	...	...	...	B(M–S)
Typewriter	S	E	B(E–M)	B(E–M)	B(VS–EX)
Large Attaché Case	M	...	E	...	B(M–S)
Projector	B(M–S)	...	B(E–M)	E	VS
Small Attaché Case	...	...	...	...	E

... | indicates the reciprocal of its transpose

	Eigenvector for Each Scale				λ_{max}	RMS	MAD
1)	0.136	0.340	0.180	0.257	0.086	5.130	0.036	0.037
2)	0.116	0.374	0.178	0.273	0.058	5.122	0.017	0.015
3)	0.104	0.395	0.170	0.281	0.050	5.094	0.015	0.005
4)	0.088	0.405	0.183	0.286	0.038	5.161	0.013	0.015
5)	0.069	0.472	0.153	0.278	0.028	5.363	0.045	0.020
6)	0.067	0.472	0.151	0.286	0.024	5.387	0.046	0.032
7)	0.062	0.470	0.158	0.289	0.021	5.486	0.045	0.023
8)	0.058	0.481	0.154	0.288	0.019	5.436	0.051	0.025
9)	0.051	0.518	0.136	0.279	0.016	5.681	0.069	0.033
10)	0.039	0.556	0.120	0.273	0.011	5.965	0.088	0.033
11)	0.011	0.726	0.054	0.207	0.002	9.124	0.172	0.026
12)	0.095	0.387	0.191	0.284	0.043	5.120	0.008	0.006
13)	0.113	0.344	0.207	0.279	0.057	5.059	0.023	0.004
14)	0.139	0.287	0.228	0.265	0.081	5.058	0.054	0.014
15)	0.181	0.207	0.249	0.233	0.131	5.233	0.102	0.042
16)	0.260	0.080	0.237	0.138	0.284	6.417	0.200	0.169
17)	0.171	0.256	0.205	0.232	0.136	5.008	0.082	0.066
18)	0.125	0.356	0.181	0.256	0.083	5.224	0.029	0.020
19)	0.097	0.422	0.161	0.263	0.057	5.528	0.024	0.020
20)	0.079	0.470	0.145	0.263	0.042	5.837	0.045	0.013
21)	0.141	0.305	0.205	0.256	0.093	5.040	0.049	0.036
22)	0.029	0.566	0.118	0.283	0.005	5.680	0.094	0.047
23)	0.009	0.681	0.068	0.243	0.001	6.650	1.150	0.052
24)	0.002	0.763	0.038	0.196	0.000	8.212	0.191	0.034
25)	0.001	0.824	0.021	0.154	0.000	10.572	0.221	0.059
26)	0.101	0.380	0.193	0.284	0.042	5.093	0.008	0.011
27)	0.120	0.344	0.201	0.274	0.060	5.058	0.024	0.016
28)	0.172	0.259	0.209	0.239	0.121	5.008	0.077	0.063
	0.100	0.390	0.200	0.270	0.040	= Actual Vector Solution		

TABLE 13-9: U.S. BEVERAGE CONSUMPTION EXAMPLE (A NEW MATRIX)

	Coffee	Wine	Tea	Beer	Soft Drink	Milk	Water
Coffee	E	B(VS–EX)	S	B(M–S)	B(E–M)	M	...
Wine	...	E	...	...	...	...	...
Tea	...	S	E	...	...	...	...
Beer	...	B(S–VS)	B(M–S)	E	...	...	...
Soft Drink	...	VS	S	M	E	M	...
Milk	...	B(S–VS)	S	B(E–M)	...	E	...
Water	M	EX	VS	S	M	M	E

... indicates the reciprocal of its transpose

	Eigenvector for Each Scale						λ_{max}	RMS	MAD
1)	0.196	0.050	0.076	0.105	0.162	0.162	0.249	7.543	0.031	1.026
2)	0.216	0.034	0.058	0.096	0.163	0.163	0.270	7.573	0.022	0.008
3)	0.215	0.025	0.048	0.088	0.168	0.165	0.291	7.590	0.018	0.016
4)	0.217	0.019	0.038	0.074	0.170	0.140	0.342	7.988	0.025	0.010
5)	0.225	0.013	0.028	0.056	0.163	0.142	0.373	8.811	0.039	0.019
6)	0.230	0.012	0.026	0.056	0.161	0.140	0.376	8.792	0.040	0.019
7)	0.225	0.010	0.023	0.049	0.161	0.125	0.407	9.214	0.051	0.015
8)	0.224	0.008	0.021	0.048	0.161	0.126	0.413	9.221	0.053	0.013
9)	0.228	0.007	0.018	0.039	0.155	0.127	0.425	10.024	0.059	0.012
10)	0.227	0.004	0.012	0.029	0.149	0.119	0.459	11.281	0.072	0.016
11)	0.227	0.000	0.002	0.004	0.129	0.105	0.533	27.564	0.103	0.026
12)	0.214	0.022	0.043	0.080	0.172	0.142	0.328	7.797	0.020	0.011
13)	0.205	0.030	0.054	0.097	0.175	0.145	0.294	7.437	0.013	0.010
14)	0.189	0.044	0.073	0.120	0.178	0.147	0.248	7.144	0.027	0.011
15)	0.160	0.075	0.107	0.157	0.175	0.144	0.181	7.054	0.061	0.045
16)	0.087	0.183	0.195	0.216	0.138	0.111	0.071	8.215	0.137	0.125
17)	0.170	0.008	0.108	0.130	0.159	0.152	0.192	7.097	0.058	0.039
18)	0.203	0.051	0.070	0.093	0.162	0.158	0.262	7.777	0.028	0.027
19)	0.220	0.034	0.050	0.071	0.160	0.157	0.308	8.595	0.024	0.010
20)	0.229	0.024	0.037	0.056	0.157	0.155	0.341	9.443	0.033	0.017
21)	0.191	0.056	0.080	0.112	0.168	0.153	0.241	7.300	0.033	0.021
22)	0.215	0.002	0.007	0.024	0.144	0.097	0.512	11.114	0.091	0.025
23)	0.185	0.000	0.001	0.006	0.120	0.066	0.622	18.969	0.135	0.029
24)	0.167	0.000	0.000	0.001	0.107	0.048	0.688	39.091	0.161	0.033
25)	0.134	0.000	0.000	0.000	0.097	0.035	0.733	90.362	0.180	0.038
26)	0.213	0.018	0.045	0.092	0.175	0.126	0.332	7.142	0.018	0.022
27)	0.205	0.029	0.059	0.105	0.175	0.135	0.292	7.088	0.012	0.014
28)	0.178	0.074	0.097	0.129	0.168	0.142	0.211	7.052	0.0478	0.024
	0.200	0.010	0.040	0.120	0.180	0.140	0.300	= Actual Vector Solution		

TABLE 13-10: OPTICS EXAMPLE

	Chair 1	Chair 2	Chair 3	Chair 4	Brightness Ratio
Chair 1	1	5	6	7	0.61
Chair 2	1/5	1	4	6	0.24
Chair 3	1/6	1/4	1	4	0.10
Chair 4	1/7	1/6	1/4	1	0.05
				λ_{max} =	4.39
				C.I. =	0.13
				C.R. =	0.14

13-9. Comments on Cost/Benefit Analysis

Often, the alternatives from which a choice must be made in a choice-making situation have both costs and benefits associated with them. In this case it is useful to construct separate costs and benefits hierarchies, with the same alternatives on the bottom level of each. Thus one obtains both a costs-priority vector and a benefits-priority vector. The benefit/cost vector is obtained by taking the ratio of the benefit priority to the costs priority for each alternative, with the highest such ratio indicating the preferred alternative. In the case where resources are allocated to several projects, such benefit-to-cost ratios or the corresponding marginal ratios prove to be very valuable.

For example, in evaluating three types of copying machines, one represents in the benefits hierarchy the good attributes one is looking for, and one represents in the costs hierarchy the pain and economic costs that one would incur in buying or maintaining the three types of machines. Note that the criteria for benefits and the criteria for costs need not be simply opposites of each other but may be totally different. Also note that each criterion may be regarded at a different threshold of intensity and that such thresholds may themselves be prioritized according to desirability, with each alternative evaluated only in terms of its highest-priority threshold level.

13-10. The Axioms of the Analytic Hierarchy Process

As we have seen, three principles guide one in problem solving using the AHP [5]: *decomposition*, *comparative judgment*, and *synthesis of priorities*.

The *decomposition* principle is applied by first structuring a simple problem with elements of each level independent of those on succeeding levels, and then working downward from the goal on the top level through criteria bearing on the goal on the second level to subcriteria on the third level, and so on, going from the more general (and sometimes uncertain) to the more particular and concrete. Saaty [5] makes a distinction between two types of dependence which he calls functional and structural. The former is the familiar contextual dependence of elements on other elements in performing their function, whereas the latter is the dependence of the importance or priority of elements on the priority itself and on a number of other elements. Absolute measurement, sometimes called scoring, is used when it is desired to ignore structural dependence between elements while relative measurement is used in the alternative case.

The principle of *comparative judgment* is applied to construct pairwise comparisons of the relative importance of elements on some given level with respect to a shared criterion or property on the level above, giving rise to the kind of matrix we have already encountered and to its corresponding principal eigenvector.

The third principle is *synthesis of priorities*. In the AHP, priorities are synthesized from the second level down by multiplying local priorities by the priority of their corresponding criterion on the level above, and weighting each element on a level according to the criteria it affects. (The second-level elements are multiplied by unity, the weight of the single top-level goal.) This gives the composite or global priority of that element, which is then used to weight the local priorities of the elements on the level below, and so on, repeating this procedure to the bottom level.

Four axioms govern the Analytic Hierarchy Process and utilize the notion of paired comparisons as a primitive [10].

Let $\mathfrak{A}$ be a finite set of n elements called alternatives. Let $\mathfrak{C}$ be a set of properties or attributes with respect to which elements in $\mathfrak{A}$ are compared. We will refer to the elements of $\mathfrak{C}$ as criteria. A criterion is a primitive.

We perform binary comparisons on the elements in $\mathfrak{A}$ according to a criterion in $\mathfrak{C}$. Let $>_C$ be a binary relation on $\mathfrak{A}$ representing "preferred to" with respect to a criterion C in $\mathfrak{C}$. Let $\sim_C$ be the binary relation "indifferent with respect to" a criterion C in $\mathfrak{C}$. Hence, given two elements $A_i, A_j \in A$, either $A_i >_C A_j$ or $A_j >_C A_i$ or $A_i \sim_C A_j$ for all $C \in \mathfrak{C}$. We use $A_i \succcurlyeq_C A_j$ to indicate "preferred to or indifferent with respect to." A given family of binary relations $>_C$ with respect to a criterion C in $\mathfrak{C}$ is a primitive.

Let $\mathfrak{P}$ be the set of mappings from $\mathfrak{A} \times \mathfrak{A}$ to $\mathbf{R}^+$ (the set of positive reals). Let $f:\mathfrak{C} \to \mathfrak{P}$. Let $P_C \in f(C)$ for $C \in \mathfrak{C}$. P_C assigns a positive real number to every pair $(A_i, A_j) \in \mathfrak{A} \times \mathfrak{A}$. Let $P_C(A_i, A_j) \equiv a_{ij} \in \mathbf{R}^+$, $A_i, A_j \in \mathfrak{A}$. For each $C \in \mathfrak{C}$, the triple $(\mathfrak{A} \times \mathfrak{A}, \mathbf{R}^+, P_C)$ is a *fundamental* or *primitive* scale. A fundamental scale is a mapping of objects to a numerical system.

Definition: For all $A_i, A_j \in \mathfrak{A}$ and $C \in \mathfrak{C}$

$$A_i >_C A_j \text{ if and only if } P_C(A_i, A_j) > 1,$$
$$A_i \sim_C A_j \text{ if and only if } P_C(A_i, A_j) = 1$$

If $A_i >_C A_j$ we say that A_i dominates A_j with respect to $C \in \mathfrak{C}$. Thus P_C represents the intensity or strength of preference for one alternative over another.

Axiom 1 *(the Reciprocal Axiom):* For all $A_i, A_j \in \mathfrak{A}$ and $C \in \mathfrak{C}$

$$P_C(A_i, A_j) = 1/P_C(A_j, A_i)$$

This axiom says that the comparison matrices we construct are formed of paired reciprocal comparisons, for if one stone is judged to be five times as heavy as another, then the other must perforce be one fifth as heavy as the first. It is this simple but powerful relationship that is the basis of the AHP.

Let $A = (a_{ij}) \equiv (P_C(A_i, A_j))$ be the set of paired comparisons of the alternatives with respect to a criterion $C \in \mathfrak{C}$. By the definition of P_C and Axiom 1, A is a positive reciprocal matrix. The object is to obtain a scale of relative dominance (or rank order) of the alternatives from the paired comparisons given in A.

We will now show how to derive the relative dominance of a set of alternatives from a pairwise comparison matrix A. Let $R_M(n)$, be the set of $(n \times n)$ positive reciprocal matrices $A = (a_{ij}) \equiv (P_C(A_i, A_j))$ for all $C \in \mathfrak{C}$. Let $[0,1]^n$ be the n-fold cartesian product of $[0,1]$ and let $W: R_M(n) > [0,1]^n$ for $A \in R_M(n)$. $W(A)$ is a n-dimensional vector whose components lie in the interval $[0,1]$. The triple $(R_{M(n)}, [0,1]^n, W)$ is a derived scale. A derived scale is a mapping between two numerical relational systems.

An important aspect of the AHP is the idea of consistency. If one has a scale for a property possessed by some objects and measures that property in them, then their relative weights with respect to that property are fixed. In this case there is no judgmental inconsistency (although if one has a physical scale and applies it to objects in pairs and *then* derives the relative standing of the objects on the scale from the pairwise comparison matrix, it is likely that inaccuracies will have occurred in the act of applying the physical scale, and again there would be inconsistency). But when comparing with respect to a property for which there is no established scale or measure, we are trying to derive a scale through comparing the objects two at a time. Since the objects may be involved in more than one comparison and we have no standard scale but are assigning relative values as a means of judgment, inconsistencies may well occur. In the AHP consistency is defined in the following way (and we are able to measure inconsistency; see Section 13–4).

Definition: The mapping P_C is said to be consistent if and only if

$$P_C(A_i, A_j) P_C(A_j, A_k) = P_C(A_i, A_k) \text{ for all } i, j, k$$

Similarly, the matrix A is consistent if and only if $a_{ij} a_{jk} = a_{ik}$ for all i, j, and k.

We now turn to the hierarchic Axioms 2, 3, and 4 and related definitions.

In a partially ordered set, we define $x < y$ to mean that $x < y$ and $x \neq y$. y is said to cover (dominate) x. If $x < y$ then $x < t < y$ is possible for no t. We use the notation $x^- = \{y | x \text{ covers } y\}$ and $x^+ = \{y | y \text{ covers } x\}$, for any element x in an ordered set.

Let H be a finite partially ordered set. Then H is a hierarchy if it satisfies the following conditions:

(a) There is a partition of H into sets L_k, $k = 1, \ldots, h$ for some h where $L_1 = \{b\}$, b a single element.

(b) $x \in L_k$ implies $x^- \in L_k+1$ $k = 1, \ldots, h-1$.
(c) $x \in L_k$ implies $x^+ \in L_{k-1}$ $k = 2, \ldots, h$.

The notions of fundamental and derived scales can be extended to $x \in L_k$, $x^- \subseteq L_{k+1}$ replacing C and A respectively. The derived scale resulting from comparing the elements in x^- with respect to x is called a local derived scale or the local priorities for the elements in x^-.

Definition: Given a positive real number $\rho \geq 1$, a nonempty set $x^- \subseteq L_{k+1}$ is said to be ρ-homogeneous with respect to $x \in L_k$ if for every pair of elements $y_1, y_2 \in x^-$, $1/\rho \leq P_C(y_1, y_2) \leq \rho$. In particular the reciprocal axiom implies that $P_C(y_i, y_i) = 1$.

Axiom 2 (the Homogeneity Axiom): Given a hierarchy H, $x \in H$ and $x \in L_k$, $x^- \subseteq L_{k+1}$ is ρ-homogeneous for $k = 1, \ldots, h - 1$.

Homogeneity is essential for meaningful comparisons, as the mind cannot compare widely disparate elements. For example, we cannot compare a grain of sand with an orange according to size. When the disparity is great, elements should be placed in separate clusters of comparable size, or on different levels altogether.

Given $L_k, L_{k+1} \subseteq H$, let us denote the local derived scale for $y \in x^-$ and $x \in L_k$ by $\psi_{k+1}(y/x)$, $k = 2, 3, \ldots, h - 1$. Without loss of generality we may assume that $\psi_{k+1}(y/x) = 1$. Consider the matrix $\psi_k(L_k/L_{k-1})$ whose columns are local derived scales of elements in L_k with respect to elements in L_{k-1}.

Definition: A set A is said to be *outer dependent* on a set C if a fundamental scale can be defined on A with respect to every $C \in C$.

The process of relating elements (e.g., alternatives) on one level of the hierarchy according to the elements of the next higher level (e.g., criteria) expresses the dependence (what is called *outer dependence*) of the lower elements on the higher so that comparisons can be made between them. The steps are repeated upward in the hierarchy through each pair of adjacent levels to the top element, the focus or goal.

The elements on a level may also depend on one another with respect to a property on another level. Input-output of industries is an example of the idea of *inner dependence*, formalized as follows:

Definition: Let $\mathfrak{A}$ be outer dependent on $\mathfrak{C}$. The elements in $\mathfrak{A}$ are said to be *inner dependent* with respect to $C \in \mathfrak{C}$ if for some $A \in \mathfrak{A}$, $\mathfrak{A}$ is outer dependent on A.

Axiom 3: Let H be a hierarchy with levels $L_1, L_2, \ldots, L_h$. For each L_k, $k = 1, 2, \ldots, h-1$,
 (1) L_{k+1} is outer dependent on L_k,
 (2) L_{k+1} is not inner dependent with respect to all $x \in L_k$,
 (3) L_k is not outer dependent on L_{k+1}.

Axiom 4 (the Axiom of Expectations):

$$C \subset H - L_h, A = L_h$$

This axiom is merely the statement that thoughtful individuals who have reasons for their beliefs should make sure that their ideas are adequately represented in the model. All alternatives, criteria, and expectations (explicit and implicit) can and should be represented in the hierarchy. This axiom does not assume rationality. People are known at times to harbor irrational expectations and such expectations can be accommodated.

Based on the concepts in Axiom 3 we can now develop a weighting function. For each $x \in H$, there is a suitable weighting function (whose nature depends on the phenomenon being hierarchically structured):

$$w_x : x^- \to [0,1] \text{ such that } \sum_{y \in x^-} w_x(y) = 1$$

Note that $h = 1$ is the last level for which x^- is not empty.

The sets L_i are the levels of the hierarchy, and the function w_x is the priority function of the elements on one level with respect to the objective x. We observe that even if $x^- \neq L_k$ (for some level L_k), w_x may be defined for all of L_k by setting it equal to zero for all elements in L_k not in x.

The weighting function is one of the more significant contributions toward the application of hierarchy theory.

Definition: A hierarchy is complete if, for all $x \subset L_k$, $x^+ \subset L_{k-1}$.

We can state the central question:

Basic Problem: Given any element $x \in L_\alpha$, and subset $S \subset L_\beta$, $(\alpha < \beta)$, how do we define a function $w_{x,S} : S \to [0,1]$ which reflects the properties of the priority functions on the levels L_k, $k = \alpha, \ldots, \beta - 1$? Specifically, what is the function $w_{b,L_h} : L_h \to [0,1]$?

In less technical terms, this can be paraphrased thus: given a social (or economic) system with a major objective b, and the set L_h of basic activities, such that the system can be modeled as a hierarchy with largest element b and lowest level L_h, what are the priorities of the elements of any level and in particular those of L_h with respect to b?

From the standpoint of optimization, to allocate a resource to the elements, any interdependence may take the form of input-output relations such as, for example, the interflow of products between industries. A high-priority industry may depend on flow of material from a low-priority industry. In an optimization framework, the priority of the elements enables one to define the objective function to be maximized, and other hierarchies supply information regarding constraints, e.g., input-output relations.

We now present the method for solving the Basic Problem. Assume that $Y = \{y_1, \ldots, y_{m_k}\} \subset L_k$ and that $X = \{x_1, \ldots, x_{m_{k+1}}\} \subset L_{k+1}$. Without loss of generality we may assume that $X = L_{k+1}$, and that there is an element $z \in L_k$ such that $y_i \in z^-$. Then consider the priority functions

$$w_z : Y \to [0,1] \text{ and } w_{y_j} : X \to [0,1] \, j = 1, \ldots, m_k$$

Construct the priority function of the elements in X with respect to z, denoted

w, $w: X \to [0,1]$, by

$$w(x_i) = \sum_{j=1}^{m_k} w_{y_j}(x_i) w_z(y_j), \quad i = 1, \ldots, m_{k+1}$$

It is obvious that this is no more than the process of weighting the influence of the element y_j on the priority of x_i by multiplying it by the importance of x_i with respect to z.

The algorithms involved will be simplified if one combines the $w_{y_j}(x_i)$ into a matrix B by setting $b_{ij} = w_{y_j}(x_i)$. If one further sets $w_i = w(x_i)$ and $w_j' = w_z(y_j)$, then the above formula becomes

$$w_i = \sum_{j=1}^{m_k} b_{ij} w_j' \quad i = 1, \ldots, n_{k+1}$$

Thus, one may speak of the priority vector w and, indeed, of the priority matrix B of the $(k + 1)$st level; this gives the final formulation

$$w = Bw'$$

The following is easy to prove:

Theorem: Let H be a complete hierarchy with largest element b and h levels. Let B_k be the priority matrix of the kth level, $k = 1, \ldots, h$. If w' is the priority vector of the pth level with respect to some element z in the $(p - 1)$st level, then the priority vector w of the qth level $(p < q)$ with respect to z is given by:

$$w = B_q B_{q-1} \ldots B_{p+1} w'$$

Thus, the priority vector of the lowest level with respect to the element b is given by:

$$w = B_h B_{h-1} \ldots B_2 b_1$$

if L_1 has a single element, $b_1 = 1$. Otherwise, b_1 is a prescribed vector. We note that the pairwise comparison process takes into consideration nonlinearities. Such nonlinearities are captured by the composition weighting process.

Network Systems

Often alternatives depend on criteria and criteria on alternatives, and there should be a cycle connecting the two which is more accurately studied with the network-feedback approach. The AHP has been generalized to deal with feedback as shown below, although people generally prefer to simplify and arrange their thinking in terms of a linear hierarchy even if the answers are only approximate.

A network is a set of nodes (each of which consists of a set of elements) and a set of arcs that indicate the order of interaction among the components. The priorities of the elements in each node are components. The priorities of the elements in each node are components of the principal eigenvector of the matrix of pairwise comparisons of the relative impact of these elements with respect to an element or node with which they interact. The interaction is indicated by an arc of the network. All such eigenvectors define what is known as a supermatrix of impact priorities. By weighting the eigenvectors corresponding to each component by the priority of that component in the system, the supermatrix is transformed into a stochastic matrix. The limiting impact priorities are obtained by computing large powers of this matrix. Formally we have:

Definition: A partially ordered set S is a network system if:

(a) There is a partition of S into sets C_k, $k = 1, \ldots, s$.

(b) There is an ordering on C_k, $k = 1, \ldots, s$ such that $x \subseteq C_k$ either implies x^- or x^+ is in C_{k_j} for some k_j or both $x^- \subseteq C_{k_j}$, $x^+ \subseteq C_{k_j}$ for one or more k_j.

(c) For each $x \in S$, there is a suitable weighting function $w_x : x^- \to [0,1]$ such that $\sum_{y \in x^-} w_x(y) = 1$ and for $C_k \subseteq S$, $k = 1, \ldots, s$ there is a weighting function

$$w_(k) : C^-{}_k \to [0,1]$$

where $C^-{}_k = \{C_h | C_k \text{ covers } C_h\}$.

We shall now turn to the calculation procedures for the weights and for the inconsistency index.

13-11. The Eigenvector Solution for Weights and Consistency

There is an infinite number of ways to derive the vector of priorities from the matrix (a_{ij}). But emphasis on consistency leads to an eigenvalue formulation.

If a_{ij} represents the importance of alternative i over alternative j and a_{jk} represents the importance of alternative j over alternative k then a_{ik}, the importance of alternative i over alternative k, must equal $a_{ij}a_{jk}$ for the judgments to be consistent. If we do not have a scale at all, or do not have it conveniently as in the case of some measuring devices, we cannot give the precise values of w_i/w_j but only an estimate. Our problem becomes $A'w' = \lambda_{\max}w'$ where $\lambda_{\max}$ is the largest or principal eigenvalue of $A' = (a'_{ij})$ the perturbed value of $A = (a_{ij})$ with $a'_{ji} = 1/a'_{ij}$ forced. To simplify the notation we shall continue to write $Aw = \lambda_{\max} w$ where A is the matrix of pairwise comparisons.

The solution is obtained by raising the matrix to a sufficiently large power, then summing over the rows and normalizing to obtain the priority vector $w = (w_1, \ldots, w_n)$. The process is stopped when the difference between components of the priority vector obtained at the kth power and at the $(k + 1)$st power is less than some predetermined small value.

An easy way to get an approximation to the priorities is to normalize the geometric means of the rows. This result coincides with the eigenvector for $n \leq 3$. A second way to obtain an approximation is by normalizing the elements in each column of the judgment matrix and then averaging over each row.

We would like to caution the reader that for important applications one should use only the eigenvector derivation procedure because approximations can lead to rank reversal in spite of the closeness of the result to the eigenvector [14]. It is easy to prove that for an arbitrary estimate x of the priority vector

$$\lim_{k\to\infty} \frac{1}{\lambda_{\max}^k} A^k x = cw$$

where c is a positive constant and w is the principal eigenvector of A. This may be interpreted roughly to say that if we begin with an estimate and operate on it successively by $A/\lambda_{\max}$ to get new estimates, the result converges to a constant multiple of the principal eigenvector.

A simple way to obtain the exact value (or an estimate) of $\lambda_{\max}$ when the exact value (or an estimate) of w is available in normalized form is to add the columns of A and multiply the resulting vector by the vector w. The resulting number is $\lambda_{\max}$ (or an estimate). This follows from

$$\sum_{j=1}^{n} a_{ij} w_j = \lambda_{\max} w_i$$

$$\sum_{i=1}^{n}\sum_{j=1}^{n} a_{ij} w_j = \sum_{j=1}^{n} \left(\sum_{i=1}^{n} a_{ij}\right) w_j = \sum_{i=1}^{n} \lambda_{\max} w_i = \lambda_{\max}$$

The problem is now, how good is the principal eigenvector estimate w? Note that if we obtain $w = (w_1, \ldots, w_n)^T$, by solving this problem, the matrix whose entries are w_i/w_j is a consistent matrix which is our consistent estimate of the matrix A. The original matrix A itself need not be consistent. In fact, the entries of A need not even be transitive; i.e., A_1 may be preferred to A_2 and A_2 to A_3 but A_3 may be preferred to A_1. What we would like is a measure of the error due to inconsistency. It turns out that A is consistent if and only if $\lambda_{\max} = n$ and that we always have $\lambda_{\max} \geq n$. This suggests using $\lambda_{\max} - n$ as an index of departure from consistency. But

$$\lambda_{\max} - n = -\sum_{i=2}^{n} \lambda_i \, ; \, \lambda_{\max} = \lambda_1$$

where λ_i, $i = 1, \ldots, n$ are the eigenvalues of A. We adopt the average value $(\lambda_{\max} - n)/(n-1)$, which is the (negative) average of λ_i, $i = 2, \ldots, n$ (some of which may be complex conjugates).

It is interesting to note that $(\lambda_{\max} - n)/(n-1)$ is the variance of the error incurred in estimating a_{ij}. This can be shown by writing

$$a_{ij} = (w_i/w_j)\, \varepsilon_{ij}, \, \varepsilon_{ij} > 0 \text{ and } \varepsilon_{ij} = 1 + \delta_{ij}, \, \delta_{ij} > -1$$

and substituting in the expression for $\lambda_{\max}$. It is δ_{ij} that concerns us as the error component and its value $|\delta_{ij}| < 1$ for an unbiased estimator. Normalizing the principal eigenvector yields a unique estimate of a ratio scale underlying the judgments.

The consistency index of a matrix of comparisons is given by C.I. = $(\lambda_{\max} - n)/n - 1$. The consistency ratio (C.R.) is obtained by comparing the C.I. with the appropriate one of the following set of numbers each of which is an average random consistency index derived from a sample of size 500 of randomly generated reciprocal matrix using the scale 1/9, 1/8, ..., 1, ..., 8, 9. See if it is about 0.10 or less (0.20 may be tolerated but not more). If it is not less than 0.10, study the problem and revise the judgments.

n	1	2	3	4	5	6	7	8	9	10
Random Consistency Index (R.I.)	0	0	.58	.90	1.12	1.24	1.32	1.41	1.45	1.49

The consistency index for an entire hierarchy is defined by

$$C_H = \sum_{j=1}^{h} \sum_{i=1}^{n_{i_{j+1}}} w_{ij}\mu_{i,j+1}$$

where $w_{ij} = 1$ for $j = 1$, and $n_{i_{j+1}}$ is the number of elements of the $(j + 1)$st level with respect to the ith criterion of the jth level.

Let $|C\bar{k}|$ be the number of elements of C_k^-, and let $w_{(k)(h)}$ be the priority of the impact of the hth component on the kth component, i.e., $w_{(k)(h)} = w_{(k)}(C_h)$ or $w_{(k)} : C_h \to w_{(k)(h)}$.

If we label the components of a system along lines similar to those we followed for a hierarchy, and denote by w_{jk} the limiting priority of the jth element in the kth component, we have

$$C_S = \sum_{k=1}^{s} \sum_{j=1}^{n_k} w_{jk} \sum_{h=1}^{|C\bar{k}|} w_{(k)(h)} \mu_{k(j,h)}$$

where $\mu_k(j,h)$ is the consistency index of the pairwise comparison matrix of the elements in the kth component with respect to the jth element in the hth component.

Since the author developed the Analytic Hierarchy Process using the principal eigenvector and compared it with other methods already in use, several other people have suggested these very same methods and a few new ones. The old ones are the method of least squares, logarithmic least squares, weighted least squares, and so on; the new ones are about variations of the eigenvalue process.

It should be noted that any method that yields nonunique solutions, such as the method of least squares, is not a good contender. We are aware of various arguments given in favor of one or the other of these methods. For several reasons we believe that the eigenvector is the only procedure to use. Recall that a decision theory must put its house in order before it can help people make choices. Thus if one method leads to a different rank order than another, we must examine the underlying problems and show which approach really addresses all the problems that arise.

The reader might reflect on the following problem. Given two irregular objects, a small one and a large one, how many copies of the small object can be combined to make the larger one? Conversely, into how many parts equal to the smaller object can the larger object be decomposed?

In theory these two questions should yield the same answer, particularly when the comparison involves tangible objects. In practice, the problem requires considerable work by the individual and as he applies himself he is likely to settle for the easier problem of combining copies of the smaller object to make the larger one and not bother with the converse to check on the answer. Even after deep reflection, when the elements being compared are intangibles for which no method of measurement is available, the comparisons are not likely to give the same answer in any case. The individual will usually settle for synthesizing copies of the smaller element to make up the larger one. Thus, by habit, people find it easier to ask for the dominance of the larger object over the smaller than to answer the questions arising from the opposite situation. Besides, people do not seem to care about whether the general problem of the equivalence of dominating and dominated has an answer, or how best to obtain it. Since the two results can in fact differ considerably, and since it is the purpose of a decision theory to measure people's preferences rather than legislate corrections that do not get incorporated into people's thinking to make an impact, the derived scale of measurement should only represent what is given and not what some philosopher or analyst thinks should be or what he thinks is desired from a technical standpoint. If he insists on making the correction, the result may distort the outcome even though the individual believes that he has done his best to represent the problem as he perceives it.

The right-eigenvector solution is the correct way to derive a scale from dominance judgments. (A proof of this fact is given below.) To this vector corresponds a left eigenvector. Left and right eigenvectors are reciprocals when the matrix is consistent but they can also be reciprocals when it is not. In general if the matrix is inconsistent it can be written as the elementwise product of a consistent matrix and a reciprocal perturbation matrix. Xu Shubo proved that left and right eigenvectors are reciprocals if and only if the row sums of the perturbation matrix are the same as corresponding column sums.

Information contained in the left eigenvector can be used to measure consistency and hence to decide on whether the information is sufficiently reliable for making a decision. If we write $a_{ij} = (w_i/w_j)\,\varepsilon_{ij}$ and also $a_{ij} = (v_j/v_i)\,\theta_{ij}$ where w and v are the right and left eigenvectors respectively, and ε and θ are the corresponding perturbation matrices, then we have:

$$\frac{\lambda_{\max}-n}{n-1} = -1 + \frac{1}{n(n-1)}\sum_{i<j}\left[\left(\frac{w_j v_j}{w_i v_i}\varepsilon_{ij}\theta_{ij}\right)^{1/2} + \left(\frac{w_j v_j}{w_i v_i}\varepsilon_{ij}\theta_{ij}\right)^{-1/2}\right]$$

Some have assumed that the problem of the left- and right-eigenvector relation should be written off by adopting a procedure that essentially ignores the presence of left and right eigenvectors. One simply assumes that they are always reciprocals of one another, artificially forcing the equivalence of the concepts of "dominating" and "dominated." As we have already said, such symmetry does not exist in the judgment of people, and assuming that it does dilutes the accuracy of what they express through dominance. However, the answer does not take into account inconsistency in general *because the reciprocal relation between left and right eigenvectors does not imply the consistency of the judgment matrix.*

It is precisely because of inconsistency that the two answers in general are not the same, and the right eigenvector is used to capture dominance. As we have seen above, the product of left and right eigenvectors puts a bound on inconsistency and the need to improve it to obtain a valid decision. Consistency is the most critical criterion because the greater the inconsistency of a matrix and the closer its entries to being randomly chosen, the less is the reliability of its information content. In decision making the object is not so much to approximate judgments by mathematically minimizing differences as to determine if the information is sufficiently reliable so that a decision can be based on it. One must then compute the priorities in such a way as to capture the dominance expressed in the judgments. It seems to us that the eigenvector is a descriptive rather than a normative approach to the problem.

The foregoing argument shows that the eigenvector captures the way people in fact represent their preferences and does not legislate by an optimization criterion what should be best for them to do. Now for the proof that the eigenvector captures dominance when the matrix is inconsistent.

If $A = (a_{ij})$, $a_{ij} > 0$, $i,j = 1, \ldots, n$, Perron proved that A has a real positive eigenvalue $\lambda_{\max}$ (called the principal eigenvalue of A) that is simple and $\lambda_{\max} > |\lambda_k|$ for the remaining eigenvalues of A. Furthermore the principal eigenvector $w = (w_1, \ldots, w_n)$ that is a solution of $Aw = \lambda_{\max}w$ is unique to within a multiplicative constant and $w_i > 0$, $i = 1, \ldots, n$. We can make the solution w unique through normalization. We define the norm of the vector w, $\|w\| = ew$ where $e = (1, 1, \ldots, 1)$, and to normalize w we divide it by its norm. We shall always think of w in normalized form. My purpose here is to show how important the principal eigenvector is in determining the rank of the alternatives through dominance walks.

There is a natural way to derive the rank order of a set of alternatives from a pairwise comparison matrix A. The rank order of each alternative is the relative proportion of its dominance over the other alternatives. This is obtained by adding the elements in each row in A and dividing by the total over all the rows. However A only captures the dominance of one alternative over each other alternative in one step. But an alternative can dominate a second by first dominating a third alternative, which then dominates the second. Thus, the first alternative dominates the second in two steps. It is known that the result for dominance in two steps is obtained by squaring the pairwise comparison matrix. Similarly dominance can occur in three steps, four steps and so on, the value of each obtained by raising the matrix to the corresponding power. The rank order of an alternative is the sum of the relative values for dominance in its row, in one step, two steps, and so on, averaged over the number of steps. The question is whether this average tends to a meaningful limit.

We can think of the alternatives as the nodes of a directed graph. With every directed arc from node i to node j (which need not be distinct) is associated a nonnegative number a of the dominance matrix. In graph-theoretic terms this is the intensity of the arc. Define a k-walk to be a sequence of k arcs such that the terminating node of each arc except the last is the source node of the arc that succeeds it. The *intensity of a* k-*walk* is the product of the intensities of the arcs in the walk. With these ideas, we can interpret the matrix A^k: the (i, j) entry of A^k is the sum of the intensities of all k-walks from node i to node j.

Definition: The dominance of an alternative along all walks of length $k \leq m$ is given by

$$\frac{1}{m}\sum_{k=1}^{m} \frac{A^k e^T}{eA^k e^T}$$

Observe that the entries of $A^k e^T$ are the row sums of A^k and that $eA^k e^T$ is the sum of all the entries of A.

Theorem: The dominance of each alternative along all walks k, as $k \to \infty$, is given by the solution of the eigenvalue problem $Aw = \lambda_{max} w$.

Proof: Let

$$s_k = \frac{A^k e}{e^T A^k e}$$

and

$$t_m = \frac{1}{m}\sum_{k=1}^{m} s_k$$

The convergence of the components of t_m to the same limit as the components of s_m is the standard Cesaro summability. Since

$$s_k = \frac{A^k e^T}{eA^k e^T} \to w \text{ as } k \to \infty$$

where w is the normalized principal right eigenvector of A, we have

$$t_m = \frac{1}{m} \sum_{k=1}^{m} \frac{A^k e^T}{eA^k e^T} \to w \text{ as } m \to \infty$$

The essence of the principal eigenvector is to rank alternatives according to dominance in terms of walks. The well-known logarithmic least-squares method (LLSM)—find the vector $v = (v_1, \ldots, v_n)$ which minimizes the expression

$$\sum_{i,j=1}^{n} \left(\log a_{ij} - \log \frac{v_i}{v_j} \right)^2$$

sometimes proposed as an alternative method of solution—obtains results (the geometric mean) which coincide with the principal eigenvector for matrices of order two and three, but deviate from it for higher order and can lead to rank reversal. In a certain application in ranking five teachers, the eigenvector ranking was D, B, C, A, E; whereas the LLSM solution ranked them as B, D, C, A, E [14].

The LLSM minimizes deviations over all the entries of the matrix log a_{ij}, which in the first place is not a_{ij}. The principal eigenvector does not attempt to minimize anything, but maximizes information preserved from all known relations of dominance.

The solution is obtained by raising the matrix to a sufficiently large power, then summing over the rows and normalizing to obtain the priority vector $w = (w_1, \ldots, w_n)^T$. The process is stopped when the difference between components of the priority vector obtained at the kth power and at the $(k + 1)$st power is less than some predetermined small value.

Linearity and Nonlinearity

The formal structure of the AHP utilizes linear algebra to develop priorities. However, the components of the vectors derived from paired comparisons are nonlinearly related. For each set of judgments a different numerical relationship is derived for the components of the eigenvector. Thus, for the range of values of each element on a level, the vector is nonlinear in the variables which represent its components. Unlike the values taken on Cartesian axes, one does not treat one value as being as likely as any other. In addition, hierarchic and feedback composition give rise to multilinear and nonlinear expressions.

13-12. Rank Preservation and Reversal [12]

As it should be, rank is unaffected when only one criterion is involved and the judgments are consistent. More generally, one can show that, with consistency, the rank order of two alternatives is unaffected when the judgment values of one dominate those of the other in every pairwise comparison matrix under the criteria. However, the final rank can change even when the judgments are consistent when an alternative dominates another under one criterion but is dominated by it under another. The following example illustrates the idea of rank reversal due to structural criteria.

In an investment decision let us use only two criteria, *Return* and *Low Risk*, to determine where it is best to invest. Let us for simplicity assume that they are equally important so that their priorities are .5 and .5. We first take the two alternatives *Tax-Free Bonds* (A) and *Securities* (B), pairwise compare them according to preference first under *Return* and then under *Low Risk*, and obtain their composite weights. We have:

RETURN	A	B	Priority
A	1	3	.75
B	1/3	1	.25

LOW RISK	A	B	Priority
A	1	1/2	.33
B	2	1	.67

The final ranks are:

A: .5 × .75 + .5 × .33 = .54;   B: .5 × .25 + .5 × .67 = .46

and A is preferred to B. On observing that the priorities of the alternatives under each criterion are obtained by adding the elements in either column and dividing by the total, which is the normalization factor, we may rewrite the above arithmetic operations to indicate that normalization is applied to rescale the priorities of the criteria and use them to weight the priorities of the alternatives before they are normalized. Thus on writing .75 = 3/4, .25 = 1/4, .33 = 1/3, .67 = 2/3 and noting that 4 is the normalization factor for *Return* and 3 is the normalization factor for *Low Risk* the above can be written as:

$$A: (.5 \times 1/4)3 + (.5 \times 1/3)1 = .54$$
$$B: (.5 \times 1/4)1 + (.5 \times 1/3)2 = .46$$

In other words, normalization may be regarded as an operation that transforms the weights of the criteria from the original scale of (.5, .5) to the new scale (.5/4, .5/3), which when normalized becomes:

[.125/(.125 + .167), .167/(.125 + .167)] = (.43, .57)

which are the rescaled priorities of the criteria. That is, relative measurement may be regarded as an operation that always introduces a new criterion that operates on the existing criteria by modifying their priorities. Earlier we called this criterion a *structural criterion*. It is known in traditional decision theory that rank reversal can occur when a new criterion is introduced and here we have encountered a fundamentally new kind of criterion that is always present when we perform relative measurement. Now let us see how rank reversal occurs here by introducing the new alternative *Savings Accounts* (C) with the resulting paired comparisons shown in the following two matrices:

RETURN	A	B	C	Priority		LOW RISK	A	B	C	Priority
A	1	3	1/2	.30		A	1	1/2	4	.31
B	1/3	1	1/6	.10		B	2	1	8	.62
C	2	6	1	.60		C	1/4	1/8	1	.08

As before, but now without the details, we have:

A: $.5 \times .3 + .5 \times .31 = (.5 \times 1/10)3 + (.5 \times 1/13)4 = .30$
B: $.5 \times .1 + .5 \times .62 = (.5 \times 1/10)1 + (.5 \times 1/13)8 = .36$
C: $.5 \times .6 + .5 \times .08 = (.5 \times 1/10)6 + (.5 \times 1/13)1 = .34$

Thus the presence of alternative C has caused rank reversal between A and B and now B is preferred to C which is preferred to A. In this case we would have to ignore the first ranking and use the second. Let us note that this process of rank reversal does not contradict any existing proven fact and we have no historical precedent in some well-developed multicriteria theory to check it against. In fact the only known theory which is concerned with rank reversal under even a single criterion is Utility Theory. There when rank reversal occurs under one criterion the new alternatives that cause it are called relevant, otherwise they are called irrelevant. The relative measurement approach of the AHP requires that in order for elements to be comparable in pairs, it is essential that they all be homogeneous, which means that they must be relevant to begin with. There is no escaping the fact that to deal with myriads of intangibles we need relative measurement and that when we use it rank reversal may occur. It is likely that rank reversal due to structural criteria happens in real life without being recognized. *There is greater uncertainity about rank optimality with increasing rate of change of information.* As we mentioned earlier, rank reversal does not occur in cases where absolute measurement is used for the alternatives.

13-13. Group Judgments

When one is dealing with group judgments, Saaty has proposed that any rule to combine the judgments of several individuals should also satisfy the reciprocal property. A proof that the geometric mean, which makes no requirement regarding who should vote first, satisfies this condition was later generalized in [1].

Group judgments can be debated through a consistency check. When several people propose radically different judgments in certain positions of the matrix these can be tested with other judgments on which there is wide agreement by solving the problem separately for each controversial judgment and measuring the consistency. That judgment yielding the highest consistency in the overall problem is retained. The following consistency comparison for each individual's judgments with those of the scale vector w derived from group judgments has been proposed.

$$\sum_{i,j=1}^{n} b_{ij} w_j/w_i - n^2 \sim .1$$

Probabilistic judgments have been studied extensively [15]. In particular L. G. Vargas has shown that when the judgments are given by a gamma distribution the derived vector belongs to a Dirichlet distribution with a beta distribution of each component.

13–14. Topics for Investigation

There are a number of areas that need further investigation and others in which ground needs to be broken. Examine AHP literature for applications and theoretical developments in most of these areas.

1. Generalization of the hierarchy and systems networks into manifolds.
2. Deeper and more extensive research on continuous judgments.
3. Testing of different group decision-making approaches to the same problem and searching for common elements. Development of A, B, C guidelines for group participation in decision making.
4. Investigation of the relationship of the principal eigenvector to the Weber-Fechner Power Law.
5. Development of applications of the AHP in Game Theory, particularly with respect to negotiation.
6. Investigation of the relationship of the AHP to optimization. Can the general-optimization problem be solved using the AHP?
7. Implementation of psychological studies to show how people's strength of feeling can be adequately represented by numerical scales.
8. Studying of the sensitivity of priorities to the number of criteria and, more generally, to the size of the hierarchy.
9. Sampling of opinions on how satisfied clients are with AHP outcomes.
10. Formulation of more cases using the AHP in resource allocation, planning, cost-benefit analysis, and conflict resolution.
11. Answering the question: Is there power in hierarchic formulation and judgments to make better predictions? How can it be tested?
12. The AHP and Risk Analysis: Putting forth a definitive theory about the use of scenarios in risk analysis.
13. Investigation of the relationship between AHP and Artificial Intelligence.
14. Development of communication and causal languages using the AHP.

REFERENCES

There are five books in English on the AHP, some of which have been translated into other languages as indicated in the references. There have been issues of two journals solely devoted to articles on the AHP: *Mathematical Modelling*, vol. 9, nos. 3–5, 1987, and the December 1986 issue of *Socio-Economic Planning*. This chapter has been particularly enriched by the article "The Analytic Hierarchy Process—What It Is and How It Is Used," prepared by Rozann W. Saaty for the *Mathematical Modelling* issue. F. Zahedi has published a comprehensive survey article on the AHP [18] with up-to-date references on its literature. Xu Shubo of Tianjin University, China, has also published a complete reference list on the AHP [17]. One book has been written on the subject by a Japanese author and two by Chinese authors.

The subject has also been presented in a chapter in the book by S. I. Gass, "Decision Making, Models and Algorithms, A First Course," John Wiley and Sons, New York, 1985; and in a chapter in the book by D. Anderson, D. Sweeney, and T. Williams, "Quantitative Methods for Business," 3d ed., West Publishing Company, St. Paul, Minn., 1986.

There is a software program available, Expert Choice [2], that runs on the IBM PC, the IBM XT, the IBM AT, and compatible computers. It requires 256K memory and one double-sided disk drive. There is also a version of Expert Choice available for the Japanese IBM personal computer (with the Kanji keyboard).

1. Aczél, J., and T. L. Saaty: Procedures for Synthesizing Ratio Judgments, *Journal of Mathematical Psychology*, vol. 27, no. 1, pp. 93–102, March, 1983.
2. Expert Choice, software package. Decision Support Software, McLean, Va.
3. Hämäläinen, Raimo T., and Timo O. Seppäläinen: The Analytic Network Process in Energy Policy Planning, *Socio-Economic Planning Sciences*, vol. 20, no. 6, pp. 399–405, 1986.
4. Harker, Patrick T.: Alternative Modes of Questioning in the Analytic Hierarchy Process, *Mathematical Modelling*, vol. 9, pp. 353–360, 1987.
5. Saaty, Thomas L.: "The Analytic Hierarchy Process." (287 pp.) McGraw-Hill International, New York, 1980. (This book has been translated into Chinese by Xu Shubo, He Jinsheng, et al. Information is available from them at the Institute of Systems Engineering, Tianjin University, Tianjin, China. A translation into Russian by Rivaz Vachnadze is currently under way.) This book was updated and republished in 1988 by RWS Publications, 4922 Ellsworth Avenue, Pittsburgh, PA 15213 (Price $22.50 at time of writing).
6. Saaty, Thomas L., and Luis G. Vargas: "The Logic of Priorities, Applications in Business, Energy, Health, Transportation." (299 pp.) Kluwer-Nijhoff Publishing, 1981.
7. Saaty, Thomas L.: "Decision Making for Leaders." (291 pp.) Wadsworth, Belmont, California, 1982. (Also reprinted as a paperback and available from the author at 322 Mervis Hall, University of Pittsburgh, Pittsburgh, PA 15260. Available in French under the title *Décider face à la complexité*, Enterprise Moderne d'Edition, Paris, 1984.)
8. Saaty, Thomas L., and Kevin P. Kearns: "Analytical Planning." (208 pp.) Pergamon Press, Oxford, 1985.
9. Saaty, Thomas L., and Joyce Alexander: "A New Logic for Conflict Resolution." In preparation.
10. Saaty, Thomas L.: Axiomatic Foundation of the Analytic Hierarchy Process, *Management Science*, vol. 32, no. 7, pp. 353–360, 1986.
11. Saaty, Thomas L.: Absolute and Relative Measurement with the AHP: The Most Livable Cities in the United States, *Socio-Economic Planning Sciences*, vol. 20, no. 6, pp. 327–331, 1986.

12. Saaty, Thomas L.: Rank Generation, Preservation and Reversal in the Analytic Hierarchy Process, *Decision Sciences*, July, 1987.
13. Saaty, Thomas L.: Rank According to Perron, *Mathematics Magazine*, 1987.
14. Saaty, Thomas L., and L. Vargas: Inconsistency and Rank Preservation, *Journal of Mathematical Psychology*, vol. 28, no. 2, pp. 205–214, 1984.
15. Saaty, Thomas L., and L. Vargas: Stimulus-Response with Reciprocal Kernels: The Rise and Fall of Sensation, *Journal of Mathematical Sciences*, 1987.
16. Vargas, L.: Reciprocal Matrices with Random Coefficients, *Journal of Mathematical Modelling*, vol. 3, no. 1, pp. 69–81, 1982.
17. Xu Shubo: "References on the Analytic Hierarchy Process." Institute of Systems Engineering, Tianjin University, China, June, 1986.
18. Zahedi, F.: The Analytic Hierarchy Process—A Survey of the Method and Its Applications, *Interfaces*, vol. 16, no. 4, pp. 96–108, 1986.

INDEX

A separate Index to Chapter 13 follows.

Accuracy of estimate, 310
Action, proper mode of, 392
Action phase, 25, 38–39
Allocation of effort, 155
Alternation, exclusive and nonexclusive, 53
Alternative syllogism, 60
Analysis, 36–37
   of variance, 317
Analytic propositions, 23
Antecedent, 54
Approximation, 108
   Newton's method, 101, 106
   to Poisson, by binomial, 261
   by normal distribution, 262
   Sterling's, 238
A priori assumptions, 236, 238, 243
Arithmetic mean, 113
Artificial vectors, 180
Autocorrelation function, 271–272
Average deviation and fractional average deviation, 228
Axioms for events with countable outcomes, 238

Backward equations, 271
Basic feasible solution, 180
Basis, 180
Bayes' theorem, 243, 244, 307
Bernoulli coefficients, 51
Bernoulli polynomials, 51
Bernoulli's differential equation, 67
Bessel function, 286, 340
Bid evaluation, 172
Bienayme-Tchebycheff inequality, 265
Binomial distribution, 258
   negative, 264
Birth and death process, 273
Blockade, 9
Blotto game, 223
Boolean algebra, 335
Brainstorming, 393

Cable problem, 407
Canonical form, reduction to, 92

Caterer problem, 172
Cauchy distribution, 249
Cauchy formula, 79
Cauchy gradient method, 106–109
Cauchy-Schwarz inequality, 112
Causal relation, 408
Central-limit theorem, 261, 305
Chain rule, 60
Chance variable, 245
Chapman-Kolmogorov equation, 270
Characteristic function, 251
Chess board, 404
Chi-square distribution, 257, 263, 311, 315
Chi-square table, 324
Chi-square test, 315–317, 367
Clairaut's differential equation, 67
Classical method of model formation, 65
Coalition, 209
Combinations, 236–237
Communication, 4, 17, 23
Competitive bidding strategy, example, 282
Complete additivity, 238
Compound probability, 239
Comprehension, 390
Computers, 293
Concave function, 130–131, 150, 154
Conditional, 54
Conditional distribution, 248
Conditional probability, 239
   example, 240
Confidence interval, 310
Conflicting interests, 210
Conjecture, 45
Conjunction, 52
Consequent, 54
Consistent estimator, 306
Constant holding time, 347
Constant input, 354
Constraint region, 142
Constraints as inequalities, 99
Constructive proofs, 48
Continuous distribution, 246
Contradiction or *reductio ad absurdum*, 47, 61

Contrapositive of conditional, 54
Convergence in probability, 266
Converse of conditional, 54
Convex function, 130–131
Convex payoff function, 222
Convex quadratic function, 198
Convex region, 195
Convex set, 115
Convolution, 85
Coordinate system, 140
Correlation, 319
   multiple, 321
   partial, 321
   significant, 320–321
Cost of a system, 291
Counterexample, 45–46
Counterfeit coin, 403
Covariance, 202
Cramer's rule, 104
Craps game, 240
Creativity, 5, 381
   of creative scientist, 393
   and discovery (eureka, eureka), 389
   enterprise, 387
   excitement, 388
   hesitation in, 387
   motivation for, 387
   and productivity, 384
   from re-creation, 389
   serendipity, 390
   stimulation of, 383
   successful use of, 387
Crout's method, 104
Cumulative distribution function, 245–246

Dams, efficiency in operation, 174
Data, 4
   grouping of, 316
Deadlock, single answer, 394
Decision, executive, 3
Decision maker, 38
Decision process, 38
Deductive method, 24–25
Defectives, 259
Degeneracy, 185
Degrees of freedom, 257
Density function, 245
Derivative, left- and right-hand, 135
Deterministic techniques, 2
   models, 34, 99
Deviation, 228
   standard, 228, 249
Diet problem, solution by simplex process, 180–184

Difference equations, nth-order linear, 77
   examples and solutions, 77–84
   homogeneous, 77
Differential difference equations, 338
Differential equations, 66–77
   application to conflict, Lanchester's law, 71
   characteristic equation of, 68
   first-order, existence and construction of solutions, 66
     seven classical types, 67
     systems of n, 70, 73
   homogeneous, 67, 68
   matrix methods of solution, 76
   nth order, 70
   nth order linear, 71
   second-order, general linear, with constant coefficients, 68
Dimensional analysis, 88, 340
   in linear programming, 176
Diminishing returns, 143
Directional of derivative, 107
Discrete distribution, 246
Distribution, binomial, 258, 264
   Cauchy, 249
   chi-square, 257, 263, 311, 315
   conditional, 248
   discrete, 246
   effort, 149
   exponential, 85, 264, 296, 336
   F test, 317
   failure, 84
   gamma, 263
   hypergeometric, 260
   log-normal, 263
   marginal, 248
   multinomial, 260
   normal (*see* Normal distribution)
   Pascal or negative binomial, 264
   Pearson Type-III, 263
   Poisson, 260
   Pollaczek-Geiringer, 264
   Student's t, 257, 315
   uniform or rectangular, 258
Dual problem of linear programming, 118, 142, 176
   solution of, 184
Dual simplex method, 193
Du Bois-Reymond lemma, 278
Dynamic programming, 155
   example of, 289

Edison, T. A., 7
Efficient combinations, 203
Effort-distribution example, 149
Electronic equipment, 401

Element standby, 290
Elimination method, 119
Encounters on line segment, 223
Entropy, 92, 157
Epsilon technique, 185
Equal likelihood, 238
Equilibrium, 250, 338
Equivalence, logical, 61
Ergodic state, 270
Erlang, A. K., 6
Erlangian, 334
Erlangian holding time, 347
Error, standard, 320
  trial and, 5, 22, 388
  types of, 311
Error function, 229
Essay, 380
Estimate, 405
  accuracy of, 310
Estimation and estimators, Bayes' formula, 307
  confidence-interval method, 310
  consistent, 306
  maximum-likelihood method, 307–309
  of parameters, 306
  unbiased, 306
Euler's equation, 132, 133, 157, 160
Euler-Maclaurin formula, 51
Evaluation, 5
  bid, 172
  (*See also* Validity)
Even function, 76
Expected payoff, 210
Expected profit, 127
Expected return, 210
Experiment, 401
Experimental verification, 368
Experimentation, 35
Exponential distribution, 85, 264, 336
  sum of, 296
Exponential holding time, 346
Expression, verbal and other, 392
Extreme point, 126
Extremum or extreme value, 125, 127
  on boundary, 132, 198
  on closed surface, 134
  to an integral, 132
  necessary condition for, 128
    Euler's equation as, 132, 133
  in region, 132
  sufficient conditions for, 128

F test, 317
Fallacy, 408
  affirming the consequent, 60
  denying the antecedent, 60

Feasible region, 125
Feasible solution, 179
Fictitious-play method, 212
Finite and infinite games, 209
Fisher's maximum likelihood, 307
Fisher's transformation, 320
Fitting of data, 89–92
Forecasting, economic, 405
Forward equations, 271
Fredholm's equation, 84
Free wheeling, 393
Freedom of speech, 408
Frequency distribution, 245
Functional equation, 156
Fundamental theorem of algebra, 100

Games, 209
  against nature, 225
  Blotto, 223
  expectation, 210, 222
  minmax theorem, 211
  nonzero-sum, 209
  relation to linear programming and solution, 220
  value of, 211
  zero-sum two-person, 209
    methods of solution, fictitious-play, 212
    geometric, 217
    kernel, 217
    other, 214
    simplex, 218
Gamesmanship, 386
Gaming, 292
Gamma distribution, 263, 297
Gamma function, 258, 297
  incomplete, 298
Gauss' integral, 76
Gauss-Jordan elimination method, 104
General programming problem, 135
Generating functions, 79
Geometric mean, 113
Geometric methods, 161
Geometric and physical probability, 282–288
Gibbs, J. W., 150
Goodness of fit, 316
Gradient, 106
Gradient conditions, 230
Gradient method, constant, 125
  exponential, 124
  infinitesimal, 125
Gradient vector, 199
Graphical solutions of root problems, 102–103

Harmonic mean, 113
Heaviside's translation theorem, 86
Histogram, 367
Hölder's inequality, 112
Homogeneous differential equations, 67, 68
Horse-race bets, example, 149
Hypergeometric distribution, 260
Hyperplane, 115, 122, 195
Hypotheses, 4, 31–32
  making of, 25
  testing of, 311
  verification of, 36–37

Idealist, 22
Imagery, 391
Imagination, 402
Impatient customers, 353, 366
Income of community, 272
Incomplete gamma function, 298
Independent random variables, 248
Induction and inductive inference, 24–25, 38, 235
Industrial-operations research, 10–13
Industry, 5, 15
Inequalities, and absolute values, 111
  Cauchy-Schwarz, 112
  elementary relations, 110
  on harmonic geometric and arithmetic means, 113–114
  Hölder's, 112
  homogeneous system of, 116
  inhomogeneous system of, 115
  Jensen's theorem, 113
  Minkowski's, 112
  use of, as constraints, 109
  useful relations, 111
Inference, 56
Information theory, application of, 92–94, 157
Initial feasible solution, 199
Initial probabilities, 270
Input-output relations, 165
Integer solutions, 145
  remarks, 184
Integral equations, 84, 342
  application and solution of, 84–86
  Fredholm's, 84
  Volterra's, 34
  Wiener-Hopf type, 343
Interactions, 91, 394
  (*See also* Correlation)
Intuitionist, 41
Intuitive method, 51
Inventory, 159, 279, 282
Inverse of conditional, 54

Isoperimetric problem, 161
Iterations, number of, 177

Jacobian, 46–47, 255, 256
Jensen's theorem, 113
Judgment phase, 25–30, 391

Kernel method, 217
Kolmogorov-Smirnov test, 328

Lagrange differential equation, 67
Lagrange method, 79, 82
Lagrange multiplier, 135, 137
Lagrangian, 137, 157, 276
Lanchester, F. W., 6
Landing of aircraft, 365
Laplace transform, 85, 88, 298
Laplace's method, 79
Large numbers, 294
  application of, 294
  law of, strong, 267
  weak, 266
Laws of thought, 63
Learning theory, 270
Least squares, method of, 89
Le Chatelier's principle, 193
Levinson, H. C., 7
l'Hôpital's rule, 252
Lifemanship, 386
Likelihood function, 308
Limited source, 353
Linear combination, 181
Linear differential equations (*see* Differential equations)
Linear form, 167
Linear independence, 140, 179
Linear programming, 154
  applications, balancing of production, 170
  bid evaluation, 172
  caterer problem, 172
  diet, 165, 168–169
  dimensional analysis, 176
  efficiency of operating system of dams, 174
  geometry of, 125
  mathematical statement, 167
  in matrix notation, 168
  optimum estimation of executive compensation, 173
  other applications, 174
  personnel assignment, 171
  policy of firm, 174
  reduction to game, 218

Linear programming, applications,
    smoothing production, 171
    textile factory, 174
    transportation, 169
    trim problem, 174
    verbal statement, 167
  parametric, 193–196
Loading of ships, 365
Log-normal distribution, 263

Machiavelli, 385
Marginal cost, 159
Marginal distribution, 248
Markov chain, 269
Markovian process, 270, 335
Matrix, application to solution of differential equation, 76
  augmented, 103, 179
  characteristic roots of, 76
  coefficient, 103, 115
  convariance, 253–254
  determinant of, 104, 180
  exponential function of, 77
  of first partial derivatives, 142
  inverse of, 104
  linearly independent columns, 115
  operations on a matrix, 215–216
  payoff, 209, 210
  of quadratic form, 129
  rank of, 103
  skew-symmetric, 219
  transition, 269
  transpose of, 197
Maximum, 125–126
  absolute, 125
  in the large, 125
  local, 125, 143
  necessary condition for, 277
  strict, 125
  (*See also* Extremum)
Maximum entropy, 157
Maximum feasible solution, 179
Maximum likelihood, 307
Maximum returns, 210
Maximum total utility, 277
Mean, 202, 248
Mean number waiting, 339
Measures of effectiveness, 27–29, 337
Median, 250
Medical services, 409
Memory, 391
Minimum (*see* Maximum)
Minkowski's inequality, 112
Minmax theorem, 14, 211, 222
Missiles, reliability of, 288
Mode, 250

Models, 32–35
  classical method for formation, 65
  deterministic, 235
Modus ponens, 60
Modus tollens, 60
Moment-generating function, 250, 306
Moments, 248
Monkey, 405
Monotone functions, 130, 159, 247
Monte Carlo method, 293–295, 367–369
Moral structure, 396
Multinomial distribution, 260
Multivariate normal distribution, 309
Municipal affairs, 409

n-person game, 209
N-square law, 6, 7
Negative, 54
Negative binomial distribution, 264
Newton's root-approximation method, 101, 106
Neyman-Pearson theory, 311
Nondegeneracy, 179
Nonparametric methods, 322
Nonstationary process, 272
Normal distribution, 157, 261
  log-, 263
  multivariate, 253, 309
  table of, 322
Normal line, 140
Normalization, 121
Null state, 270

Objective functions, linear and quadratic, 165, 197
Objectives, 3, 126, 382
  and values, 26
Observation, 35
Offensive patrols, 8
Operation, 3, 22, 26, 126
Operational problem, 384, 401
Operationalism, 22–25
Operations analyst, 3, 16, 21
Operations research, applications, 7–13
    future, 16
  competitive strategy, 14
  definitions, 2–4
  equipment and system, 15
  groups, 9–11
  journals, 12
  methods, 1
  post-World War II, 9
  prediction, 14
  scope, 13
  societies, 12

Operations research, tactical optimization, 15
  team work, 17
  universities, 12
  useful features, 39
Operators, difference, 87
  differential, 86
  shifting, 87
Optimization, 125, 130
  with equality constraints, 135
  equivalence with saddle-value problem, 153
  examples, 137–139
  with inequality constraints, 139, 148
  of integral subject to constraints, 156
  inventory example illustrating corner condition, 159
  necessary conditions for, 139
  of nonconstrained functions, 127
  in operations research, 99
  and probabilities examples, 275–282
  special cases, 149
  subject to constraints, 134
  tactical, 15
  use of geometric methods, 161
Optimum, 125
  on the boundary, 134, 198
  of several functions, 135
Optimum estimation of executive compensation, 173
Optimum operation level in textile industry, 174
Optimum policy, 155
Optimum stock example, 281
Optimum strategy, 280
Optimum vertex, 195
Ordering cost, 281
Organization, talent for, 390

Parabolic path of projectile, 105
Paradoxes, 408
Parametric linear programming, 193–196
Partial differential equation, 370
Pascal distribution, 264
Payoff, 209, 210
Payoff function, 202
Payoff matrix, 210
Pearson Type-III distribution, 263
Periodic state, 270
Permutations, 236–237
Personnel assignment, 171
Perspectives, 392
Piecewise continuous derivative, 159
Piecewise linear functions, 152
Pistol duel, 222
Pivot element, 200

Platonist, 22
Play, 209
Player, 209
Poisson distribution, 260
Poisson process, 273, 334, 335
Policy, 155
  of firm, 174
Pollaczek-Gciringer distribution, 264
Polyhedron, 177, 404
  regular, 43
Polynomial, 100
Population, 302
  mean, 305
  normal, 308
  variance of, 305
Positivist, 22
Prediction, 2, 14, 27, 37
Preemptive service, 350
Premises, 56
Primal, 176
Principles, of identity, excluded middle and contradiction, 60
  of indifference, 236
Priority, 348, 350, 351, 361
Probability, application, 78, 240, 243, 270, 275–299, 331–371
  autocorrelation function, 272
  axiomatic approach, 238
  Bayes' theorem, 243–244
  birth-death process, 273, 331–371
  bivariate case, 247
  characteristic function, 25–252
  compound, 239
  conditional, 239, 240
  convergence in, 266
  deduction, 235
  density function, 245
  distribution, conditional, 248
    continuous, 246
    cumulative, 245–246
    discrete, 246
    frequency, 245
    of sums, products, and quotients, 255–258, 306
  distributions, 258–264
  element of, 246
  equal likelihood, 238
  geometric and physical, 282–288
  independent variables, 248
  induction, 235
  initial, 270
  joint frequency function, 256
  law of, addition, 239
    large numbers, 266
  limit theorems, 265
  marginal distribution, 248
  Markov chains, 269

INDEX 457

Probability, Markov process, 270
  mean, 248
  median, 250
  mode, 250
  models, 34
  moment-generating function, 250
  moments, 248–249
  occurrence of an event, 238
  optimization and, 275–282
  Poisson process, 273
  pure birth process, 272
  queues, 331
  random variable, 244
  relative frequency, 238
  of speaking truth, 407
  standard deviation, 249
  stationary process, 271
  stochastic process, 267, 335
  total, 238
  transformation of variables, 254
  transition matrix, 264
  useful inequalities, 265
  variance, 249
Probable error, 294
Problem, hypothetical, 30
  in operation, 29–30
  optimization, 29
  prediction, 29
  remedial, 29
  transference, 29
Production, 144
  balancing and smoothing of, 170–171
  flow in, 367
Profit, expected, 127
Profit function, 143
Programming, 165
Propositions, indeterminate, 23
  meaningless, 23
  undecidable, 46
Pure birth process, 272, 335
Pure strategy, 210

Quadratic form, 128, 131, 197
  positive definite, 129, 198
  semidefinite, 129
Quadratic programming, 197
Quantitative, quantification, 3, 51, 408
Queueing theory, applications, 364–367
  channels, in parallel, 359
    in series, 357
    single, 345
  impatient customer, 353
  input, 333–334
    constant, 354
    limited, 353, 358, 363
    Poisson, 345

Queueing theory, input, unlimited, 358
  measures of effectiveness, 337
  models, 338–345
  Monte Carlo, 367–369
  notation, 332
  priorities, 348, 350, 351, 361
  queue-length dependence, 354
  random selection, 353
  service time, 333–334
    arbitrary, 354
    exponential, 357
  time dependence of parameters, 356
  time independent, 80

Railroad classification yards, 366
Random number, tables, 327
Random selection for service, 353
Random variables, 244, 247
Ranking, 328
Rare events, 261
Rational framework, 403
Realist, 22
Recommendation, 8, 11, 38
Re-creation, 389
Recurrent state, 270
Region of concavity, example, 131
Regression, 320–322
  partial, 321
Relative frequency, 238
Relaxation method, 121
Reliability, effect of switch on, 291
  of systems, 288
Renewals, 84, 295
Research phase, 25, 30–38
Return function, 149
Riccati's differential equation, 68
Robot, 402
Root-mean-square deviation, 228
Rouché's theorem, 81

(s,S) policy, 281
Saddle point, 153, 210
Saddle-value problem, 153
Sales campaign, example, 149
Sample and sampling, corrected variance, 305
  judgment, 305
  mean, 305
  random, 304
  stratified, 304
  systematic, 304
  variance, 305
Scheduling in clinics, 365
Scientific method, 1, 2, 5, 21, 25
  framework, 22

Scientific method, in operations research, 25–39
  in statistics, 302
  supplements decisions, 39
Search theory, 277, 286, 290
Separable payoff function, 222
Separable variables, 67
Sequential analysis, 314
Serendipity, 390
Series expansion, 83–84
  linear and quadratic terms of, 92
Significance, 329
Significance levels, 317
Simplex process, 125, 178, 199
Simplex tableau, 200
Simulation, 292
Simultaneous equations and inequalities and methods of solution, 88, 103, 105, 109, 114, 119, 121, 124, 125, 139
Slack variable, 125, 180
Social constraints, 396
Solution, 35, 42–51
  approximation to, 43, 50–51
  basic feasible, 180
  bounds on, 196
  characterization of, 42
  comparison, 99
  conditions for, necessary and sufficient, 42, 135, 139–143, 150–153
  construction of, 42, 149, 178, 296
  convergence to, 42, 49, 124, 139
  economic consideration in, 127
  errors in, 37, 42
  existence of, 41, 45–46, 116
  mortally urgent, 388
  order of magnitude, 196
  successive choice of, 139
  time and, 127
  uniqueness and nonuniqueness of, 41, 193, 296
  upper bounds for number and value of, 42–45, 178
Solution vectors, 149
Spare-part ordering, 275
Spherical cap, 110
Square array, 402
Standard deviation, 228, 249
Standard error, 320
Statement, connectives of, 52
Statesman, consequences of acts of, 1
Stationary ridge, 92
Stationary, 271, 335
Statistical mechanics, example form, 286
Statistics, analysis of variance, 317
  application to queues, 367
  Bayes' formula in, 307
  central-limit theorem, 30^r

Statistics, confidence intervals, 310
  correlation, 304
  estimation, 304
  histogram, 303
  inductive and, 235
  maximum likelihood, 307
  nonparametric methods, 322
  population, 302
  regression and correlation, 319
  sampling, 304
  sequential analysis, 314
  tables, 322–327
  testing of hypotheses, 311
  Wald's theory, example, 314
Steady state, 270, 338
Steepest ascent along fitting curves, 91, 292
Steepest descent or saddle-point method, 106
Steiner's proof, 161
Sterling's formula, 238
Stieltjes integral, 251
Stochastic process, 267, 335
Stochastic variable, 245
Stocking, 145, 275, 279, 401
Strategy, 209
  mixed, 211
  pure, 210
Strictly convex function, 203
Student's t distribution, 257
  table, 323
Suboptimization, 126
Success, rules for, 385
Sufficiency tests, 128
Sum of functions, 152
Supply demand, examples, 127, 276
Symbolic logic, 51
Symmetry argument, 224
Synthetic propositions, 23
System standby, 290
Systems, 15

Tables, 322–327
Tactical weapon, 5
Target, detection of, 277
Task and its effect, 149
Tautology, 61
Taxi stand, waiting at, 366
Taylor, F. W., 6
Taylor's formula, 128
Tchebycheff's inequality, 265
Team, 17
Telephone conversations, 365
  duration of, 245
Television, 403
Time series, 271
Total-cost minimum, 159

Total-effort constraint, 152
Total probability, 238
Trajectory of projectile, 73
Transcendental functions, 101
Transformation of variables, 254
Transient equation, 338
Transient state, 270
Transition matrix, 269
Transportation model, 169
Transportation problem, solution by simplex process, 185–193
Transposition theorem, 116
Trial and error, 5, 22, 388
Trim problem, 174
Truncation, 51
Truth, 21–25
 operational, 23
 speaking, probability of, 407
Truth tables, 53–63
 application of, 56
 for complex expressions, 55

Unbiased estimator, 306
Uniform or rectangular distribution, 258
Upper and lower bounds, 35, 92, 406
 (*See also* Solution)

Utility function, 99
Utilization factor, 333, 339

Validation (*see* Models; Validity)
Validity, 21–25, 57
Variance, 249, 317
 analysis of, 317
 in queues, 340
 ratio table, 325–326
Variate, 245
Verification, 4, 24
 experimental, 368
 (*See also* Validity)
Volterra's equation, 84

Waiting time, 333
War, 2, 5
Weights and weighted average, 127, 135
Wilcoxon test, 328
World War II and operations research, 7–9

Zero-sum two-person game, 209
 (*See also* Games)

INDEX TO CHAPTER 13

Algebra, linear, 444
Analytic Hierarchy Process,
 axioms of, 416, 425, 432–437
 general, 415–416, 425
Arcs, 437, 443
Artificial Intelligence, 447

Basic Problem, 436
Benefits and costs, 416, 428, 432, 447
Beverage consumption, 428, 431

Cartesian axes, 444
Causal languages, 447
Cesaro summability, 443
Comparative judgment, 418, 424, 432, 433
Comparisons, paired, 415–418, 420, 421, 423–426, 433–435, 437, 438, 440, 442–445
Composition of priorities, principle of, 421
Conflict resolution, 415, 447
Consistency, 415, 416, 425–427, 434, 439–442, 446
Consistency index, 440
Consistency ratio, 440
Consistent matrix, 441

Continuous judgments, 447
Costs and benefits, 416, 428, 432, 447

Decision theory, 441
Decomposition, 418, 419, 425, 432
Dependence, 416, 421, 432, 433, 435
 functional, 432, 433
 inner, 435
 outer, 435
 structural, 432, 433
Derived scale, 434, 435
Dirichlet distribution, 447
Dominance, 416, 428, 434, 441–445

Eigenfunctions, 416
Eigenvalues, 438–440, 442, 443
Eigenvectors, 416, 419, 424, 437–444, 447
 left, 441, 442
 right, 441–443
Employees, evaluating for raises, 423, 424
Expectations, Axiom of, 435, 436
Expert Choice, 417

Fechner, Gustav Theodor, 425, 426, 447
Feedback, 416, 437

Feedback-network structures, 416, 437
Fredholm Operators, 416
Fundamental scale, 433, 435

Game Theory, 447
General-optimization problem, 447
Group judgments, 416, 446–447

Hierarchies, 418, 419, 421, 423, 425, 427, 428, 434–436, 447
Homogeneity Axiom, 435
House, choosing the best, 418–423

Inconsistency index, 438
Input-output relations, 436
Interdependence, 421, 422, 436
Inverse-square law of optics, 429

Judgment, comparative, 418, 424

k arcs, 443
k-walk, intensity of a, 443
k-walks, 443

Least squares, method of, 440
Linearity, 444
Logarithmic least squares, method of, 440, 444

Measurement, absolute, 416, 417, 423, 433
  relative, 416–418, 433
Multilinear expressions, 444

Negotiation, 447
Network-feedback approach, 416, 437
Networks, 437, 447
Nodes, 437, 438, 443
Nonlinearities, 437, 444
Normalization, 417, 422, 424, 438, 439, 443, 445

Optics, 428, 429, 432
  inverse-square law of, 429

Optimization, 415, 436, 442, 447

Perturbation matrix, 441
Planning, 415, 447
Primitive scale, 433
Priorities, 417–419, 421–424, 427, 432, 433, 436–438, 440, 442, 445–447
Probabilistic judgments, 447
Psychology, 416, 447

Rank preservation, 416, 444
Rank reversal, 416, 439, 444–446
Ratings, 417
Reciprocal Axiom, 433–435
Resource allocation, 415, 447
Risk Analysis, 447

Saaty, Thomas L., 432, 446
Scale comparisons, 428–432
Scaling, 416, 420, 425, 426, 428, 433–435, 441, 447
Scoring, 417, 433
Stochastic matrix, 438
Supermatrix of impact priorities, 438
Synthesis of priorities, 421, 432, 433
Structuring, 418, 427, 428, 445

Topics for investigation, 447

Utility Theory, 446

Vargas, L. G., 447

Weber-Fechner, psychophysical law of, 426, 447
Weber's law, 425
Weight comparison, 428, 430
Weighted least squares, method of, 440
Weighting function, 436–438

Xu Shubo, 441

A CATALOG OF SELECTED
DOVER BOOKS
IN ALL FIELDS OF INTEREST

A CATALOG OF SELECTED DOVER BOOKS IN ALL FIELDS OF INTEREST

DRAWINGS OF REMBRANDT, edited by Seymour Slive. Updated Lippmann, Hofstede de Groot edition, with definitive scholarly apparatus. All portraits, biblical sketches, landscapes, nudes. Oriental figures, classical studies, together with selection of work by followers. 550 illustrations. Total of 630pp. 9⅛ × 12¼.
21485-0, 21486-9 Pa., Two-vol. set $25.00

GHOST AND HORROR STORIES OF AMBROSE BIERCE, Ambrose Bierce. 24 tales vividly imagined, strangely prophetic, and decades ahead of their time in technical skill: "The Damned Thing," "An Inhabitant of Carcosa," "The Eyes of the Panther," "Moxon's Master," and 20 more. 199pp. 5⅜ × 8½. 20767-6 Pa. $3.95

ETHICAL WRITINGS OF MAIMONIDES, Maimonides. Most significant ethical works of great medieval sage, newly translated for utmost precision, readability. Laws Concerning Character Traits, Eight Chapters, more. 192pp. 5⅜ × 8½.
24522-5 Pa. $4.50

THE EXPLORATION OF THE COLORADO RIVER AND ITS CANYONS, J. W. Powell. Full text of Powell's 1,000-mile expedition down the fabled Colorado in 1869. Superb account of terrain, geology, vegetation, Indians, famine, mutiny, treacherous rapids, mighty canyons, during exploration of last unknown part of continental U.S. 400pp. 5⅜ × 8½. 20094-9 Pa. $6.95

HISTORY OF PHILOSOPHY, Julián Marías. Clearest one-volume history on the market. Every major philosopher and dozens of others, to Existentialism and later. 505pp. 5⅜ × 8½. 21739-6 Pa. $8.50

ALL ABOUT LIGHTNING, Martin A. Uman. Highly readable non-technical survey of nature and causes of lightning, thunderstorms, ball lightning, St. Elmo's Fire, much more. Illustrated. 192pp. 5⅜ × 8½. 25237-X Pa. $5.95

SAILING ALONE AROUND THE WORLD, Captain Joshua Slocum. First man to sail around the world, alone, in small boat. One of great feats of seamanship told in delightful manner. 67 illustrations. 294pp. 5⅜ × 8½. 20326-3 Pa. $4.95

LETTERS AND NOTES ON THE MANNERS, CUSTOMS AND CONDITIONS OF THE NORTH AMERICAN INDIANS, George Catlin. Classic account of life among Plains Indians: ceremonies, hunt, warfare, etc. 312 plates. 572pp. of text. 6⅛ × 9¼. 22118-0, 22119-9 Pa. Two-vol. set $15.90

ALASKA: The Harriman Expedition, 1899, John Burroughs, John Muir, et al. Informative, engrossing accounts of two-month, 9,000-mile expedition. Native peoples, wildlife, forests, geography, salmon industry, glaciers, more. Profusely illustrated. 240 black-and-white line drawings. 124 black-and-white photographs. 3 maps. Index. 576pp. 5⅜ × 8½. 25109-8 Pa. $11.95

CATALOG OF DOVER BOOKS

THE BOOK OF BEASTS: Being a Translation from a Latin Bestiary of the Twelfth Century, T. H. White. Wonderful catalog real and fanciful beasts: manticore, griffin, phoenix, amphivius, jaculus, many more. White's witty erudite commentary on scientific, historical aspects. Fascinating glimpse of medieval mind. Illustrated. 296pp. 5⅜ × 8¼. (Available in U.S. only) 24609-4 Pa. $5.95

FRANK LLOYD WRIGHT: ARCHITECTURE AND NATURE With 160 Illustrations, Donald Hoffmann. Profusely illustrated study of influence of nature—especially prairie—on Wright's designs for Fallingwater, Robie House, Guggenheim Museum, other masterpieces. 96pp. 9¼ × 10¾. 25098-9 Pa. $7.95

FRANK LLOYD WRIGHT'S FALLINGWATER, Donald Hoffmann. Wright's famous waterfall house: planning and construction of organic idea. History of site, owners, Wright's personal involvement. Photographs of various stages of building. Preface by Edgar Kaufmann, Jr. 100 illustrations. 112pp. 9¼ × 10. 23671-4 Pa. $7.95

YEARS WITH FRANK LLOYD WRIGHT: Apprentice to Genius, Edgar Tafel. Insightful memoir by a former apprentice presents a revealing portrait of Wright the man, the inspired teacher, the greatest American architect. 372 black-and-white illustrations. Preface. Index. vi + 228pp. 8¼ × 11. 24801-1 Pa. $9.95

THE STORY OF KING ARTHUR AND HIS KNIGHTS, Howard Pyle. Enchanting version of King Arthur fable has delighted generations with imaginative narratives of exciting adventures and unforgettable illustrations by the author. 41 illustrations. xviii + 313pp. 6⅛ × 9¼. 21445-1 Pa. $5.95

THE GODS OF THE EGYPTIANS, E. A. Wallis Budge. Thorough coverage of numerous gods of ancient Egypt by foremost Egyptologist. Information on evolution of cults, rites and gods; the cult of Osiris; the Book of the Dead and its rites; the sacred animals and birds; Heaven and Hell; and more. 956pp. 6⅛ × 9¼. 22055-9, 22056-7 Pa., Two-vol. set $21.90

A THEOLOGICO-POLITICAL TREATISE, Benedict Spinoza. Also contains unfinished *Political Treatise*. Great classic on religious liberty, theory of government on common consent. R. Elwes translation. Total of 421pp. 5⅜ × 8½. 20249-6 Pa. $6.95

INCIDENTS OF TRAVEL IN CENTRAL AMERICA, CHIAPAS, AND YUCATAN, John L. Stephens. Almost single-handed discovery of Maya culture; exploration of ruined cities, monuments, temples; customs of Indians. 115 drawings. 892pp. 5⅜ × 8½. 22404-X, 22405-8 Pa., Two-vol. set $15.90

LOS CAPRICHOS, Francisco Goya. 80 plates of wild, grotesque monsters and caricatures. Prado manuscript included. 183pp. 6⅜ × 9⅜. 22384-1 Pa. $4.95

AUTOBIOGRAPHY: The Story of My Experiments with Truth, Mohandas K. Gandhi. Not hagiography, but Gandhi in his own words. Boyhood, legal studies, purification, the growth of the Satyagraha (nonviolent protest) movement. Critical, inspiring work of the man who freed India. 480pp. 5⅜ × 8½. (Available in U.S. only) 24593-4 Pa. $6.95

CATALOG OF DOVER BOOKS

ILLUSTRATED DICTIONARY OF HISTORIC ARCHITECTURE, edited by Cyril M. Harris. Extraordinary compendium of clear, concise definitions for over 5,000 important architectural terms complemented by over 2,000 line drawings. Covers full spectrum of architecture from ancient ruins to 20th-century Modernism. Preface. 592pp. 7½ × 9⅝. 24444-X Pa. $14.95

THE NIGHT BEFORE CHRISTMAS, Clement Moore. Full text, and woodcuts from original 1848 book. Also critical, historical material. 19 illustrations. 40pp. 4⅝ × 6. 22797-9 Pa. $2.50

THE LESSON OF JAPANESE ARCHITECTURE: 165 Photographs, Jiro Harada. Memorable gallery of 165 photographs taken in the 1930's of exquisite Japanese homes of the well-to-do and historic buildings. 13 line diagrams. 192pp. 8⅜ × 11¼. 24778-3 Pa. $8.95

THE AUTOBIOGRAPHY OF CHARLES DARWIN AND SELECTED LETTERS, edited by Francis Darwin. The fascinating life of eccentric genius composed of an intimate memoir by Darwin (intended for his children); commentary by his son, Francis; hundreds of fragments from notebooks, journals, papers; and letters to and from Lyell, Hooker, Huxley, Wallace and Henslow. xi + 365pp. 5⅜ × 8.
20479-0 Pa. $5.95

WONDERS OF THE SKY: Observing Rainbows, Comets, Eclipses, the Stars and Other Phenomena, Fred Schaaf. Charming, easy-to-read poetic guide to all manner of celestial events visible to the naked eye. Mock suns, glories, Belt of Venus, more. Illustrated. 299pp. 5¼ × 8¼. 24402-4 Pa. $7.95

BURNHAM'S CELESTIAL HANDBOOK, Robert Burnham, Jr. Thorough guide to the stars beyond our solar system. Exhaustive treatment. Alphabetical by constellation: Andromeda to Cetus in Vol. 1; Chamaeleon to Orion in Vol. 2; and Pavo to Vulpecula in Vol. 3. Hundreds of illustrations. Index in Vol. 3. 2,000pp. 6⅛ × 9¼. 23567-X, 23568-8, 23673-0 Pa., Three-vol. set $37.85

STAR NAMES: Their Lore and Meaning, Richard Hinckley Allen. Fascinating history of names various cultures have given to constellations and literary and folkloristic uses that have been made of stars. Indexes to subjects. Arabic and Greek names. Biblical references. Bibliography. 563pp. 5⅜ × 8½. 21079-0 Pa. $7.95

THIRTY YEARS THAT SHOOK PHYSICS: The Story of Quantum Theory, George Gamow. Lucid, accessible introduction to influential theory of energy and matter. Careful explanations of Dirac's anti-particles, Bohr's model of the atom, much more. 12 plates. Numerous drawings. 240pp. 5⅜ × 8½. 24895-X Pa. $4.95

CHINESE DOMESTIC FURNITURE IN PHOTOGRAPHS AND MEASURED DRAWINGS, Gustav Ecke. A rare volume, now affordably priced for antique collectors, furniture buffs and art historians. Detailed review of styles ranging from early Shang to late Ming. Unabridged republication. 161 black-and-white drawings, photos. Total of 224pp. 8⅜ × 11¼. (Available in U.S. only) 25171-3 Pa. $12.95

VINCENT VAN GOGH: A Biography, Julius Meier-Graefe. Dynamic, penetrating study of artist's life, relationship with brother, Theo, painting techniques, travels, more. Readable, engrossing. 160pp. 5⅜ × 8½. (Available in U.S. only)
25253-1 Pa. $3.95

CATALOG OF DOVER BOOKS

AMERICAN CLIPPER SHIPS: 1833–1858, Octavius T. Howe & Frederick C. Matthews. Fully-illustrated, encyclopedic review of 352 clipper ships from the period of America's greatest maritime supremacy. Introduction. 109 halftones. 5 black-and-white line illustrations. Index. Total of 928pp. 5⅜ × 8½.
25115-2, 25116-0 Pa., Two-vol. set $17.90

TOWARDS A NEW ARCHITECTURE, Le Corbusier. Pioneering manifesto by great architect, near legendary founder of "International School." Technical and aesthetic theories, views on industry, economics, relation of form to function, "mass-production spirit," much more. Profusely illustrated. Unabridged translation of 13th French edition. Introduction by Frederick Etchells. 320pp. 6⅛ × 9¼. (Available in U.S. only)
25023-7 Pa. $8.95

THE BOOK OF KELLS, edited by Blanche Cirker. Inexpensive collection of 32 full-color, full-page plates from the greatest illuminated manuscript of the Middle Ages, painstakingly reproduced from rare facsimile edition. Publisher's Note. Captions. 32pp. 9⅜ × 12¼.
24345-1 Pa. $4.95

BEST SCIENCE FICTION STORIES OF H. G. WELLS, H. G. Wells. Full novel *The Invisible Man,* plus 17 short stories: "The Crystal Egg," "Aepyornis Island," "The Strange Orchid," etc. 303pp. 5⅜ × 8½. (Available in U.S. only)
21531-8 Pa. $4.95

AMERICAN SAILING SHIPS: Their Plans and History, Charles G. Davis. Photos, construction details of schooners, frigates, clippers, other sailcraft of 18th to early 20th centuries—plus entertaining discourse on design, rigging, nautical lore, much more. 137 black-and-white illustrations. 240pp. 6⅛ × 9¼.
24658-2 Pa. $5.95

ENTERTAINING MATHEMATICAL PUZZLES, Martin Gardner. Selection of author's favorite conundrums involving arithmetic, money, speed, etc., with lively commentary. Complete solutions. 112pp. 5⅜ × 8½.
25211-6 Pa. $2.95

THE WILL TO BELIEVE, HUMAN IMMORTALITY, William James. Two books bound together. Effect of irrational on logical, and arguments for human immortality. 402pp. 5⅜ × 8½.
20291-7 Pa. $7.50

THE HAUNTED MONASTERY and THE CHINESE MAZE MURDERS, Robert Van Gulik. 2 full novels by Van Gulik continue adventures of Judge Dee and his companions. An evil Taoist monastery, seemingly supernatural events; overgrown topiary maze that hides strange crimes. Set in 7th-century China. 27 illustrations. 328pp. 5⅜ × 8½.
23502-5 Pa. $5.95

CELEBRATED CASES OF JUDGE DEE (DEE GOONG AN), translated by Robert Van Gulik. Authentic 18th-century Chinese detective novel; Dee and associates solve three interlocked cases. Led to Van Gulik's own stories with same characters. Extensive introduction. 9 illustrations. 237pp. 5⅜ × 8½.
23337-5 Pa. $4.95

Prices subject to change without notice.

Available at your book dealer or write for free catalog to Dept. GI, Dover Publications, Inc., 31 East 2nd St., Mineola, N.Y. 11501. Dover publishes more than 175 books each year on science, elementary and advanced mathematics, biology, music, art, literary history, social sciences and other areas.